The Praying Mantids

The Praying Mantids

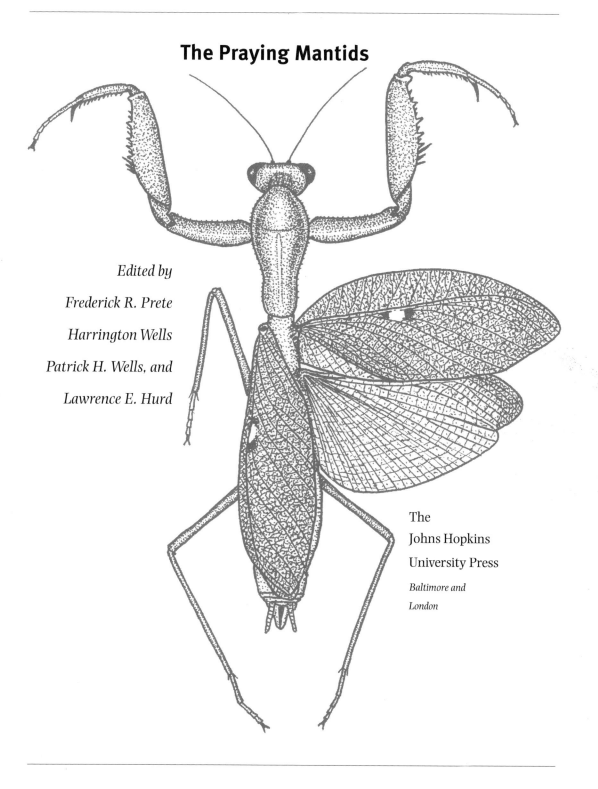

Edited by
Frederick R. Prete
Harrington Wells
Patrick H. Wells, and
Lawrence E. Hurd

The
Johns Hopkins
University Press

*Baltimore and
London*

© 1999 The Johns Hopkins University Press
All rights reserved. Published 1999
Printed in the United States of America on acid-free paper
9 8 7 6 5 4 3 2 1

The Johns Hopkins University Press
2715 North Charles Street
Baltimore, Maryland 21218-4363
www.press.jhu.edu

Library of Congress Cataloging-in-Publication Data
will be found at the end of this book.

A catalog record for this book is available from the British Library.

ISBN 0-8018-6174-8

To Gregory Marshall, Elizabeth Sarah, and Heather Anne

F. R. P.

Contents

Contributors ix
Foreword *John Alcock* xi
Introduction *Frederick R. Prete* xiii

History, Morphology, and Taxonomy

1. The Predatory Behavior of Mantids: Historical Attitudes and Contemporary Questions *Frederick R. Prete, Harrington Wells, Patrick H. Wells* 3

2. Morphology and Taxonomy *Roger Roy* 19

Ecology and Mating Behavior

3. Ecology of Praying Mantids *L. E. Hurd* 43

4. The Ecology and Foraging Strategy of *Tenodera angustipennis* *Toshiaki Matsura, Tamiji Inoue* 61

5. Mating Behavior *Michael R. Maxwell* 69

Hearing and Vision

6. Hearing *David D. Yager* 93

7. Binocular Vision and Distance Estimation *Karl Kral* 114

8. Prey Recognition *Frederick R. Prete* 141

Motor Behaviors

9. Flight and Wing Kinematics *John Brackenbury* 183

10. Prey Capture *Frederick R. Prete, Kristy Hamilton* 194

11. The Hierarchical Organization of Mantid Behavior *Eckehard Liske* 224

Defensive Behavior

12. Ontogeny of Defensive Behaviors *Eckehard Liske, Kristian Köchy, Heinz-Gerd Wolff* 253

13. Ethology of Defenses against Predators *Malcolm Edmunds, Dani Brunner* 276

Techniques

14. Rearing Techniques, Developmental Time, and Life Span Data for Lab-Reared *Sphodromantis lineola* *Frederick R. Prete* 303

15. Comparative Aspects of Rearing and Breeding Mantids *David D. Yager* 311

16. Mantids in Ecological Research *L. E. Hurd* 318

17. Histological Techniques for Mantid Research *David D. Yager* 322

References 327

Index 355

Color illustrations for chapters 5 and 13 follow p. 282.

Contributors

John Alcock, Department of Zoology, Arizona State University, Tempe, Arizona 85287-1501 USA

John Brackenbury, University of Cambridge, Department of Anatomy, Cambridge, CB2 3DY Cambridge, England

Dani Brunner, New York State Psychiatric Institute, 722 West 168th Street, New York, New York 10032, USA

Malcolm Edmunds, Department of Biological Sciences, University of Central Lancashire, Preston PR1 2HE, England

Kristy S. Hamilton, Department of Psychology, Indiana University, Bloomington, Indiana 47405, USA

Lawrence E. Hurd, Department of Biology, Washington and Lee University, Lexington, Virginia 24450, USA

Tamiji Inoue, Center for Ecological Research, Kyoto University, Shimosakamoto 4, Otsu 520-01, Japan

Kristian Köchy, Zoologisches Institut Physiologie, Technische Universität, Pockelsstraße 10a, Braunschweig, Germany

Karl Kral, Institut für Zoologie, Karl-Franzens-Universität, Universitätplatz 2, A-8010 Graz, Austria

Eckehard Liske, Zoologisches Institut Physiologie, Technische Universität Pockelsstraße 10a Braunschweig, Germany

Toshiaki Matsura, Department of Biology, Kyoto University of Education, Fushimi-ku, Kyoto 612, Japan

Michael R. Maxwell, Marine Biological Laboratory, Woods Hole, Massachusetts 02543, USA

Frederick R. Prete, Department of Biological Sciences, DePaul University, Chicago, Illinois 60614, USA

Roger Roy, Muséum National d'Histoire Naturelle, Entomologie, 45 rue Buffon, Paris 75005, France

Harrington Wells, Department of Biology, The University of Tulsa, 600 S. College Avenue, Tulsa, Oklahoma 74104, USA

Patrick H. Wells, Department of Biology, Occidental College, Los Angeles, California 90041, USA

Heinz-Gerd Wolff, Zoologisches Institut Physiologie, Technische Universität Pockelsstraße 10a Braunschweig, Germany

David Yager, Department of Psychology, University of Maryland, College Park, Maryland 20742, USA

Foreword

Most people feel an affinity of one sort or another with the dog, the cat, the grizzly bear, and the lion but believe that insects exist to be ignored—except when one or another bug makes a nuisance of itself, and then we get out the fly swatter or the insect spray and consign the nuisance to oblivion. Although the typical attitude about insects ranges from indifference to antipathy, a handful of species have managed to overcome our pro-mammalian bias and stimulate our admiration, or at least generate a certain fascination. The ladybird beetle comes to mind as well as the honeybee, forgiven in part for its ability to sting because of its beneficial economic effects. But perhaps the premiere harmless insect in terms of making an impression on people is "the" praying mantis and all because among the two thousand or so species of mantids are some whose females have been observed feasting on their copulatory partners. As John and Helen Philbrick say in *The Bug Book*, "because the female devours the male at a certain stage of their life cycle, [the praying mantis] has always received a good deal of enviable publicity."

Whether enviable or not, this aspect of mantid biology clearly amazes people who hear about it. As a result, considerable ink has been spilt on the subject by nonscientists and research biologists alike. One of the many strengths of *The Praying Mantids* is that it provides a complete review of "sexual cannibalism" from a number of vantage points. In the course of this review, the authors thoroughly debunk a number of myths associated with the phenomenon while providing a sober analysis of the "de-mythified" reality that remains, namely, that females do, indeed, sometimes consume their mates.

The contributors to this book also clear up a number of other misunderstandings about mantids, such as the widespread belief that they are designed to feed exclusively on flies and the allied view that mantids are robotic automatons totally lacking in behavioral flexibility. In addition, the notion that mantids are big-eyed diurnal animals whose lives are dominated by vision to the exclusion of other senses turns out to be quite wrong. Readers of this book will learn that the actual capabilities of mantids are really much more interesting than the ones they are commonly but incorrectly believed to possess.

The various contributors to *The Praying Mantids* share an eagerness to describe mantis biology as it really is because they have a great enthusiasm for mantises as they really are. It is my impression that scientists generally select research projects for one of two reasons: either a deep aesthetic curiosity about an organism that has managed to catch their imagination or a conviction that the species in question just happens to be well suited to help solve a theoretical issue of interest. When an emotional involvement with an organism combines with a desire to resolve an important theoretical puzzle, the resulting research is more likely to be exceptionally satisfying for both the scientist and those who read about the research.

This book is a case in point because not only are mantids wonderful creatures but they also offer much for those who would resolve such intriguing matters as the evolutionary significance, if any, of sexual cannibalism, or the relation between a species' ecology and the evolved design of its perceptual equipment, or the role of generalist predators in structuring the ecological communities to which they belong, or the means by which animals organize their behavioral repertoire so as to avoid incapacitating conflicts, among others.

A key feature of the book that further increases its value is the willingness of the authors to identify the many unresolved questions and puzzles that remain about mantid biology. This makes the book an important resource for other scientists, perhaps especially for graduate students in biology who sometimes get the impression that everything worth studying has already been studied. The fact that the biology of such appealing creatures as mantids is still so incompletely known is both sobering and encouraging, a reflection of how much remains to be done and an indicator of the many pleasures of discovery that lie ahead. This book helps ensure that the road to discovery is well marked, which is good news for those of us with a fondness for the world's marvelous mantids.

JOHN ALCOCK
Arizona State University

Introduction

Frederick R. Prete

After finishing two years of graduate research in the Sleep Research Laboratory at the University of Chicago, I found myself with the doldrums. I felt quite lucky to be at the U. of C., and I enjoyed doing research and writing, but I had not discovered a project that excited me enough to pursue it for my dissertation. Then, early on a bright autumn morning, while on my way to the science library, I saw a creature that would change my life. On the sidewalk just at the foot of the stairs stood a large, elegant insect that assumed a stance which has been described like this: "The wing-covers open, and are thrust obliquely aside. . . . [They are like] parallel screens of transparent gauze, forming a pyramidal prominence which dominates the back; the end of the abdomen curls upwards. . . . The murderous forelimbs . . . open to their full extent, forming a cross with the body. . . . The object of this mimicry is evident; the Mantis wishes to terrorize its powerful prey, to paralyse it with fright" (Fabré, 1912, 74–75). Well, quite frankly, Fabré was not far from the truth. I was not quite "demoralized with fear," but I did pick up the creature gingerly. I carefully slipped it into my briefcase and literally ran back to my office to look more closely at this marvelous little animal.

Soon would I learn that what I had found was a female *Tenodera aridifolia sinensis*. However, before I knew its scientific name, before I had seen any scholarly papers about mantids, before I turned an analytical eye to its behavior, it was simply thrilling! After a day with my mantid, my future research agenda was set. I went straight to the chair of my committee (which is like a department at the U. of C.) and asked his permission to take over a small laboratory in the basement of Green Hall—a laboratory which had once belonged to a rather well-known ethologist but which now sat empty, locked, and abandoned for more than a decade. After I assured the chair that I would ask for no departmental money and that I would do all of the necessary cleaning and refurbishing myself, he agreed. I was elated. I supported my research by teaching part-time at several local universities, and I borrowed old, unused laboratory equipment from almost everyone. Over the next two years I collected enough data on the psychophysics of prey recognition and the kinematics of the predatory strike to assemble a dissertation. Although my dissertation was accepted and I passed my final defense in August 1990, I stayed an additional year to continue my research on this fascinating insect. These three years were wonderful.

I learned quite a bit in the few years that I had my first lab. I learned about resourcefulness, about research, about universities, about invertebrate perception, about mantids, and, most important, about science. In particular, I learned how enjoyable and rewarding science is when it is collaborative, cooperative, and collegial, and that is one of the reasons that I initiated this book. A second motivating factor was my desire to promote a scientific interest in praying mantids. To that end, my colleagues and I have included some chapters at the end of the book that deal with certain technical aspects of mantid research. This information is available nowhere else, and we were anxious to share it.

From the beginning of this project, all of the authors were encouraged to put their personal stamps on their contributions. The result is a lively and at times wonderfully idiosyncratic look at the topics that they discuss. For instance, Roger Roy calls for a thorough reevaluation of mantid taxonomy (chap. 2), and Frederick Prete asks us to completely rethink the ways in which we approach the questions of mantid prey recognition and prey capture (chaps. 8 and 10). Interestingly, in spite of the fact that mantids have fascinated people for millennia (chap. 1), little is known about their basic biology, and some of the information in this volume has never been published before. Hence, some chapters present ground-breaking data on topics not previously explored with mantids: Toshiaki Matsura and Tamiji Inoue (chap. 4) give us an interesting look at the foraging behavior of *Tenodera angustipennis*, an animal about which virtually nothing else is known; David Yager reviews his work on mantid hearing, a sensory system that he discovered in mantids (chap. 6); Karl Kral explains his innovative work on the use of binocular vision as a cue to distance estimation by mantids (chap. 7); and Eckehard Liske gives us a wonderfully detailed experimental analysis of two previously unexplored aspects of mantid behavior, its hierarchical organization and the ontogeny of defensive reactions (chaps. 11 and 12). In all cases, the chapters in this book have been written by the people who have done the pioneering work in their particular field of mantid research, and their contributions reflect their strong intellectual commitment to their areas of expertise. This is particularly evident, for instance, in the thorough analyses of mantid ecology by L. E. Hurd (chap. 3), mating behavior by Michael Maxwell (chap. 5), mantid flight by John Brackenbury (chap. 9), and defense against predators by Malcolm Edmunds and Dani Brunner (chap. 13).

Regretfully, I must acknowledge one omission from this book. I would very much have liked to have had Dr. Samuel Rossel write a chapter about his work on binocular vision in mantids. Without question, he is an important and highly respected member of the mantid community. Unfortunately, Dr. Rossel was simply too busy to accept my invitation, but he applauds our efforts and wishes us success.

It goes without saying that a project such as this requires the goodwill and collegial cooperation of a large number of people. These include, of course, all of the contributors to this book, especially my co-editors. I am deeply indebted to each and every one of these people. Their friendships have been rewarding, their expertise has been invaluable, and I have learned much from each of them. I also thank Robert Harington and Ginger Berman of the Johns Hopkins University Press for their patience, support, and guidance. In addition, I am particularly grateful to our copyeditor, Sherry Hawthorne, and to those anonymous referees and reviewers who gave up their valuable time in order to offer thoughtful comments, criticisms, and suggestions regarding the original prospectus and various versions of the final product. The project benefited from their input. It goes without saying that any errors that remain are mine.

History, Morphology, and Taxonomy

1. The Predatory Behavior of Mantids: Historical Attitudes and Contemporary Questions

Frederick R. Prete, Harrington Wells, Patrick H. Wells

Everyone who writes about mantids seems to have an interesting story to tell, and we hope that we do, too. Our story has two interconnected plots. The first is about how mantids came to be known by Europeans and Americans, that is, who wrote about mantids and what they said about them. The second, more complicated plot, has to do with how a number of preexisting beliefs and ideas shaped early writers' understandings of mantid behavior and, in turn, shaped the scientific questions that came to be asked about mantids. Until recently, just one topic has dominated discussions about mantids, their remarkable predatory behavior. The mantid's dramatic method of capturing prey is thrilling to watch, often distracting one from the other interesting things that this insect does to make a living. We argue here that, early on, a fundamental misconception about mantid predatory behavior became widespread. That misconception is that mantids engage in two separate kinds of predatory behavior: "Normal" predatory behavior was thought to involve the identification and capture of small prey items such as flies. In fact, a considerable amount has been written over the last forty years explaining the (putative) morphological and neurological adaptations that make the mantids "ideally" suited to that end (Prete, 1995; Prete and Wolfe, 1992). The other kind of behavior, regarded as "abnormal," is predatory behavior in which mantids capture very large prey, such as small vertebrates or same-sized conspecifics. Historically, this behavior is seen as strange, anomalous, or aberrant and consequently in need of some special explanation.

Early on, instances of mantids capturing very large prey were dismissed as products of their pugnacious or evil temperament. More recently, however, theorists have largely ignored instances of mantids capturing large prey, except in the cases of adult females capturing and eating adult males. These cases are usually imagined to be instances of that rather nebulous behavioral category of sexual cannibalism, irrespective of the conditions under which they occur, and are rarely perceived as normal predatory behavior. The historical reasons that mantids and mantid predatory behavior have come to be viewed in this way make a fascinating story that embraces almost all of what has been written about these insects, from the ancient Egyptian Book of the Dead to relatively recent scientific papers.

The Asian Mantis

Insects have occupied a more prominent place in the Chinese culture, and have done so for a longer period of time, than they have in most other cultures (Pemberton, 1990; Tsai, 1982).

Much of this story appeared originally in F. R. Prete and M. M. Wolfe's 1992 article "Religious supplicant, seductive cannibal, or reflex machine? In search of the praying mantis," *Journal of History of Biology* 25: 91–136. We thank the copyright holders for their kind permission to reprint that material.

For instance, sericulture and silk technology began in China some 4500 to 7000 years ago, commercial beekeeping was prospering 1800 years ago, the first pest control officer was installed 2230 years ago, and insects have been used for medicine and entertainment for at least for 2000 years (Tsai, 1939). The praying mantids, some of which are indigenous to China, have been subjects of scientific study, frequent subjects of art and literature, and objects of entertainment for at least a thousand years (e.g., Tsai, 1982; Kevan, 1983).

The mantid's seemingly pious demeanor did not deceive Asian observers as it did those in other countries. In China (and later Japan), the mantid became a symbol of strength, courage, and boldness extending sometimes even to foolhardiness. This conceptualization, based on the willingness of mantids to attack creatures as large as or larger than themselves, was popularized in a Chinese proverb that says that although a mantis is not strong enough to stop a bullock cart, it can be brash enough to give it a try (Bodenheimer, 1928; Kevan, 1983, 1985). So well established was the mantid's reputation for foolhardiness that it received this advice in "The Praying Mantis Also Gets a Warning" by the ninth-century Chinese poet Liu Yü-Hsi:

> Your neck indeed, my friend, looks strong;
> Armed with huge battle axes you convey a fierce expression.
> But do not presume on your arrogant ways and swallow up your kind;
> Be on your guard for the yellow bird, as he may open his mouth and finish you!
>
> (Kevan, 1983, 455)

Interestingly, the mantid's boldness was put to the test in the Chinese gaming arenas. Beginning in the tenth century A.D., several species of orthopteroid insects, including crickets and mantids, were captured or bred for fighting. The sport eventually became widespread and supported an extensive trade in exquisitely made insect cages and accouterments (Gressitt, 1946; Daufer, 1927; Gressitt, 1969).

Tales of the mantid's seemingly pugnacious personality (both in and out of the gaming arena) were brought home by European explorers much to the astonishment, disbelief, and revulsion of those who read them. Readers' reactions were so strong because the stories clashed forcefully with traditional European beliefs about the mantid's demeanor. In contrast to the harsh (but more realistic) light in which mantids were viewed in Asia, peoples of the Middle East, Africa, and Europe saw these insects as gentle, magical, or divine.

. . . In Juxtaposition

Insects have always played important roles in the daily lives of people in the Middle East and Africa. They can be, of course, unrelenting competitors for space and resources, often preying upon both crops and people. However, in addition to having practical and economic import, insects have also played important symbolic roles. In Egypt, for instance, the sign of the oriental hornet (or bee, depending on whom you read) was associated with royalty at least since the First Dynasty (Bodenheimer, 1973; Fraser, 1931), and the scarab beetle is so common in the artifacts of early Egypt that it has been called the virtual personification of ancient Egypt (Harpaz, 1973).

Mantids, too, played an important symbolic role in Egypt, appearing first in the Eighteenth Dynasty (1555–1350 B.C.) in what has become known in the West as The Book of the Dead. In this text, the mantis appears as a minor deity (the "Bird-Fly") whose function is to conduct the souls of the dead to the netherworld in order that they might meet the greater divine spirits (Kevan, 1978):

> I have made my way into the Royal Palace, and it was the Mantis who brought me thither.
> Hail to thee, who fliest up to Heaven, to give light to the stars and protect the White Crown which falleth to me.
> Stable art thou, O mighty god, for ever, Make thou for me a path upon which I may pursue my course.
>
> (Renouf, 1894, cited in Kevan, 1978, 391)

Mantids with supernatural powers also appear in what is probably the first zoology textbook (Harpaz, 1973). A collection of cuneiform texts assembled by the Assyrian king Ashurbanipal (669–626 B.C.) and stored in the royal library of Nineveh includes a Sumero-Akkadian lexicographical dictionary known as Har-ra=Hubulla. Compiled during the ninth century B.C., the texts contain systematically organized lists of Sumerian names with their corresponding Akkadian translations. Animal names are arranged in zoological groups, each of which is characterized by a common prefix. Among the list of the Orthopteran insects (Sumerian prefix *buru*), are two Sumerian names for mantis: "buru.EN.ME.LI." and "buru.EN.ME.LI.a sha(g).ga," which roughly translate, respectively, as *necromancer* (a fortune-teller who communicates with the dead) and *soothsayer* of the field. The taxonomic groups of the Har-ra=Hubulla are similar to those used by Pliny and were the basis for much of the zoology in the Talmud and for some medieval Arab writings (Harpaz, 1973). Influence of the Sumerian classification was persistent; even Linnaeus grouped the mantids with the locusts in his early writings.

Although there are a number of references to *mantis* in ancient Greek literature, most mean simply prophet or soothsayer and do not refer to the insect (Prete and Wolfe, 1992). After a time, however, the insect did come to acquire the name *mantis* due to its seemingly pious demeanor and purported ability to point the way home to lost travelers (Kevan, 1978). Of all of the references to the mantis in early Greek writings, one is particularly noteworthy because it would later appear in the two most influential entomology texts of the seventeenth century. The reference is an allusion, in the tenth *Idyll* of Theocratus (ca. 270 B.C.), to the mantid's raptorial forelegs. In lines 17 and 18, Theocratus draws an analogy between the arms of a young woman and those of the insect: "Lord! thy sin hath found thee out. Thou'dst wished and wished, and now, 'faith, thou'st won. There'll be a mantis to clasp thee all night long" (Edmunds, 1912, cited in Kevan, 1978).

Compendia and Bestiaries

Indian, Hebrew, and Egyptian legends and folklore were absorbed into Greek and Roman folklore, poetry, and art, ultimately being included in Alexandrian handbooks of paradoxology and medical-magical treatises. From these, writers such as Pliny and Aelian passed the legends to the Christian world (Curley, 1979; Cook, 1919). Now infused with religious and moral teachings, the legends were collected by an unknown author in a popular zoological text known as *The Physiologus*, the original Greek version of which dates from the fourth century. This book was the most widely distributed zoological work of the time, marking the transition, some have argued, from antiquity to the Middle Ages (Morge, 1973).

The Physiologus is a collection of Christian allegories revolving around a number of animals, stones, and trees. The insects discussed in various versions include the ant, fly, bee, scarab beetle, wasp, moth, and grasshopper or locust; but the mantis, as such, seems not to have been included in any version of *The Physiologus* of which we are aware (e.g., see McCulloch, 1960). However, the suggestion has been made that the myth behind the use of the locust as a symbol representing conversion to Christianity (or triumph over evil) may actually have originated in observations of mantids subduing their prey (Kevan, 1985). This hypothesis seems plausible in that the original myth involves a locust (of which mantids were considered a type), seizing a snake (which represents Satan, or opposition to the church) and killing it by biting its neck. If this story has an empirical basis, it is unlikely that the original observation could have been of any insect other than a mantid (Wilkinson, 1984).

By the end of the twelfth century the natural history compendium had begun to evolve into a new genre of popular nature book that would come to be known generically as the bestiary. These texts could be voluminous and grew to absorb all extant animal legends and lore, including those contained in the various versions of *The Physiologus* (Curley, 1979). Arguably, the three most important of the early natural history texts are *Liber de Natura Rerum* by Thomas Cantipratanus (b.1201, d. between 1263 and 1293), *De Animalibus* by Albert Magnus (1193–1280), and a codex of animals written about 1460 by Petrus Candidus Decembrus (1399–1477). Each of these works included some information about locusts. Cantipratanus and Decembrus mentioned the mantis specifically (Bodenheimer, 1949, as explained in Harpaz, 1973), and Magnus described the mysterious snake-killing orthopteroid (Opimacus) to which we referred above (Scanlan, 1987).

European biology in general and entomology in particular were the last to become free of superstition and religious dogma (Beier, 1973).

Until about the mid-sixteenth century, entomology was concerned primarily with practical considerations and was frequently included in medical texts, because many students of medicine were also avid entomologists (e.g., Edward Wotton, 1492–1555, author of *De Differentiis Animalium Libri Decem*, 1552). During the latter half of the sixteenth century, however, entomology took a significant step forward. This was made possible both by a growing independence of the natural scientists–philosophers from the ancient authorities and a reorganization in the universities such that medical/zoological studies were no longer restricted to strictly pragmatic concerns. The first of the two most influential entomologists of this period was the philosopher-physician Ulisse Aldrovandi (1522–1605). Among the hundreds of manuscripts Aldrovandi produced is *De Animalibus Insectis Libri VII*, first published in 1602 (Aldrovandi, 1602). As had others, Aldrovandi considered mantids to be large grasshoppers and in doing so confused the organism's structure-function relations (Bodenheimer, 1928; Prete and Wolfe, 1992).

Although Aldrovandi's work was generally influential, historically it did not fare as well as did that of his contemporary Konrad Gesner (1516–65). Gesner became an established scholar of Hebrew, Greek, botany, and zoology and chief physician of Zurich before succumbing to the plague at the age of forty-nine. Upon his death he left an unfinished manuscript dealing with the natural history of insects, which eventually passed into the hands of Thomas Moffett (also spelled Moufet or Muffet, 1553–1604). Moffett added both descriptions and drawings to the text. His *Inectorum sive Minimorum Animalium Theatrum* was published posthumously in 1634; it was the first entomology book to appear in England (Gillispie, 1970; Moffett, 1634) and was subsequently translated from Latin into English as part of Topsell's *History of Four-Footed Beasts and Serpents* (1608, 2d ed.).

Moffett's book introduced to a broad readership the accumulated information about a wide range of insects, including European beliefs about mantid behavior. In chap. 16, Moffett described the locusts, which he divided into two broad groups, the winged and the wingless. The former he again divided into two groups, the ordinary and the rare, the latter of which included the mantids. Moffett claimed to have seen but three kinds of these rare "locusts" in the meadows and pastures of France and Britanny, "i.e., *Italian, Greek,* and *African:* They are called *Mantes, foretellers,* either because by their coming . . . they do shew the Spring to be at hand, So *Anacreon* the Poet sang; or else they foretell death and famine, as . . . *Theocratus* observed. Or lastly, because it alwaies holds up its forefeet like hands praying" (emphasis in original). Moffett continued by recounting the myth that if a traveler is lost, the mantis will point the way home: "So divine a creature is this esteemed, that if a childe ask the way to such a place, she will stretch out one of her feet, and shew him the right way, and seldome or never misse" (Moffett, 1634, 982–83).

Tales of Voracious Cannibalism

Moffett's characterization notwithstanding, the mantis's reputation for pious helpfulness was about to fade. Unflattering accounts were offered by the well-known German painter and naturalist J. A. Rösel von Rosenhof (Rosenhof, 1746–61). In attempting to study the mantid's habits, Rösel placed several in a glass jar for observation. To his amazement, he found that both young and old battled viciously with each other, the victor always devouring the vanquished. Rösel likened these ferocious battles to those of Huzzars, fighting to the death with razor-sharp sabers. But even more important to our story is the fact that Rösel claimed that his mantids did not cease to attack, kill, and eat each other even when their hunger had been satisfied by an abundant supply of other nourishing food (Prete and Wolfe, 1992; Cuvier, 1832; Gosse, 1857).

A second, unflattering account of mantid behavior appeared in a widely read travel documentary about China written by John Barrow (1764–1848). In his discussion of Chinese gaming practices, Barrow described the "cruel and unmanly amusement" of animal fights, shamefully, he noted, also practiced in Europe. However, in China the sport not only included the cock and the quail, but extended even "into the insect tribe, in which they have discovered a species of gryllus, or locust that will attack each other with such ferocity as seldom to quit their hold without bringing away at the same time a limb of their antagonist." These insects were almost certainly mantids (Barrow, 1805, 106–8; Prete and Wolfe, 1992).

An equally shocking account of mantid ferocity was given in James Smith's popular travelog *A Sketch of a Tour on the Continent*. In describing the unusual plants and animals around Montpellier, Smith included that "singular" insect, *Mantis religiosa*, which gets its name from the "perpetual erection of its forepaws," and will, according to Moffett, point the way home to lost travelers. However, Smith's accompanying entomologist offered a contradictory characterization, which, Smith noted, "savors little of divinity" (Smith, 1807, 168–69). After putting a captured male and female together in a bottle, Smith and his colleague witnessed a series of events not unlike those reported by Rösel: First, the pair mated, after which "the larger and stronger [female] . . . devoured the head and upper part of the body of her companion. But the most wonderful circumstance is, that a subsequent union took place; the life and vigor of the male being unimpaired by the loss of his head, as that part is not in insects the seat of the brain: this was no sooner concluded than his insatiable mate ate up the rest of his body" (Smith, 1807, 170).

The tales of voraciousness offered by Rösel, Barrow, and Smith rapidly found their way into popular natural history and the general scientific literature, and soon it was the mantid's predatory behavior (especially its cannibalistic propensities) that became the primary focus of people's interest. Rösel's, Barrow's, and Smith's accounts were repeated by virtually everyone who wrote about mantids, often being placed in juxtaposition to other, kinder characterizations (e.g., Shaw's *General Zoology* of 1806).

As the mantid's notoriety increased, so did people's apparent outrage about its cannibalism, and, consequently, an interesting dichotomy began to develop regarding the mantid's eating habits (Prete and Wolfe, 1992). That is, those writing about mantid behavior at the time were unable to accept the idea that cannibalism could be a "normal" part of the mantid eating behavior. The general distaste with which people viewed mantid cannibalism is evident, for instance, in the very widely read *Introduction to Entomology* written by the natural theologians William Kirby (1759–1850) and William Spence (1783–1860). Kirby and Spence recognized that, by feeding on other species, insects help keep their total numbers in check: "I cannot doubt that you will recognize the goodness of the Great Parent in providing such an army of counterchecks to the natural tendency of almost all insects to incalculable increase." But as good as it may be to eat superfluous, unrelated insects, it is bad, indeed, to eat your own kind: "But before I quit this subject I may call your attention to what may be denominated Cannibal Insects, since in spite of those disclaimers who would persuade us that man is the only animal that preys upon his own species, a large number of insects are guilty of the same offence." And among the guilty of course is the "cowardly and cruel" mantis who seems to engage in the practice of cannibalism out of sheer wantonness even "*when in no need of other food*" (emphasis added, Kirby and Spence, 1815, 267–70). Now note here that cannibalism is described yet again as an unusual behavior disconnected from hunger.

Kirby and Spence relied on earlier accounts for their information on mantid behavior, and these same accounts (along with others) were repeated ad infinitum in both the scientific and the popular literature of the time. For instance, in the fifteenth volume of *Animal Kingdom*, the French naturalist Georges Cuvier (1769–1832) reiterated the reports of Shaw, Moffett, Aldrovandi (even quoting the tenth *Idyll* of Theocratus), and Rösel, as well as an account of mating-related cannibalism written by the French naturalist Poiret (Cuvier, 1832). In addition, Cuvier embellished his description with three important comments, which accurately reflect the consensus of his contemporaries regarding mantid behavior. The first of these comments is that mantids are in some sense abnormal insects: "The mantes may be called truly anomalous insects." The second is that mantids have a basically evil disposition: "The mantis is as cowardly as it is cruel, for it will fly away from the ant, though it will destroy abundance of helpless flies." Third, and most important from our point of view, is Cuvier's reiteration of Rösel's claim that mantid cannibalism is a product of the insect's disposition, not a part of its normal eating behavior: "They are so cruel and carnivorous that (even as hatchlings) they kill and eat one another without being compelled to do so by hunger" (Cuvier, 1832, 188–90).

So, by the time Cuvier's comments were published, mantid predatory behavior had come to be seen not just as unusual, but as actually consisting of two separate components, each corresponding to one of the two sides of the mantid's character. Rooted in the commonplace juxtaposition of pious helpfulness versus cruel voraciousness, mantids had come to be seen as unusual animals that eat small prey (such as flies) for nutrition (a manifestation of their good-natured, or normal, side) and cannibalize large prey (such as conspecifics) out of savage wantonness, not hunger (a manifestation of their cruel, or abnormal, side). As explained in *The Natural History of Insects*, when prey is of appropriate size (such as a blue bottlefly) and "not too large, it is curious to remark how cunningly [the mantis] endeavors to entrap its prey." However, even when well fed, "They never ceased to attack, kill, and eat each other when they met" (*The Natural History of Insects*, 1835, 165–69.). Of course, these cannibalistic attacks appeared particularly abhorrent when considered in juxtaposition to the mantid's pretensions to divinity (e.g., see *Domestica*, 1867, 265).

But What Does a Mantis Really Eat?

The belief that mantids eat flies or small, fly-sized insects for nutritional reasons and engage in cannibalism not from hunger but, rather, from savage wantonness presupposes that mantids really do maintain themselves on small, fly-sized prey. And, indeed, this was the common presumption. However, the idea that mantids maintain themselves on small prey did not become entrenched because there was an absence of disconfirming data; it became entrenched for two other reasons (Prete and Wolfe, 1992).

The first reason is the constant repetition of Rösel's claim that mantids "will destroy [an] abundance of helpless flies." This line, or words to its effect, were consistently included in all types of descriptions of mantid behavior. Further, the implications of this claim were bolstered by the fact that flies became the food of choice for everyone who was keeping mantids and writing about them. Hence, it came to be "clear" to everyone that flies (or at least fly-sized prey) are the mantid's normal diet (Denny, 1867, 144; Figuier, 1869, 288–89; Duncan 1870, 338; Newman, 1874, 188; Bellamy, 1890, 530; Brighten 1895, 204; Kellog, 1905, 129; Harmer and Shipley, 1922, 274).

The second reason that the idea of mantids maintaining themselves on flies or fly-sized prey became so firmly entrenched was simply that people continued to dismiss as anomalous any evidence to the contrary. Reports that mantids will capture a variety of large prey never became integrated into the model that people had about mantid predatory behavior, and to some degree the data still remain unintegrated (Prete, 1995; Prete and Wolfe, 1992).

The "Anomalous" Data

Contrary to the strongly held notion that flies or fly-sized insects are the mantid's normal diet, reports began to appear that mantids are actually opportunistic predators that will eat a wide range of prey, including very large arthropods and even small vertebrates. These reports began appearing around the mid-eighteenth century and continued into the late twentieth century but were either not cited in articles on mantid predatory behavior or dismissed as so unusual as not to be of general importance. Had these data been integrated into people's thinking about mantid predation, it would have been impossible to maintain the belief that cannibalism (i.e., capturing same-sized conspecifics) is in any way abnormal.

The earliest published account of a mantis eating a vertebrate seems to have been that of C. A. Zimmerman, which was first reported in the German literature in 1838 by his colleague, the well-known entomologist H. C. C. Burmeister (1807–92): "It devoured as before, several dozen flies daily, sometimes also robust grasshoppers, then some young frogs and even a lizard . . . three times as long as the insect" (Burmeister, 1838, cited in Kevan, 1985, 3). (Again note the reliance on flies as the mantid's primary food.) Zimmerman's report appeared again in German in 1843, (Erichson, 1843) and then in English in *The Annals and Magazine of Natural History* (1844). The fact that the normal eating behavior of mantids was thought to be limited to small insects is evidenced by the fact that the latter was a replication of Zimmerman's first account in an attempt to refute the skepticism of the editor who originally published it: "Instead of the striped lizard . . . I made use of a species of newt. . . . One newt after the other was seized, and to a greater or less extent devoured. I send you the very specimen of mantis with which these experiments were performed" (Tulk, 1844, 78).

Reports such as Zimmerman's continued to appear periodically and included instances of mantids catching small birds, lizards, frogs, mice, snakes, and turtles (e.g., Kingsley, 1884, 176; Johnson, 1976; Ridpath, 1977; Nickle, 1981; reviewed in Kevan, 1985, and Prete and Wolfe, 1992). The original reports, and numerous references to them in articles and textbooks, appeared in English, German, and French from the mid-eighteenth century on, with virtually no impact on subsequent descriptions of what constitutes the mantid's "normal" fare: that is, the reports were dismissed as anomalies. So, by the turn of the century, the belief that the mantid's normal diet consists of flies or fly-sized prey was ubiquitous, and the presumption came to play a key role in shaping both the subsequent experimental paradigms in which mantids were used and the interpretations of the data derived from such paradigms. (But more of that later.)

The interesting historical question is, of course, why did this model persist, especially in light of the anecdotal evidence that mantids are, in fact, generalized predators that will eat anything they can hold onto, regardless of its size? There are, we believe, four reasons. The first has been revealed by the history presented so far. In Asia, people had long observed mantids. Perhaps due to their prevalence, certainly due to the ancient sport of insect fighting, it was generally recognized that mantids are quite bold, ready to challenge a variety of other creatures. However, in the Middle East, Africa, and Europe a general unfamiliarity with mantid behavior allowed a variety of myths to flourish, all of which were based on the idea that mantids are gentle, helpful, and "pious." When a general interest in nature was rekindled and the sciences began to evolve, people turned their attention back toward nature itself. What they found in the case of the mantis, however, was that this insect's apparent cruelty and insatiable voracity clashed horribly with their (preferred) belief in the insect's gentle piousness and their strong revulsion to cannibalism in general.

Much of the subsequent scientific literature on mantids published from the mid-nineteenth to the early twentieth century narrowed its focus to taxonomy, descriptions of various species and their habitats, and the like. However, there were a few notable exceptions that would be critical in maintaining the conceptual dichotomy between "normal" mantid eating behavior and mantid cannibalism. Of these exceptions, the most popular and influential voice was that of Jean Henri Fabré (1823–1915). In his widely read major work, *Souvenirs Entomologiques* (1879–1907), Fabré included observations on mantids (for references see Prete and Wolfe, 1992). In spite of his gift of careful observation, however, Fabré (whose poetic style is wonderful to read) bolstered several misunderstandings about mantids.

Fabré seems to have recognized correctly that mantids will regularly capture and eat prey as large as or larger than themselves: "These particular captures are destined to show me just how far the vigor and audacity of the Mantis will lead it. They include the large grey cricket . . . which is larger than the creature that devours it" (Fabré, 1912, 73; also see Didlake, 1926). However, he never made the connection to the capture of same-sized conspecifics. Like many of his predecessors, Fabré attributed mantid cannibalism to less than noble causes, the first of which was the quick temper of gravid females. Here is how he described the goings-on in his colony: "At the outset matters did not go badly. . . . But this period of concord was of brief duration. . . . The swelling of the ovaries perverted my flock, and infected them with an insane desire to devour one another. . . . Ferocious creatures! It is said that even wolves do not eat one another. The mantis is not so scrupulous; she will eat her fellows when her favorite quarry, the cricket, is attainable and abundant" (Fabré, 1912, 81).

But even more appalling to Fabré was the act of cannibalism when associated with sexual behavior: "These observations reach yet a more revolting extreme [for which there is not] the excuse of hunger. . . . The male Mantis, a slender and elegant lover, . . . throws himself timidly on the back of his corpulent companion. . . . Finally the two separate, but they are soon to be made one flesh in a much more intimate fashion. . . . Here we have no case of jealousy, but simply a depraved taste. . . . I have seen the same Mantis treat seven husbands in this fashion. She admitted all to her embraces, and all paid for the nuptial ecstasy with their lives." And, in Fabré's opinion, it gets even worse: "The custom of eating the lover after the consummation of the nuptials, of making a meal of the exhausted pygmy, who is henceforth good for nothing, is not so difficult to understand . . . but to devour him during the act surpasses anything that the most morbid mind could imagine. I have seen the thing with my own eyes, and I have not yet recovered from my surprise" (Fabré, 1912, 84).

Fabré saw the latter, most distasteful acts of cannibalism as being the product of primitive, undeveloped, instincts:

> I do not deny that the limited area of the cage may favor the massacre of the males; [they have nowhere to flee], but the cause of such butchering must be sought elsewhere. It is perhaps a reminiscence of the carboniferous period. . . . The Orthoptera . . . are the firstborn of the insect world. Uncouth, incomplete in their transformation, they wandered amidst the arborescent foliage, already flourishing when none . . . sprung of more complex forms of metamorphosis were as yet in existence. . . . Manners were not gentle in those epochs, which were full of lust to destroy in order to produce; and the Mantis, a feeble memory of those ancient ghosts, might well preserve the customs of an earlier age. (Fabré, 1912, 84–85)

As we have discussed, the dichotomy between normal eating behavior and cannibalism was maintained because reports of mantids capturing large prey were never completely integrated into people's understanding of mantid predation. Attacks on conspecifics were attributed to the dark, undeveloped side of the mantis, its innate wantonness, greed, and unbridled voracity. However, if a belief in the existence of these traits were the only force maintaining the idea that mantids have two separate predatory strategies, then the dichotomy should have disappeared along with the ascription of the traits to the insect. That is, after Fabré, mantids should have come to be seen as generalized predators that regularly eat large prey, including conspecifics. But this did not occur.

New Alibis

Fabré marked the last important attempt to claim that mantid cannibalism is distinct from other predatory behaviors due to the mantid's innate, despicable character. However, without the alibi of an ill-temperedness, how might one account for such behavior? This question is best answered, we believe, by asking a more fundamental question: Why does one need to account for mantid cannibalism? Why could it simply not be dismissed by saying "The mantid was hungry"? As we have pointed out, one key reason that it needed a special explanation is that capturing large prey—especially conspecifics—was not seen as a part of the mantid's normal eating behavior; it was not seen as a product of hunger. A second reason is that cannibalism could not be ignored theoretically as could instances of mantids eating other large prey (such as newts): Cannibalism was too common and had received too much bad press. A third reason that cannibalism could not be dismissed is that when cannibalism happens to be of the adult-female-on-adult-male variety, and the male is decapitated prior to or during copulation, he can still manage to mate. If the headless male simply ran around for a few moments like a decapitated chicken, the whole scenario would be of little interest, scientifically or otherwise. To the best of our knowledge, no one has sought to determine the biological advantage to the headless chicken's undignified display. (This analogy is borrowed from Liske and Davis [1984, 1987].) Unlike the chicken, however, the headless mantis can sometimes manage that most important evolutionary task. Surely, this could not be the case by chance: Headless mating cannot be analogous to the frenetic, undirected behavior of the decapitated chicken. (Or can it?) A fourth reason is that, even in science, once a proposition is accepted as "fact" it is perniciously persistent (Wenner and Wells, 1990). "Verifications" (in this case cannibalistic decapitations) are regularly reported; "refutations" (amicable matings) go unreported.

Contemporary ideas about mantid behavior were shaped by two basic notions. As we have explained, the first is the well-entrenched belief that the mantid's normal fare is flies, or fly-sized prey. The second has to do with the functional organization of the insect central nervous system (CNS). The general plan of the insect CNS is that there is a series of discrete ganglia (groups of functionally related nerve cells) along the length of the insect, connected by two parallel, ventral nerve cords (or longitudinal connectives). The largest aggregation of nerve cells is in the insect's head and is sometimes referred to as a brain (although this term is somewhat misleading). Ventral to the brain, and also in the head, is the subesophageal ganglion, with which we will be concerned in a moment. Thereafter follows a series of ganglia, the number of which varies from species to species. In most insects, including mantids, the three thoracic ganglia remain separate, each innervating the muscles and sense organs of the segment in which it is located. Likewise, a varying number of abdominal ganglia innervate the structures in the abdomen (see, e.g., Chapman, 1982; Gillott, 1980). What is important to note is that because each ganglion contains motor neurons, sensory neurons, and some capacity to integrate information, all of the necessary neural machinery to emit a complex behavior may be present in a single ganglion. For instance, the ability to learn to avoid a shock has been demonstrated in a single, isolated cockroach ganglion and its corresponding leg (Einstein and Cohen, 1965). Functionally, this means that an insect may be able to walk or to mate apparently normally—even without a head.

The understanding that mantid behaviors are to a greater or lesser extent discretely organized in and controlled by individual ganglia was first derived from anecdotal observations of damaged mantids emitting complex, seemingly normal behaviors. Of particular influence was a brief observation made in 1784 by the French naturalist Poiret (cited in Hennegyy, 1904).

As had others, Poiret noted that a male mantis will still mate if partially cannibalized. The implications of this observation were examined in the four subsequent articles, and their importance requires citing several lines from three of the articles. From the first:

> Duges has seen that the [separated] posterior section of the body of the praying mantis . . . could still stand . . . resist impulses, raise itself, and regain its balance when overturned. . . . Duges concluded that this single thoracic ganglion senses the fingers which apply pressure . . . recognizes the point at which it is grasped, wishes to disencumber itself, and directs the appendages which it animates. [My confirmation of these facts has revealed behaviors] as well coordinated as prior to the removal of the head. (Dubois, 1893, 205–207; also see Moore, 1901)

> On Thursday November 2 [1911], in order to study the mouthparts, I had decapitated a female [mantis]. The next day I found . . . the insect still alive, . . . her posture had remained remarkably similar to that of normal individuals. [On] November 6, . . . the insect fashioned a perfectly constructed ootheca, of entirely normal shape. . . .

This fact demonstrates that, despite the suppression of the nerve centers, insects can accomplish certain actions which are in appearance very complicated. . . . [My] assistant at the laboratory informed me that she died only on November 21. (L. Chopard, 1914, 481–82)

And finally,

Having introduced a male into a cage where a female had resided for several days, the latter immediately seized the former [and] after a few moments the decapitation was complete. The remainder of the body was nevertheless not reduced to immobility and, . . . the abdomen executed a series of movements exactly comparable to those which a male executes during the preliminaries to mating. . . . Does not the important phenomenon lie in the execution of complex movements in the absence of cerebral ganglia?

To this account Etienne Raubad added the following theoretical note:

Under normal conditions, attraction as a result of sensory influences, and copulation as a result of sensory excitation form a whole. In all probability, the cerebral ganglia directly influence the copulatory reflex, by exercising an inhibitory action on them, to such an extent that a male may remain on a female for hours without making the slightest movement. Liberated from this influence, the reflex undergoes no modification in form or rapidity, but it occurs under the most singular conditions. (57–59)

Unfortunately, these comparatively sophisticated analyses were never translated or reprinted. (As far as we know, the first translations, done by Cathryn Easterbrook, appear in Prete and Wolfe [1992].) The fact that they remained sequestered in the early French entomological literature may have been part of the reason that the papers had no apparent impact on contemporary views of mantid behavior (except, as we will explain, that Ken Roeder usurped Raubad's neurological model). In contrast to these sophisticated reports, it was a comparatively brief, rather simplistic five-hundred-word anecdote that (unfortunately) came to be known as the earliest account of the decapitated male's "remarkable" abilities (i.e., Howard, 1886). This brief account has been called erroneously the beginning of "the modern version of the decapitation myth" (Brown, 421), and ironically, "the first account . . . of an all-time favorite among nature's curious facts" (Gould, 1984, 10), despite the fact that Howard, himself, cited three well-known English language texts that had reported the very same events!

In his report, Howard echoed the long-held belief that female cannibalism is independent of hunger:

Not until she had eaten all of his thorax except about three millimeters, did she stop to rest. All this while the male had continued his vain attempts to obtain entrance . . . and he now succeeded. . . . The female was apparently full fed when the male was placed with her, and had always been plentifully supplied with food. The extraordinary vitality of the species which permits a fragment of the male to perform the act of impregnation is necessary on account of the rapacity of the female, and it seems to be only by accident that a male ever escapes alive from the embraces of his partner. (Howard, 1886, 326)

Six years later, in a brief (half page), second-hand account, Riley and Howard modified Howard's original claim only slightly. "The nonchalance with which the male devoted himself to the sacrifice . . . indicated that the male has no serious objection to this method of suicide" (Riley and Howard, 1892, 145; also see Lampa, 1894; Baltchley, 1896). Howard only echoed the well-established fact that cannibalism does occur and appended his thought that the male must be capable of mating without a head to compensate for the female's rapacity.

And now we come to a historical turning point. By this time, anthropomorphic explanations for cannibalism could not be taken seriously; wanton voraciousness was no longer an acceptable alibi, but an alibi was needed nonetheless. Further, cannibalism could not be explained simply as a part of the mantid's normal eating behavior; it was now well established that "normal" prey recognition meant recognizing fly-sized prey, not newt-sized prey. So, the only acceptable alibi for cannibalism stemmed from the "promising" idea that it was linked in some way to sexual behavior. This alibi made perfect sense, of course, if one accepted Raubad's hypothesis that the cerebral ganglia "directly influence the copulatory reflex, by exercising an inhibitory action on them"; under these circumstances, partial cannibalism of the male would "release" his copulatory reflex.

This transition from the wanton voraciousness alibi to the sexual behavior alibi is clearly evidenced in the fact that when the former alibi faded from the literature so did all references to cannibalism between sexually immature and same sex conspecifics, because these instances did not fit the sexual behavior model. In fact, discussions of mantids hunting any large prey subsided about this time (Prete and Wolfe, 1992). The reason, of course, is that if one blames cannibalism on wanton voraciousness, then accounts of male-on-male and female-on-female cannibalism and of sexually immature mantids cannibalizing each other are of great interest (e.g., Rösel's or Fabré's accounts). On the other hand, if one uses the idiosyncracies of mantid mating behavior as the context for cannibalism, then only instances of adult female-on-male or male-on-female cannibalism are of potential interest. However, because adult female mantids are generally larger, stronger, and heartier eaters than are males, male-on-female cannibalism seldom occurs among adults. However, if male-on-female cannibalism occurred with any frequency, we are sure that a story would have sprung up claiming that the female has evolved such that decapitation releases the egg-laying response from cerebral inhibition as a protection against a male's attack—recall Chopard's account.

We come now to a watershed event, the publication of Ken Roeder's paper, "An Experimental Analysis of the Sexual Behavior of the Praying Mantis (*Mantis religiosa*, L.)" (Roeder, 1935; also see Roeder, 1963). Roeder was, of course, aware of much (if not all) of the literature that we have presented here. In his references he listed, for instance, Aoki and Takeishi (1927), Chopard, Fabré, Howard, Riley and Howard, and Raubad. In addition, however, he cited a fascinating article, the historical importance of which has been discussed only once (Prete and Wolfe, 1992). This article, which appeared in *Transactions of the Academy of Science, St. Louis*, in 1913, was neither poetic nor anecdotal (as were Fabré's accounts), nor was it based on single observations (as were the accounts of Chopard, Howard, Howard and Riley, and Raubad). It is an extensive, "modern" study undertaken by Phil and Nellie Rau on the biology of *Stagmomantis carolina*.

The Raus began their article by reminding their audience of just how fascinating mantid cannibalism is: "The fact that the female mantis *almost always* devours her mate while the pair are in copula, and the male unresistingly clings while he is slowly being eaten, makes the mating habits of this species arouse more than ordinary interest (emphasis added)" (Rau and Rau, 1913, 29). The Raus' fascination with this event is evident in both their text and the accompanying photos. In fact, their fascination is so great that, as one reads the article, one comes gradually to a realization that would have been impossible for accounts based on just one or two observations: that the observations were dramatically shaped by the authors' fascination with cannibalism itself. In fact, the Raus' opening claim that "the female mantis almost always devours her mate" was not even remotely supported by the behaviors of the mantids that they observed: In only three of the more than thirty matings that they watched were males cannibalized, and of those few cases, at least one had nothing to do with sexual behavior!

Despite their lack of evidence, however, the Raus concluded: "The practice of the female devouring her mate may be one of the little economical devices of Nature. . . . Why should he not go to help nourish the female while she goes through the function of egg-laying?" (p. 39). To dramatize their point, the Raus included a full-page photograph of half-devoured males, still in copula and pinned to their mates, as were the specimens sent to Riley and Howard two decades before. (By the way, the Raus included the Riley and Howard paper among their references.)

So now we have a plausible explanation for a behavior that originally had been seen as strange or abnormal: Cannibalism was one of the "little economical devices of Nature." There might have been a problem with this explanation if anyone remembered that the earliest accounts of mantid cannibalism did not usually involve sexual partners. In fact, most were reports of sexually immature nymphs or of females eating one another. There might also have been problems if anyone took seriously the accounts of mantids eating other large prey such as newts, or if anyone stopped to wonder why so few of the Raus' males actually were eaten while mating.

In spite of the stark inconsistency between the Raus' data and their belief in the regularity of mating-related cannibalism, claims such as "[the] female generally devours the male after copulation" (Lefroy, 1923, 49) appeared ever more frequently in both the scientific and the popular literature after the Raus' paper was published, and cannibalism gradually came to be seen not as a product of the mantid's pugnaciousness, not as a part of its normal eating behavior, but as a regular—sometimes even necessary—part of the mating ritual. As the Harvard entomologist William Wheeler explained, "the nutritive meaning of the male to the female is clearly revealed, for both Fabré, Howard and the Raus have shown that he himself is actually devoured piecemeal by his spouse after copulation" (Wheeler, 1928, 160). This was an amazing leap from anecdote to grand theory, but the leap seemed so completely reasonable that even articles that presented well-balanced accounts of the mantid's eating behaviors could not resist the inevitable comment, "the female almost invariably devours her mate" (Teale, 1935, 25).

Two decades after the Raus published their report, Ken Roeder wrote his classic article on mantid sexual behavior (Roeder, 1935). In it, he actually argued against the position that males are regularly cannibalized by females during mating, claiming both that the event is strictly fortuitous and that most males remain unscathed. Further, he reiterated Fabré's claim that mantids capture a variety of large insects and he argued, as did Fabré, that cannibalism is exacerbated by the close quarters in which mantids are generally kept. However, the work had little impact for two important reasons: First, as in the Raus' paper, there are critical inconsistencies in Roeder's portrayal of mantid predatory behavior, and these inconsistencies allowed a variety of misinterpretations to flourish. Second, Roeder offered plausible experimental evidence that appeared to support the hypothesis (originally proposed by Raubad) that decapitation releases the mantid's copulatory reflex from the inhibitory influences of the cerebral ganglia.

The inconsistencies in Roeder's paper involve his claims about the size of the mantid's prey. Although he claimed that male mantids are eaten simply because their movements attract the female's attention (i.e., there is nothing unusual about a female mantid capturing a large insect) and even noted a case of a mantis eating a lizard (Morgue, 1909), he also claimed that mantids will eat only those insects that are "not too large" (Roeder, 1935, 205). He then tried to reconcile these contradictory claims about the size of prey that a mantid will capture by reiterating the old belief that it is unusual circumstances (i.e., overcrowding or an unusually high level of hunger) that lead to the capture of large conspecifics. Hence, even Roeder's elegant analysis begins with the longstanding assumption that the capture of large prey is in some way unusual behavior and, therefore, needs special explanation.

The second reason that Roeder's paper did not change the general misconceptions that people had about mantid predatory behavior is that Roeder not only offered experimental evidence that removal of the male's head disinhibits the copulatory reflex, he also reintroduced Raubad's model of mantid brain organization to explain the phenomenon. (Note that although Roeder cited Raubad's paper, he did not give Raubad credit for originally coming up with this idea.) Roeder's experiments demonstrated that removal of the subesophageal ganglion (SEG) either through decapitation or by severing the ventral nerve cords between the SEG and the first thoracic ganglion yielded copulatory movements which (he thought) appeared normal even in sexually immature adults. It is particularly noteworthy, however, that Roeder also found that this operation led to variety of other not-so-normal behaviors, such as a general reduction in muscle tone, spontaneous locomotor movements occurring with the copulatory movements, and immediate grasping of any elongated object accompanied by "violent attempts" at copulation (Roeder, 1935, 212–13). Unfortunately, these behaviors have been subsequently ignored.

Based on his observations, Roeder hypothesized that "the nerve center responsible for . . . copulatory reflexes, is situated in the last abdominal ganglion" (Roeder, 1935, 216). Of course, the idea that complex behaviors were organized within individual ganglia was already well established (Baldi, 1922; Bethe, 1898; W. von Buddenbrock, 1921; Ewing, 1904; Kopec, 1912). Roeder went on to explain that the SEG contains a "center which is antagonistic to the last abdominal center, inhibiting copulatory movements except when the male is in contact with the female" (Roeder, 1935, 217–18). Hence, contact with a female or extirpation of the SEG removes the inhibitory influences to the last abdominal ganglion, and copulatory behavior is released.

So, on the one hand, Roeder argued that mantid cannibalism is simply a fortuitous event that is not a normal part of mantid predation or mantid mating: "Insects up to the size of the mantis may be caught and eaten, including other mantids of the same species. This fortuitous cannibalism is probably minimized in nature by the inactive and cryptic habits of mantids. [A] cannibalistic attack may take place during the male's [courtship] approach, directly after mounting, or as the couple separates [but] it is not inevitable" (Roeder, 1963, 204–9). Then, on the other hand, he attempts to make mating-related cannibalism look like a well-designed behavior by claiming (erroneously) that "the head and prothorax of an approaching male are naturally most exposed to the female's attack, and are therefore eaten first" (Roeder, 1963, 136). The implication is, of course, that the remainder of the male stays intact and can complete the mating.

Roeder's claim that male mantids are neatly decapitated in conjunction with his seductive neurological model made an irresistible combination. Although he argued against the idea that mantid cannibalism occurs with any frequency in the wild, Roeder ended his 1935 article with this comment: "Taken together, the continuous copulatory and lateral locomotor movements constitute a very beautiful means for securing fertilization of the female, should the preliminary courtship be unsuccessful and the male captured" (Roeder, 1935, 218). He never did explain, however, how or with what frequency the decapitated rear half of the male managed to escape from the female's clutches, walk around to her back end, mount her, hang on tightly, and successfully mate. That would be quite a feat for a severely mangled male!

Thus, one could cite Roeder selectively, and still argue for the regularity (or necessity) of mating-related cannibalism without obviously contradicting or challenging Roeder's work. For instance, Roeder's findings are presented in some detail in the textbook *Mechanisms of Animal Behavior* (1967). In this text, the authors leave the reader with the unmistakable impression that decapitation of the male, if not absolutely necessary, certainly "facilitate[s] the occurrence of sexual behavior" (Marler and Hamilton, 1967, 208–10). Similarly, in a 1980 symposium on Orthopteran mating systems, it was suggested that male mantids may be engaging in a "strategy of 'investing all' in the female and/or offspring by being eaten . . . (especially if the female is undernourished)." Here, one of Roeder's strongest claims is used ("the male's head is eaten first"), and it is suggested that this may be "an evolved tactic by which the male ensures copulation . . . if the female begins feeding on him" (Gwynne, 1983, 346). From this hypothesis it is a very short step to the following claim, which appeared in an ethology textbook: "In some species of mantis the males are themselves the food offering. . . . In the grisly ceremony . . . the male's head [is] eaten [and] the remainder of his body mates. . . . After mating . . . the female will consume the remainder of the male" (Gould, 1982, 365). Finally, consider this claim: "Females commonly attack and devour males either before or after copulation. . . . Destruction of the [subesophageal] ganglion allows copulation to proceed" (Vickery and Kevan, 1985, 89). Interestingly, but unfortunately, this is precisely what many of us were taught in one biology class or another. As it was explained to one of us by a senior graduate student: "The mantis has two brains so the female has to rip his head off to get him to mate."

When Roeder (inadvertently) gave people the means to link cannibalism tightly to sexual behavior (i.e., he popularized a neurological model which made female-on-male cannibalism biologically understandable), he left unexplained all other types of cannibalism (male-on-female, female-on-female, and that between sexually immature mantids). Further, he left unanswered the questions of how or why females recognize male mantids, or other large creatures, as prey. Unfortunately, until very recently, no one has tried to answer these questions. Let us explain why.

As we have noted, the earliest accounts of cannibalism referred only to mantid pugnaciousness or voraciousness, not to their gender. Neither did gender enter into subsequent, popular accounts of mantid cannibalism. Characteristic is this description: "If two of these insects be shut up together . . . they deal each other blows with their front legs, and do not leave off fighting until the stronger has succeeded in eating off the other's head" (*Popular Science Monthly*, 1874, 711). Virtually the same account appears in *The Insect World* in 1869, with the matter-of-fact addendum that "the male being smaller than the female, is often its victim" (Figuier, 1869, 290). In fact, voraciousness was attributed equally to both sexes: "The winner, that is to say the survivor, generally consummates his victory by devouring the body of his slaughtered foe" (Wood, 1871, 485). What happened to these types of anecdotes? In other words, after Roeder, why were there no doctoral dissertations on the ecological or evolutionary significance of same sex cannibalism? Why were there no early– to mid–twentieth-century experimental papers describing the ways in which mantids recognize conspecifics or newts as prey? We believe that these omissions were the product of three interacting forces.

The first, as we have explained, was the general repulsion to animal cannibalism. Early on, cannibalism could be explained away by accusing the mantid of being psychologically disturbed, that is, of having an evil, primitive, or undeveloped temperament. Later, cannibalism could be described as an evolutionarily sound strategy by which males "invest" in their offspring. This model is given credence by the fact that, under just the right circumstances, the half-eaten male can still mate. The second force responsible for the omissions noted above is that it is nearly impossible to believe that the ability of a half-eaten male to successfully mate could be simply a fluke of nature, especially in the face of Roeder's seductively straightforward neurological explanation of the release of the copulatory reflex. But as powerful as these two factors were, there was a third, even more influential, force that maintained the conceptual dichotomy between mantid predation on small versus large insects. This force was the belief that the mantis is designed, evolutionarily speaking, to be the perfect fly-catching machine. Obviously, the latter belief does nothing but strengthen the long-held contention that predation on large organisms like newts and conspecifics is not part of the mantid's normal predatory repertoire and, consequently, needs a special explanation.

Born to Catch Flies

As we have explained, the mantids were introduced to a broad readership through the works of Aldrovandi and Moffett, and during the eighteenth century authors continued to rely on these sources for their information about these unusual "locusts." However, natural history expeditions, the collecting and trading of specimens, and an array of often beautifully illustrated natural history texts led to the dissemination of increasingly more firsthand information about mantids and their behaviors (for references see Prete and Wolfe, 1992). One outcome of this dissemination was the recognition of the strikingly diverse morphologies among mantid species. Some mantids resemble phasmids (i.e., leaf insects), which led to the mantids being grouped for a time with the phasmids rather than the locusts. However, it was quickly realized that there are two critical differences between phasmids and mantids, which led to yet another taxonomic reclassification.

In 1797 a paper by Anthony A. H. Lichtenstein, read before the Linnean Society of London, proposed that the praying mantids be placed in their own genus (Lichtenstein, 1802; also see Stoll, 1787). In characterizing the morphological and behavioral traits that set the mantids apart, Lichtenstein captured much of what would remain the defining characteristics of the group for the next two centuries. In fact, two of the morphological traits that he cited are precisely those that elicit so many anthropomorphisms from virtually everyone who sees a mantid: keen eyes set in a highly mobile head and dexterous hands. The critical, defining behavioral trait that Lichtenstein noted was, of course, that mantids prey on living creatures. Historically, this latter trait has elicited as much approbation and moral condemnation as the former characteristics have elicited identification. Indeed, it is the "humanlike" qualities of the mantis that make its "uncivilized behavior" so appalling to so many.

Given the consensus that mantids feed on flies or fly-sized prey, and observations of the amazing speed and accuracy with which mantids capture their prey, it was inevitable that late nineteenth- and early–twentieth-century aficionados would conclude that the mantid's forelegs are remarkably well evolved for the task of capturing flies: "the peculiar structure of their fore-legs [is] marvelously adapted for seizing flies and other insects on which they feed" (*Saturday Review*, 1890, 735); and "Their wonderful raptorial forelegs . . . are amazingly rapid and dexterous, often capturing an insect as it flies past" (Bateson, 1913, 174).The belief that the mantid's forelegs are specifically adapted to capture fly-sized prey was allowed to develop precisely because the mantids' capturing of larger (even vertebrate) prey was seen as highly anomalous. Certainly, mantids could not have evolved specifically to capture fly-sized prey if, in fact, they regularly feed on much larger prey. So, from this perspective, too, the fact that mantids capture (large) conspecifics had to be accounted for by a special explanation.

The convenient separation of mantid predatory behavior into two categories (normal eating versus cannibalism) in the early twentieth century was simply the continuation of a century-old dichotomy. In the light of continued (though sporadic) reports of mantids eating large prey, the dichotomy might have eventually disappeared had mantids not been used as an experimental model system by some early ethologists (Thorpe, 1979). The early ethologists focused their attentions on innate, or instinctive, behaviors, the belief being that even the most complex of these behaviors could be broken down into discrete components or subroutines, each of which could, in theory, be acted upon by natural selection. This way of viewing innate behavior fit well with prevailing assumptions as to the generally mechanistic nature of invertebrate (including insect) behavior (e.g., von Uexküll, 1909; Lorenz, 1950, 1981; Tinbergen, 1951). As luck would have it, several of the mantid's interesting behaviors made it an appealing experimental animal on which to test these assumptions.

The ability to conceptualize and explain many seemingly complex behaviors in terms of chains of simpler, discrete components was aided by ethology's timely convergence with control theory, or cybernetics (e.g., Wiener, 1948; Hoyle, 1984). An insightful and creative pioneer in this early merger, as well a student of insect behavior, was Horst Mittelstaedt. As a student, Mittelstaedt had been included by his mentor, and Europe's premier biologist, Erich von Holst, in the small, close-knit group (which included Lorenz and Tinbergen) that would become the kernel of the modern ethological movement (Thorpe, 1979). Mittelstaedt's early work included the enthusiastic application of cybernetics to the visual tracking behavior of the mantis (Mittelstaedt, 1957, 1962).

In order to pursue his interest in the control of visual tracking behavior in mantids, Mittelstaedt collaborated with a well-known American ethologist who was considered the expert on mantid neurophysiology at the time. That American ethologist was Ken Roeder—the very same person whose work we discussed earlier. As an aid to their studies, one of Roeder's graduate students, Susan Rilling, made a series of paper lures that were used to elicit tracking and striking behavior by the mantids. Although this aspect of Rilling's work was not central to the senior investigators' project, she pressed for publication of her findings on the effects of various lure configurations on mantid predatory behavior (Prete and Wolfe, 1992). This first, systematic analysis of mantid prey recognition, coauthored by Rilling, Mittelstaedt, and Roeder, was published in 1959, well after Roeder's original work on mantid sexual behavior (Roeder, 1935).

Rilling's mantids—which, by the way, she had raised on flies—were presented with both flies and a series of paper dummies hanging from strings that were twirled and/or swung back and forth. The results of her experiments led Rilling to the conclusion that the strongest stimuli for eliciting prey-catching behavior were dummies that mimicked the appearance and movements of flies. What is critical here is that Rilling constructed no dummies that were other than flylike. That is, all were small ovals or rectangles with legs or wings affixed, and all were dangled in the air. None resembled the other things that mantids were known to capture: none looked like other mantids, newts, mice, caterpillars, butterflies, or the like. As I am sure you have guessed, the assumption behind Rilling's choice of stimuli must have been the belief that the most potent releasing stimulus for prey capture would turn out to be flylike.

Rilling's article became the final word on mantid prey recognition for three decades in spite of the fact that a study published five years later demonstrated that at least one species of mantid strikes most frequently at elongated stimuli almost twice as large as Rilling's (Holling, 1964). The impact of Rilling's article was so strong that subsequent experimental studies were generally oriented toward verifying the hypothesis that both mantid behavior and morphology have evolved specifically to capture fly-sized prey (Prete, 1995; Prete and Wolfe, 1992).

The results reported by Rilling et al. completed the picture of mantid predatory behavior begun two centuries earlier. Their results, in combination with previous findings on mating behavior, appeared to support the general belief that mantids have two separate predatory strategies. The strongest eliciting stimulus for prey capture (i.e., the eliciting stimulus to which hungry mantids respond innately) is a fly or a flylike object. The fact that same-sized (or near same-sized) conspecifics are sometimes eaten is a product of something other than normal eating behavior: It is due to "abnormal" living conditions (e.g., overcrowding), to an "abnormally" high hunger level, or to a special mantid reproductive strategy (e.g., an investment strategy by the male, or the decapitation-induced release of the inhibitory controls over the male's copulatory reflex).

What, then, became the explanation for the capture of large prey when it clearly wasn't an instance of so-called sexual cannibalism; that is, when cannibalism was male-on-female, between members of the same sex, or between immature mantids? Interestingly enough, accounts of these instances simply disappeared from the literature after Roeder (Prete, 1995; Prete and Wolfe, 1992). And what about accounts of mantids eating newts and the like? Until relatively recently, instances of mantids capturing very large prey was considered too anomalous to be explored experimentally.

Epilogue

Happily, the study of mantids has changed dramatically. The mantid's method of capturing its prey no longer dominates the field, although it remains an important area of research for a number of reasons. Contemporary research on mantids now spans a variety of disciplines, and research in each one of them is exciting. The creativity, insight, and candor of those who have contributed to this book are evidence of that fact. To some extent, the growing interest in all aspects of mantid behavior is a product of a much larger change in people's attitudes toward insects in general. The character of this new vision was summed up best by colleagues who study another orthopteroid insect, the grasshopper. In their words: Insects display "an impressive array of mechanisms for behavioral flexibility and give lie to the idea that [they] are simple, hardwired automata" (Simpson and White, 1990, 512). And, in our (perhaps subjective) opinion, mantids display the most impressive array of all.

2. Morphology and Taxonomy

Roger Roy, translated by Cathryn Easterbrook

The Position of Mantodea within the General Classification of Living Beings

The Mantodea are, first of all, part of the animal kingdom, now restricted to multicellular animals or metazoa (Margulis and Schwartz, 1982). Obviously, they are put among the Invertebrata, a heterogeneous assemblage that commonly gathers all animals without vertebrae, unlike the monophyletic group of Vertebrata. More specifically, the mantids are considered triploblastic due to the formation of three germinal layers, Coelomates due to the appearance of coelomic chambers, and Protostomes because the mouth is derived from the blastopore. These factors place them among the large phylum Arthropoda, which is characterized by articulated appendages carried by a metamerized body formed of consecutive segments.

The presence of antennae and mandibles defines the mantids as Antennata as opposed to Chelicerata, and the simple character of the appendages places them within the Uniramia rather than Biramia (i.e., the Crustacea). The six locomotor appendages attached to the thorax define them as Hexapoda in contrast to the Myriapoda, and the arrangement of the mouth parts defines them as Ectognathata (Insecta *stricto sensu*) rather than Entognathata.

Within the Insecta, mantids can be categorized successively as members of the Euentomata, the Dicondylata, the Pterygota, the Opisthopterata, the Neopterata, the Polyneopterata, the Orthopterodida, the Blattiformida, the Dictyopterida, the Cursorida, and the Blattarida (Boudreaux, 1979). This last subdivision, which for this author is at the rank of subtercohort, groups together Isoptera (termites), Blattaria (cockroaches), and Mantodea (mantids), a relationship recognized by all authors. There is not, however, unanimous agreement on the ranking of these insects at the level of order: Some think that the three constitute a single order, some identify two orders by grouping the Blattaria and Mantodea within Dictyoptera and leaving the termites alone as Isoptera, and others believe that each of the three groups belongs to a distinct order (Fig. 2.1).

Here, I opt for the middle ground which I feel is most consistent with the biological reality. Members of Dictyoptera (or Dictuoptera) and Isoptera both have heads with compound and simple eyes, primitive grinding-type mouthparts, four wings with the forewings overlapping the hindwings on the dorsal surface when at rest, an abdomen with ten segments that typically terminates in multiarticulated cerci and uniarticulated styles, and a holopneustic respiratory apparatus with two pairs of thoracic stigmata and eight pairs of abdominal stigmata. The majority of these characteristics are also found in other orders of Polyneoptera. However, Isoptera are clearly distinguished from Dictyoptera by a typically prognathous and virtually immobile head, fairly short and moniliform antennae, poorly developed pronotum, a generally reduced number of tarsal segments, virtually identical membra-

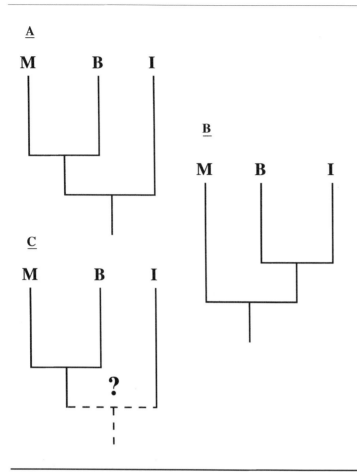

Fig. 2.1 Three recent hypotheses for the relationships among the Mantodea (M), the Blattodea (B), and the Isoptera (I): *A*, Thorne and Carpenter (1992) proposed that the mantids and roaches are sister groups and that the termites are a sister group to the mantids+roaches. *B*, Kristensen (1995) suggested that the roaches may be more closely related to the termites than to the mantids. *C*, Using molecular techniques, Kambhampati (1995) found strong support for the closer relationship of mantids and roaches, but could neither confirm or refute the sister group status of the termites. In this paper, the order Dictyoptera is taken to comprise two suborders, Blattodea and Mantodea; Isoptera is considered a separate, but closely related, order.

nous wings with a reduced anal area and a basilary suture (which permits autotomy), the occasional presence of styles in females, and a reduced or often absent genital armature in the males, ovipositor in the females, and a relatively small number of Malpighian tubes (most often two or four pairs, but sometimes six or eight). Further, their eggs are not grouped in oothecae, and they are always social with differentiated castes.

In contrast, the Dictyoptera have a very mobile head that is rarely prognathous, antennae typically filiform and elongated, a well-developed and always evident pronotum, tarsi that usually have five segments (but only four in regenerated legs), dissimilar wings, forewings that are thickened in the tegmina with a still apparent venation, and hindwings that are membranous with a highly developed anal area, styles only in the males, an asymmetrical genital armature in males, typically a ventral hypophallus and two epiphalli (right and left), a symmetrical ovipositor in the female, numerous Malpighian tubes, and eggs that are grouped in oothecae. Finally, the Dictyoptera never demonstrate eusociality or castes. It also should be noted that in Dictyoptera the wings often show various degrees of reduction that may render them completely nonfunctional, especially in the females, but sometimes the wings may be missing even in the males.

Mantodea are essentially distinguished from Blattodea—considered here as suborders—by the following characteristics: Their head is always very evident, typically orthognathous, and sometimes more or less prognathous or opisthognathous while that of Blattodea is always opisthognathous and more or less hidden by the front of the pronotum. Mantodea always have three simple eyes, while Blattodea only have two. Mantodea display an often substantial sexual dimorphism in their antennae that is almost nonexistent among the Blattodea. The postclypeus of Mantodea is almost always well defined and forms a frontal sclerite, while this structure is much less clear in Blattodea. The pronotum, generally several times longer than it is wide, is divided into a prozone and metazone by a supracoxal groove, while that of the Blattodea is always without separation and usually wider than it is long. The mantid's forelegs are raptorial with elongated coxae and femora and the presence of opposed rows of spines on the femora and tibiae. In Mantodea, only the first segment of the tarsi is very elongated, while in the Blattodea the last one is elongated as well. The tegmina have a pterostigma that is generally very apparent.

The oothecae of Mantodea have a foamy outer layer of varying thickness, which hardens shortly after laying, while the oothecae of Blattodea have no foamy layer and may be carried for a considerable time by the females before being deposited. Life is always solitary and more or less heliophile for the Mantodea, while Blattodea often live in groups and are generally nocturnal and occasionally burrowing. The diet of Mantodea is always that of a predatory carnivore, with frequent cannibalism (see Maxwell, this volume, chap. 5; Hurd, this volume, chap. 3), while Blattodea are saprophage or detritivore, with some being additionally guanivore. Finally, it is to be noted that the distribution of mantids is more restricted than that of Blattodea and more limited to warm regions (as is that of the Isoptera).

The adaptations of the forelegs for predation are always critical to the identification of Mantodea. However, it is important to note that these raptorial forelegs are not exclusive to Mantodea. In fact, they are also found in Reduviidae Emesinae (Heteroptera), Mantispidae (Neuroptera), and Empididae Hemerodromiinae (Diptera). Interestingly, it is the Mantispidae that show the greatest overall morphological similarity to the Mantodea.

Main Features of Mantodea and the Limits of Variability

Mantodea are generally fairly large in size and more or less elongated in shape (e.g., Figs. 2.2–2.5). The smallest species is without doubt *Mantoida tenuis,* of the neotropical forest, in which some males barely reach 1 cm in length, while the largest is incontestably *Ischnomantis gigas,* from the savannas of Western Africa, whose females may grow to over 17 cm. In terms of weight, two genera from the tropical forest regions spring to mind: *Macromantis* in America and *Plistospilota* in Africa, each of which have females weighing approximately 5 g (precise weights remain to be determined for living specimens).

The Head

In mantids, the head is generally wider than it is long, more or less triangular in shape, and held in a typically orthognathous position. In certain genera, however, such as *Compsothespis, Orthoderella,* and *Calamothespis* (each of which are in different subfamilies), the head is longer than it is wide and has a tendency toward prognathism. In general, the vertex of the head is most often straight or slightly convex, possibly irregular, and only rarely concave. In the latter case, the concavity may be great, such as it is in *Schizocephala bicornis,* the lone representative of the Schizocephalinae (e.g., Fig. 2.6). Well-individuated lateral tubercles, which may be rounded or angular, are found in some taxonomically disparate genera such as *Amorphoscelis, Hoplocorypha, Danuria,* and *Achlaena.* A median prolongation of the head is found regularly in some groups of mantids (e.g., Sibyllinae and Empusidae; see Fig. 2.7) and

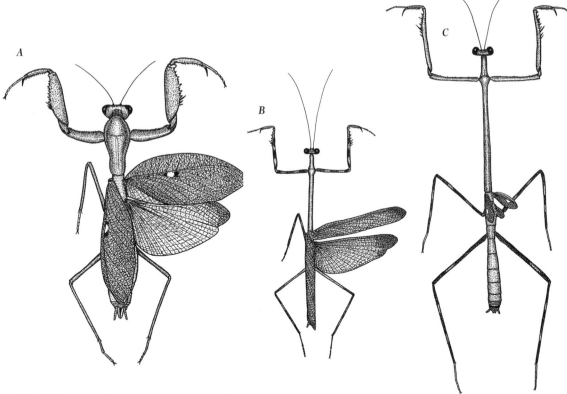

Fig. 2.2 *A*, Mantidae: *Sphodromantis lineola* (Burmeister) female. *B, C*, Mantidae: *Euchomenella heteroptera* (de Haan), male, female, respectively. Reprinted from Bragg (1997) with the kind permission of Phil Bragg.

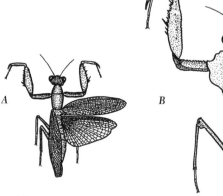
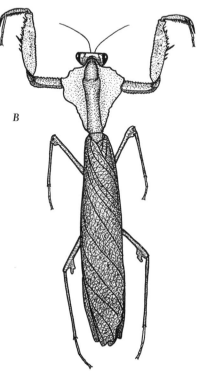

Fig. 2.3 *A*, Hymenopodidae: *Odontomantis micans*, female. *B*, Mantidae: *Deroplatys desiccata*, male. Reprinted from Bragg (1997) with the kind permission of Phil Bragg.

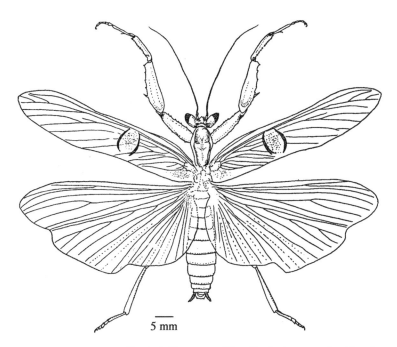

Fig. 2.4 Hymenopodidae: *Creobroter* sp., male. Adapted from Mukherjee et al. (1995) with the kind permission of Associated Publishers.

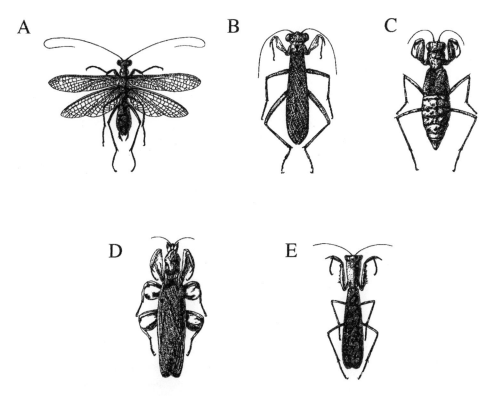

Fig. 2.5 *A*, Mantoididae: *Mantoida brunneriana*, female. *B*, Amorphoscelidae: *Amorphoscelis annulipes*, female. *C*, Eremiaphilidae: *Eremiaphila denticollis*, female. *D*, Hymenopodidae: *Hymenopus coronatus*, female. *E*, Mantidae: *Orthodera ministralis*, female. Adapted from Chopard (1949).

frequently in other groups (e.g., Tarachodinae, Vatinae, Hymenopodidae), but is never found in the remaining groups. This prolongation may be more or less developed, simple or triangular, bifid (sometimes with lateral denticulations), irregular and asymmetrical, or even sometimes foliaceous. The presence of both a median prolongation and lateral tubercles is infrequent; the most remarkable instance of such an occurrence is in *Stenophylla cornigera*, the lone representative of the Stenophyllinae.

The compound eyes are always well developed in mantids and may contain up to ten thousand ommatidia each. They are usually globular but sometimes slightly conical, and they are generally elongated when the head is longer than it is wide. The surface of the compound eyes may show a nonvisual elongation that is most often

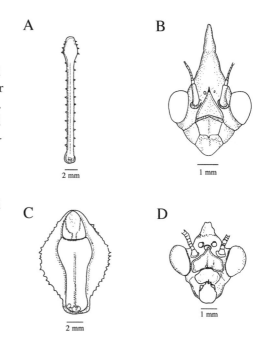

Fig. 2.7 *A, B,* Empusidae: *Empusa spinosa* (Kraus), female; pronotum, head, respectively. *C, D,* Empusidae: *Blepharopsis mendica* (Fabricius), female; pronotum, head, respectively. Adapted from Mukherjee et al. (1995) with the kind permission of Associated Publishers.

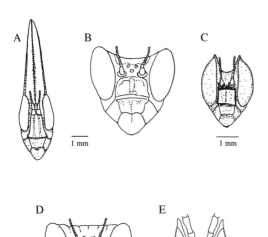

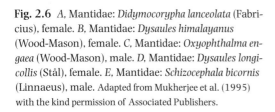

Fig. 2.6 *A,* Mantidae: *Didymocorypha lanceolata* (Fabricius), female. *B,* Mantidae: *Dysaules himalayanus* (Wood-Mason), female. *C,* Mantidae: *Oxyophthalma engaea* (Wood-Mason), male. *D,* Mantidae: *Dysaules longicollis* (Stål), female. *E,* Mantidae: *Schizocephala bicornis* (Linnaeus), male. Adapted from Mukherjee et al. (1995) with the kind permission of Associated Publishers.

pointed. Such a lateral elongation may be small as in *Calamothespis* or large as in *Oxyothespis*. An oblique elongation is found in *Toxodera*, *Pseudacanthops*, and some *Heterochaeta*; an anterior elongation is found in *Episcopomantis*. Tubercles may be found in the border of the eye: a single small one in *Sphodromantis*, two in Paraoxypilinae, and three in Amorphoscelinae. Mantids always have three simple eyes (ocelli) arranged in the shape of an isosceles triangle whose angle at the vertex varies from 75 to 120 degrees. The simple eyes are always globular, and they are generally smaller in the female than in the male. In two cases—Paraoxypilinae and some of Vatinae—they are carried by ocelligerous tubercles.

The antennae of mantids are most often filiform and fine and consist of a large number of segments. They may be glabrous or garnished with setae, which can be fairly long and abundant in the case of the males of *Gonypetella* and *Promiopteryx*. The antennae are bulbous over the length of their basal region in *Schizocephala* and *Brunneria*, but in other cases are moniliform, serrulated, pectinated, or even bipectinated as in male Empusidae.

The postclypeus is differentiated by a frontal sclerite, which is usually well defined with a surface that may be flat, convex, or concave. It may also be bordered, possibly with granules, striations, ridges, and crests. Further, it may be transverse as in *Solygia*, or more or less elevated with a rounded or angular superior border, which is sometimes prominent as it is in Empusidae. The clypeus and the labrum cover the typical grinding mouth parts, which include quite powerful mandibles, maxillae with five-segmented palpi, and a labium with three-segmented palpi. The surface of the head may be smooth and shiny or dull, or more or less irregular and marked with numerous granulations, which may be affixed to the various extensions described above. This leads to a shape of great complexity in *Metoxypilus spinosus* and *Pseudacanthops spinulosa*.

The Pronotum

The mantid's pronotum is always divided into a prozone and a metazone by a supracoxal sulcus, with the prozone always being the shortest (e.g., Figs. 2.8B, 2.9B, and 2.10D). It is seldom wider than it is long—a configuration that occurs only in Amorphoscelinae (e.g., Fig. 2.10A) and Eremiaphilidae— and it is never wider by much. The pronotum is approximately as wide as it is long in Perlamantinae and Mantoididae and a bit longer than it is wide in Chaeteessidae and Metallyticidae.

Most of the time the pronotum is clearly longer than it is wide: In general it is two to three times longer in Amelinae, Iridopteryginae, and Hymenopodinae; four to eight times longer in the majority of species; and more than ten times longer in a few groups such as Angelinae and Schizocephalinae. The record is apparently a pronotum twenty times longer than wide (80 × 4 mm) for a female of *Leptocola stanleyana* (personal observation). When the pronotum is short, its widest point is generally in the prozone. For most species this corresponds to a supracoxal enlargement of varying degree, which is generally rounded but may be angular or even spiny. The lateral borders of the pronotum may be smooth, granular, or more or less strongly dentated as in *Stauromantis* and *Catoxyopsis*. In some species one sees a lamellar enlargement, especially in *Phyllocrania paradoxa*, *Brancsikia*, and *Deroplatys*, *Idolomantis diabolica*, and chiefly in *Choeradodis*, in which the pronotum has become regularly wider than it is long (e.g., Fig. 2.3B). A particularly remarkable case is that of *Pnigomantis medioconstricta*, whose pronotum is shrunken between two enlargements. Generally, the axis of the pronotum is straight; however, it is bent convex dorsally in *Toxodera*. The surface of the pronotum is often

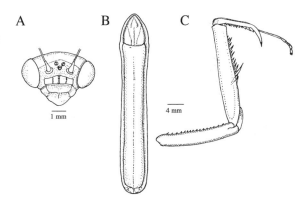

Fig. 2.8 A–C, Mantidae: *Mesopteryx robusta* (Wood-Mason), male; head, pronotum, left foreleg, respectively. Adapted from Mukherjee et al. (1995) with the kind permission of Associated Publishers.

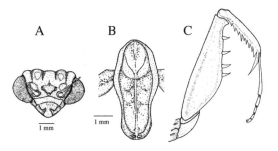

Fig. 2.9 A–C, Mantidae: *Eomantis guttatipennis* (Stål), female; head, pronotum, right foreleg, respectively. Adapted from Mukherjee et al. (1995) with the kind permission of Associated Publishers.

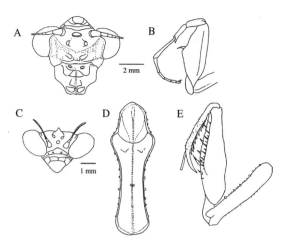

Fig. 2.10 *A, B,* Amorphoscelidae: *Amorphoscelis annulicornis* (Stål); dorsal head and pronotum, left foreleg, respectively. *C–E,* Hymenopodidae: *Acromantis insularis* (Giglio-Tos), male; head, pronotum, right foreleg, respectively.

smooth but may be granular, tuberous, or spiny in various ways, with protuberances that may be bilaterally symmetrical or positioned medially. Such protuberances are particularly uneven in *Junodia* and *Metoxypilus*. Often the metazone presents a medial, longitudinal carina that is generally low and regular, but may be scalloped as in *Enicophlebia*, lobed as in Paraoxypilinae, or spiny as in *Haania*, or have large foliated expansions as in *Paratoxodera cornicollis*.

The prosternum does not generally demonstrate particular morphological characteristics, except for the occasional spines (known as *acetabular*) near the insertion of the anterior prothoracic coxae.

The Prothoracic Legs

The raptorial prothoracic legs, so characteristic of mantids (Fig. 2.11), begin with very elongated coxae, which are generally simple in shape, and end on the ventral (inner) side with two apical lobes, which may be short and separated but which are most often well developed and divergent as in *Solygia*. In some genera (e.g., *Pseudomantis*) these apical lobes are close together, and one may be adjacent to or cover the other as in *Macromantis* or *Brancsikia*, respectively. The lobes are particularly large among the Empusidae, and their development is at its maximum in *Gongylus*. The apical region of the coxae can be dilated, and this is the case among Thespinae and some of Vatinae, mainly *Danuria*. The anterior edge and less frequently the posterior edge may have granules, tubercles, or spines, possibly bulbous or compressed, with the most spectacular case being that of *Cardioptera squalodon*, which bears large spines on the anterior edge of their coxae that are reminiscent of shark's teeth. The ventral side of the coxae sometimes carries a limited number of callused spots in Mantinae, in particular *Sphodromantis* and especially *Paramantis*. Finally, the only known case in which the anterior coxae are lamellar is that of *Idolomantis diabolica*, in which the coxae have two widened edges.

The anterior trochantera, always adjoined to the femora, vary little in either shape or size. On the other hand, the anterior femora are always longer than the coxae and are variously proportioned, sometimes very thin, sometimes bulbous with a lamellous anterior edge in some Empusidae and Hymenopodidae, especially in the genus *Hestiasula* (Fig. 2.12). It also happens that the anterior edge of the femur may have a small subapical lobe as in *Zoolea* and *Popa* or a subbasal lobe as in *Pseudacanthops*. The widest point is found most often at the level of a small depression on the posterior edge of the femur, which is called the claw-

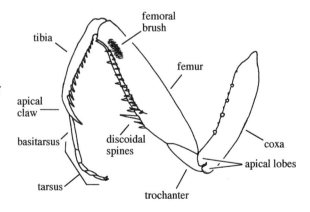

Fig. 2.11 Right foreleg (ventral view) of *Sphodromantis lineola* (Burmeister). After Ragge and Roy (1967), modified.

groove and is used to receive the terminal claw of the tibia when it is bent against the femur. Depending upon the case, this groove may be close to the base of the femur as in *Tarachodes*, in the middle of the femur as in *Mantis*, or fairly close to the apex as in Oligonychinae and some Angelinae such as *Leptocola*.

Except for two subterminal spines, one dorsal and one ventral, the posterior edge of each femur normally bears three series of spines: discoidal spines near the claw-groove, dorsal or external spines (they are on the exterior side when the mantis is "praying"), and ventral or internal spines. The dorsal and ventral series of spines occupy the largest portion of the distal region of the femur. Between these two rows of spines, the posterior surface of the femur is wider than the anterior surface, and it is particularly so in *Chaeteessa*. The spines in the three series have various shapes and dimensions. Although generally pointed, femoral spines may be blunted as they are in some of the internal spines of Perlamantinae, or sometimes very slender as in *Chaeteessa* and *Stenophylla*.

The discoidal spines are absent only in Metallyticidae; there is only one in Perlamantinae and Amorphoscelinae (e.g., Fig. 2.10B); there are two in Chaeteessidae, *Astape* (Haaniinae), and in only some Paraoxypilinae, Compsothespinae, and Iridopteryginae, subfamilies for which the normal number is three. As a rule, there are also three discoidal spines near the claw furrow in Mantoididae, Eremiaphilidae, Photininae, Toxoderinae, and Acontistinae, the spine in the middle being the longest. By far the most frequent number of these discoidal spines is four, with the third being the longest. In many (but not all) cases, these spines are aligned, and often they are preceded by a variable number of small tubercles. Four is the normal maximum number seen, but five discoidal spines have been noted in Empusinae (Giglio-Tos, 1927, 633; Beier, 1934, 3) and *Brancsikia* (personal observation).

External spines are absent in Amorphoscelinae and Perlamantinae. A single external spine in this category is exceptional and is found only in *Metoxypilus* (Paraoxypilinae) and *Thesprotia* (Oligonychinae). The presence of two spines has

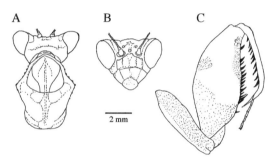

Fig. 2.12 A–C, Hymenopodidae: *Hestiasula castetsi* (Bolívar), female; pronotum and head, head, left foreleg, respectively. Adapted from Mukherjee et al. (1995) with the kind permission of Associated Publishers.

probably never been observed. The presence of three is rare and has been cited only for *Cliomantis* and *Myrmecomantis*. In contrast, four is the most frequent number among the Chaeteessidae, Metallyticidae, Mantoididae, and Eremiaphilidae, and it is also the rule for the majority of subfamilies of Mantidae and various Hymenopodidae (e.g., 2.9C). The presence of five external spines is also relatively widespread, in particular among Tarachodinae, Photininae, Toxoderinae, and Acromantinae; further, it is the rule among Empusidae. Six external spines is clearly less frequent but is seen in the genera *Macromantis*, *Calamothespis*, *Toxodera*, *Acanthops*, and *Gongylus*, among others. The maximum number normally seen is seven, as in *Metilia*, *Paratoxodera*, and various *Calamothespis*, and it may be strictly anomalous that there are sometimes eight among certain species where the usual number is seven, or sometimes even six. The external spines are all usually about the same size and may be fairly long but sometimes are relatively short as, for example, in *Acanthops*. Among Metallyticidae, the first external spine is much longer than the three others while in Schizocephalinae the two median spines are longer than the other two spines. Otherwise, the interval between the two first spines is generally smaller than between the others. In some cases, as in *Miomantis*, *Cilnia*, and *Acanthops* and in the Empusidae, granulations are found between the spines, but this is infrequent.

The internal row of femoral spines is missing in Amorphoscelinae in which the armor of the anterior legs is thus reduced to a single discoidal spine. They are also missing in the Compsothespinae. It appears that there is never a case where only a single internal spine is present: *Paraoxypilus armatus* has two, *Gyromantis* and *Phthersigena* have three, Perlamantinae have four that are quite unequal in length, *Triaenocorypha* has five, and greater numbers up to ten are present among various genera. But it is the numbers from ten to sixteen that are the most frequently observed in many groups. Seventeen internal femoral spines is the maximum seen in *Popa*, *Omomantis*, *Dactylopteryx*, and *Theopropus*, and eighteen is the maximum number seen in *Macromantis*, *Stenophylla*, and *Metilia*. In *Chaeteessa*, *Eremiaphila*, *Empusa*, and *Hypsicorypha* there can be more than twenty internal femoral spines, and overall maximum number of thirty has been counted in *Gongylus gongylodes* and *Idolomantis diabolica*. It should also be noted that the more numerous the spines, the more important are the limits of variation.

The internal spines are rarely all of approximately equal length, although this is the case in the Metallyticidae. The two first and the last two are larger than the others in Eremiaphilidae. The first are also larger in Mantoididae and the last ones are larger among some Amelinae, for instance, *Ligaria* and *Macracanthopus*, for which the last internal spine is enormous. However, the most frequent arrangement is that large and small spines alternate for the complete length of the row (with the first usually being somewhat misaligned from the rest of the row), and sometimes supernumerary small spines are present between the last ones. It is only in the Empusidae that we find an arrangement in which one large spine alternates with three small spines, except for the most basal for which one small spine is situated between two large. Here, too, there may be small supernumerary spines in the distal area, and the small spines may not have the same length. Although the discoidal and external spines described above are always spaced out at their base, this is not always the case for the internal spines, especially when they are numerous.

The femoral brush, which is always present, is at the level of the last internal spine. The brush consists of short, robust bristles, often flattened into a spatula shape, which are used for cleaning the head.

The prothoracic tibiae are always shorter than the femora (up to 4.5 times less), and depending on the particular case, they may be either longer or shorter than the coxae. Except for the genus *Chaeteessa*, they terminate in a sturdy apical claw, and they are often slightly arched. As in the femora, the anterior edge is thin and the posterior edge is comparatively thick; the latter bears two series of spines, one dorsal or external and the other ventral or internal. These spines face the corresponding series of the femoral spines when the tibiae are closed.

The internal and external spines are generally of equal importance, with the external spines most often numbering less than the internal ones. In a general way, the size of the spines increases from the base toward the apex of the tibia, but not always in a regular way. Again, depending on the case, the spines may or may not be contiguous at the base, and are more or less pointed, seldom blunt, and infrequently as slim as are the femoral spines. Further, the number of tibial spines is usually more variable than is the case for the femoral spines.

The armor of the anterior tibiae is reduced to only the apical claw in Amorphoscelinae, Perlamantinae, and Compsothespinae. The situation is the same for Paraoxypilinae if one disregards a small subapical row of internal granules, which face a similar row of granules on the femora. In Oligonychinae the spines of the anterior tibiae—which are always short and bulging—are few in number, particularly the external spines, which sometimes may be missing (as in *Thesprotiella*) or reduced to only one (as in *Thesprotia* and *Oligonyx*), and sometimes some of the internal spines may take an anterior position. The number of external and internal spines is regularly less than ten in Metallyticidae, Mantoididae, Oxypilinae, Schizocephalinae, in the majority of Thespinae, and also in certain Tarachodinae, Iridopteryginae, Amelinae, and Liturgusinae. There are fewer external than internal spines in Chaeteessidae,

Eremiaphilidae, and Sibyllinae, as well as in some of the Angelinae (e.g., *Angela*) and Toxoderinae (e.g., *Calamothespis*). However, the most frequent case among many of the Mantinae and Vatinae is the presence of seven to eleven external spines and ten to fifteen internal spines. The numbers of spines are regularly higher in the Photininae, in many of the Hymenopodidae (for which the external spines are most often short and leaning against one another), and especially in certain Toxoderinae (*Paratoxodera Stenophylla*) and Empusinae. Once again, the record is held by *Gongylus gongylodes*, which has twenty-eight to thirty-four external spines and twenty-seven to thirty-three internal spines. It is an interesting contrast that there can be a total of as many as one hundred spines on each predatory foreleg of *Gongylus gongylodes* but only two spines in Amorphoscelinae!

The anterior tarsi are inserted in an anterodorsal position on the tibiae. They are generally inserted closer to the apex of the claw than to the base of the tibia, but they can be inserted in the middle of the tibiae as they are in Amorphoscelidae or closer to the base than to the apex as in Compsothespinae and *Thesprotia* (Oligonychinae). With some exceptions, the tarsi have five segments. There are only four in certain species of *Heteronutarsus* and in regenerated legs. The penultimate segment is the shortest and the first, known as the *metatarsus* but better termed the *basitarsus*, is the longest segment, often longer than all the others together. The last segment ends in two claws, which always appear to be equal in size and between which one does not find any arolium.

The Mesothorax and Metathorax

The mesothorax and the metathorax are shorter than the prothorax in almost all instances (e.g., Fig. 2.4). Only in Mantoididac does one find the prothorax of equal length to or a bit shorter than the mesothorax. In some cases the mesothorax is longer than the metathorax, but more often it is the opposite. The mesonotum and the metanotum have very similar shapes, with a smooth surface that is a bit raised on the median line. The mesosternum and the metasternum also are similar in appearance, except that the metasternum shelters the unpaired auditory organ between the posterior coxae (see Yager, this volume, chap. 6).

The median and posterior legs are little modified in contrast to the prothoracic legs. They are very similar in shape, with the posterior usually a bit larger than the median. Depending on the group, there is a large variation in the relative length of these legs in relation to the length of the body. For instance, they are especially short in Toxoderinae (especially in *Calamothespis*) and very long in Eremiaphilidae. The absolute record for length is held by one female *Leptocola stanleyana* (personal observation) that has posterior legs 137 mm long (femur = 51 mm; tibia = 63 mm).

The mesothoracic and metathoracic coxae are relatively elongated and are approximately as a sawn-off cone in shape. They are sometimes dressed with foliated lobes as in some Empusidae, *Paratoxodera*, and *Phyllocrania*. The trochanters are always of a reduced size and without special characteristics. The femora are most often elongated and simple in shape and without foliate expansions. The latter, however are fairly common in Hymenopodidae, Empusidae, and various subfamilies of Mantidae. In many cases these adornments are only subapical lobules that are more or less rudimentary on the posterior side as, for example, in female *Stenopyga* and certain species of *Cardioptera*. The corresponding subfamilies of Angelinae and Photininae never have larger expansions, but there are comparable lobules in various Vatinae such as *Chopardiella* and *Heterovates* and in many of Hymenopodidae such as *Epaphrodita* and *Chlidonoptera*; in these latter groups many species display very large expansions. A much rarer case exists in the *Hestiasula* and *Pseudacanthops*, which have a femoral lobule in a basal rather than subapical position. There are a number of cases in which one finds large lamellose expansions that display a degree of diversity. For example, there are semicircular expansions in the females of *Hymenopus coronatus*, expansions with sinuous contours that overhang the femurs in front and in back in *Toxodera* and *Phyllocrania*, both basal and subapical lobes in the *Sibylla*, three successive lobes in *Vates multilobata*, and three subapical lobes situated in various planes in

Gongylus. It is interesting to note also that in some cases, as in *Macrodanuria elongata*, the median femora are the only ones with lobes.

Each of the four femora end with a dorsal and a ventral genicular lobe, which are generally short and rounded but sometimes pointed or crenelated and occasionally very elongated (e.g., in *Belomantis*). On the ventral side of the femora there is usually a genicular spine that varies in length and sometimes may be absent, which is the case, for instance, in *Mantis*. This spine may be either straight or slightly bent and is greatly bent in *Paratoxodera cornicollis*.

The tibiae may be shorter or longer than the corresponding femora. They generally have a smooth surface like that of the femur but are carinate in Vatinae, and in certain cases such as in *Phyllovates* they are arched. The tibiae may also display foliated expansions, but this occurs less often than in the femora and it usually involves smaller adornments. In some species of *Acanthops* and *Hagiotata hofmanni*, one finds small expansions on the tibiae but none on the femora. At their distal extremity the tibiae always display two spines, generally of comparable size, though in certain groups they differ, with the external (dorsal) spine being then the longer as in Metallyticidae, Eremiaphilidae, Tarachodinae, Oxyothespinae, Toxoderinae, and Photininae. These spines differ the most in the species *Paratoxodera cornicollis* in which the external spine is very elongated and wide.

The mesothoracic and metathoracic tarsi are similar to the prothoracic tarsi but may consist of only three segments in some *Heteronutarsus*. The tarsi terminate in two claws, which are generally a bit stronger than those of the prothoracic tarsi. The claws are especially thickened and a bit unequal in *Eremiaphila typhon*, with the ventral being the heaviest; this may be adaptation for walking on sandy soil. Such inequalities in the terminal claws are even greater in other species of the same genus, such as *Eremiaphila numida*.

The Wings

The tegmina are very variable in size according to whether the individuals are macropterous, more or less brachypterous, micropterous, or apterous. The tegmina demonstrate a costal area of varying widths between the costal vein and the radial vein (the *radiale antérieure* of Giglio-Tos, 1927), a generally wider discoidal area between the costal area and the posterior cubital vein (the *ulnaire postérieure* of Giglio-Tos), and finally an anal area connected to the mesothorax by a jugal field. In fact, the nomenclature used for the veins and wing areas has varied from author to author, which poses obvious problems, some of which are not yet solved. With some changes, I have followed Beier (1968), whose work seems to be widely accepted. It should be noted that the wing venation is simplified and often modified, sometimes in a manner that is not easy to interpret as in the case of extreme brachyptery (e.g., for Eremiaphilidae among others).

The costal vein C corresponding to the anterior edge (Fig. 2.13) is always single and more or less convex toward the front and sometimes sinuous in its apical part, for instance, in *Acanthops*. The radial vein R, which limits the costal area toward the rear, is almost always stronger than the former, and it corresponds, in general, to the ridge of a obtuse fold of the tegmen. It seems to end at the apex of this "fold," often after having emitted a bifurcation in the discoidal area (i.e., radial sector, the significance of which is perhaps an anterior median). It may rarely yield two or three rays, as in the case of *Idolomantis*, or not split at all, as in *Liturgusa*. The subcostal vein Sc (the *médiastine* of Giglio-Tos) is found in the costal area, always single and incomplete; it is a low vein situated most often closer to the radial than to the costal vein, but also found almost equidistant from the two, but only when the costal area is narrow. In fact, when the costal area is wide, it is always the portion between C and Sc that is dilated, for example, in the females of *Chopardiella latipennis*. The transverse veins are always single and straight between Sc and R, and they may also be simple and straight between C and Sc, but they are most often arranged in a more complex fashion, that is, oblique and branched out in a more or less dense net substituting or superadding to the original arrangement.

In the discoidal area, just after and often very close to the radial vein is the median vein M (*radi-*

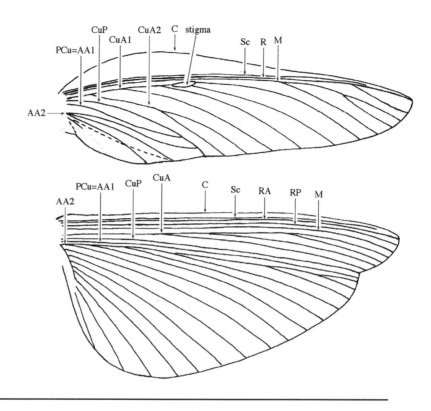

Fig. 2.13 Right wings (dorsal view) of *Sphodromantis lineola* (Burmeister). Abbreviations: AA - anterior anal vein, C - costal vein, CuA - anterior cubital vein, CuP - posterior cubital vein, M - median vein, PCu - postcubital vein, R - radial vein, RA - anterior radial vein, RP - radial posterior vein, and Sc - subcostal vein.

ale postérieure of Giglio-Tos), which can be single or show in its apical part a sector with one or two branches (as in *Metallyticus*). Next, the anterior cubital vein CuA (the *ulnaire antérieure* of Giglio-Tos) always branches from its base, usually forming three to eight branches (but up to twelve or more in *Metallyticus*) constituting, generally, two groups (CuA1 and CuA2, the latter being the *cubitale postérieure* of Chopard, 1949). The posterior cubital CuP (the *ulnaire postérieure* of Giglio-Tos; *anale* of Chopard) which marks the end of the discoidal area is a low vein that is always single, is in general more or less concave toward the rear, and that ends on the posterior edge of the tegmen most often between one-third and one-half way from the base. One of the rare exceptions in which it ends sooner occurs in *Empusa*.

More often it ends after the midpoint as in *Chaeteessa, Mantoida, Iridopteryx,* and *Majunga*. Among these longitudinal veins are found the transversal veins arranged in a more or less regular fashion, sometimes forming networks as in *Orthodera* and *Pseudostagmatoptera*.

On the level of the discoidal area there is a thickened formation, the stigma (pterostigma), which is sometimes very subtle but always present. It may be elongated or short, thin or wide, oval or triangular, of the same or a different color (lighter or darker) than the rest of the tegmen, and sometimes bicolored. Its most usual position is between the median and the first branches of the anterior cubital, but it can also be found positioned against the median (as in *Compsothespis*) or only on the level of the branches of the anterior cubital. It is longest in *Enicophlebia pallida*, obliquely crossing almost the entire discoidal area, while in *Enicophlebia hilara* it is much shorter and only at the level of the posterior branches of CuA, in a very unusual position.

In the anal area of the tegmen one first finds the postcubital (Pcu) or first anterior anal vein (AA1) (the *vena dividens* of Giglio-Tos; *first anal* of Ragge, 1955), which is more or less parallel to CuP. In some cases it is connected to this vein (as in *Mantis religiosa*) and may go to the edge of the tegmen or be interrupted before it does. It is single in the vast majority of cases, but may be bifurcated (as in *Metallyticus*). Then comes the second anterior anal vein (AA2) (the *vena plicata* of Giglio-Tos; *veines axillaires* of Chopard, 1949), which is always branched at the base, most often with three but seldom with only two (e.g., Mantoididae and Perlamantinae) or with four or even five branches (e.g., Metallyticidae, Orthoderinae), each shorter than the previous one. The most anterior branch may be connected to the postcubital (as in *Danuria*, for example). A plicature of the anal area marks the limit between an anterior anal area and a posterior anal area (the *champ jugal* of Chopard, 1949) in which one may find prolongations of the last branches of the second anterior anal vein and sometimes a badly individualized posterior anal vein, but often this area only has one reticulation, which is slightly tightened. This posterior anal area is articulated by another plicature at the jugal field (in the strict sense).

The posterior wings are generally shorter than the tegmina, sometimes quite a bit so. However, they are often wider than the tegmina due to the great development of their anal area. It is only in a few cases, as in the males of *Cliomantis*, and the females of *Brancsikia* (and in the enigmatic species *Astollia chloris*), that the wings are longer than the tegmina, while they are reduced to their most simple expression in the females of *Coptopteryx*, which are brachypterous for the tegmina. One distinguishes for the wings the same areas as ones does the tegmina, and the nomenclature used for their veins poses the same problems. However, the variability of the observed structures is much smaller. The costal area of the wings is always narrow, with a subcostal vein that does not reach the apex. The radial vein is generally single, but it may be branched toward its apical extremity. One qualifies it as anterior radial (RA) because it occurs just before a low vein, which is always single, the radial posterior vein (RP). The latter is the first of the discoidal area. (It is the *veine radiale médiane* of Giglio-Tos, for whom the *radiale postérieure* is that which one now calls the median.)

The median vein is also usually single while the anterior cubital (*veine discoïdale* or *ulnaire antérieure* of Giglio-Tos) is, in contrast, always branched once (as in *Stagmomantis*) or, most commonly, two or three times, and up to five times in *Metallyticus*. Next is the posterior cubital vein, which is always single and is longer and straighter than the corresponding vein of the tegmina and which ends the discoidal area. The anal area, most often the largest, consists of a single postcubital vein (*vena dividens*: first anterior anal) and a series of other anal veins (the *veines axillaires* of Chopard, 1949) of which only the first one is branched. Usually this part of the wing is folded into a fan at rest, but this is not the case in Eremiaphilidae (Grandcolas, 1994) or in other cases in which the wings are very shortened.

The Abdomen

The abdomen always has ten tergites, with nine sternites visible in the males but only seven in females (e.g., Fig. 2.4). The abdomen is most often simple in shape and thinner in the males, but may show lateral, midventral, or middorsal prolongations. The latter occurs usually at the level of the portion of the abdomen not covered by shortened or absent wings (e.g., Thespinae females). The lateral or medioventral prolongations are particularly long and ribboned in *Presibylla elegans*, while the sixth sternite carries two strong spines in the females of Eremiaphilidae. The tenth and final tergite constitutes the supra-anal plate, which is most often transverse, but frequently is more or less elongated, with a rounded posterior edge, or either truncated, arched (ribbed), or triangular, with smooth or denticulated lateral edges (as in female *Episcopomantis congica*). It is sometimes carinate as in

Hoplocorypha and even has a foliaceous median dorsal lobe in *Epaphrodita*. The record length is 26 mm for the supra-anal plate of an *Ischnomantis gigas* male.

The abdomen terminates in a pair of cerci. These are somewhat hairy, generally consist of ten to twenty joints, and are quite variable in length and shape: For instance, they are very short in Eremiaphilidae and very long in Chaeteessidae. In cross section they are most often circular, but they can be flattened, or foliated as in *Angela*. The last segment is often the smallest (e.g., in Mantinae), but it can also be the largest (e.g., in *Caudatoscelis*, *Oxyothespis*, and *Belomantis*). The last segment may also be either narrower or wider than the preceding segment, simple in shape or with a bifid apex, toothed or truncated in a characteristic fashion.

The last (ninth) sternite in the males is the subgenital plate, which is generally fairly large and externally convex, rounded, truncated, or indented on its posterior edge. It is usually symmetrical, but may be asymmetrical as in *Polyspilota* and *Chlidonoptera*. The subgenital plate carries the styles, which always appear to be symmetrical and simple in shape, sometimes very small (as in *Chloroharpax*, *Eremiaphila*), generally fairly long, and particularly elongated in *Theopompella* where they are sinuous and as long as 4 mm.

The male genitalia, situated in the concavity of the subgenital plate, are always strongly asymmetrical and consist of three principal parts or phallomeres: a ventral hypophallus and right and left dorsal epiphalli. The hypophallus can be very simple in shape with a rounded posterior edge, or it may show varied and diversely placed prolongations. For instance, there may be lobes, denticulations, points, or diverticulations. The latter may be curved or sinuous, and rounded or pointed at the apex. When there is only one prolongation it is most often situated on the right side on the posterior edge.

The right epiphallus is most often roughly triangular with an enlarged base and a narrowly rounded apex. Toward its middle there is a predominant sclerified formation that is more or less curved, the apophysis, accompanied by an equally sclerified thickening, the contra-apophysis. In Tarachodinae, among others, the shape is often more complex, with supplementary formations and an apex that may be pointed or hooked. The left epiphallus is generally fairly narrow, with the lateral edge fairly regularly convex. It has most often a sclerified phallic apophysis (also called the pseudophallus), fairly variable in shape, accompanying the membranous penis. This epiphallus frequently also carries an additional prolongation, the titillator, which is always curved toward the front with a simple or sometimes double apex (as in *Mantis religiosa*).

The whole of these parts is sclerified and pigmented to varying degrees—*Anasigerpes* is one case in which the sclerification and pigmentation are slight. They may be glabrous or provided with long or short setae, fine or inflated, sparse or localized, or sometimes forming brushes or tufts.

In the females the seventh and last sternite also forms a kind of subgenital plate, analogous with that of the males, which is indented in the rear on the level of the ovipositor. It carries no appendix, but in *Rivetina* it is armed with two strong spines, which are remarkably similar to those in *Eremiaphila*. The ovipositor varies little, is composed of six valves, two dorsal and two ventral, and two shorter internal valves.

Coloration

The general coloration of mantids is most often in harmony with the environment in which they live. So, for instance, ground-dwelling Eremiaphilidae range from beige to brown; tree-dwelling Liturgusinae are marbled brown, like bark; grass-dwelling *Pyrgomantis* are the color of straw; *Brancsikia* and *Phyllocrania* have the appearance and coloration of dead leaves; and many Mantinae living in bushes are green, like the leaves. Often, especially in this latter subfamily, members of the same species may be green or brown. Moreover, sometimes one can find specimens that are almost black in the burned savannas. In some cases the green and the brown are juxtaposed; thus, for example, the males of *Polyspilota aeruginosa* are often brown, with the front half of their body and the costal area of the tegmina green. Coloration systems with dark stains on a light background still exist in certain groups: one can thus list the genera *Maculatoscelis, Oxyelaea,* and *Astape* and also the remarkable *Calospilota pulchra,* all of different subfamilies.

Another system of definite coloration is found in flower-dwelling species such as *Pseudocreobotra* in tropical Africa and *Theopropus* in Southeast Asia, which are green, yellow, and black. Finally, there are also rare metallic colorations in mantises, generally in Metallyticidae but also in the Iridopteryginae (specifically, *Nemotha metallica*).

Certain elements of coloration are ordinarily hidden and do not become apparent until the secondary postures of defense (Edmunds and Brunner, this volume, chap. 13). Essentially this is a case of ornamentation of the prosternum, the ventral (internal) face of the predatory legs, and the wings, which are normally folded under the tegmina. These elements, when they exist, have the effect of enhancing the impressive appearance of the insect when it faces you and displays these unusual structures. But for many species, the ventral surface shows no particular detail of coloration, and their wings are hyaline. The particular ornamentations of the prosternum consist of black stains of various shapes generally spread out, in one or two transversal bands or points symmetrically arranged as in certain species of *Prohierodula*. The ornamentations of the predatory legs are more diversified. They are found especially on the level of the coxae and femora with stains that may be callused and diversely colored, sometimes ocellated, with transversal or longitudinal bands (as on the coxae of *Oxypilus*) or with dark or light points of varying number. The wings may be irised, partially or completely smoky (as in *Thesprotia infumata* males), with portions colored in yellow or red (sometimes in association with black), brown, deep violet-blue with metallic reflections, pink, or, in light, turquoise as in *Amorphoscelis elegans* females. In some cases the wings' transversal veins are sharply underlined, while in other cases the transparent parts subside in hyalin windows on the colored background. A remarkable curiosity is *Tithrone roseipennis,* whose wings are a bright pink, as is the one tegmen that is covered at rest, while the other tegmen is green. This is truly one of the rare cases, if not the only case, where there is regularly asymmetrical coloration in mantids.

Sexual Dimorphism

Sexual dimorphism is important in mantids and sometimes exists to such an extent that it is difficult to clearly match the two sexes, and errors on this subject have been committed many times in the past. In general, females are larger, stockier, and therefore stronger than the males. The female's abdomen is notably larger than the male's due to the voluminous ovaries. The largest disproportion in size is found in *Hymenopus coronatus,* in which the males average 25 mm in length and the females average 50 mm in length; taking into consideration the difference in bodily proportions the female must be heavier than the male by a factor of ten or more. However, by virtue of the great individual variability in some species, especially in *Mantis religiosa*, on occasion the females can be notably smaller than the males, especially if one compares different populations. In any case, the females generally have smaller simple eyes, shorter and finer antennae (with the exception of *Schizocephala*), and sometimes simpler antennae, as in Empusidae.

In the females the wings are frequently more colored and less transparent than in the males, often also shorter, with apterism being much more frequent than in the males (in fact, it is the rule in Thespinae). In contrast, the costal area of the tegmina is often wider in females, with possibly a more accentuated apical shrinking. The prolongation of the vertex, the tubercles of the pronotum, and the expansions of the abdomen are also in general more developed in the female.

Finally, it seems that in some genera there are two types of females differing by a few morphological peculiarities while the males are always of the same type. In terms of chromosomes, the males are always heterogamous with two possibilities, both of which are widely distributed: XO and X_1X_2Y. The haploid number, n, may vary from eight to twenty, with fourteen being most frequent as in many Mantinae, and also in Vatinae, Toxoderinae, Hymenopodinae, and Empusinae (Hughes-Schrader, 1950).

Internal Anatomy

The internal anatomy of Mantodea was discussed by Beier (1964, 1968) and, as I have no additional information on this rarely tackled subject, I will not discuss it here. I am sure, however, that there is valuable work to be done in this area.

Preliminary Data for a Rational Classification of Mantodea

The classification of mantids has varied greatly since Linnaeus (1758) identified just ten species as *Mantis* within the genus *Gryllus,* all of which, by the way, were not really mantids.

As the number of known species increased, a variety of authors made attempts at classification. The most important of these are Audinet-Serville (1831, 1839), Burmeister (1838), Saussure (1869 and the years following), Stål (1873, 1877), Westwood (1889), Brunner von Wattenwyl (1893), Kirby (1904), Giglio-Tos (1917, 1919, 1927), Beier (1934, 1935), Chopard (1949), and Beier again (1964 and 1968) (Fig. 2.14). Beier proposed an exhaustive classification of mantids worldwide, which has been widely used since, especially by Brown Jr. in the *Synopsis and Classification of Living Organisms* (1982). In recent years several modifications to mantid taxonomy have been proposed, chiefly within the scope of regional studies. These include Vickery and Kevan (1983), Roy (1987), Wang (1993), Terra (1995), and Kaltenbach (1996).

In spite of the large amount of effort directed toward mantid taxonomy, many problems remain. Unfortunately, all of the necessary data have not been collected as yet, which leaves too much room for subjective opinions based on intuition rather than fact. Under these circumstances, it would be premature to propose a classification system that attempts to reconcile all existing disagreements. Nevertheless, along with the obvious uncertainties, there is an undeniable body of knowledge. If we attend to that knowledge and clear up the uncertainties, we get a better picture of the overall relationship between the various mantids. And it is only by doing so that an acceptable classification system will emerge.

In order to arrive at a rational classification for Mantodea, it would be helpful, first, to take an inventory of various useful morphological characteristics and evaluate their relative importance and, then, to determine for each which is primitive (plesiomorph) and which is derived (apomorph) by calculating which changes (probably) arose only once and which could have arisen several times.

It is also necessary to consider the fact that major variations do not necessarily mean a distant phylogenetic relationship and that striking resemblances do not necessarily translate into a close relationship. That is why *Mantis* and *Paramantis,* on the one hand, and *Tenodera* and *Epitenodera,* on the other, must be clearly separated, despite the fact that they have long been combined (Roy, 1967, 1973). Conversely, *Iris* and *Paroxyophthalmus,* although sometimes placed in different subfamilies, have been revealed as being very closely related (Roy, 1971). In practice, a group of conformable, even minimal, characteristics is more important to take into consideration than one single, very obvious criterion, which may have arisen simply by convergence. The morphological characteristics that are the most immediately apparent are obviously the best known, but they are not enough, in and of themselves, to clearly determine phylogenetic affinities.

A head that is wider than it is long is clearly primitive, with a simple vertex, rounded compound eyes, a transverse frontal shield with minimal depth, and filiform antennae. The elongation of the head appears in several lines, as do various elongations of the vertex or other areas in various ways. The shape of the eyes constitutes a criterion of relative importance, since it is not well fixed at the level of certain genera and even for certain species. The characteristics of the antennae seem to be more important to consider, especially in the male.

The most primitive mantids have a short prothorax, without supracoxal enlargement, and with a smooth surface and regular borders. The prothorax, having remained short, may have acquired a surface that is more or less undulating

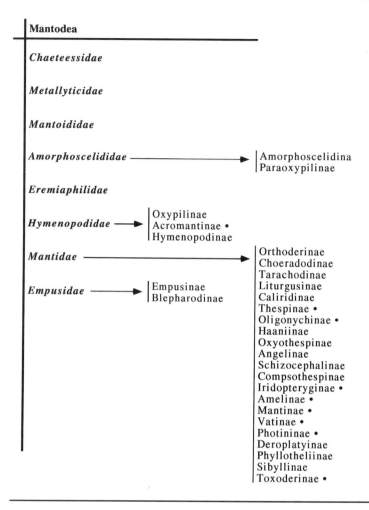

Fig. 2.14 The families and subfamilies of the suborder Mantodea as proposed by Beier (1964, 1968). Subfamilies denoted by black dots are further subdivided into tribes. Although Beier did not present this as a phylogenetic hypothesis, he felt that these were the most natural groupings given the information available at the time. Note that 21 of the 28 subfamilies were placed in the family Mantidae. More recent work suggests that this large family requires revision and that some subfamilies should be elevated to family status.

with irregular lateral borders. The elongation must have appeared gradually, no doubt in a unique manner initially, and must have differed between groups, in some cases leading to a hypertelic development and in others leading to a secondary shortening. This elongation was accompanied by the formation of a very clear supracoxal enlargement, rounded initially, but which may have become angular, while the surface and lateral borders acquired, in some species, rough patches or various diversely located dilatations, which appear in several instances.

The coxae of the predatory forelegs, always elongated, are primitive with simple edges and short terminal lobes that are slightly divergent; evolution has produced more elongated lobes, which were eventually unequally developed and which could get nearer, whereas the edges acquired particular variations, and all of this occurred several times independently.

The anterior femora, also elongated from the beginning, should typically carry four external spines and a larger yet undetermined number of internal spines. The discoidal spines, typically four in number, probably appeared at a later time as did differences in the sizes of the internal spines. Proportions (i.e., more squat or more elongated) vary between lines, while the various categories of spines vary in number and in dimension independently of the former. Similarly, granulations and various expansions have appeared at several times.

The anterior tibiae, primitively only a bit shorter than the femora, were able to become much shorter. They end typically in a spine and carry an initially indeterminate number of external and internal spines, which may increase, decrease, or eventually disappear in different lines. The migration of some internal spines to the dorsal position occurred only once.

The mesothoracic and metathoracic legs, simple initially, have acquired in several instances and in various ways a diverse expansion of shapes, and the genicular spine that is typically present has disappeared in several instances.

Primitively, the flight organs are well developed, without noticeable differences according to sex; the longitudinal veins are relatively numerous, but the costal area of the tegmina is narrow. On several occasions, evolution has caused a sometimes irregular dilatation of the costal area, simplifications in the venation, and reduction of the importance of the flight organs, most often in the females, which has led in some instances to flightlessness (see Yager, this volume, chap. 6).

The abdomen is primitively elongated and simple in shape, slightly larger in the female. Diversely placed elongations have appeared in several instances, and in the females a major dilatation may occur. The supra-anal plate, transverse initially, has been elongated in several lines by taking various shapes and occasionally by acquiring a keel. The cerci, which are primitively rounded in cross section, have often become flattened, with the terminal joint being extremely variable in its shape. The subgenital plate in the males, normally regular and symmetrical, has become indented in certain cases and, independently, asymmetrical in others. In addition, the styles, which were probably average in length initially, may vary greatly in their development.

The primitive type of male genitalia is very similar to the primitive one of Blattodea: the various phallomeres are fairly simple in shape and not very sclerified, the phallic apophysis has no basal elongation, and the titillator has a rounded apex. Numerous complications as well as simplifications have intervened in the various lines. However, knowledge is still insufficient, since many genera have not yet been examined from this point of view, nor have sufficient species been studied, which means that one still cannot obtain a precise idea of all the variations that exist.

Besides the primitive XO chromosomal arrangement, an X_1X_2Y type has evolved, for sure, only once; this condition has never been noticed in the groups with short pronotums. But then again, and in a manner still more obvious than for the genitalia, enormous gaps still exist in our knowledge, since a large number of families and subfamilies have not been examined yet in this regard. It would be desirable if more effort were made in this area.

Our knowledge of behavior is still too rudimentary to be reliable, with the exception of specific cases that have provided valuable data. It would seem that the breeding of mantises, now more and more common, will serve as a basis for more specific work in this area, in addition to results concerning biological cycles.

Finally, the molecular biology of mantids remains an immense area for exploration.

A Critical Review of the Different Groups

The groups with a short pronotum, with little or no supracoxal enlargement, are certainly the most primitive, and it is remarkable that among the groups, now universally treated as families (Metallyticidae, Chaeteessidae, Mantoididae, Amorphoscelidae, and Eremiaphilidae), the vertex is always without median elongation, the eyes are without spines, the antennae are filiform, the legs are without lobes, and the supra-anal plate is transverse: These are all characteristics that are plesiomorphic. The chromosome type, known only for Mantoididae and Amorphoscelidae, is XO for the males.

The Metallyticidae, with only *Metallyticus* in the eastern region, are probably the most primitive of mantids today. They are the only ones without discoidal spines on the anterior femurs, which have a very characteristic bulging shape with a very strong primary external spine and internal spines all of the same size. In terms of flight organs, the cubital anterior of the tegmina has the maximum number of branches, while the anterior anal area of the same tegmina is also very branched. The wings, which have an anal area very scarcely spread out, demonstrate similar branching. Their unusual metallic coloration and their way of life, which is reminiscent of the cockroaches on which they feed, contributes again to making them a distinct group that is considered a superfamily by Vickery and Kevan (1983).

The family of Chaeteessidae, currently limited to one South American type (*Chaeteessa*), but present in Europe at the beginning of the Tertiary (Nel and Roy, 1996), is also very remarkable, in part because the anterior tibiae lack the strong terminal spine which is never missing in the other mantids. In its place only a kind of setigerous tubercle is found. The other categories of spines of the anterior legs are present, thin and tapered, and rather weak. The filiform cerci have the particularity of being elongated, carrying up to twenty-two joints.

In these two families the flight organs are equally developed in both sexes, with a narrow costal area of the tegmina. This is also the case for the family of Mantoididae, which one can place, consequently, only with the neotropical genus *Mantoida*, which contains the smallest species of the group and which have among other peculiarities antennae much longer than their body. Their exterior aspect is very similar to that of *Chaeteessa*, but their forelegs have normal armature such as that found in Mantodea. Further, the venation of the tegmina and wings demonstrates only minor differential characteristics in relation to those of the preceding family. The two genera, *Mantoida* and *Chaeteessa*, have long been grouped more closely, especially by Giglio-Tos, who grouped them in the Mantoidea, and one can ask if this grouping will not explain relatively narrow affinities, all the more so since the two genera are found in the same biogeographic zone and their genitalia present some similarities.

The superfamily Chaeteessoidea, created in 1983 by Vickery and Kevan for the Chaeteessidae alone, could well apply to the entirety of the two families.

After these three monogeneric families, one can consider the Amorphoscelidae (or Amorphoscelididae), another universally recognized taxon, which is diversified throughout most of the Ancient World and is justifiably considered by Wang (1993) as very likely constituting a superfamily its own right. In this group, the head is very characteristic, with the vertex provided with lateral tubercles behind the eyes. The consistently short pronotum may present blunt or spiny tubercles and a median carina that is eventually relieved by a crest in the metazone, and there may be a supracoxal enlargement. The anterior legs have a particular type of armature, with various reductions, which have served as the basis for the distinction of three well-defined subfamilies: Amorphoscelinae, Perlamantinae, and Paraoxypilinae. These three subfamilies have very marked and specialized characteristics in terms of the armature of the anterior legs and of the surface of the pronotum. None of these subfamilies would be considered primitive in relation to the others, but the first two (Amorphoscelinae and Perlamantinae) have more in common with each other than with the third. But, all things considered, they do not seem to have more points in common than do the Chaeteessidae and the Mantodidae. It should be noted again that the equal development of the flight organs, constant in the preceding families, is not found in the Amorphoscelidae, whose females may show reductions that are always discrete in terms of the Amorphoscelinae and Perlamantinae but often major in Paraoxypilinae.

The well-defined family Eremiaphilidae, in whom the pronotum is always short, is clearly different from those discussed above, which would justify a superfamily. Members of this group are located in regions with bare soil, mainly desert, in northern Africa and western Asia, and they form a group that is completely homogenous and wherein only two genera have been distinguished. Their principal peculiarities are their globular head with slightly protruding eyes and reduced antennae, their trapezoid pronotum with a bumpy surface, their strong predatory forelegs armed with short spines, and mesothoracic and metathoracic legs that are long and thin, which allow then to move quickly in an open environment. Their wings are always shortened and nonfunctional, and they have aberrant venation. Lastly, the females have two strong spines on their sixth abdominal sternite.

All the other mantids, that is, the vast majority, have a relatively elongated pronotum with a clear supracoxal enlargement, and they can be easily grouped in the superfamily Mantoidea, which is used here in a more restrictive sense than in Vickery and Kevan (1983). The separation of the Mantoidea into families and subfamilies is not obvious and is still the object of major controversy. For the three families which Beier defined in 1964, the only one that is unanimously accepted is Empusidae, which almost everyone agrees to place at the end of the list. However, this taxon encompasses only a relatively limited number of genera and species living in the south of Europe, in Africa, in Madagascar, and throughout western Asia up to China. The members of this taxon are essentially characterized by the bipectinated antennae of the males and the facts that the large internal spines of the anterior femora usually carry three small spines between them, that there is always an elongation of the vertex, and that lobes on the legs are frequent. The grouping is clearly monophyletic, with three sets treated as groups by Giglio-Tos, while only two subfamilies, Empusinae and Blepharodinae, are distinguished by Beier.

The family of Hymenopodidae, represented in all the tropical countries with the exception of Australia, is also widely accepted but it should be noted that its boundaries are uncertain and it is not obviously monophyletic. The basic distinguishing criterion for the group is the fact that the external spines of the anterior tibia are rather numerous, thick, short, squat, close together, and layered one upon another in a characteristic manner. However, this arrangement is not identical in all the genera, and some genera that are not closely related to one another have been placed in this family without demonstrating these characteristics. Such is the case for *Oxypilus, Amphecostephanus,* and *Tithrone,* which were placed in different tribes by Beier in 1968. On the other hand, *Parablepharis* and *Brancsikia,* which have the distinguishing arrangement of spines, were excluded from Hymenopodidae in the same publication and placed in the Mantidae beside *Deroplatys,* forming the subfamily of Deroplatyinae. The three subfamilies of Hymenopodidae that were distinguished by Beier—Acromantinae, Hymenopodinae, and Oxypilinae—represent but one possible approach to the phylogeny of this group. It will be necessary to review these classifications and, perhaps, make both internal and external revisions based on critical studies of other subfamilies that he placed among the Mantidae. Only the Hymenopodinae must be homogenous, and it seems that the Acontistini would be better placed in their own group than among the Acromantinae. As for the Epaphroditinae (or Acanthopinae) they probably deserve to be grouped separately, but the *Amphecostephanus* and *Antemna* do not seem to be properly grouped with them.

The remaining mantids, approximately two-thirds of the genera and four-fifths of the species, were grouped by Beier in the large family Mantidae, which encompasses twenty-one subfamilies of unequal importance, some of which are subdivided into tribes. Many of these subfamilies have been considered, rightly or wrongly, as complete families. Hence, one finds in the literature mention of Orthoderidae, Choeradodidae, Deroplatyidae, Thespidae, Iridopterygidae, Toxoderidae, Sibyllidae, Liturgusidae, and Vatidae. Other subfamilies no doubt also deserve to be elevated to the rank of family, such as Tarachodinae and Amelinae, but various revisions remain to be made, taking into account the maximum number of criteria.

The carinated character of the median and posterior tibiae, placed in the forefront as a characteristic to separate the Vatinae, appears to have only minor value. If it is more or less evident in the genera that are usually placed within Vatinae, it is also found among genera that are classified among the Toxoderinae and Epaphroditinae. Further, the characteristic has led to grouping more closely genera as different as *Danuria, Heter-*

ochaeta, Vates, and *Stagmatoptera.* This is clearly an artificial group that should be divided into subfamilies, which is what Terra (1995) has begun to do. The lamellar character of the pronotum, used to distinguish the Choeradodinae and the Deroplatyinae, but equally present among the *Rhombodera* and the *Rhomboderella* (Mantinae), as well as among the *Phyllocrania* (Hymenopodidae) and the *Idolomantis* (Empusidae), has clearly appeared several times in different lines and should be used with caution.

The shape of the eyes, the modifications of various parts of the body, the elongation of the prothorax, the supra-anal plate, the cerci, and the shortening of the flight organs should also not constitute absolute criteria for defining taxa, especially if they are considered in isolation. On the other hand, the genitalia and chromosomal formula, and certain behaviors can surely furnish good criteria, but gaps in our knowledge of these areas do not yet allow us to use these criteria systematically.

The Mantidae should no doubt be restricted to groups with a chromosomal formula of X_1X_2Y, with a certain number of well-distinguished subfamilies such as the Orthoderinae, Compsothespinae, Mellierinae, Antemninae, Photininae, Vatinae, Stagmatopterinae, and surely several others, besides Mantinae limited in their scope. Further, several other subfamilies should be retired in order to establish families, such as the Tarachodinae, Amelinae (perhaps in a modified sense), and the Iridopteryginae, which would be placed as families alongside the Thespidae and Liturgusidae. Finally, the subfamilies of Caliridinae, Sibyllinae, Toxoderinae, and Stenophyllinae, among others, should find their place as they become better known, but a great deal of work remains to be done before we will arrive at an appropriately detailed and entirely satisfactory taxonomy.

Ecology and Mating Behavior

3. Ecology of Praying Mantids

L. E. Hurd

> Organizations of species and individuals . . . exist in a state of interdependence.
> August Möbius (1877)

> A struggle for existence inevitably follows.
> Charles Darwin (1859)

The disciplines of ecology and evolutionary biology converge on a common problem: to explain the distribution and abundance of life on Earth. The study of evolution has a venerable history in Western thought, while as a quantitative science, ecology is relatively young. The development of animal ecology from descriptive natural history to controlled experiments in natural systems is less than four decades old, arguably beginning with Connell's (1961) manipulation of a marine invertebrate assemblage that demonstrated the importance of competition to barnacle distribution. Since then, we have seen the rise and fall of many hopeful paradigms as the field has moved from big-picture predictions to reductionism, and at least part way back again.

Mantids are not among the organisms which, like birds or lizards, have helped define much of modern animal ecology. In fact, the contribution of mantid studies to ecology up to the present time falls far short of its potential. As generalist arthropod predators, mantids have much in common with spiders; however, the ecological literature concerning spiders is considerably richer (Wise, 1993). At least part of the discrepancy may be explicable by the substantially greater numbers and diversity of spiders in temperate regions, which is where most ecological studies have been conducted. In any case, as with the other authors in this volume, part of my goal is to stimulate future research with an inherently interesting group of creatures.

In this chapter I will discuss research with mantids on the levels of population and community ecology. Because other contributors to this volume (in particular, my colleague Frederick Prete) have far more experience and expertise than I on the subject of behavior, my treatment of that aspect of mantid biology will be limited to the impact of some behaviors on population dynamics.

The Niche of the Praying Predator

> Since the environment is a complex of many factors, every animal lives surrounded by and responds to a complex of factors. . . . Can a single factor control distribution?
> V. E. Shelford (1911)

As its name—*the prophet*—implies, the mantis was considered by ancient naturalists to possess mystic wisdom (Prete and Wolfe, 1992). Other subjective attributes, which have been exploited by writers of dramatic fiction, include hunting prowess (Fox, 1979) and evil cunning (Johnson, 1937). Stripped of the mythology, however, the mantis belongs to one of a number of predaceous arthropod taxa that inhabit many kinds of complex terrestrial ecosystems, and like most other predaceous arthropods, it is a generalist predator (Hassell, 1978). Hence, studies of mantid niches should yield insights into the general ecology of arthropod predators. The term *niche* used here (*sensu* Hutchinson, 1957) is the *n*-dimensional hypervolume resulting from the confluence of the suite of biotic and abiotic environmental conditions (i.e., niche axes) necessary for survival of an organism. The niche thus defines the evolutionary/ecological strategy of a species, which is subject to change by natural selection.

The nineteen hundred or so species of mantids run the gamut of patterns of occurrence and adaptive strategies. Some are quite rare, such as *Galapagos solitaria* (Scudder, 1893), which appears to be limited to three islands in the equatorial Galapagos archipelago (Linsley and Usinger, 1966). In contrast, the temperate species *Tenodera aridifolia sinensis* is distributed widely throughout the north-temperate zones of both the eastern and western hemispheres (Gurney, 1950; Yan et al., 1981). Most mantids are ambush predators, but species such as the diminutive *Yersiniops sophronicum* run down their prey (Helfer, 1963). Most mantids are sexually dimorphic, but *Brunneria borealis* is a totally female, parthenogenetic species (Helfer, 1963, and personal observations). Mantids of seasonal temperate zone habitats generally experience no overlap in generations, but some tropical species such as *Cardioptera brachyptera* exhibit parental care (Terra, 1992). There are both winged and wingless species, which relates to interhabitat dispersal ability, and there are large- and small-bodied species, which influences the breadth of potential prey and potential enemies within arthropod assemblages. Almost nothing is known about how these factors affect mantid ecology. We are just beginning to appreciate the fact that a number of other aspects of mantid biology have ecological implications, for instance, the role of cuticular hydrocarbons in water conservation and, perhaps, as sex attractants (Jones et al., 1997).

At present, most ecological work per se has involved a few north temperate species, especially the Chinese mantis *Tenodera a. sinensis*. This is a large-bodied species which is, arguably, the most widely distributed of any extant mantid. It is univoltine and semelparous: It undergoes a complete life cycle in the space of a growing season, dying shortly after oviposition in the fall, and its eggs overwinter to hatch the following spring. *T. a. sinensis* is a generalist with respect to habitat as well as to prey, occupying old fields with a wide range of vegetation types and successional ages.

Perhaps the most salient feature (i.e., niche axis) of any mantid's niche is its food, given that the bulk of research indicates that predatory arthropods are generally food limited in the wild. Food limitation sometimes is interpreted as energy limitation, but for many animals quality of nutrition is more limiting than quantity of energy. For instance, a specific nutrient such as nitrogen (White, 1993) may be the limiting factor for many animals, and the kind of food needed during the reproductive phase may be different from that required just to sustain an individual (Hurd, 1985b). Prey availability has been shown to have significant impact on many factors influencing fitness of arthropod predators, including their behavior, growth, survivorship, and fecundity (Lawton, 1971; Anderson, 1974; Fox, 1975; Benke, 1976; Takafuji and Chant, 1976; Formanowicz, 1982; Eisenberg et al., 1981; Rypstra, 1983; Baars and Van Dijk, 1984; Folsom and Collins, 1984; Wise, 1984; Lenski, 1984; Hurd, 1989; Matsura and Murao, 1994; and many others).

Mantid nymphs are efficient processors of food: growth efficiency of first stadium *T. a. sinensis* (Hurd, 1991) and of its congener *T. angustipennis* (Matsura et al., 1984) is between 35 and 40% when they have been fed as much as they can possibly eat. However, at lower than maximum prey level, growth efficiency actually increases to almost 60% for *T. a. sinensis*, indicating that beyond a moderate prey density, nymphs actually move food through the gut faster than they can most efficiently process it. This can be interpreted as prey wastage, which argues against mantids being optimal foragers (Hurd, 1991). In other words, for habitually food-limited animals, perhaps there is no selection against eating too fast because the level of prey in the environment never gets high enough for such an animal to do so.

For euryphagous predators such as mantids, which apparently can choose from such a wide menu of potential prey, it may at first seem strange that food should be limiting. However, food limitation is a consequence of the life history of *T. a. sinensis*, and probably is so for many other temperate zone mantids as well. To begin with, the egg cases of *T. a. sinensis* (and of *Mantis religiosa*, for that matter) are spatially aggregated in the field (Eisenberg and Hurd, 1990). This dispersion is not explained by multiple ovipositions per female, since it regularly occurs even in portions of the geographical range that do not have sufficiently long seasons to permit females the time to produce more than a single ootheca (Eisenberg et al., 1981). Therefore, the females themselves must be contagiously dispersed. One possible explanation for this is the clonal nature of plants like asters and goldenrod, the stems of which are favored oviposition sites. However, oothecae are not uniformly distributed even among these plants. Another possible but yet untested explanation is that female *T. a. sinensis* are attracted to pheromones released by other females; that is, for mantids, sex pheromones are also aggregation pheromones. The rationale for this hypothesis is that multiple females would produce a higher concentration of pheromones than would any single individual, and in turn that would increase the probability that males would respond. Females rarely eat each other, and males can mate with more than one female (Bartley, 1982; Hurd et al., 1994), so the idea is not as farfetched as it might first appear for these reputed cannibals.

However it comes about, aggregated dispersion of oothecae can result in locally very high abundances of hatchlings in the following spring. In two Delaware old fields, 49 to 71% of *T. a. sinensis* oothecae were located in the same 2 m^2 area as was at least one other ootheca, and 29 to 61% of these were paired within a single 1 m^2 area (Eisenberg and Hurd, 1990). Mean ootheca density in both of these fields was only about 0.1/m^2, but nymphal emergence could well exceed 780/m^2 based on prediction from the mass of two large egg cases (Eisenberg and Hurd, 1977). Most of the eggs in any single ootheca hatch simultaneously, and emergence is synchronous to within about two weeks among oothecae. Because, at the time of emergence early in the spring, prey scarcity is apt to cause massive starvation (Hurd et al., 1978; Hurd and Eisenberg, 1984a), it would appear that mantid mothers do their offspring no favor by aggregating egg cases.

Two questions arise from these observations: (1) Why don't female mantids produce two or more small oothecae spaced far apart instead of one large egg mass; and (2) why do nymphs hatch so early in the spring when food supply is low and there is danger of mortality from a late frost?

As to the first question, well-fed *T. a. sinensis* females in the protected environment of the laboratory can produce four to six viable oothecae, with an interval of 8 to 10 days between successive ovipositions (Butler, 1966; Eisenberg et al., 1981; and personal observations). Therefore, given enough time, a single female potentially could spread her offspring among several different oviposition sites within a field before she is killed by autumn frost. The key here is time: Although the season is only long enough to permit one oviposition by females in Delaware and Maryland, in North Carolina adult females often survive long enough to oviposit twice (personal observations over the past 15 years), and the contiguous range of *T. a. sinensis* in eastern North America extends into part of South Carolina (Rooney et al., 1996) where the growing season may be long enough to allow females to realize their full reproductive potential. This raises yet another question: Why is *T. a. sinensis* limited to temperate zone distribution? Unlike *Mantis religiosa* (Salt and James, 1947), *T. a. sinensis* eggs do not require a period of low ambient temperature to break diapause, so its distribution in more tropical climes is not curtailed by climate. A native American species, *Stagmomantis carolina*, also has discrete (i.e., winter-terminated) generations throughout most of its range in the United States but has overlapping generations in the vicinity of Fort Lauderdale, Florida (personal observations).

The answer to the question of why *T. a. sinensis* nymphs emerge so early in the spring could be as simple as a trade-off between the danger of starvation (and late frost) from hatching early and the low probability of maturing in time to oviposit before an autumn frost if hatching is delayed (Hurd et al., 1995). The difficulty with accepting this simple, physiological-limitation hypothesis lies in the intersection of the niche food axis with another important axis, intraseasonal temperature. Food limitation retards development (Hurd and Eisenberg, 1984a), resulting in shorter adult females with lower potential fecundity (Dussé and Hurd, 1997), but rate of development even for well-fed mantid nymphs is highly temperature dependent (Fig. 3.1). Both temperature and prey level are higher later in the spring than when mantid eggs normally hatch.

Shepherd et al. (forthcoming) tested the hypothesis that delaying egg hatch of *T. a. sinensis* until the weather is warmer and prey is more plentiful could allow these insects to "catch up" with their normal maturation schedule. The results indicated that delaying egg hatch as much as two months did not prevent maturation at the normal time in a Virginia field population, which is precisely what was predicted from temperature-controlled laboratory cohorts. Therefore, the reason for early egg hatch may be ecological rather than physiological, that is, avoidance of predation or competition from other members of the community. (I will discuss this point later.)

Of course, just because mantid physiology can compensate for normal seasonality does not mean that these animals are immune to weather anomalies. Though single oviposition is the norm for *T. a. sinensis* in Delaware and Maryland, even this level of fecundity is not assured if the weather does not cooperate. For example, the summer of 1992 was abnormally cool, probably due to the atmospheric infusion of volcanic aerosols from the eruption of Mt. Pinatubo in the Philippines the previous year (Minnis et al., 1993). Oviposition in an unmanipulated mantid population in Maryland was reduced 94% (just three oothecae from 281 adult females) from the mean of the previous two years, and there were still nymphs in the population three weeks beyond normal maturation time (Hurd et al., 1995). This could not be attributed to food limitation, because experimentally fed females from this population that were maintained in field enclosures also failed to oviposit before killing frost. This

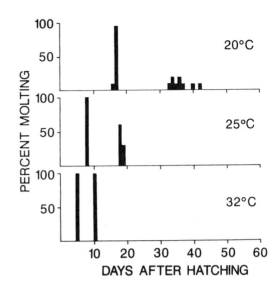

Fig. 3.1 Relationship between ambient temperature and rate of development for the first two stadia of *Tenodera aridifolia sinensis* raised in laboratory incubators. The percentage of cohort that molted on any given day becomes both earlier and more synchronous at progressively higher temperatures. Data from Hurd et al., 1989.

raises the specter of local or even regional extinction of these mantids from weather anomalies as short as a single season. For longer-term climatic changes such as global warming, the effect is likely to be much more severe for this species, as well as for other seasonal insects with low vagility (Rooney et al., 1996).

In the field, most *T. a. sinensis* nymphs die from starvation, predation, and (probably) desiccation during the first stadium. Those that survive to molt disperse within the habitat, which decreases both local density and aggregation within the habitat (Hurd and Eisenberg, 1984a). Growth rates are not uniform within a cohort due to slight differences in hatch times among oothecae and to differential success at hunting among nymphs. Differential success in capturing prey can be caused by the patchiness of the resource, but even among nymphs fed equally in the laboratory, predation efficiency and development can be variable (Hurd and Rathet, 1986). Developmental asynchrony produces body-size differences within a cohort that can promote cannibalism of smaller nymphs by larger ones (Hurd, 1988; Fagan and Odell, 1996). Cannibalism can be a mechanism that regulates populations of cannibalistic predators through negative density-dependent feedback on survivorship (Fox, 1975; Heessen and Brunsting, 1981; Polis, 1981). For *T. a. sinensis*, the percentage of cannibalism among cohorts of experimentally starved first instars in laboratory cages was found to be density independent by Hurd and Eisenberg (1984a), but Woodson and Hurd (forthcoming) found that among well-fed mixed laboratory cohorts of first- and second-stadium nymphs, cannibalism increased with increasing initial density. The extent to which cannibalism is density-dependent may therefore be a function of feeding condition, life stage, or an interaction between these two factors, at least among laboratory cohorts.

In the field, where feeding condition and life stage is apt to vary within a population, intraseasonal regulation of these insects appears to occur through density-dependent dispersal within the habitat rather than through mortality (Hurd and Eisenberg, 1984a). When different initial densities of *T. a. sinensis* were experimentally established in a large-plot, open-field study, dispersal increased disproportionately with increasing density (Fig. 3.2), which by itself accounted for regulatory convergence of final population densities. In a study of *M. religiosa*, Fagan and Hurd

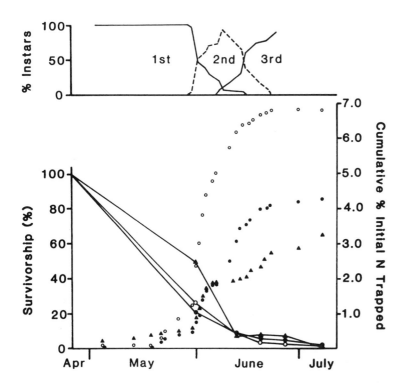

Fig. 3.2 Relationships between survivorship, emigration rate, and time for three established densities of *T. a. sinensis* nymphs in open-field experimental plots. Most mortality (solid line in lower graph) occurs during first stadium (upper graph). Most emigration from plots (sigmoid curves), judged by individuals captured in sticky trap dispersal barriers, occurred during second stadium. Proportional emigration was greater in successively higher-density treatments: triangles - density 1x; closed circles - density 3x; open circles - density 9x. Therefore, within-habitat dispersal was density dependent and, in this case, responsible for intraseasonal regulation of the populations over the course of the season. Modified from Hurd and Eisenberg, 1984a, with the kind permission of Blackwell Science, Ltd.

(1994) found that a combination of mortality and dispersal produced density-dependent convergence of experimental cohorts that were introduced to the field at different density levels. In these, as in all our field studies, experimental densities were well within the range of naturally occurring population densities.

Both mortality and dispersal decline markedly as the mantid population attains a carrying capacity or K in the Verhulst logistic model. (K equals the maximum density of a population that is sustainable on the limiting resource [Hutchinson, 1978].) In mantids, carrying capacity is dictated by ambient food level (Hurd and Eisenberg, 1984a). Carrying capacity can, of course, vary between seasons and among habitats in a given geographical area, so population densities of mantids can differ among fields and years. Although most of the mortality occurs in the first stadium, nymphs that survive this initial period do not find themselves free of resource constraints. Continued food limitation during the middle and late developmental periods affects development rate (Hurd and Eisenberg, 1989a), adult body size (Eisenberg et al., 1981), and oviposition (Eisenberg et al., 1992b). During this later period of development, Ilyse Rathet sat in a field up to nine hours at a stretch watching and timing the behavior of individual *T. a. sinensis* nymphs (Rathet and Hurd, 1983). She found that feeding occupied only a little more than 4% of their time, whereas more than 93% of the time they were entirely motionless (the rest of the time was spent grooming or changing location). This was not because nymphs passed up opportunities to strike at passing insects, but because very few insects passed by. My field notes gathered over the past two decades indicate that only about 5% of the mantids that I have encountered in the field were feeding. What this means is that, in spite of the fact that mantids in the laboratory can learn to prefer one prey item over another (Gelperin, 1968; Bartley, 1983), they rarely get a chance to exercise choice in nature because encounters with even a single prey item are infrequent. Rejection of suboptimal prey (Stephens and Krebs, 1986) or formation of some optimal attack strategy (Charnov, 1976) seems unlikely to apply to animals in the real world who cannot afford to pass up any opportunity to feed.

For Adults Only

The child is father of the man.
   William Wordsworth (1807)

The nymph is father of the imago, in the sense that the potential for success among adults is determined by the environmental milieu of developing juveniles (Fagan and Hurd, 1991b). If maturation is late, the adult will not reproduce (Hurd et al., 1995); if food level is low for juveniles, adult body size and consequent maximum potential egg production will be reduced even if prey are plentiful during oogenesis (Eisenberg et al., 1981). For species in temperate seasonal communities, there is a race to produce and lay eggs before the season closes, as noted above. This is really a two-edged sword for *T. a. sinensis*: If oviposition occurs too early, eggs may hatch before winter. In the piedmont region of North Carolina during late October it is not uncommon to find hatchling nymphs frozen in the act of emerging *en masse* from an egg mass, perhaps triggered by the fickle warmth of an Indian summer morning. Since females in this region routinely produce two oothecae, perhaps the earlier of the two can be viewed as a bet-hedging strategy (*sensu* Partridge and Harvey, 1988) against the chance of unusually early winter that would preclude a second attempt.

During the race with the end of the season, the food requirement of male and female mantids is unequal: males eat little and gain almost no mass following maturation; females must double their body mass to produce an ootheca that weighs about one-third as much as they do (Eisenberg et al., 1981). All of this occurs in a community that is winding down its productivity, with arthropod numbers dropping as resident species are depleted by predation or just complete their life cycles and die. Evidence from field studies indicates that, at least for three common temperate zone species (*T. a. sinensis, T. angustipennis,* and *Mantis religiosa*), food limitation may be the rule for adult mantids (Eisenberg et al., 1981; Matsura and Marooka, 1983; Hurd, 1989; Lawrence, 1992;

Hurd et al., 1994, 1995, but see Maxwell, 1995, for contrary results in *Stagmomantis limbata*). At such a time, late season flowers can be critical for female mantids. Asters and goldenrods act as magnets for flower foragers such as wasps, bees, and butterflies, which become concentrated on their blooms (Sholes, 1984). Mantids located on flowers often eat bees laden with pollen, which raises the question as to the potential value of vegetable sources of protein and other nutrients to their diet. Smith and Mommsen (1984) demonstrated that pollen grains caught in the webs of juvenile spiders can be an important food source, and Pollard (1993) reported that crab spiders can feed on the nectar of Queen Anne's lace. For that matter, since mantids eat many folivorous insects that have vegetable matter in the gut, might not indirect herbivory be both routine and important to these predators?

Female mantids that are located on flowering plants are more likely to oviposit, and to produce larger oothecae, than those without this advantage (Hurd, 1989). It remains to be tested whether female mantids exert choice in where they forage, as do some other predators, for example, crab spiders (Morse and Fritz, 1982). However, casual observations suggest that *T. a. sinensis* do not remain in a flower patch any longer than they do in nonflowering vegetation: Foraging perch for these predators probably is more a matter of chance than of choice.

Female *T. a. sinensis* do exert another kind of choice, however: How to respond to males, that is, whether to mate with them, eat them, or both. The cannibalism of male mantids by females is firmly rooted in biomythology, about as commonly believed as the much-rumored legal fine for killing or collecting a praying mantis. (I almost never have given a public mantid lecture at the end of which someone did not ask me how I got a permit to collect mantids "legally." In fact, mantids are protected only in Florida, which in 1979 declared "the praying mantid" to be the state insect.) That intersexual cannibalism happens is indisputable, but how frequently it happens and under what conditions are open questions for most species (e.g., Prete and Wolfe, 1992; Prete, 1995). There are two ways to look at this behavior: adaptive suicide of males or prey supplementation by females (Newman and Elgar, 1991). The first hypothesis is a male-choice paradigm that requires that a male, who is going to die anyway when the season ends, should allow himself to be eaten by the mother of his offspring as a form of parental investment. This requires a high degree of certainty on the part of the victim that the cannibal is in fact carrying eggs that he fertilized (Maynard Smith, 1977). Since both males and females of *T. a. sinensis* can and do copulate with more than one partner, this may be problematic, and one can argue that multiple matings would do more for a male's fitness than a single suicide (Birkhead et al., 1988).

The alternative hypothesis is a female-choice paradigm, deriving from a consideration of the state of the environment during the critical period of oogenesis. Food supply is diminishing (unless she stays on a flower), she must gain substantial mass, and male mantids are the largest potential prey items on the menu. All she has to do is attract and mate with a single male—which she may or may not eat during or after copulation—after which she can continue to attract and eat as many more males as are required to produce oothecae. This is the more parsimonious hypothesis, since there is no assumption of the victim's paternity, only that the male cannot tell whether the pheromone trail he has just followed will end in love or death (Hurd et al., 1994).

The initial sex ratio of adult cohorts of *T. a. sinensis* typically is male biased (>60% [Hurd et al., 1994, 1995]), which seems fortuitous in view of the extra wear and tear on the male component of the population. This also is true of *T. angustipennis* (Matsura, personal communication). This ratio may be reversed by sexual cannibalism by the end of the season (Hurd et al., 1994), probably depending upon the degree to which females are food limited. I have been censusing a population of *T. a. sinensis* in Rockbridge County, Virginia, for the past four years in which the sex ratio has remained male biased until nearly the end of the season. I have seen very little cannibalism in that population, and body mass of females indicated that they were well fed in that field. In any case, the reason for the lopsided adult sex ratio appears to be disproportionate mortality among juvenile females, because the hatchling sex ratio is 50:50 (Moran and Hurd, 1994a). The reason for higher female mortality among nymphs in the field remains to be elucidated, but it may be that the larger-bodied females require more prey than males and thus are more prone to starvation in a food-limited environment. An alternative hypothesis, that males may more frequently cannibalize females than the other way around while they are nymphs, was expressed as follows by my student Erika A. Woodson:

> There once was a mantid who preyed,
> On some of his cohort each day;
> But when it came time to mate,
> He had so cleared his plate,
> That the poor prayer couldn't get laid.

A curious feature of *T. a. sinensis* is its wide distribution and abundance in the face of low adult vagility. Nymphs can disperse over short distances within a habitat, but normally cannot be expected to colonize new habitats. Adults of both sexes are winged, but only the males are normally observed flying (females quickly become too heavy during oogenesis), and males cannot by themselves colonize new territory. Eisenberg et al. (1992a) made an intensive study of movement within a large open-field population of these mantids and found no difference between net distance traveled by females over ground and males through the air. In fact, few individuals moved more than 20 m from the time they were first marked until the end of the life cycle, and most moved less than 10 m. From these data it would at first seem that local mantid populations are virtually closed, that is, not subject to metapopulation dynamics involving migratory replenishment of sink demes by source demes (Pulliam, 1988). Because this species has been in the United States only since 1896 (Laurent, 1898), or about 100 generations, how did it achieve its present broad geographical distribution here? The answer, as with the Norway rat, is people power.

Humans have distributed egg masses purposefully, for the dubious value of mantids as pest control agents in "organic" gardens (Hurd, 1985a; Maxwell and Eitan, forthcoming). However, I suspect that this has been less important than the inadvertent distribution of eggs on transplanted nursery stock (which is how they got here in the first place) and on various modes of human transportation. Unintentional anthropogenic transport is not limited to *T. a. sinensis*. I first found *Mantis religiosa* in northern Delaware (so far as I know, the farthest south this species occurs) in fields along a railroad track in 1980. Subsequently, I have on a number of occasions found oothecae of this mantid attached to the underside of freight cars. Mother mantis played unintentional roulette with her offspring, gambling that they would be in a good place to live when they hatched. I would be greatly surprised if vehicular dispersal, from bicycles to buses, has not been responsible for an appreciable fraction of the distribution of mantids in this country.

Guilds of Contentious Consumers

> The struggle for existence among animals is a problem of the relationships between the components in mixed growing groups of individuals.
> G. F. Gause (1934)

In any reasonably diverse animal assemblage in nature, there will be groups of species that make their living exploiting the same resources. Such groupings are called *guilds* (Root, 1967). The concept of guild is founded on the most prevalent and persistent paradigm in ecology: interspecific competition is the organizing force in natural communities of organisms (Elton, 1946; Schoener, 1982). This is based on a logical construct, the principle of competitive exclusion (Hardin, 1960), which holds that more than one species cannot long exist on exactly the same limiting resource. The intensity of competition thus is determined by the degree of resource niche overlap: "competition is keenest when the individuals are most similar" (Gause, 1934). Subdivision of the resource portion of the niche can be based on differences in temporal or spatial exploitation patterns, as exemplified by the several guilds of arthropods utilizing scotch broom (Waloff, 1968). Often, very slight differences in morphological characters cause differences in resource usage among coexisting species (e.g., the beaks of finches that coexist on islands of the Galapagos Archipelago [Weiner, 1994]). The assumption, at least tacitly, is that such character divergence evolved in response to historical competition.

The term *guild* originally was intended to apply "without regard to taxonomic position" (Root, 1967), though most authors have applied it to species related at least at the level of family. Thus, in addition to Root's foliage-gleaning bird assemblage, we have taxonomically narrow guilds of everything from plethodontid salamanders in streams (Gustafson, 1993) to cellular slime molds in forest soil (Eisenberg et al., 1989). However, often it is more useful to consider guilds as functional units apart from taxonomic composition, for example, the fish and shrimp species that share the same prey assemblage are components of a guild of mobile predators that forage in marine benthic communities (Evans and Tallmark, 1985). Sometimes large taxonomic groupings are divided into functional guilds, for example, beetles in successional old fields can be divided among predator, herbivore, and decomposer guilds (Waliczky, 1991). The many species of cursorial spiders can be considered members of a guild within an arthropod assemblage, but this guild also can be subdivided by foraging habit and morphology. Vagrant web-spinners, running spiders, crab spiders, and wolf spiders all can be considered separate guilds-within-a-guild of cursorial spiders (Hurd and Fagan, 1992).

Arthropod species assemblages in complex terrestrial ecosystems contain taxonomically diverse predators that can be considered members of the same guild (e.g., Rosenheim et al. 1993). Though food-limited competition may not be common among folivorous insects (Lawton and Strong, 1981), among terrestrial arthropod predators the preponderance of data suggests that they may be subject to food competition generally, as predicted by Hairston et al. (1960). With the exception of parasitoids, most predators are generalists with respect to prey (Hassell, 1978), which complicates the question of what controls guild structure. Generalist arthropod predators such as mantids typically are bitrophic consumers: They simultaneously occupy the third and fourth trophic levels by virtue of feeding both on herbivores and on each other (Hurd and Eisenberg, 1990a). Within a guild of generalist predators, then, interspecific predation, cannibalism, and interspecific and intraspecific competition for prey all may occur (Spiller, 1984, 1986; Hurd, 1988; Polis et al., 1989; Anholt, 1990; Wise and Wagner, 1992). The important question is whether competition or predation is the defining relationship among bitrophic predators.

The trade-off between intraguild predation and competition can depend upon differences in phenology and ontogeny among predator populations (Wissinger, 1992), as can the outcome of interspecific competition (Lawler and Morin, 1993; Werner, 1994). In seasonal communities such as temperate old fields, timing of egg hatch and resultant ontogenetic differences among species can determine the outcome of both competition and predation for univoltine, semelparous predators such as mantids. Interspecific differences in phenology therefore constitute an important niche parameter for predator guilds in seasonal communities.

Within the guild of bitrophic arthropod predators inhabiting many north temperate old fields, there is a component guild consisting of three species of praying mantids: *T. a. sinensis*, *T. angustipennis*, and *M. religiosa*. Since they are all generalists, they may be expected to share the general pool of arthropod prey, and this broad niche overlap might be expected to generate competitive relationships among them. According to traditional guild theory, we ought to be able to find ways in which this common prey resource pool is divided up to minimize or avoid competition. As it happens, the three mantid species differ in three important ways: (1) egg hatch phenology (relative timing of nymphal emergence in the spring), (2) optimal prey size, and (3) vertical placement in the habitat. These differences can be expressed as regions of nonoverlap on three niche axes (Fig. 3.3).

Niche difference is most pronounced between the two species of *Tenodera*: *T. a. sinensis* always hatches before *T. angustipennis* (Rathet and Hurd, 1983; Hurd and Eisenberg, 1989b; Iwasaki, 1996; Matsura, personal communication). As a

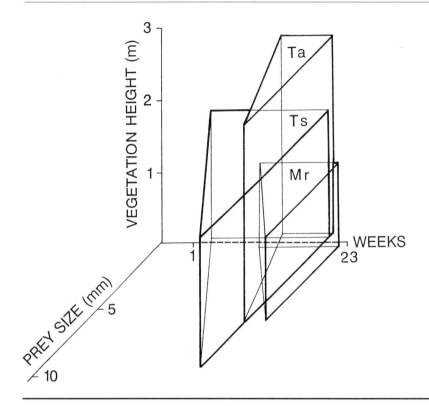

Fig. 3.3 Three-dimensional niche volumes of three mantid species in a guild: Ts = *T. a. sinensis*; Ta = *T. angustipennis*; Mr = *M. religiosa*. Areas of nonoverlap on one or more niche axes potentially promote coexistence by reducing competition among guild members, according to conventional theory (data from Hurd and Fagan, forthcoming). Ta oviposits highest in the vegetation; Ts hatches earlier and eats larger prey; Mr occurs lowest in the vegetation. Regardless of differences in egg hatch phenology, all three species die at the end of the season.

consequence, *T. a. sinensis* would seem to have an advantage, both by not having to share the habitat with a potential competitor for the first two or three weeks of life and by being further along in development (i.e., larger) by the time its rival hatches. This latter feature results in a niche difference, because mantids of different body sizes have different optimal prey sizes (Bartley, 1983). The difference in body size persists through the developmental period (Hurd and Eisenberg, 1989a). Thus, we might expect that these two predator species exploit different portions of the available prey, thereby avoiding competitive exclusion. Further, *T. a. sinensis* might be expected to have an advantage in this two-species interaction, because larger predators generally are able to exploit a wider range of prey size than smaller predators (Cohen et al., 1993). However, as previously pointed out, hatching early entails risks: greater food limitation because prey are scarcer, and mortality from late frost.

In any case, looking for competition avoidance between these two species may be following a red herring, because the evidence suggests that predation is more important. To begin with, the interspecific body-size difference that arises as a result of egg hatch phenology encourages predation. In a laboratory experiment (Hurd, 1988), *T. a. sinensis* and *T. angustipennis* nymphs were paired according to developmental stages that would co-occur as a result of normal egg hatch phenology. These pairs of nymphs were provided *Drosophila melanogaster* prey, upon which both species subsist quite well in the laboratory through the third stadium. However, in spite of a plentiful supply of flies, when *T. a. sinensis* was more than 30% larger than its congener, the smaller mantid became prey for the larger. According to conventional competition theory, linear morphological character differences (in this case, body length) of 10 to 40% should promote coexistence between potential competitors (Hutchinson, 1959; but see Simberloff and Boecklen, 1981; Strong and Simberloff, 1981). The apparent advantage *T. a. sinensis* has in this interaction was borne out by two field experiments in which interspecific predation, rather than competition, was the key interaction (Moran, 1995; Snyder and Hurd, 1995). Thus, selection may have favored early egg hatch in *T. a. sinensis* to gain a predatory advantage over *T. angustipennis*, which may compensate for the drawbacks of an early spring environment by adding one more prey item to the menu. The fact that the timing of egg hatch for each species is retained in habitats where only one or the other is present does not disprove this hypothesis, as suggested by Iwasaki (1996): Once evolved, there is no reason such a trait might not persist in a regional population even if selection is absent in some demes.

Egg hatch asynchrony occurs (to a lesser extent) within, as well as between, species. The same size ratio that encouraged interspecific predation in this experiment led to cannibalism among nymphs hatched at different times, as subsequently corroborated by Woodson and Hurd (forthcoming). Cannibalism resulting from intraspecific asynchrony and resultant size differences with a cohort was also found in dragonfly larvae by Hopper et al. (1996).

The egg phenology of *M. religiosa*, the third member of this guild, is midway between and overlaps with that of the other two species (Hurd and Eisenberg, 1989b). In the southern part of its range, this species also has lower hatch success than *Tenodera* spp. (numbers of emerging nymphs compared with numbers of eggs in an ootheca [Eisenberg et al. 1992b; Eisenberg and Hurd, 1993]). Eggs of *M. religiosa* are adapted to more northerly climates (Salt and James, 1947). However, the most important niche difference for this species is that *M. religiosa* occurs lower in the vegetation than the other two (Rathet and Hurd, 1983) and thus may be expected to avoid at least predation from the larger *T. a. sinensis*, if not competition from either or both *Tenodera* spp. Even here, the relatively slight difference in egg hatch phenology between *M. religiosa* and *T. a. sinensis* is important. When egg hatch was experimentally synchronized between these two species in a field enclosure experiment (Hurd and Eisenberg, 1990b), competition (not predation) from *T. a. sinensis* reduced both survivorship and growth rate of *M. religiosa*, while *T. a. sinensis* was unaffected by the interaction. Such one-sided, asymmetric exploitative competition appears to be more common than mutual interference among insects (Lawton and Hassell, 1981).

Interactions between mantids and other members of the broader generalist predator guild, and with other species in natural ecosystems, are not yet well defined. The most important such relationships may be with cursorial spiders, with which mantids have much in common. Mantids are not as similar to spiders as they are to each other, so by Gausean reasoning insect-arachnid competition should not be as intense as it might be among mantid species. Mantids and spiders eat each other, but the larger species of mantids can escape spider predation by outgrowing prey status and becoming top carnivores (Hurd and Eisenberg, 1990a). Whether by predation or competition, mantids have been shown to affect spider numbers in arthropod assemblages. One such effect, rarely documented in predator guild experiments, is on spider behavior. In two open-field experiments, small spiders emigrated disproportionately from plots to which mantid nymphs (*T. a. sinensis*) had been added (Moran and Hurd, 1994b; Moran et al., 1996). This was interpreted as avoidance of predation on the part of these spiders.

In another study (Moran and Hurd, 1997), experimentally increasing prey levels (*Drosophila virilus*) in experimental field plots enhanced growth rate of mantids, but actually decreased survivorship. This was because emigration from plots of cursorial spiders large enough to eat mantid nymphs declined relative to controls. Here, mantid nymphs were intermediate predators, and spiders were top predators. This behavioral response of spiders to increased prey levels caused greater mantid mortality through intraguild predation. This contrasts with the findings of Anholt and Werner (1995), in which increasing algal food of tadpoles reduced mortality from their odonate predators. This suggests that, when the food source is the same for both predators and prey (as with added flies in the mantid-spider system), increasing food level can have detrimental effects on intermediate predators and the benefit may accrue only to top predators.

In the Garden of the Prophet: The Community Niche

> The concept of community as something more than a mere chance assemblage of species is at the focus of interest.
>
> A. Macfadyen (1963)

> It is possible to deal objectively and quantitatively with big and complex structures.
>
> R. Margalef (1963)

The most difficult conceptual framework, by virtue of complexity, in which a biologist can work is the ecology of communities and ecosystems. Whether there really is a science of ecology at this level depends upon the whole being more than the sum of its component populations (i.e., reductionism versus holism). The reductionist view of communities as stochastic assemblages of species' populations derives from the individualistic hypothesis of Gleason (1926), whereas the holistic notion that species' interactions define a unit of organization beyond that which can be learned from studying the population ecology of component organisms derives from the community-unit hypothesis of Clements (1936) (McNaughton and Wolf, 1979). Community integrity even has been compared with that of organisms (e.g., Allee et al., 1949). This dichotomy has a broader context in modern biology: whether studying the behavior of molecules and genes is sufficient to describe and predict Mother Nature or whether we must get our hands dirty with the muck and mire of whole organisms and their interactions. While I am skeptical of such megaholistic notions as the Gaia hypothesis (Lovelock, 1988), I am not cynical enough to be an explanatory reductionist (*sensu* Mayr, 1982). An intelligent but naive person presented with a spark plug might just possibly glean enough physics to imagine internal combustion, but that same person could not from the same premise of information predict the consequences of a traffic jam in Quito, Ecuador, during rush hour. (I exempt, of course, those wonderfully imaginative people who write science fiction.)

One problem with which community ecologists must contend is that of scale: whether the structure of natural species assemblages is influenced more by interactions among organisms within a local habitat or by the regional environment afforded by proximity to other communities. This is an ongoing debate (e.g., Shmida and Wilson, 1985; Ricklefs, 1987; Compton et al., 1989; Lawton, 1990; Cornell and Lawton, 1992). However important regional phenomena may be, predation, parasitism, competition, and symbiotic relationships do occur within localized food webs. This argues for the value of a holistic, multitrophic approach to studying species interactions on a local scale.

How important are generalist predators such as mantids to the structure and function of natural communities? The influence of predators on species diversity often has been discussed theoretically and documented in a variety of ecosystems (Connell and Orias, 1964; Pianka, 1966; Spight, 1967; Clarke and Grant, 1968; Maguire et al., 1968; Harper, 1969; Paine, 1969; Dayton and Hessler, 1972; Addicott, 1974; Vance, 1979; Smedes and Hurd, 1981; Pacala and Roughgarden, 1984, 1985; Spiller and Schoener, 1990, 1994; and others). More than three decades ago, Hairston et al. (1960) proposed the broad generalization that predators control herbivores in terrestrial ecosystems. More recently, predators have been implicated as major selective agents in the evolution of phytophagous insects (Lawton and Strong, 1981; Bernays and Graham, 1988). In an influential review, Price et al. (1980) argued that predators and parasites, as enemies of herbivores, should be thought of as mutualists of plants. However, some predators have a potentially negative effect on plants by feeding on pollinators (Louda, 1982). This may apply to mantids, which can obtain a significant portion of their prey from flower-foraging arthropods late in the growing season (Hurd, 1989). In any case, if predators exert control on herbivores, and herbivores are important to the process of plant succession (McBrien et al., 1983; Brown, 1985), then these predators well may exert important effects on all trophic levels of natural terrestrial communities.

Community-level studies of mantids have revealed several key effects: (1) simple, direct depression of prey; (2) nonadditive and indirect effects; and (3) a trophic cascade. The first two of these depend upon the power of resolution to which the community is subjected. In the earliest controlled experimental study of the effect of three *T. a. sinensis* densities on old-field community structure, no effect of mantids was detected on the rest of the arthropod assemblage (Hurd et al., 1978). The level of resolution in that study was biomass and numerical abundance of carnivore and herbivore trophic levels; plants were not examined. As a small historical note, I now believe that the failure to detect a significant impact of mantids on the arthropods in that study was a function of the limited replication (only two plots per treatment) and low sample number, the consequently low power of statistical testing, and the relative naiveté of the investigators. After all, mantids must eat, and therefore must depress their prey populations, as Mook and Davies (1966) showed for *M. religiosa* feeding on grasshoppers in the field.

The failure to detect mantid impact in Hurd et al. (1978) was not likely due to the experimental densities' having been too low. In subsequent studies that have detected significant community-level effects, experimental densities have all been lower than the higher of the two in Hurd et al. (1978). Ironically, Hairston (1989) stated that the highest experimental density employed by Hurd and Eisenberg (1984b), which was less than half that of the earlier study, was "unrealistic." This was an unfortunate misreading on Hairston's part: As we stated in that paper, and as has been true of all subsequent work, experimental densities have been within the range of naturally occurring mantid densities (usually on the low side).

The experiment reported in Hurd and Eisenberg (1984b) yielded the first two of the three key results enumerated above: Overall abundance and biomass of arthropod herbivores and carnivores were depressed by the highest of the three mantid densities. However, some body-size groups within those trophic levels and some taxa actually increased in abundance. Since then, other studies have confirmed both of these general results (Hurd and Eisenberg, 1990a; Fagan and Hurd, 1991a, 1994). However, there is not enough consistency among experimental results as yet to predict with confidence the effect of mantids on some taxa in multispecies prey assemblages. For example, although mantids have reduced abundance of leafhoppers and their allies (Homoptera) in all studies, cricket (Gryllidae) numbers were enhanced by *T. a. sinensis* in an open-field experiment (Hurd and Eisenberg, 1984b), but unaffected in a field enclosure study (Hurd and Eisenberg, 1990a). Grasshopper abundance (Acrididae) was unaffected by mantids in the former study, but reduced in the latter. On the other hand, *M. religiosa* reduced abundance of crickets both in an open-field (Fagan and Hurd, 1994) and in an enclosure study (Fagan and Hurd, 1991a).

Another experiment can be used to illustrate that here, as with guild interactions (see above), timing can determine the impact of predators on other arthropods in the assemblage. This experiment consisted of two replicate tests in which realistic numbers of *T. a. sinensis* hatchlings were added to an old-field arthropod community early and late in the spring. We set up ten replicated enclosures, constructed of Lumite Saran mesh over a 1 m^3 PVC frame, in an early successional field in northern Delaware during the first three weeks of May (normal hatch) and an identical set during the first three weeks of June (delayed hatch, accomplished by refrigerating oothecae). These enclosures were open at the bottom, containing the vegetation and associated arthropods on 1 m² of ground. At the start of each test, we introduced one hundred mantid nymphs from fresh egg hatches into each of five randomly selected enclosures. The remaining five enclosures served as controls without mantids. At the end of each test, all enclosures were exhaustively sampled with a D-Vac sampler and subsequently searched by hand. Because mantids are size-dependent predators, arthropods were divided into categories of body length as well as of taxa.

The principal effect of mantids on the arthropod assemblage during the May test was a reduction in numbers of individuals less than 2 mm in body length. The majority (>90%) of these were aphids, which were reduced about 80% from control levels (Fig. 3.4). Flies, which represented the only other abundant taxon at this time of the season, also were reduced about 80% from control. Arthropods in the largest size category were not significantly affected by mantids.

The same experiment in June produced quite different results. As in May, the principal effect was among arthropods in the smallest size cat-

egory, but the direction of impact was reversed: mantids elevated, rather than depressed, total arthropod abundance. The majority (77%) of these smaller arthropods were aphids, the numbers of which rose to almost three times control level, while flies were again reduced (Fig. 3.4). The increase in aphid abundance can be explained by mantid predation on a group of arthropods that were virtually absent in May: small (<4 mm) spiders were reduced 61% in treatment enclosures (Fig. 3.4). Aphids were the most abundant suitable prey for these spiders, while spiders are more mobile, attractive prey items than aphids for mantids. Thus, overall predation on aphids was reduced in June because mantids fed on a different trophic level than in May, that is, on the predators of aphids rather than on aphids themselves.

This indirect enhancement of aphids need not be attributed solely to increased aphid survival in treatment enclosures. The prodigious reproductive potential and short generation time of these parthenogenetic insects mean that death of a few individuals at any given time can have a large effect on later population size. In any case, the positive effect of mantids on aphids in this experiment more than compensated for the depression in numbers of flies and other insects.

The community effects of mantids and other members of the bitrophic guild are not always in accord. Though cricket abundance was reduced by *M. religiosa* in the Fagan and Hurd (1991a) study, it was enhanced by experimentally augmented cursorial spider density in that same experiment. Sometimes the effects of mantids and spiders are in the same direction, but their combined effects are not predictable from their separate influences (i.e., they are not additive). Mantids and cursorial spiders eat each other commonly, though mantids can eventually escape predation by becoming too large for most spiders to eat. In a study by Hurd and Eisenberg (1990a), late stadium *T. a. sinensis* nymphs were top predators that depressed numbers of large wolf spiders (*Hogna rabida*). Both the large spiders and the mantids reduced numbers of smaller spiders in the field, but their combined effect was no greater than the individual effect of either predator.

This mixture of positive and negative effects on

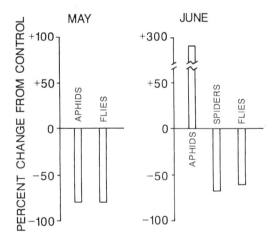

Fig. 3.4 The percentage increase (+) or decrease (-) in arthropod abundance in treatment (mantid addition) enclosures compared with control enclosures in the May and June replicate tests. Negative changes represent direct effects of mantid predation; the positive change in aphids during June represents an indirect enhancement due to trophic switching by mantids from aphids to spiders as prey.

the prey assemblage is a consequence of the bitrophic niche of mantids. The combination of competition and predation within a guild of bitrophic predators complicates their impact on the rest of the community. Some of these complications have been outlined in detail elsewhere, most recently by Holt and Lawton (1994). As a consequence, interpretation of the results of predator manipulation on a target community often can be difficult. Decline of a target population might be caused either by predation from the experimental predator or by competition with some other species on the same trophic level that has benefited from the predator's presence. The first cause is direct; the second is indirect, because it is mediated by a change in some other (third) population (Abrams, 1987). Enhancement of a target population by a predator is usually seen as indirect and generally ascribed to release from competition with other species that may be depressed because they are favored prey. This is the essence of the "keystone" effect of Paine (1966).

Indirect effects of predation have been found in a number of experimental studies, especially of aquatic communities (Morin, 1986; Kerfoot and Sih, 1987; Kneib, 1988; Morin et al., 1988). Whether these effects are the exception or the rule depends on how selective one is at reviewing the literature. Indirect effects were found in the majority of predation experiments examined by Sih et al. (1985). However, in a much more limited survey (Diehl, 1993) only two of twenty-two studies documented significant positive effects of predators on target populations. In any case, these kinds of results underscore the nonadditivity of population dynamics within communities (see also Holt, 1977; Vandermeer, 1980; Bender et al., 1984; Wilbur and Fauth, 1990; Bowers and Sacchi, 1991). Indirect effects are just beginning to be incorporated into predictive theory (e.g., Bertness and Callaway, 1994; Wootton, 1994a). Such results are important to ecological theory in the context of holism, or "emergence" (Mayr, 1982). Sih et al. (1985) used the term *unexpected* to describe these experimental outcomes, which means that they could not have been predicted from a consideration of the population biology of individual species or of pairwise species interactions. Indirect effects therefore represent an emergent property of communities.

The third key result enumerated above, that is, trophic cascade, comes from recent experimental studies with *T. a. sinensis* (Moran et al., 1996; Moran and Hurd, 1998). The context of these studies is the controversy between "bottom-up" and "top-down" control in ecosystems (Hunter and Price, 1992; Power, 1992; Strong, 1992). The top-down model of trophic cascade suggests that predators can, by virtue of their negative influence on herbivores, positively affect plants. This is consistent with the suggestion that predators are mutualists of plants (Price et al., 1980). This kind of effect has been demonstrated for some vertebrate predators (Altegrim, 1989; Marquis and Whelan, 1994; Spiller and Schoener, 1990, 1994; McLaren and Peterson, 1994; Dial and Roughgarden, 1995). It also has been demonstrated for a specialized hymenopteran parasite (Gomez and Zamora, 1994) and for spiders in a garden test system (Riechert and Bishop, 1990). However, the results of Moran et al. (1996) are the first demonstrating that a generalist arthropod predator can enhance plant productivity in a complex natural ecosystem.

In that study, although herbivore biomass was depressed by mantid predation, the effect was spread out among arthropod taxa such that no single group of arthropod prey exhibited a statistically significant reduction in density. Herbivore load (a ratio of herbivore to plant biomass [Root, 1973]) was reduced by mantids, thus reducing herbivory, which reduced loss of plant biomass to herbivorous arthropods. By analogy, we can examine predator load, the proportional contribution of predator biomass to total arthropod biomass. Here, predator load was initially elevated by mantid addition, but this effect quickly abated as mantids died and emigrated from experimental plots. However, the initial boost to predator load resulted in a sustained effect on the lower two trophic levels through the season until the experiment was terminated. Evidently, herbivores did not recover from the initial mantid onslaught, either through reproduction or immigration in this open-field experiment, sufficiently to compensate for their initial reduction in numbers.

Predator load may provide an insight both to population regulation and community effect of generalist predators such as mantids. Hurd and Eisenberg (1984b) found that adding *T. a. sinensis* to a naive old field established a higher predator load in experimental than in control plots, which remained above control level throughout the season (Fig. 3.5). Fagan and Hurd (1994) found that predator load for open-field plots in which three experimentally established densities of *M. religiosa* had been established converged to a common value, several times higher than that of control plots. The reason that predator load, chiefly a function of mantid biomass in both cases, could be sustained in spite of substantial mortality is that surviving mantids grew. Thus, it is quite possible to find the same biomass in a large population of first instars as in a reduced number of sixth instars.

The convergence of predator load among three disparate population densities (Fagan and Hurd, 1994) suggests that carrying capacity (K) for mantids can be thought of as a compensation of growing biomass for declining numbers, that is, an ontogenetically adjustable K (OAK). The fact that predator load in treatment plots remained higher than control levels in both of the above studies suggests that the background community was not saturated with predators. However, the fact that control levels remained constant over time in both studies calls this into question: If there is room for more predators, why aren't there more eventually through numerical response (Holling, 1959)? A possible explanation is that large-bodied, extreme generalists such as mantids can find a place at table even when more specialized forms perceive the environment as full. Through OAK, they can exploit the broad spectrum of prey biomass in a way that smaller bodied, shorter-lived forms cannot, that is, achieve a higher carrying capacity than specialists in a given environment. Though their effect on the rest of the arthropod assemblage may be diffuse and sometimes hard to measure, yet they are capable of benefiting the plant community by reducing herbivory.

Future Research

> Around and around and around we spin,
> With feet of lead and wings of tin.
> Bokonon (Kurt Vonnegut, 1963)

In this chapter, I have tried to indicate some major questions and identify some loose ends that might be worth pursuing. Many of these are not specific to mantid biology. This reflects the fact that my primary research interest is not mantids per se, but rather in using mantids as tools to gain insight into the broader problem of the community niche of generalist predators. Conversely, what we learn about this niche from studies of other predaceous taxa may well give us insight into the biology of mantids.

Although we have learned much about a very few species of this diverse group of insects, we know almost nothing about how the different life history strategies represented within the taxon

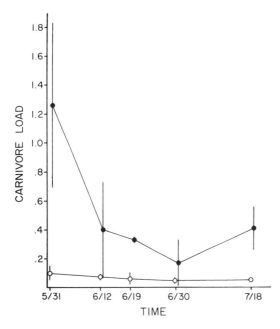

Fig. 3.5 Carnivore load, the ratio of carnivore to herbivore biomass (analogous to predator load in text) within an old-field arthropod assemblage for control plots without mantids (open circles) and treatment plots (closed circles) to which a cohort of *T. a. sinensis* was established as oothecae in the spring. Growth of individual mantids partially compensates for declining population density. After Hurd and Eisenberg, 1984b, with the kind permission of Blackwell Science, Ltd.

affect population biology, guild membership, or community impact. That which seems to be true of a few semelparous, univoltine, temperate zone, ambush species may not hold for tropical species, or those with higher vagility or with overlapping generations or that are active hunters. How do all of these factors affect population dynamics and community interactions? Another facet of mantid biology about which we know very little is parasitology; for example, how important are egg parasites to mantid host population dynamics over successive generations? Clearly, there is plenty of work to go around among mantidologists and general ecologists.

If *T. a. sinensis* populations are, in the short run at least, virtually closed because of low interhabitat dispersal ability, are there other mantids that exhibit regional metapopulation dynamics among source and sink demes? Could such vagile generalist predators reduce the tendency for habitat fragmentation to reduce predator load and release herbivores from control (Kreuss and Tsharntke, 1994)? The phenomenon of habitat fragmentation is part of a larger consideration of the role of disturbance in community dynamics. Even supposedly equilibrial ecosystems such as rainforests and coral reefs are, to a larger extent than tradition has held, subject to and dependent upon disturbance (Karlson and Hurd, 1993). Old-field ecosystems, the home of many mantid species, are secondary successional systems that exist only as a result of disturbance. They are temporary habitats, and residence in them for mantids is temporally limited. How do the adaptive strategies of old-field mantids differ from those inhabiting ecosystems of longer duration, such as climax forests and deserts?

What about mantids as agents of biological pest control? The finding that mantids can significantly reduce herbivory suggests that these predators can be good for plants. However, we must remember that pests and their crop hosts are by their nature in specialized relationships. The diffuse effects of mantids on arthropods does not necessarily bode well for tracking specific prey populations, nor is their numerical or functional response to changing prey density likely to be strong enough to keep up the pressure on offending insects (Hurd, 1985a). This is especially true because pest species are by definition numerically out of control and require close tracking by their enemies. An additional potential drawback is the propensity of mantids to feed on other predators and on pollinating insects. However, some evidence suggests that generalists such as mites (Sandness and McMurtry, 1972) and spiders (Haynes and Sisojevic, 1966) can exert a regulatory functional response on prey. Further, even highly polyphagous predators may be able to regulate prey if their life cycles are out of synchrony (DeAngelis et al., 1975; Post and Travis, 1979; Murdoch et al., 1985). Thus, biocontrol by mantids is as yet an open question.

A larger question is: How do generalist predators contribute to the integrity of ecosystems? If ecosystems are functional entities beyond their constituent populations, how do such predators contribute to the repeated patterns of food web dynamics (e.g., Pimm et al., 1991) that appear to lend order to natural communities of organisms? This question derives from a long-standing debate about what factors are responsible for community stability. The notion that the number of species in a community contributes to its resistance to change was first tested in a successional old-field ecosystem, and the results were measured for the producer and two consumer trophic levels (Hurd et al., 1971; Hurd and Wolf, 1974; Mellinger and McNaughton, 1975). In that experiment, diversity and stability were positively related only for plants; the greatest instability occurred among carnivorous arthropods the year following a fertilizer perturbation. Recently, Tilman (1996) confirmed that diversity can stabilize overall biomass of a plant assemblage, but not abundances of the individual populations.

It appears that stability might be an emergent property of ecosystems, but we are as yet far from building a predictive model. As a graduate student, I read Paine's papers on keystone species and decided that mantids might be good keystone predators, holding the fabric of arthropod assemblages together. Results from some of the work cited here (e.g., Fagan and Hurd, 1991a) suggest they indeed might be, at least in some communities.

ACKNOWLEDGMENTS

The work reported here from my laboratory has been supported over the years by funds from a number of sources: NSF Grants 77-11141 and BSR 8506181; Sea Grants from NOAA; UNIDEL and UDRF Grants from the University of Delaware (U. D.); John Glenn and R. E. Lee Grants from Washington and Lee University (W & L). Rob Eisenberg provided companionable collaboration for two decades. Several generations of graduate and undergraduate students from the Experimental Ecology course at U. D. and the General Ecology course at W & L contributed to data collection. Matt Moran, Mike Maxwell, and Trina Welsheimer made useful comments on earlier drafts of this chapter. I am grateful to my allergist, C. R. Donoho Sr., M.D., for making it possible to work in old-field habitats. Finally, I acknowledge my graduate advisor, Larry L. Wolf, for attempting to teach me how to ask the right questions. This chapter was written on the one-hundredth anniversary of the introduction of *Tenodera a. sinensis* into the United States.

4. The Ecology and Foraging Strategy of *Tenodera angustipennis*

Toshiaki Matsura, Tamiji Inoue

Praying mantids are sit-and-wait predators: They patiently wait for prey to approach their ambush sites, from which they may harvest those that come too close. Ambush sites are temporary and require no energetic investment on the part of the mantids. Since prey density is variable, and fecundity of predacious insects depends upon prey consumption (Hassell, 1978), movement from poor to better hunting sites will enhance a mantid's fitness.

We begin this chapter by describing the foraging strategy of the grassy-field mantid, *Tenodera angustipennis* (Saussure) and discussing experiments designed to test what we call "the hunger hypothesis." Second, we estimate mantid feeding levels in natural habitats to assess the successfulness of their foraging strategy. Finally, we compare their behavior to that of ant lion larvae, a more sedentary predator that has an energetic investment in its ambush site.

Foraging Strategy

We have for some time been interested in the factors that correlate with the change of ambush sites by *Tenodera angustipennis*. It has been suggested that a reduced rate of prey encounter or a reduced number of captures per unit time would induce movement to new hunting places (Formanowicz, 1987). However, we believe that it is unreasonable to think that mantids use either of these criteria. What predators must ultimately increase in order to raise their overall fitness is not the *number* of prey but the total *amount* (i.e., biomass) of prey that is eaten.

The praying mantids are polyphagous predators; they capture various sizes of prey from small flies weighing less than 10 mg wet mass to large grasshoppers weighing more than 1000 mg. For encounter or capture rates to be effective assay criteria for mantids, they would have to remember a catalog of prey sizes and retain a temporally indexed memory of hunting activity. Such sophisticated intellectual abilities would be surprising in a "simple" insect predator.

This being the case, we hypothesize that hunger level is the most likely criterion by which mantids assess the quality of their hunting grounds and that insufficient food intake induces movement to new ambush sites. We do know that hunger level governs capture behavior in the mantid *Hierodula crassa* (Holling, 1966); perhaps it also induces movement to a new hunting area.

Movement Patterns in Small Cages

In our initial experiments, we made two enclosures (90 x 135 x 90 cm), which had walls covered by nylon mesh to enable the mantids to walk around easily. No prey were supplied to the mantids in cage 1, and a constant density of prey was supplied to those in cage 2. The prey level in cage 2 was maintained at ninety houseflies (*Musca domestica*).

We introduced nine fifth-stadium nymphs of *T. angustipennis* into each cage. Four of each set of nymphs were food deprived for 2 days (hunger level [HL] = 2), and five were given houseflies *ad libitum* until the day of the experiment (HL = 0). All of the mantids were marked individually with small labels. We observed their daytime behavior and recorded their positions every 15 minutes for 4 days. Day length, temperature, and humidity were not controlled (Inoue and Matsura, 1983).

Figure 4.1 shows changes in the cumulative distance that each mantid moved in cage 1 (a, b), and cage 2 (c, d). Since the nymphs rarely moved at night, only the diurnal movement pattern is shown. Satiated nymphs (HL = 0) in cage 1 rarely changed their ambush sites until the morning of day 4, when they began active movement. In contrast, all of the hungry nymphs (HL = 2) except #1 moved around actively from the beginning of the experiment (b). The inactivity of mantid #1 may have been due to molting, however. Unfortunately, on day 2, nymph #2 ate nymph #7, so we added a new nymph to the HL = 2 condition (cage 1). On the morning of day 4, nymph #2 cannibalized a second victim (#3), after which it remained in one position.

In cage 2, where there were prey, there was not a great difference in the cumulative distance moved between the satiated and hungry nymphs (c versus d). On the first day, however, movement velocity of the hungry mantids was more than twice that of the satiated ones, but as they consumed prey, they became more sedentary.

Results of this experiment convinced us that nymphs of *T. angustipennis* begin to walk around when they have gone without food for longer than a certain period of time. In the case of these fifth instars, two or three days of fasting motivated them to change their ambush sites. But if enough prey were captured, they remained quite close to their original site.

Spatial Distribution of Prey and Foraging Behavior

In the previous experiment we observed the behavior of mantids in small cages, and we did not manipulate spatial distribution of the prey. In the field, however, prey are usually distributed more or less patchily. So the next issue that we examined was how the mantids respond behaviorally to spatial differences in prey density.

In these experiments, we cultivated fifty similarly sized pepper plants (about 80 cm high) that were planted in a 10 x 5 array within a semicylindrical frame (11 x 5 x 3 m) covered with vinyl sheeting (Fig. 4.2). Adult female grasshoppers, *Atractomorpha bedeli* (Bolivar), with an average wet body weight of 550 mg, were used as prey. These grasshoppers were found mainly on soybean plants, and we often observed adult *T. angustipennis* eating them. To obtain a heterogeneous spatial distribution of prey, grasshoppers were tethered with 30 cm cotton threads, eight per designated pepper plant. Grasshoppers that died or were killed were replaced with fresh ones every 2 hours during the daytime in order to keep prey density constant. Sixteen pepper plants containing eight grasshoppers each were arranged as shown in Figure 4.2, and no prey were put on the other thirty-four plants. All other potential prey had been killed with pesticide three weeks before the start of the experiment.

Adult *T. angustipennis* that were to be used for the experiment were collected in the field and fed on houseflies *ad libitum* for 3 days. Half of them were then deprived of food for 3 days (HL = 3), and the other half continued to receive houseflies *ad libitum* (HL = 0). Twenty hungry and twenty satiated mantids (ten of each sex per group) were released at random at night, and their ambush positions were recorded hourly during the daytime for the next 5 days.

Interestingly, this experiment revealed a significant difference in movement patterns between the sexes: Males were more active than females. The average total distance moved by females during 5 days was 11.2 m, while that traversed by males was 19.6 m ($p < .05$, F-test). Regardless of food status or hunger level, males frequently changed their ambush positions. Among females, satiated mantids (HL = 0) moved less than hungry (HL = 3) mantids; the former seldom moved around within prey-rich patches as they often did in prey-poor areas (Inoue and Matsura, 1983; Matsura, 1982; Matsura et al. 1975).

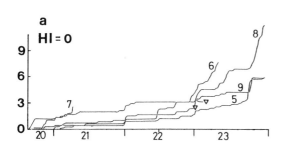
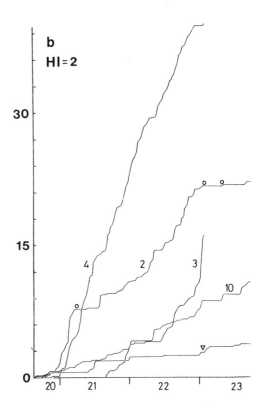
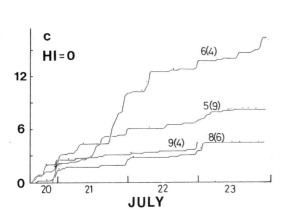
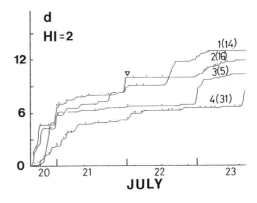

Fig. 4.1 Cumulative distances moved by (*a*) satiated and (*b*) hungry mantids in Cage 1 in which no additional prey was provided, and cumulative distances moved by (*c*) satiated and (*d*) hungry mantids in Cage 2 in which a constant density of prey was available. ▽, ○, and ● show molting, cannibalism, and feeding on a housefly, respectively.

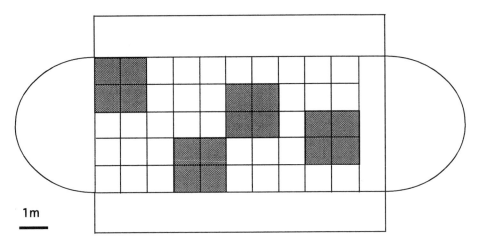

Fig. 4.2 Spatial arrangement of 50 pepper plants, 16 with prey (shaded, 8 grasshoppers each) and the rest without prey, in an experimental enclosure into which 40 mantids were released.

In summary, hungry adult female *T. angustipennis* changed ambush sites more frequently in patches without prey than in patches with prey, and, overall, females moved less frequently than males. Males were more peripatetic, and prey densities had little influence on their movements. The reasons for this will be clarified in the next section.

Field Studies

After these experiments, we reasoned that, although we had released our mantids in a large semicylindrical arena, their behavior might have been affected by the enclosure. So we carried out a field survey to examine the mantids' distribution and behavior under more natural circumstances.

The research area was approximately 50 × 110 m and consisted of rice paddies, goldenrod (*Solidago altissima*) fields, and soybean fields. It was surrounded by 5-m-wide roads, and we observed no migration of mantids in or out of the area. A census was taken twenty-three times (i.e., every 1 to 3 days) from 10 September to 24 October. We walked a patrol route that covered the peripheral 2 m of paddy fields. Mantids captured for the first time in each habitat type were marked individually and then released where they had been captured. We recorded mantid capture positions and identity numbers on a research map.

Figure 4.3 shows the seasonal change in number of adult mantids, estimated by the Jolly-Seber method (Jolly, 1965; Seber, 1965) and the sex ratio (males/total). We reasoned that the increase in number of individuals before mid-September would be due to recruitment by adult eclosion. Thereafter, the population remained relatively constant until early October and then decreased rapidly after mid-October. This decrease correlated with increased male mortality (note the sex ratio trend in Fig. 4.3). Males of *T. angustipennis* have a shorter physiological longevity than females (Matsura, personal observation). Although the possibility of cannibalism by females was not excluded, such incidents were rare in our caged mantid experiments (Inoue and Matsura, 1983). Similar seasonal changes in sex ratio are seen in the congeneric species *T. sinensis* (Hurd et al., 1994).

Seasonal changes in density of mantids captured on each type of vegetation (rice paddies, goldenrod fields, and soybean fields) are shown in Figure 4.4. Until about 7 October, there were remarkable differences in mean densities between paddy fields and other vegetation types ($p < .0001$, Fisher's PLSD test). A good food supply of grasshoppers (*O. japonica*) in the soybean fields coincided with the abundance of mantids there. Females tended to stay longer in the soybean patches; they consumed a large amount of prey, and hunger levels apparently seldom exceeded the threshold for movement to other hunting sites.

In the paddy fields prey density was very low, and we observed few mantids eating prey on rice plants. Rice plants may be too flexible to provide good footholds for prey capture, as Iwasaki (1995) demonstrated with *Tenodera aridifolia sinensis*. This was especially the case on windy days

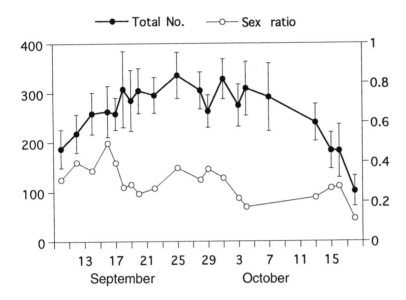

Fig. 4.3 Seasonal change in estimated number of adult mantids and sex ratio (males/total) in an open research area.

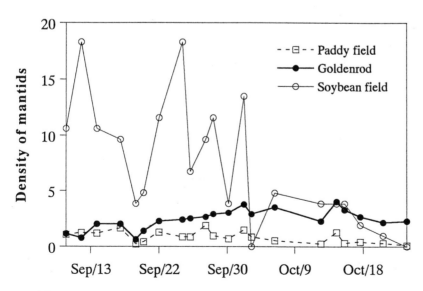

Fig. 4.4 Seasonal change in density of mantids (number/m^2) in different field types within the research area.

during which mantids seemed to conceal themselves and few were found on the foliage surface.

More mantids were found in the goldenrod patches than in the paddy fields. In addition to goldenrod, various plants such as *Cosmos bipinnatus* and *Aster subulatus* grew at our site, and hover flies coming to the flowers provided a rich secondary prey supply. After mid-October mantid density in goldenrod fields was higher than was observed on other vegetation. The reaping of rice and soybean plants had begun by this time, and female mantids were migrating to goldenrod fields in search of oviposition sites; females prefer solid stems such as goldenrod for oviposition sites.

Unlike the females, males moved around widely in the study area, independent of vegetation type. They are much lighter than the females (less than one-third of the female's weight), and often flew several meters over the vegetation. Figure 4.5 shows seasonal changes in the daily movement of males (A) and females (B) and number of matings observed (C). Daily movement was calculated by dividing the distance in meters between two successive capture points by the number of intervening days. The net movement of males was much greater than that of females during late September and early October, when the frequency of matings was the highest.

Matsura et al. (1984) reported that mature males of *T. angustipennis* consume only one-third to one-tenth as much food as do females. That result suggests that adult males maximize the number of matings by moving around widely, while females behave so as to acquire the maximum amount of prey. A clear positive correlation between prey consumption and fecundity has been shown to exist for *T. angustipennis*: Females that consumed more prey during the adult stage laid more and larger oothecae (Matsura and Morooka, 1983). Although the proximate objective for searching (i.e., moving around) within their habitat is different for males and females, the ultimate result for both is enhancement of fitness.

Evaluation of the Foraging Strategy of *T. angustipennis*

Female *T. angustipennis* seem to switch tactics between ambush and active search, using hunger

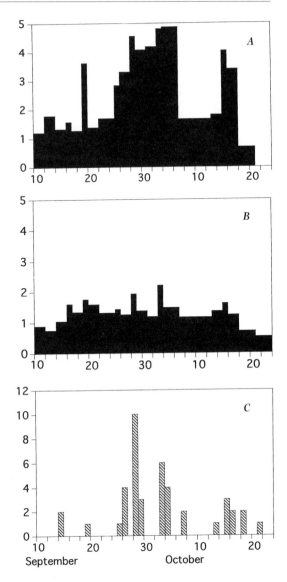

Fig. 4.5 Seasonal change in daily movement (m/D) of (*A*) males and (*B*) females and (*C*) in the number of matings observed.

level as the criterion for staying or moving. If this foraging strategy is efficient, they should consume an adequate amount of prey in their natural habitats. To evaluate that assumption, we first estimated prey consumption during a long period in the natural habitat, and then we compared that estimate with the maximum prey consumption obtained under laboratory conditions.

Prey Consumption

A positive correlation between food intake and growth rate has been found in various predatory insects (e.g., dragonfly larvae, Baker, 1982; larvae of tiger beetles, Pearson and Kinsley, 1985; ant lion larvae, Matsura et al., 1991). We found that a similar relationship existed for *T. angustipennis* (Matsura et al., 1975) when we fed houseflies to mantids from the fourth stadium until they became adults. When we varied the number of prey given to each mantis, we found a linear relationship between the number consumed and the increase in body length during the study period (Fig. 4.6). Although both male and female sixth and seventh instars developed, the data points were well described by the linear equation: $y = 0.10x + 26.70$, where x is the number of houseflies consumed and y (in millimeters) is the increase in body length.

Using this equation, we estimated the prey consumption of mantids in the field during a period from fourth instar until adult emergence. We then determined that the body length of fourth-instar nymphs in the field was 30 mm (29.87 ± 0.18 s.d., $n = 429$). Hence, $Y - 30$ is the average increase in the body length from fourth instar to adult eclosion, where Y is the body length of adult mantids collected in the field. Substituting $Y - 30$ for y in the previous equation, the prey consumption (Ce) of these mantids is estimated as $Ce = (Y - 56.70)/0.10$.

Samples of adult mantids were collected from three different habitats near the campus of the Kyoto University of Education in Kyoto City, and prey consumption was estimated for the fourth-instar to adult interval (Matsura, 1982). Results varied a little among habitats, but were in the range of 70 to 90% of maximum consumption measured in the laboratory, which was an average of 212.5 and 313.8 houseflies for males and females, respectively ($n = 6$ each).

Analysis

A comparison of the percentage of maximum prey consumption for mantids to those estimated for other arthropods is instructive. Using an index similar to ours, Lawton (1971) estimated prey consumption by damselfly larvae (*Pyrrhosoma nymphula*) living in a pond. The values obtained were lower in winter (20–30%) than during summer (70%), with the annual mean consumption less than 50% of maximum. Anderson (1974) found that a combination of widths of the abdomen and cephalothorax in the spiders *Lycosa lenta* and *Filistata hibernalis* is a good indicator of nutritional status. Comparing spiders reared under "starved" or "satiated" conditions with those collected in the field, he concluded that in nature these spiders are often exposed to starvation. Even though food supplies for mantids may not be stable in the wild, and they may experience food shortages, our estimated feeding levels suggest that the foraging strategy adopted by *T. angustipennis* is more successful than those of some spiders and damselflies.

Even so, the estimated prey consumption levels of wild mantids were less than those measured for laboratory animals fed *ad libitum*. That may be because the mantids use hunger level as a criterion for leaving an ambush site; that is, they will remain at a prey-poor site until a threshold

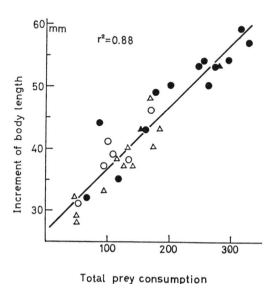

Fig. 4.6 Relationship of body weight increment to prey consumption during development from fourth instar to adult in *Tenodera angustipennis*. Circles - females; triangles - males: open - sixth instar, solid - seventh instar.

hunger level is reached. Unlucky mantids that move from one poor site to another poor site may even starve to death. Mantid behaviors seem not to follow the "marginal value theorem" of Charnov (1976), which assumes that, by prior sampling, a predator knows the average prey density in its habitat. Optimal predators decide how long they should spend foraging in a patch by comparing the present-patch prey density with the average prey density of the habitat. This model would be useful for wide-searching foragers such as great tits (Covie, 1977) or bumblebees (Whitham, 1977) but may not be applicable to sit-and-wait predators, whose sedentary ambush style of hunting precludes general sampling of the habitat. Inoue (1983a, b) generalized the foraging strategy of sit-and-wait predators as models of predators having limited information about prey distribution. Those models of learning and decision making have been developed further by various authors (reviewed by Krebs and Kacelnik, 1991).

Ant Lions

Among sit-and-wait predators there is a gradation in the degree to which they remain sedentary at the hunting site, that is, the lengths of time before they "give up" and move on vary between animals. Although mantids—which have no energetic investment in their site—often change ambush sites, pit-building ant lion larvae—which have a considerable investment in their pit—do not change sites (Matsura, 1987; Matsura and Murao, 1994). For example, fifth-instar *T. angustipennis* begin to move around after they have been starved for about 3 days, while residence times for antlion larvae under starvation conditions range from 25 to 72 days (e.g., *Myrmeleon bore,* 72 days; *M. formicarius,* 54 days; *Hagenomyia micans,* 25 days [Matsura and Murao, 1994]). Moreover, ant lion larvae can tolerate starvation for a considerable length of time. For example, third-stadium *Myrmeleon bore* can survive 84 days, *M. formicarius* can survive 147 days, and *Hagenomyia micans* can survive 55 days without food. On the other hand, fifth-instar and adult male *T. angustipennis* can survive only 15 days and females only 27 days without food (Matsura, 1981). There is an interesting difference in physiological responses to starvation between ant lion larvae and mantids: Ant lions decrease their respiratory rates dramatically during fasting to only 16 to 24% of the rate at satiation (Matsura and Murao, 1994). In contrast, the respiration rates of fasting mantids remain at 84% of the satiated rate (Matsura, 1981).

These factors suggest that *T. angustipennis* may have evolved under more favorable feeding conditions than were experienced during ant lion evolution. Hence, starvation tolerance was not a strongly favored trait for this mantid. Praying mantids do not pay a great energy cost for their ambush sites, whereas ant lion larvae must expend considerable energy to construct and maintain pitfall traps (Lucas, 1985). The foraging strategy adopted by *T. angustipennis*—switching their tactic between ambush and active search in response to prey population density—is found in various other arthropod predators, such as clerid beetles (Frazier, 1981), rove beetles (Richards, 1983), diving beetle larvae (Formanowicz, 1982), and scolopendrid centipedes (Formanowicz, 1987). We expect that this flexible foraging strategy will be found in a wide range of other arthropod predators.

5. Mating Behavior

Michael R. Maxwell

Sexual reproduction involves some degree of coordination between male and female behaviors. Biologists used to view this process as a harmonious act designed to ensure the survival of the species (Holmes, 1916; Tinbergen, 1951). Williams (1966), however, argued convincingly that natural selection acts on the individual rather than the group. Thus, interactions between members of a population—including mating encounters—are shaped by an ongoing competition to pass the maximum number of genes possible into the next generation. On top of this level of competition, Bateman (1948) and Trivers (1972) noted an important difference between the sexes: females typically invest more time and energy per gamete than do males. Hence, males recover more quickly from poor mate choices and are generally served best by maximizing the number of females with which they mate. Females, on the other hand, should do best by "holding out" for males that are of superior genetic quality or can provide other, direct, benefits such as food. The mating process, then, is akin to a joint business venture, with each partner following a separate, rather selfish agenda.

In mantids, the conflict between these agendas takes a dramatic form when a female cannibalizes a male. This grim feature of mantid mating behavior has fascinated scholars for several centuries (reviewed in Prete and Wolfe, 1992; Prete et al., this volume, chap. 1). Modern accounts stem from Poiret's 1784 report in the *Journal de Physique* of a male *Mantis religiosa* that continued to copulate following decapitation (quoted in Henneguy, 1904). A century later, Howard's (1886) vivid description of cannibalism and copulation in *Stagmomantis carolina*, along with Fabré's (1897b, 1914) poetic descriptions of *M. religiosa*, captured the attention of the scientific community. Subsequent studies examined the neurobiology of mantid behavior, with an emphasis on the male's ability to copulate following decapitation (Dubois, 1892; Chopard, 1914; Rabaud, 1916; Aoki and Takeishi, 1927). These investigations led ultimately to the well-known experiments of Ken Roeder (1935, 1967).

The mating behavior of mantids in the wild went largely unstudied, however, until the second half of this century (Phipps, 1960; Quesnel, 1967; Edmunds, 1975; Norsgaard, 1975; Robinson and Robinson, 1979). The relatively recent emergence of theoretical (Buskirk et al., 1984) and laboratory (Liske and Davis, 1987; Birkhead et al., 1988) work on mantid mating behavior has motivated a number of field studies on mating ecology (e.g., Lawrence, 1992; Hurd et al., 1994; Maxwell, 1998a).

My goals in this chapter are to review the state of knowledge about mantid mating behavior and to stimulate questions for further research. To this end, I have divided the chapter into four sections. The first section discusses when sexual activity occurs at the level of the individual and of the population. The following sections discuss mating behavior in terms of the chronology of its component behaviors, beginning with the mechanisms of mate attraction in the second section. This discussion continues in the third section, in which I include my thoughts on the possible proximate and ultimate reasons for female-on-male cannibalism and on how the risk of cannibalism might affect male behavior. The final section considers copulation and fertilization.

When to Mate
Sexual Maturation

Adults of several mantid species have been observed to copulate within the first few days following eclosion. For example, Rau and Rau (1913) reported seeing a newly eclosed female *S. carolina* copulate, and then go on to copulate five more times before ovipositing 13 days after eclosion. In the same study, a male of this species mated just 30 hours after eclosion. Kumar (1973) claimed that 48 hours posteclosion is the usual time at which copulation occurs in *Catasigerpes occidentalis*, *Miomantis paykullii*, *Sphodromantis lineola*, and *Tenodera superstitiosa*. In *Stagmomantis limbata*, Roberts (1937a) observed copulation between a 5-day-old female and a 9-day-old male.

In other species of mantids, individuals initiate sexual behavior some time after the first week of adult life, generally beginning at 2 weeks posteclosion (Roeder, 1935; Heath, 1980; Birkhead et al., 1988; Kynaston et al., 1994). Robinson and Robinson (1979) found that adult females emit pheromones after day 8 in *Acanthops falcata*. In the laboratory, male *Tenodera aridifolia sinensis* initiate copulation around day 10 to 14 (Liske and Davis, 1987; Davis and Liske, 1988). *Mantis religiosa* females apparently do not signal their sexual receptivity until 2 weeks after eclosion. Lawrence (1992) maintained newly eclosed adult females in field cages and monitored the arrival of males on a daily basis. The youngest female that had any males on her cage was 16 days posteclosion.

In many insects, the expression of sexual behavior partly depends on the development of the gametes (reviewed in Engelmann, 1970; Thornhill and Alcock 1983), and sex differences in the time necessary for gametic development may result in different ages at sexual maturity. In several mantids, females do not have mature oocytes upon eclosion and must feed in order to begin vitellogenesis. So, in the first few weeks of adult life, they may consume enough food to increase their body mass as much as three to five times (Matsura and Morooka, 1983; Birkhead et al., 1988; Maxwell, 1995). In the laboratory, *Coptopteryx viridis* and *Stagmomantis limbata* require 20 days to develop mature oocytes, which, in the latter species, results in a clutch with a volume on the order of 300 to 500 mm^3 (Cukier et al., 1986; Maxwell, 1995). In contrast to the size of this clutch, the male's spermatophore is roughly 5 mm^3 and presumably requires less time to produce. If the emergence of sexual behavior does, indeed, depend upon gamete development, then one might expect that the male *S. limbata* would begin sexual activity at a younger age as an adult than would the female. Field observations lend support to this hypothesis. Males as young as 7 days posteclosion have been observed copulating ($n = 4$, mean age = 11 days) (Maxwell, 1998a), but no females younger than 14 days posteclosion have been observed to do so ($n = 7$, mean age = 19 days) (Maxwell, unpublished data).

Time of Year

In temperate regions, mating activity is constrained to a few months out of the year. For mantid populations in Portugal, Maryland, and northern California, adults eclose in August and September and persist until October or November (Lawrence, 1992; Hurd et al., 1994; Maxwell and Eitan, 1998). Only the eggs overwinter, to hatch between April and June. As latitude approaches the equator, generations within a given population are more likely to overlap, with adults being present throughout the year. Such is the case in western Africa and Trinidad (Ene, 1964; Quesnel, 1967; Kumar, 1973; Edmunds, 1986). Under these conditions, mating occurs over a greater portion of the year (Quesnel, 1967; Bonfils, 1967).

Time of Day

For many mantids, it seems reasonable to assume that the majority of copulations begin during the day, because finding, approaching, and mounting a female may involve a number of visual cues. In nature, *Iris oratoria* and *Stagmomantis limbata* males have been observed to mount females between the hours of 1200 and 1800 (Maxwell, 1998a). In staged trials, Rau and Rau (1913) observed that mounting occurred in *S. carolina* during all daylight hours.

Mating behavior may be constrained to a much narrower time interval in some species. In the neotropical hymenopodid *Acanthops falcata*, mounting and copulation occur primarily, if not exclusively, around dawn (Robinson and Robinson, 1979). These females release pheromones only at this time, and the males fly in search of females guided, presumably, by the pheromones. Interestingly, photoperiod manipulations in the laboratory have demonstrated that the dark-to-light switch induces both pheromone emission by female *A. falcata* and spontaneous flight by males, regardless of the actual time of day (Robinson and Robinson, 1979).

Nighttime searches, approaches, and mountings are not entirely out of the question, however, because males might pursue females by moonlight. Although actual accounts of males mounting females during the night are rare (Rau and Rau, 1913; Phipps, 1960), one should realize that this paucity undoubtedly reflects the diurnal activity patterns of the researchers as much as it does that of the insects. Furthermore, several lines of evidence suggest that there is a considerable amount of nocturnal activity by mantids. For example, Edmunds (1975) has described what might be pheromone emission by a female *Tarachodes afzelii* after sunset, and Lawrence (1992) reported that male *Mantis religiosa* arrive at cages of females during the night. Males of many species fly toward lamps at night (Tinkham, 1938; Ene, 1964; Roy and Leston, 1975; Edmunds, 1986; Cumming, 1996; personal observations), and male *Miomantis* cf. *natalica* fly substantially more by night than by day (Cumming, 1996). Studies on mantid evasive responses to bat echolocating sonar emissions suggest a long history of nocturnal predator-prey interactions (Yager and May, 1990b; Yager et al., 1990; Yager, this volume, chap. 6). Indeed, Ene (1964) reported nocturnal predation of flying male *Tarachodes afzelii* by bats, owls, and night-

jars. Thus, it seems that the "night lives" of mantids warrant further investigation.

Getting the Sexes Together

In nature, adult mantids tend to be thinly dispersed. In several populations, the density is just 0.01 to 0.30 adults/m^2 (Bartley, 1982; Inoue and Matsura, 1983; Hurd et al., 1994; Maxwell, 1998a). How might individuals in such low-density populations find members of the opposite sex?

Many insects solve this problem through methods of mate attraction based on three classes of signals: mechanical, chemical, and visual (Cade, 1985). Mechanical signals include acoustic signals, such as stridulatory calls. Some mantids produce sounds by scraping or rubbing the underwings, hindlegs, or abdomen against the outer wings, but these seem to be part of startle or threat displays, not mating calls (reviewed in Ramsay, 1990).

Mate attraction through chemical signals such as pheromones may be quite common in mantids. Indeed, the release of pheromones by females is probably the first step in bringing the sexes together in many mantid species. Female-emitted pheromones have been demonstrated in the hymenopodid *A. falcata* (Robinson and Robinson, 1979). Several researchers suspect female pheromones within the Mantidae as well (Rau and Rau, 1913; Kelner-Pillault, 1957; Bonfils, 1967; Slifer, 1968; Edmunds, 1975; Kaushik, 1985; Lawrence, 1992; Maxwell, 1998b), and it appears that female *M. religiosa* can attract males from at least 100 m away (Kelner-Pillault, 1957). In general, researchers assume that males are the active searchers and the more sedentary females emit the pheromones. For species in which only the males fly, this is most likely the case. Some studies on the movement patterns of *Tenodera* spp. support the "searching male" scenario. Bartley (1982) and Inoue and Matsura (1983) found that male *T. aridifolia sinensis* and *T. angustipennis*, respectively, move more often than females, move farther per day than do females, and, at the population level, track the movements of females through the habitat. In other field studies, however, females were found to move as far as do males during a given period of time (*Mantis reli-*

giosa, Lawrence, 1992; *T. a. sinensis*, Eisenberg et al., 1992a). In these latter studies, it was probably the search for prey and oviposition sites—not for males—that influenced movement by the females (Matsura and Inoue, this volume, chap. 4).

The use of pheromones has been most intensively studied in *A. falcata* (Robinson and Robinson, 1979). During dawn, the flightless females adopt what has been termed a "pheromone-release" posture, which involves raising the wings and curling the abdomen ventrally, exposing two shiny protuberant glands on the anterior edge of tergite VII (Fig. 5.1). Robinson and Robinson's experiments demonstrated that males will fly toward virgin females in the pheromone-release posture even if visual cues are eliminated by a cloth partition between the males and females.

A. falcata females maintain the release posture for only about 20 minutes at dawn. Edmunds (1975) reports a similar time dependency in the possible pheromone release by female *Tarachodes afzelii*. Between the hours of 1930 and 2000 on three consecutive nights, a female curled her abdomen ventrally and repeatedly exposed two gray globules between tergites V and VI for 15 minutes. In contrast, pheromone release by female *Mantis religiosa* might be on a looser time schedule. Both Kelner-Pillault (1957) and Lawrence (1992) found that males arrived at female cages throughout the day. Interestingly, male arrivals peaked at specific times in both studies: 0930 to 1130 hours (Kelner-Pillault, 1957) and 1200 to 1400 hours (Lawrence, 1992). These intervals might have been periods of peak pheromone emission, or, alternatively, they might have been either the times of day when the males were at optimal body temperature or the times when the pheromones were most volatile.

Pheromone release by females depends not only on exogenous variables such as the time of day, but also on endogenous variables such as reproductive and feeding status (reviewed in McNeil, 1991). For example, *A. falcata* females stop emitting pheromones after they have copulated and resume emission after laying one or more oothecae (Robinson and Robinson, 1979). *Mantis religiosa* females usually do not attract males for at least 3 days following copulation, although Lawrence (1992) observed a female copulate with a male just 41 minutes after she had finished copulating with a previous male. Perhaps there is a delay in the onset of the refractory period in female *M. religiosa*. Alternatively, the second male might have forced a copulation with the female; this male had clung to the female throughout her previous copulation.

Pheromone release might also be influenced by nutritional state. As suggested in the previous section, the initiation of sexual behavior in some mantids might depend upon gametic development which is, in turn, influenced by diet. It is conceivable that females of some species might not release pheromones until they have developed a certain quantity of oocytes, as is the case for armyworm moths (*Pseudaletia unipuncta*, Cusson and McNeil, 1989). Indeed, the age dependency of pheromone release in *A. falcata* (Robinson and Robinson, 1979) is consistent with the idea that pheromone release must be preceded by a period of feeding. If this is the case, then food-limited females might be physiologically unable to emit pheromones. If, on the other hand, pheromone release does not depend upon gametic development or feeding condition, then food-limited females might attract males as a foraging strategy; that is, they might lure males in order to eat them (Hurd et al., 1994). Such *femme fatale* behavior could create selective pressures on males that would lead to an ability to determine a female's intentions before he ventures too close.

In addition to cues provided by pheromones, mantids may catch sight of each other from some distance away. Norsgaard (1975) witnessed a male *Tenodera aridifolia sinensis* fly onto a female from roughly 30 cm away, but this observation does not rule out olfactory cues. Unfortunately, long-distance vision is poorly understood for mantids, although *Iris oratoria* and *Stagmomantis limbata* visually track conspecifics from at least 45 cm away (personal observations) and *Mantis religiosa* purportedly can detect females at 1 m or more (Roeder, 1935). Studies on *Tenodera aridifolia sinensis* (Corrette, 1990; Prete et al., 1990) indicate that a female's maximum strike distance is roughly one body length, or 5 to 10 cm for many mantids. Thus, initially spotting the female from beyond this distance may work in the male's favor, as it affords him a safe opportunity to assess her.

Making Contact: Cannibalism, Approaches, and Mounts

Once the male or the female is close enough to see the other, a new phase of precopulatory behavior begins when either approaches the other. This approach phase might include bending or waving of the abdomen or legs by either adult. Eventually, the male makes contact by mounting the female, typically by leaping onto her back. In many species, the threat of an attack by the female is imminent throughout this precontact phase. The male and female visually track each other's movements during these events, suggesting the importance of visual cues. To highlight this point, Rau and Rau (1913) removed one or both antennae from male *S. carolina* and found that the males could still mount females as well as could intact males. Indeed, after watching several encounters between males and females, one cannot shake the notion that each sex is studying the other.

Female-on-Male Cannibalism among Taxa

Female mantids are infamous for connubial cannibalism (Prete and Wolfe, 1992). However, in spite of the fact that they stand to gain nutritional benefits by consuming males, not all females eat their suitors. Table 5.1 summarizes the data available on cannibalism during encounters between male and female mantids, beginning with Howard (1886). Cannibalism has been observed in 76% of the species listed, either in captivity or the field, and in all but two of the subfamilies (Sibyllinae and Vatinae). The mating behavior of four species has been studied repeatedly in the field (i.e., five or more unmanipulated encounters): *Iris oratoria, Mantis religiosa, Stagmomantis limbata,* and *Tenodera aridifolia sinensis* (Bartley, 1982; Lawrence, 1992; Hurd et al., 1994; Maxwell, 1998a), and cannibalism of the male occurs in each of these species. Cannibalism in the wild has also been documented in *Orthodera novaezealandiae, Polyspilota aeruginosa,* and *Sphodromantis lineola,* although with smaller sample sizes (Phipps, 1960; Ramsay, 1990; Preston-Mafham, 1990).

As one can see from Table 5.1, the frequency of cannibalism varies both between and within species. What causes this variation? In order for cannibalism to occur, the female must be able to overpower the male. Female mantids are typically larger than males and have longer and stronger raptorial forelegs. One can predict, then, that species with higher degrees of size dimorphism between the sexes will exhibit higher frequencies of cannibalism. This idea has not been tested in mantids; Elgar (1992), however, found little support for this prediction among spider taxa. Notably, the species in Table 5.1 represent two of Beier's (1964) eight families within Mantodea. The mating behaviors of species within the other six families remain unknown. Undertaking such studies will provide more degrees of freedom when testing for interspecific variation in cannibalism.

Interestingly, size dimorphism does not appear to explain variation in cannibalism within *Iris oratoria*. In mating trials, the degree of dimorphism between paired mantids did not predict whether the males would be attacked (Maxwell, in press). It seems, then, that differences in ecological or laboratory conditions (e.g., the feeding regimes of the females) are most likely to contribute to behavioral variation within and between populations, as I will discuss below.

How Do Females Cannibalize Males?

The female may attack the male during one of three phases: before he attempts to mount, while he is mounted on her back, or after he has dismounted. When the male is not mounted (i.e., during the premount or postmount phases), the female's attack consists of a characteristic predatory strike. When the male is mounted, she can often reach back with one foreleg, grab the male by the head or anterior prothorax, and pull him to her mouth.

Both the timing and voraciousness of the female's attack can determine whether or not copulation occurs. If the female attacks and devours the male during his initial approach, mating is precluded (Edmunds, 1975; Kynaston et al.,

Table 5.1 Review of cannibalism of males by females within the Mantodea

Family	Species	Setting	Encounters[a]	Premount[b]	Mount[c]	Postmount[d]
Hymenopodidae						
Acromantinae						
	Acanthops falcata[g]	Cage	31	0	0	0
		Field	4	0	0	0
	Acontiothespis multicolor	Field	7+	0	0	0
	Acontista sp.	Field	1	0	0	0
	Catasigerpes occidentalis	Lab	—[h]	—	3	—
	Hestiasula brunneriana	Cage	2	0	1	0
		Cage	1	0	0	0
	Tithrone sp.	Field	1	0	0	0
Hymenopodinae						
	Creobater urbana	Lab	2	0	0	0
	Ephestiasula amoena	Cage	2	0	0	0
		Lab	—	0	0	0
	Euantissa ornata	Lab	6	0	0	0
Oxypilinae						
	Oxypilus hamatus	Lab	9	0	1	0
Mantidae						
Amelinae						
	Litaneutria minor	—	—	—	Yes[h]	—
Liturgusinae						
	Humbertiella similis	Cage	30	0	1	1
Mantinae						
	Hierodula sp.	Lab	—	1	0	0
	Hierodula membranacea	Lab	31	~7	~3	~11
		Lab	14	0	~3	~11
	Hierodula patellifera	Lab	—	Yes[i]	—	—
	Hierodula tenuidentata	Lab	—	Yes	—	—
	Iris oratoria	Field	7	0	1	0
		Cage	42	0	0	0
		Cage	38	2	0	4
	Mantis religiosa	Cage	—	—	Yes	—

Cannibalism[e]	Description of encounters[f]	Reference
0	Unmanipulated; colony of adults in 1 x 1 x 0.5 = m cage	Robinson & Robinson (1978, 1979)
0	Unmanipulated	Preston-Mafham & Preston-Mafham (1993)
0	Staged; at least 7 pairs observed.	Quesnel (1967)
0	Unmanipulated	Preston-Mafham (1990)
—	—	Kumar (1973)
0	Staged	Mathur (1946)
0	Staged	Ahmad et al. (1985)
0	Unmanipulated	Preston-Mafham (1990)
2	Staged	Daniels et al. (1989)
0	Staged	Loxton (1979)
1	Unmanipulated; colony of adults	Loxton (1979)
0	—	Mathur (1946)
4	Staged	Edmunds (1975)
—	—	Roberts (1937b)
4	Staged	Kaushik (1985)
0	—	Heath (1980)
0	Staged. Trials with females on High (n = 5), Med (n = 12), and Low (n = 14) diets. Twenty-one co-occurrences of copulation and cannibalism; I have divided these by category based on percentages given in article.	Birkhead et al. (1988)
0	Staged. Trials with females on Low diet that were allowed to cannibalize males. I have divided these by category based on percentages given in the article.	Birkhead et al. (1988)
—	Staged	Lawrence (1992)
Yes	—	Roeder (1967)
1	Unmanipulated	Maxwell (1998a)
0	Staged; females not starved before trials	Maxwell (in press)
1	Staged; females starved 6–8 days before trials	Maxwell (submitted-a)
—	Staged	Fabre (1897b, 1914)

Table 5.1 *(continued)*

Family	Species	Setting	Encounters[a]	Premount[b]	Mount[c]	Postmount[d]
		Cage	1	1	0	0
		Cage	—	Yes	0	0
		Cage	4	0	0	0
		Cage	—	0	0	0
		Field	13	0	4	0
		Field	17	3	0	0
	Miomantis caffra	—	—	—	Yes	—
	Miomantis paykullii	Lab	—	—	3	—
	Polyspilota aeruginosa	Lab	4	0	1	0
		Field	2	0	1	0
	Sphodromantis lineola	Field	3	0	0	1
		Large room	1	0	1	0
		Lab	—	0	0	0
		Lab	11	0	0	1
		Cage	42	0	0	0
		Cage	19	1	0	0
	Stagmomantis carolina	Cage	1	1	0	0
		Lab	1	0	1	0
		Cage	—	0	Yes	0
		Cage	32	0	7	4
	Stagmomantis limbata	—	—	—	Yes	—
		Field	45	0	4	0
		Cage	18	0	0	0
		Cage	18	0	0	0
		Cage	7	0	0	1
	Tenodera angustipennis	Cage	35	0	1	0
		Cage	24+	—	—	—
	Tenodera arid. sinensis	Cage	1	0	0	1
		Field	1	0	0	1
		—	20	0	0	0
		Field	14	0	0	0
		Cage	15	0	0	0
		Cage	6	0	1	0
		Cage	7	0	0	0
		Field	71	0	3	0
	Tenodera superstitiosa	Lab	—	0	0	0

Cannibalism[e]	Description of encounters[f]	Reference
0	Staged	Rabaud (1916)
Yes	Staged	Roeder (1935)
2	Staged	Joulin (1983)
0	Unmanipulated; colony of 43 adults in large cage	Ehrmann (1986)
0	Unmanipulated; some M cases could be Pre-M	Lawrence (1992)
1	Staged	Lawrence (1992)
—	—	Ramsay (1984, 1990)
—	—	Kumar (1973)
3	—	Heath (1980)
0	Unmanipulated	Preston-Mafham (1990)
0	Unmanipulated	Phipps (1960)
0	Unmanipulated	Phipps (1960)
0	—	Kumar (1973)
2	Staged	Edmunds (1975)
0	Staged; females not starved before trials	Kynaston et al. (1994)
17	Staged; females starved for 4 days before trials	Kynaston et al. (1994)
0	Staged	Howard (1886)
0		Riley & Howard (1892)
—	Staged	Blatchley (1896)
8	Staged; adults confined together for several days	Rau & Rau (1913)
—	—	Roberts (1937a)
5	Unmanipulated; some M cases could be Pre-M	Maxwell (1998a)
1	Staged; females on High-quality diets	Maxwell (unpub. data)
3	Staged; females on Low-quality diets	Maxwell (unpub. data)
1	Staged; females starved 3–8 days before trials	Maxwell (unpub. data)
0	Unmanipulated; 40 adults kept in 15×11×3-m cage	Inoue & Matsura (1983)
0	Staged; at least 24 pairings	Matsura & Morooka (1983)
0	Staged	Gurney (1950)
0	Unmanipulated	Norsgaard (1975)
2	—	Schauff & Jones (1978)
0	Unmanipulated	Bartley (1982)
0	Staged. Females not starved before trials; trials were terminated at intromission	Liske & Davis (1984, 1987)
4	Staged. Females starved for 3–5 days before trials; trials were allowed to continue after intromission	Liske & Davis (1984, 1987)
2	Staged. Females starved for 5–11 days before trials; trials were terminated at intromission	Liske & Davis (1984, 1987)
9	Unmanipulated; some M cases could be Pre-M	Hurd et al. (1994)
0	—	Kumar (1973)

Table 5.1 *(continued)*

Family	Species	Setting	Encounters[a]	Premount[b]	Mount[c]	Postmount[d]
Oligonychinae						
	Oligonyx insularis	—	—	—	—	Yes
Orthoderinae						
	Orthodera ministralis	Lab	—	—	—	—
	Orthodera novazealandiae	Field	—	—	Yes	—
Photininae						
	Coptopteryx viridis	—	—	—	—	—
Sibyllinae						
	Sibylla pretiosa	Lab	—	0	0	0
Tarachodinae						
	Tarachodes afzelii	Lab	18	3	2	0
Vatinae						
	Stenovates sp.	Lab	—	0	0	0

1994; personal observations). On the other hand, males are renowned for their ability to initiate copulation while being eaten. The organization of a mantid's central nervous system will allow both copulation and spermatophore transfer in the absence of descending input from the cephalic ganglia (Roeder et al., 1960; Roeder, 1967).

Whether the unfortunate male achieves intromission largely depends upon the form of the female's attack. One way by which copulation and cannibalism may co-occur begins with the female capturing the male and holding him perpendicular to her body, with her forelegs gripping his thorax (Fig. 5.2). If she chews through his prothorax, thereby cleaving him in two, and then releases the rear portions while she feeds on the anterior prothorax and head, the mesothoracic and metathoracic legs retain the ability to grasp her and position themselves so that the abdomen can initiate copulation (Roeder, 1935, 1960; Davis and Liske, 1988; personal observations; Fig. 5.4). The joint occurrence of premounting cannibalism and copulation is given as "Premount" in Table 5.1.

The female may also attack the male after he has mounted her, either before he attempts copulation, during copulation, or after copulation. Roeder (1935, 1967) claimed that female *M. religiosa* will attack a mounted male only immediately after he has mounted and while he is still shuffling around on her back, or when he is positioned poorly (e.g., when his head is very close to hers). Similarly, I have seen a female *Iris oratoria* start eating a copulating male who had positioned himself much more anteriorly than is usual for this species (i.e., his head was to the side of hers; Maxwell, 1998a). Other authors report females grabbing mounted males without describing precisely the mounted male's position (Riley and Howard, 1892; Rau and Rau, 1913; Phipps, 1960; Edmunds, 1975; Kaushik, 1985; Liske and Davis, 1987).

The female may also attack the male after he has copulated and dismounted. Confining the male and female in a cage undoubtedly increases the probability of such an event. Norsgaard (1975), however, describes an encounter in nature in which a male *T. a. sinensis* remained within 30 cm of his mate after dismounting and, after a few hours, the female approached and ate him.

Cannibalism may occur in the absence of cop-

Cannibalism[e]	Description of encounters[f]	Reference
—	—	Bonfils (1967)
4		Suckling (1984)
2	Unmanipulated	Ramsay (1990)
Yes	—	Cukier et al. (1979)
0	—	Heath (1980)
11	Staged	Edmunds (1975)
0	—	Heath (1980)

[a] Encounters: total number of male-female encounters. For staged encounters, this is the number of male-female pairings. For unmanipulated encounters, this is the number of occurrences of mounting, copulation, and/or cannibalism.

[b] Premount: copulation co-occurring with cannibalism that was initiated before a mount by the male.

[c] Mount: copulation co-occurring with cannibalism that was initiated while the male was mounted.

[d] Postmount: copulation co-occurring with cannibalism that was initiated after the male had dismounted.

[e] Cannibalism: cannibalism, with no copulation observed.

[f] *Unmanipulated* encounters refer to male-female pairings that occurred without intervention by researchers. *Staged* encounters refer to males and females that were brought together by researchers.

[g] Genera are assigned to family and subfamily according to Beier (1964). *Acontiothespis, Ephestiasula, Euantissa,* and *Stenovates* do not appear in Beier (1964) and are assigned to subfamily according to Giglio-Tos (1927).

[h] Table entries: — = behavior not reported; Yes = behavior asserted but frequency not given.

ulation. This can result from an attack by the female either before the male attempts to mount, while he is mounted but before he attempts copulation, or after he dismounts without having copulated. Authors do not always make a distinction between these forms of cannibalism, so they are grouped into one category ("Cannibalism") in Table 5.1.

As a definitional note, cannibalism of the male in a sexual context has been described as sexual cannibalism by various authors. Two definitions of *sexual cannibalism* exist in the arthropod literature. Most authors define sexual cannibalism as the joint occurrence of copulation and cannibalism within the same male-female encounter (Pollis, 1981, 1990; Buskirk et al., 1984; Forster, 1992; Lawrence, 1992; Hurd et al., 1994; Andrade, 1996). This definition does not include cases where the male sexually courts the female but is cannibalized in the absence of copulation. Hence, a second definition of sexual cannibalism is the joint occurrence of cannibalism and either courtship or copulation within an encounter (Elgar, 1992; Johns and Maxwell, 1997; Maxwell, 1998a). This definition requires one to distinguish courtship from other behaviors (Prete, 1995). In mantids, courtship behavior may involve such behavior as movements of the abdomen and legs, which I discuss below.

Why Do Females Cannibalize Males?

The cannibalism of males occurs in many arthropod species (reviewed in Elgar, 1992), including mantids. In the field, conspecific males constitute 63% and 26% of predation events by female *T. a. sinensis* and *Stagmomantis limbata*, respectively (Hurd et al., 1994; Maxwell and Eitan, 1998). Why might this be so? With regard to mantids, the question can be answered from three different perspectives. One could consider male and female morphology, one could look at the structure of intersexual encounters between adults, or one could examine the evolutionary costs and benefits associated with cannibalism. From the first perspective, one could argue that females eat males simply because they can—mantids are generalist predators, and females are typically larger and stronger than males (Prete, 1995). As discussed above, it remains to be determined whether sexual dimorphism can explain variation in cannibalism between and within species.

The topology of the encounter between male and female mantids may be a better predictor of the male's fate. An important factor in the outcome of a sexual encounter is whether the female sees the male and is in a position to attack him before he attempts to mount her (Roeder, 1935; Lawrence, 1992; Maxwel, in press). Captivity in and of itself might actually promote cannibalism either by confining the male with the female and thereby eliminating his opportunity for escape, by not allowing enough room for him to maneuver before or after mounting, by distracting him with the movements or shadows of observers, or by pairing him with an immature or unreceptive female.

Although the physical arrangement of an encounter plays a part in determining its outcome, other factors are at work because encounters in cages do not always end in the male's death (Table 5.1). An examination of the reproductive costs and benefits associated with female-on-male cannibalism provides insight into why females cannibalize males (Table 5.2). The first point to consider is that female mantids gain a direct benefit from eating a male: food to support somatic and/or reproductive efforts. Hence, one should expect cannibalism to increase as females become more food limited, and this seems to be the case. Mating trials with *H. membranacea*, *S. lineola*, and *T. a. sinensis* suggest that poorly fed or starved females are more cannibalistic than better-fed females (Liske and Davis, 1987; Birkhead et al., 1988; Kynaston et al., 1994). Within a species, variation in food abundance between

Table 5.2 Reproductive benefits and costs associated with the cannibalism of the male

Benefits	Costs
Female	
Receive nutrients for somatic effort (survival) and/or reproductive effort (produce ova)	Possibly preempt or interrupt the receipt of sperm
	Possible injury from the male
Male	
Possibly increase female's reproductive output; male benefits only if he fertilizes the female	Death
Possible paternity advantage: cannibalism may result in the male fertilizing a higher proportion of ova than an uncannibalized male	

place and time may contribute to variation in the occurrence of cannibalism, as suggested by the different degrees of cannibalism reported by Bartley (1982) and Hurd et al. (1994) for *T. a. sinensis* (see Table 5.1).

There are several ways one might explain the mechanisms that underlie female-on-male cannibalism in mantids. On the one hand, one could argue that cannibalism represents either instances of "mistaken identity" in which conspecifics are not recognized as such (Elgar, 1992) or that it represents the outcome of selective pressures for indiscriminate rapacity on the part of the developing nymphs (Arnqvist and Henriksson, 1997; also see Prete, 1995, on this point). The former hypothesis requires that one accept the assumption that a mantid will not attack a conspecific that it actually has identified as such. The latter hypothesis argues that nymphs that are more liable to attack and eat any potential prey item, including conspecifics, will grow faster and eclose as larger adults. According to Arnqvist and Henriksson, then, cannibalism by adult females might reflect selection on earlier life stages. These ideas about mistaken identity and indiscriminate rapacity have not been examined in mantids.

On the other hand, one could argue that females specifically target males. Conspecifics are probably a particularly rich food source (Polis, 1981). Presumably, their bodies contain all of the essential ingredients needed to produce viable mantids and, by implication, viable mantid eggs. Hence, females might consume males preferentially and be able to do so because the males can be lured by pheromones.

Cannibalism might also be a form of female-female competition (Polis, 1981). By cannibalizing a male, a female denies all other females access to him. At several sites in the temperate zone, mantid populations become increasingly female-biased as the season progresses (Lawrence, 1992; Hurd et al., 1994; Maxwell, 1998b). In such populations, males might be a very limited resource late in the season, and females might have to compete for access to them. Exploitation competition may occur, wherein an unmated or sperm-limited female competes with her neighbors to at-tract a male; eating the male after insemination ensures that he does not inseminate her rivals. A "sperm-satisfied" female might even continue to attract males in order to cannibalize them, thereby gaining nutrients as well as removing the males from the mating pool. Females might also actually fight over males. Rau and Rau (1913) report such interference competition between female *S. carolina*. They observed two females repeatedly leap at a copulating pair with which they were caged. The Raus felt compelled to remove the copulating pair in order to prevent further calamity. Interfemale competition for access to males remains to be studied in nature.

In some cases, it may actually be in a female's interest to prevent or interrupt the receipt of sperm from a male, either because she is sexually immature, because she has already mated a sufficient number of times, or because the male is of inferior quality. With regard to the last reason, cannibalism becomes the ultimate form of mate rejection (Rau and Rau, 1913; Kyriacou, 1987).

The cannibalism of the male involves two potential costs to the female (Table 5.2). First, females risk injury during the inevitable struggle. Although male mantids seem ill-equipped to kill the larger females, they can indeed do so (Prete, unpublished data), and I have observed some male *Iris oratoria* chew on the wings of females during attacks. The second potential cost is that the female loses a mate if she attacks the male too soon or too vigorously. Because this cost is most severe for virgin females (Newman and Elgar, 1991), one can predict that, if feeding status and male availability were held constant, unmated females will be less likely to cannibalize males before mating than will mated females. This prediction has yet to be tested.

Other Female Behaviors

Females of several species behave in an interesting way that may signal their receptivity to mating. When situated face to face with a male, some females turn 180°, such that their rears face the males (*T. a. sinensis*: Gurney, 1950; *E. amoena*: Loxton, 1979; *I. oratoria*: Maxwell, in press; *S. limbata*: Maxwell, unpublished data). In *I. oratoria* females "presented" in this way in 45% of mating trials (Maxwell, in press). These presentations seemed to correlate with the occurrence of mount attempts by males, but the trend was not statistically significant.

Movements of the abdomen occur in both sexes in some species (summarized in Table 5.3). In females, these movements are typically low-amplitude, dorso-ventral pumps (Loxton, 1979; Liske and Davis, 1987; Maxwell, 1995). Based on Table 5.3, it appears that the approach phase does not involve abdominal movements in most mantids. However, there have been so few observations made on so few species that little is known about courtship behaviors (but see, Quesnel, 1967; Edmunds, 1975; Loxton, 1979; Kynaston et al., 1994). Interestingly, Liske and Davis (1987) reported a behavior in *T. a. sinensis* that has not been described in other species. A few females approached males and touched them with their antennae before the males attempted to mount. One female even stroked the male's foreleg for several minutes. Approaches by female mantids do not always appear so conciliatory, and males sometimes run or fly away from approaching females (Edmunds, 1975; Lawrence, 1992; Maxwell, unpublished data). It is not known if specific characteristics of the female's approach triggers such escape behavior.

Sometimes females confront males with characteristic defensive (i.e., "deimatic") displays (Maldonado, 1970b), or by striking at them (e.g., *A. multicolor*, Quesnel, 1967; *S. lineola*, Edmunds, 1975). Such behaviors do not necessarily deter amorous males, however. For example, nine of forty-two well-fed *Iris oratoria* females struck at males before or just after the males mounted the females (Maxwell, in press). All of these males mounted the females, and eight copulated. These attacks did not appear to indicate female choice; attacked males did not differ significantly from nonattacked males in terms of body mass or pronotum length.

Costs and Benefits of Cannibalism for the Male

Adult female-on-male cannibalism has been documented for most of the species in Table 5.1. Thus, reproduction for many male mantids is a dangerous mating game. Over evolutionary time, the risk of cannibalism can have one of two effects on male behavior (Gould, 1984). First, in order to maximize their reproductive benefits by being cannibalized by the female with which they mate, the males might be "suicidal"; that is, they may behave in such a way so as to facilitate the co-occurrence of copulation and cannibalism. Alternatively, the costs of cannibalism might select for males that act to lower the probability that they will be eaten.

A male may gain a reproductive advantage by being cannibalized (Table 5.2). His soma may contribute to the fecundity or prolonged survival of his mate, thereby enabling her to lay more eggs. Studies using chemical tracers demonstrate that female insects manufacture ova from nutrients derived from male spermatophores or seminal fluids (reviewed in Parker and Simmons, 1989; Boggs, 1995), but such work remains to be done on mantids. Birkhead et al. (1988) found that cannibalizing a male significantly increased female reproductive output in food-limited female *Hierodula membranacea*. Maxwell (submitted-a), however, failed to detect an effect in otherwise well-fed female *Iris oratoria* that were starved before mating trials.

Obviously, in order to benefit from being cannibalized, a male must fertilize the ova of the female that eats him (Wickler, 1985; Simmons and Parker, 1989). Unfortunately, it has not been determined whether or not this actually happens in any mantid species.

Decapitated males are apparently capable of fertilizing females as well as are intact males (Roeder, 1935; Lawrence, 1991). In fact, Roeder (1967) suggested that partially cannibalized males may actually fertilize a higher proportion of eggs than intact males. In an inbred strain of *Hierodula tenuidentata*, decapitated males performed more vigorous sexual movements and, in a single mating, fertilized more eggs than intact males. It is unclear as to whether this higher rate of fertilization is due to decapitated males passing more sperm or simply providing greater mechanical stimulation to the females (Roeder et al., 1960; Davis and Liske, 1988). Interestingly, Andrade (1996) provided another explanation that may apply to male mantids. In the Australian redback

Table 5.3 Observations of abdominal movements during the approach phase

Family	Species	Female	Male	Reference
Hymenopodidae				
Acromantinae				
	Acontiothespis multicolor[a]	No[b]	Yes[c]	Quesnel (1967)
	Catasigerpes occidentalis	No	No	Kumar (1973)
	Tithrone roseipennis	No	Yes	Quesnel (1967)
Hymenopodinae				
	Ephestiasula amoena	Yes	No	Loxton (1979)
Mantidae				
Liturgusinae				
	Humbertiella similis	No	No	Kaushik (1985)
Mantinae				
	Hierodula coarctata	No	No	Loxton (1979)
	Hierodula membranacea	No	No	Loxton (1979); Birkhead & Lawrence (1988); Lawrence (1991)
	Hierodula patellifera	—	No	Lawrence (1991)
	Hierodula tenuidentata	No	No	Roeder et al. (1960); Loxton (1979)
	Iris oratoria	Yes	Yes	Maxwell (1995)
	Mantis religiosa	No	No	Fabre (1897b, 1914); Roeder (1935, 1967); Lawrence (1992)
	Miomantis paykullii	No	No	Kumar (1973)
	Sphodromantis lineola	No	No	Kumar (1973)
		No	Yes	Edmunds (1975); Kynaston et al. (1994)
	Stagmomantis carolina	No	No	Rau & Rau (1913); Quesnel (1967)
	Stagmomantis limbata	Yes	Yes	Maxwell (unpub. data)
	Tenodera angustipennis	No	No	Inoue & Matsura (1983)
	Tenodera aridifolia sinensis	Yes	Yes	Liske and Davis (1987)
	Tenodera superstitiosa	No	No	Kumar (1973)
Oligonychinae				
	Oligonyx insularis	No	No	Bonfils (1967)
Tarachodinae				
	Tarachodes afzelii	No	No	Edmunds (1975)
Vatinae				
	Stagmatoptera biocellata	No	No	Loxton (1979)

[a] Genera are assigned to family and subfamily according to Beier (1964). *Acontiothespis*, *Ephestiasula*, and *Stagmatoptera* do not appear in Beier and are assigned to subfamily according to Giglio-Tos (1927).

[b] *No:* abdominal movements not observed.

[c] *Yes:* abdominal movements observed. For females, these typically involve the pumping of the abdomen. For males, these involve dorso-ventral and lateral bendings of the abdomen.

spider (*Latrodectus hasselti*), cannibalized males remain in copula twice as long as uncannibalized males, which results in roughly a doubling of the number of offspring sired by the cannibalized males. Perhaps this also occurs in mantids.

The potential reproductive benefits of cannibalism cost the male dearly. Despite what some literature might suggest (reviewed in Prete and Wolfe, 1992), decapitation is not a necessary prerequisite to copulation, and intact males do fertilize females (Roeder, 1935; Phipps, 1960; Quesnel, 1967; Edmunds, 1975; Matsura and Morooka, 1983; Birkhead et al., 1988; Lawrence, 1991, 1992; Maxwell, in press). Moreover, male mantids are capable of copulating with more than one female (Rau and Rau, 1913; Roeder et al., 1960; Roeder, 1967; Bartley, 1982; Inoue and Matsura, 1983; Lawrence, 1992; Hurd et al., 1994; Kynaston et al., 1994; Maxwell 1995). Field studies report multiple copulations by individual males (Bartley, 1982; Lawrence, 1992; Hurd et al., 1994), as well as multiple noncopulatory encounters with females (Maxwell, 1998a).

Are Male Mantids Suicidal?

One could argue that the benefits of being cannibalized might be large enough to favor male "suicide," with the males actively offering themselves as nuptial meals to the females (Parker, 1979; Simmons and Parker, 1989). Such behavior seems to occur in certain spider species, where the males somersault into the females' mouth parts during copulation (Grasshoff, 1964; Cariaso, 1967; Forster, 1992; Andrade, 1996).

Buskirk et al. (1984) developed a model of sexual cannibalism that predicts that the benefits of cannibalism plus copulation outweigh the costs if the female substantially increases her reproductive output, provided that the male actually sires these offspring, and the probability of additional mating opportunities is low for the male. Buskirk et al.'s model notwithstanding, there are few data that could be used to argue for male "suicide" in mantids. Instances of males mounting aggressive females (Quesnel, 1967; Edmunds, 1975; Maxwell, 1998a) do not involve overt indications of male suicide, such as lunges into the females' forelegs or the subsequent cannibalism of the males. Even when males copulate beyond the time necessary to pass a spermatophore, they generally appear to remain out of the female's reach while on her back (Maxwell, 1998a). Besides, prolonged copulation may be a mechanism to prevent other males from mating with the female (reviewed in Alcock, 1994).

Many aspects of male mating behavior suggest that they actually try to avoid being eaten. These include a variety of mechanisms to reduce the risks of premounting cannibalism (e.g., displays while approaching females, approaching females slowly, mounting females from behind), of cannibalism during the mount (e.g., the nature of the copulatory posture), and of postmounting cannibalism (e.g., quick dismounts and hasty escapes). Finally, male *I. oratoria* strike back at attacking females (personal observations), suggesting that they are not willing participants in the act of cannibalism.

Males might behave to reduce the risk of cannibalism because the benefits of being eaten do not compensate for the loss of future matings. Furthermore, cannibalism might not pay off at all. After being cannibalized, the male's rivals might fertilize the female. For example, female *M. religiosa* and *S. limbata* have been observed in nature to copulate with a second male after copulating with and cannibalizing a first (Lawrence, 1992; Maxwell, 1998a). Unfortunately, measures of sperm precedence, such as P_2 values, are not published for any mantid species, although Lawrence (1991) reports first-male precedence in *Hierodula membranacea* and second-male precedence in *Hierodula patellifera*.

Male Choice: Which Female to Mount

A male's choice of a mate might depend upon factors other than the risk of cannibalism, such as female fecundity. Many male arthropods actively choose large, fecund females over smaller ones (Gwynne, 1981, 1984, 1985; Hatziolos and Caldwell, 1983; Svensson and Petersson, 1988; Gwynne and Simmons, 1990; Simmons and Bailey, 1990; Shelly and Bailey, 1992; reviewed in Andersson, 1994). In mantids, a female's fecundity seems to be negatively correlated with her propensity toward cannibalism (Liske and Davis,

1987; Birkhead et al., 1988; Kynaston et al., 1994) and positively correlated with her nutritional status (Maldonado et al., 1967; Matsura and Morooka, 1983; Birkhead et al., 1988; Hurd, 1989; Fagan and Hurd, 1991a; Maxwell, 1995). These relationships suggest that male choice for a noncannibalistic and a fecund female might be one and the same.

Of course, in order for male choice to occur, females must vary in quality, and males must be able to detect this variation. Although several field studies demonstrate the former (Eisenberg and Hurd, 1977, 1993; Eisenberg et al., 1981, 1992b; Matsura and Nagai, 1983; Hurd, 1989; Fagan and Hurd, 1991a; Lawrence, 1992; Hurd et al., 1995; Maxwell, 1998a), the latter remains an open question. However, males might be able to detect variations in female fecundity indirectly, via the silhouette of a female's abdomen; a thick abdomen indicates a fecund and, presumably, relatively noncannibalistic female (Maxwell, 1995, in press).

Given that some species satisfy requirements for the expression of mate choice by males, how would such a strategy fare in nature? That is, when can males afford the luxury of rejecting "low-quality" females? Maxwell (submitted-b) explored the ecological conditions that favor male choice through a theoretical model. In the model, a hypothetical adult male encounters a number of females described in terms of their fecundity and the probabilities of their cannibalizing the male and/or being inseminated by him. Two female phenotypes exist in this model, high-quality females that are relatively more fecund and less cannibalistic than low-quality females. Each male may follow one of two mating strategies. He may be "choosy" and mate with only high-quality females or "indiscriminate" and mate with any female. The model compares the reproductive pay-offs of the choosy and indiscriminate strategies over a range of organismal and ecological parameters.

A fundamental outcome of this model was that the risk of cannibalism alone selects for some level of male choice within a population. When the probability of cannibalism is zero for both female phenotypes, mating is risk-free, and the choosy strategy never outperforms the indiscriminate strategy. Once copulation involves the risk of cannibalism, however, the choosy strategy confers higher reproductive pay-offs as male daily survivorship, maximum male life span, or female encounter rates increase. Several empirical studies demonstrate the effect of varying encounter rate in other arthropods, with individuals becoming less choosy as mate density decreases (Manning, 1980; Everson and Addicott, 1982; Gwynne, 1984, 1985, 1990, 1993; Thornhill, 1984; Ryan, 1985; Lawrence, 1986; Shelly and Bailey, 1992). On the other hand, choosiness does poorer as the benefit of being cannibalized increases, and the indiscriminate strategy outperforms being choosy as a male grows older.

Theory aside, the key question is whether male mantids discriminate between females in the real world. They might. Lawrence (1992) found that more male *M. religiosa* arrived at field cages housing females with high body mass, but female feeding condition and pronotum length had no effect on male arrivals. Unfortunately, the actual mating behavior of the arriving males was not observed. Maxwell (in press) conducted mating experiments with *I. oratoria* and found that males were more likely to mount females of "high" feeding condition. In contrast, female responses were not related to the size of the males. In this experiment, female feeding condition was positively correlated with abdominal girth and fecundity. This study, however, did not support a negative relationship between female condition and the probability of an attack. Hence, the males' preferences for females of high feeding condition were more consistent with male choice for fecundity than with male choice for nonaggressive females.

Approaches and Mounts

If the female does not approach the male after he has arrived near her, he may initiate contact by typically approaching her on foot for some distance and then leaping onto her back. In general, once a male is near a female his antennae oscillate, suggesting the use of olfaction (Roeder, 1935; Loxton, 1979; Kaushik, 1985; Liske and Davis, 1987; Kynaston et al., 1994). He may then begin a slow approach that can take a few minutes to a few hours (Roeder, 1935; Quesnel, 1967; Edmunds, 1975; Schauff and Jones, 1978; Liske and Davis, 1987; Kynaston et al., 1994; Maxwell, 1995, in press).

In contrast to these general characteristics, other facets of the male's approach behavior vary between species. In some cases, approaches are rather cursory and without overt displays (e.g., *M. religiosa*, Roeder, 1935; Lawrence, 1992). Once the male sees the female, he freezes immediately. Then he walks forward slowly whenever the female looks away. Loxton (1979) and Kaushik (1985) report similar behavior in *Hierodula tenuidentata*, *H. membranacea*, *H. coarctata*, *Humbertiella similis*, and *Stagmatoptera biocellata*.

Males of other species appear to perform various types of displays while approaching females. Particularly conspicuous are abdominal movements involving dorso-ventral and lateral flexions (Table 5.3). In *Sphodromantis lineola*, whether a male displays depends upon the direction of his approach. Kynaston et al. (1994) found that, during the approach, 86% of the males flexed their abdomens and approached the females from any direction. The remaining males did not perform abdominal flexions and always approached females from behind, freezing whenever the females looked at them ("sneaking"). Based on their data, Kynaston et al. suggested that *S. lineola* males fall into two behavioral morphs: those that flex their abdomens and those that sneak. They observed no males that switched tactics from one trial to the next.

This link between abdominal bending and the direction of approach is not as apparent in other species. In *T. a. sinensis*, males were observed to bend their abdomens in 64% of mating trials (Liske and Davis, 1987). Most males that approached females from the front bent their abdomens, although some that approached from behind bent theirs as well (Liske and Davis, 1987).

This abdominal flexing behavior by males raises two questions. First, are the flexions signals or courtship displays, or do they simply reflect a general state of sexual arousal? Kynaston et al. (1994) noted that male *S. lineola* started to bend their abdomens when the females looked at them, which supports the signal hypothesis, although it does not rule out the alternative. On the other hand, *T. a. sinensis* males have been seen to flex their abdomens when behind the females, which is consistent with the arousal, but not the signal, hypothesis.

Second, if abdominal flexions are signals, what are they used for? One idea is that the flexions reduce female aggression (Liske and Davis, 1987; Kynaston et al., 1994). Results from mating trials, however, indicate that female nutritional status is a stronger determinant of whether cannibalism occurs. Kynaston et al. (1994) found that no well-fed *S. lineola* female cannibalized a male, regardless of the presence or absence of abdominal flexions by the males, whereas all but one of the starved females cannibalized males, again regardless of whether abdominal flexions occurred. Liske and Davis (1987) found similar results for *T. a. sinensis*.

Males move other parts of their bodies during the approach. Male *Oxypilus hamatus*, *Ephestiasula amoena*, *Acontiothespis multicolor*, and *Tenodera aridifolia sinensis* move their legs in various patterns, described as boxing, semaphore, or stamping displays (Quesnel, 1967; Edmunds, 1975; Loxton, 1979; Liske and Davis, 1987). Bonfils (1967) describes a different behavior in *Oligonyx insularis*. During the approach males move around the female while "trembling" their wings. The significance of these behaviors is not known.

As previously noted, the duration of the approach phase can vary from a few minutes to several hours, even within a species. The female's behavior may contribute to quick approaches. For example, females may approach or walk by males. When this happens, the males may mount them quickly, without any further stalking (*M. religiosa*, Roeder, 1935) or abdominal flexing (*T. a. sinensis*, Liske and Davis, 1987). Preston-Mafham (1990) suggests that, in nature, males may often forego premounting formalities and may simply fly onto unsuspecting females.

Regardless of whether or not males perform a display during the approach phase, males of many species typically attempt to mount the females by leaping onto them when they are about 4 to 10 cm away, although the distance can be as much as 30 cm (Roeder, 1935; Roeder et al., 1960; Quesnel, 1967; Norsgaard, 1975; Robinson and Robinson, 1979; Liske and Davis, 1987; Lawrence, 1992; Kynaston et al., 1994), while some males mount females by crawling directly onto them (*Iris oratoria, Stagmomantis limbata*, Maxwell, 1995). Males of several species tend to mount females from the rear (Fig. 5.3A; Roeder, 1935; Maxwell, in press) or when the females are not looking at them (Birkhead et al., 1988), a tactic that has been termed a *surprise* mount by Preston-Mafham (1990). However, mounts are not without their problems, and a misjudgment or improper positioning by the male can end in cannibalism (Roeder, 1935, 1967; Quesnel, 1967; Schauff and Jones, 1978; Liske and Davis, 1987; Lawrence, 1992; Maxwell, 1995, 1998a).

Copulation and Fertilization
Copulation

Copulation is similar among many mantids. Once the male mounts, he aligns his body with the female's such that he faces the back of her head, and his forelegs grip the posterior end of her pronotum or her anterior abdomen (Fig. 5.3B; Roeder, 1935; Quesnel, 1967; Liske and Davis, 1987; personal observations). Once in this posture, the male begins what have been called *S-bendings* of his abdomen, which allow his phallomeres and cerci to probe the tip of the female's abdomen (Roeder, 1935; Roeder et al., 1960; Quesnel, 1967; Schauff and Jones, 1978; Liske and Davis, 1987; Davis and Liske, 1988; Liske, 1991b). During these behaviors, his antennae may flagellate the female's head (Roeder, 1967; Schauff and Jones, 1978; Liske and Davis, 1987).

Copulation begins from a few minutes to several hours after the male assumes the copulatory posture (Fig. 5.5; Quesnel, 1967; Inoue and Matsura, 1983; Maxwell, 1995). For intromission to occur, the female lifts her dorsal ovipositor valvulae. The male's cerci then guide his phallomeres between the female's valvulae and subgenital plate. Davis and Liske (1988) determined that the removal of the cerci in male *T. a. sinensis* prevents intromission. It is not known if the female's cerci play a role in copulation. In *S. carolina*, removal of the cerci apparently does not affect the mating behavior of the female (Rau and Rau, 1913).

Once mounted, males do not always achieve intromission (Rau and Rau, 1913; Edmunds, 1975; Inoue and Matsura, 1983; Kaushik, 1985; Maxwell, 1998a, in press), and some males remain mounted without copulating for one or more days (Rau and Rau, 1913; Edmunds, 1975; Maxwell, 1998a). Thus, copulation appears to involve active cooperation by the female. For example, female *T. a. sinensis* can prevent intromission by raising their abdomens (Schauff and Jones, 1978), and presumably females can keep their valvulae shut in order to lock out suitors (Kaushik, 1985; Lawrence, 1991). The hooklike shape of male phallomeres suggests they could play a role in forcing copulations, but this possibility has not been tested in mantids.

In some species, the male transfers a spermatophore within 45 to 110 minutes after the initiation of copulation (*I. oratoria*, Maxwell, 1995; *Hierodula patellifera*, Lawrence, 1991). When cannibalism begins before copulation, however, *H. patellifera* pass their spermatophores sooner than when cannibalism does not occur (Lawrence, 1991). In *Ameles desicolor, I. oratoria, M. religiosa,* and *Stagmomantis limbata*, the spermatophore is spherical, being 1.5 to 3 mm in greatest diameter (Gerhardt, 1914; Lawrence, 1992; Maxwell, 1995). In *M. religiosa*, the spermatophore consists of a sperm capsule surrounded by two membranes, with a duct leading from the capsule apparently to guide the spermatozoa into the female's genital chamber (Gerhardt, 1914, translated in Mann, 1984).

Copulation lasts for about an hour or less in some species (Mathur, 1946; Bonfils, 1967; Robinson and Robinson, 1978; Maxwell, 1995, 1998a), although two or more hours is more typical (Rau and Rau, 1913; Fabré, 1914; Roeder, 1935; Roberts, 1937a, b; Quesnel, 1967; Kumar, 1973; Edmunds, 1975; Schauff and Jones, 1978; Inoue and Matsura, 1983; Liske and Davis, 1987; Lawrence, 1992; Maxwell, 1998a). The longest reported duration is 40 hours (Rau and Rau, 1913). Decapitated males are capable of maintaining copulation for several hours in captivity (Roeder, 1935; Kumar, 1973; Maxwell, unpublished data) and for at least 20 hours in nature (Maxwell, 1998a). Within minutes to hours after withdrawing his genitalia from the female, the male dismounts, often by simply dropping off of her or by flying away (Roeder, 1935; Quesnel, 1967; Bonfils, 1967; Robinson and Robinson, 1978; Schauff and Jones, 1978; Lawrence, 1992; Maxwell, 1995). Either tactic probably reduces the risk of postcopulatory cannibalism.

Fertilization

The fate of the male's spermatophore is by no means certain once it is transferred to the female. She may use his spermatozoa to fertilize her eggs, or she may transport some or all of them to her spermatheca for storage. Sperm in the spermatheca remain viable for several months. For example, in *H. membranacea*, sperm from a single mating can be sufficient to fertilize as many as seven clutches laid over a 5 month period (Birkhead et al., 1988). On the other hand, the female may eject the spermatophore presumably before the spermatozoa drain from it (Phipps, 1960; Quesnel, 1967; Lawrence, 1991), which might be a mechanism of postcopulatory or "cryptic" female choice (reviewed by Eberhard and Cordero, 1995).

Even spermatozoa that make their way into the female's genital tract do not necessarily fuse with her ova. The female may simply store the sperm in her spermatheca indefinitely, or, as in other insects, she might metabolize them (Thornhill and Alcock, 1983). Furthermore, sperm from a rival male may be present. Females of several species copulate with more than one male both in the wild and within captive colonies (Bartley, 1982; Inoue and Matsura, 1983; Lawrence, 1992; Hurd et al., 1994; Maxwell, 1998a). Subsequent matings might increase the genetic diversity of the female's offspring, increase her chances of receiving high-quality genes from at least one male, or ensure that she receives sufficient sperm to last her reproductive life (Walker, 1980). She may also gain nutrients by receiving additional spermatophores, which some females eat (i.e., *Acontiothespis multicolor, Hierodula membranacea, H. patellifera,* Quesnel, 1967; Lawrence, 1991).

When more than one male copulates with a female, the paternity of the offspring becomes unclear. Lawrence (1991) reports first-male precedence in *H. membranacea* and second-male precedence in *H. patellifera*. In another member of the genus, *H. coarctata*, the spermatheca is spherical (Sathe and Joshi, 1986), which suggests that multiple matings could lead to an equal mixing of competing sperm (Walker, 1980; Ridley, 1989; Parker et al., 1990).

Sperm competition is just one form of competition for access to females. Other interactions between males have been observed that may be forms of competition, but the data are too sparse to form any conclusions. These behaviors include "juddering" (a rapid shaking of the body) (*M. religiosa*, Lawrence, 1992), nonlethal battles (Rau and Rau, 1913; Ramsay, 1990; Preston-Mafham, 1990; personal observations), and killing of other males (*Iris oratoria*, personal observations), all done in the presence of females.

As discussed earlier, extended copulations and/or mounts may function as mate-guarding tactics (Alcock, 1994). Copulations that exceed 20 hours—as they can for *Stagmomantis limbata*—are probably longer than the time required for spermatophore transfer. Remaining in copula for so long locks out rival males, but does not necessarily deter them. In nature, two or more males have been seen clinging to a single female in several species (Quesnel, 1967; Preston-Mafham, 1990; Lawrence, 1992; Preston-Mafham and Preston-Mafham, 1993; P. M. Johns, personal communication), and five males managed to mount a single *T. a. sinensis* female (L. E. Hurd, personal communication).

Conclusion

Although the study of mantid mating behavior has a very long history, several important questions remain unanswered. This is especially true with regard to research on the timing of sexual behavior. Although copulation has been observed primarily during the day, several lines of research suggest that there is a wealth of behavioral information to be discovered at night. Males and females may differ in terms of the age at first copulation, and differences in nutritional requirements for gametic development may explain this. This question reaches beyond the world of mantids and gets at the fundamental reproductive conflict between the sexes.

We have some understanding of the influences of age and mating status on pheromone production and/or emission by female mantids, *Acanthops falcata* in particular. We know very little about the use of pheromones in other mantid species. Furthermore, the relative importance of chemical and visual cues in mate recognition remains unclear.

Interactions between males and females involve a variety of behaviors, the most notorious of which is adult female-on-male cannibalism. Cannibalism has been observed in most species studied, and the topic has recently attracted intensive investigations in the field. Why cannibalism seems to be more frequent in some species than in others has yet to be explained. Size dimorphism between the sexes might account for the occurrence of cannibalism beween and within species, as might an evaluation of the evolutionary costs and benefits associated with cannibalism. With regard to the latter, it remains to be determined if cannibalized males actually fertilize the females and if females consistently increase their reproductive output through cannibalism.

The risk of being cannibalized might influence the evolution of male behavior in at least two ways. First, males might be selected to offer themselves as nuptial meals. The behavioral evidence for male suicide, however, appears to be weak at best. Second, males might be under selection to reduce the chance that cannibalism occurs. Indeed, much of male behavior is compatible with this notion, including male approach behavior (e.g., direction of approach, possible courtship displays), mounts (e.g., the copulatory position), and dismounts. Furthermore, the risk of cannibalism can lead to male choice, with males preferentially mating with fecund and/or noncannibalistic females, and recent work supports this idea. The prospect of male choice raises the question of the cues upon which males might base their mating decisions. Abdominal girth, a reflection of female nutritional status, is a likely cue.

The male-female conflict does not end once the male mounts the female. She might eject the male's spermatophore or store sperm in her spermatheca indefinitely. While the female stores one male's sperm, she may mate with his rivals. It is in the male's interests to discourage these interlopers, but it is not clear how effective a male is in doing this. Determining patterns of precedence among competing batches of sperm will shed some light on this problem.

The largest gap in our knowledge is a matter of taxonomy. Virtually nothing is known about mating in species outside the Hymenopodidae and Mantidae. Data on these taxa will increase the power of comparative tests within the Mantodea, as well as improve our understanding of reproductive strategies in the wild.

By pointing out these and other unresolved issues, I do not mean to diminish the labors of previous researchers. Rather, my intention is to inspire present and future workers to ask new questions and to investigate new species. It is my hope that this chapter contributes toward this endeavor.

ACKNOWLEDGMENTS

My gratitude goes first to the many research assistants, faculty and staff, and fellow students at the University of California at Davis who aided my initial study of mantids. This research was funded by grants from the National Science Foundation, Sigma Xi, and the University of California. I thank Marc Mangel for reviewing this chapter, Nan Garrett-Logan of the National Marine Fisheries for steering me through various French-language articles, and Michael Robinson of the Smithsonian's National Zoological Park for contributing the photograph of *Acanthops falcata*. The National Research Council provided funding during the preparation of this chapter.

Hearing and Vision

6. Hearing

David D. Yager

Audition using tympanate hearing plays a major role in the lives of many insects (Michelsen and Larsen, 1985; Fullard and Yack, 1993; Robert and Hoy, 1995), but praying mantids would seem unlikely candidates for membership in the ranks of the auditorily competent. The historical view is of a diurnally active insect whose large eyes and excellent vision dominate its sensory world (Prete and Wolfe, 1992). Further, no one had ever reported hearing mantids make sounds in non-defensive contexts, and there was no evidence of acoustically mediated behavior—a view reinforced by many reports that courtship, a behavior commonly linked to hearing in other insects, ranges from minimal to nonexistent in mantids. Max Beier, one of the great authorities on mantis biology, drew the "obvious" conclusion in his 1968 monograph, flatly stating that tympanal organs are absent in mantids (in fact, in all Blattoidea).

The recent literature on mantids as reviewed throughout this volume makes clear how little we understand mantis biology, in part, by exposing errors in previous assumptions. Mantids—or at least *some* mantids—are active both at night and during the day. Courtship is not simple at all in these insects, and at least one nonvisual sense, olfaction, certainly plays a pivotal role. Mantids may, in fact, use intraspecific acoustic communication signals—but, as discussed below, they would be ultrasonic and beyond our own auditory range. In light of our growing knowledge of mantids, it is not particularly surprising to discover that they do have tympanal organs and that hearing is an important part of their biology.

This chapter presents the information on hearing in mantids that has accumulated in the fifteen years since its discovery. Here we deal only with tympanate hearing and do not consider possible detection of low-frequency (<2 kHz) airborne sound by subgenual organs, cerci, or other mechanoreceptors. The emphasis is on comparative and ontogenetic studies directed toward understanding the evolution of the mantis auditory system.

The Praying Mantis Auditory System

The description that follows applies to the most prevalent form of auditory system anatomy and physiology in the Mantodea. Museum studies of external ear anatomy found this form in approximately 65% of all mantids examined (Yager, 1990). As a shorthand, we refer to this pattern as the DK ear (*d*eep groove with *k*nobs; see below). At least three other major patterns of audition occur in the suborder and are discussed later in this chapter.

Structure

The highly unusual appearance of the mantis ear explains to a large extent why it was overlooked by anatomists as well as naturalists. Several authors (Levereault, 1936; Adam and Lepointe, 1948; LaGreca, 1949; Debouteville, 1952; Mat-

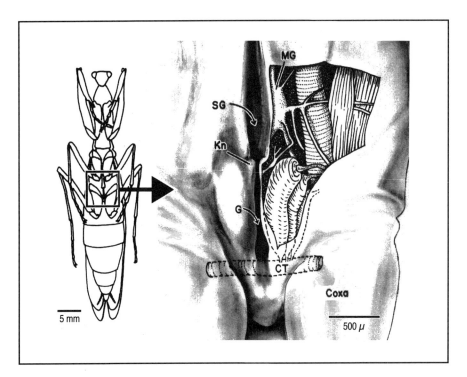

Fig. 6.1 The DK mantis ear. The schematic drawing to the left shows the location of the ear in the ventral midline of the metathorax. The detailed view to the right shows the deep groove (G) and the cuticular knobs (Kn) of the ear. A shallow groove (SG) that is not part of the ear extends from the ear anteriorly toward the mesothorax. The basisternum anteriorly with the fused pre-episternum posteriorly have been removed on the animal's left to show the tracheal sac that arises from a commissural trachea (CT) and backs the tympanum (shown here slightly pulled away). The tympanal organ is also visible attaching to the anterior end of the tympanum. The tympanal nerve carries axons from the tympanal organ to the metathoracic ganglion (MG). Modified from Yager and Hoy (1987) and used with permission.

suda, 1970) have included it in figures but apparently believed it was simply a complex muscle attachment site.

The mantis peripheral auditory system consists of a single ear located in the ventral midline of the thorax near its junction with the abdomen (Figs. 6.1, 6.12C; Yager and Hoy, 1986; Yager and Hoy, 1987). Except at its posterior end, the ear is surrounded by thick cuticle: the basisternum anteriorly and the pre-episterna laterally. Much of the cuticle near the tympana (primarily the furcasternum) is especially thick and extremely rigid; it does, in fact, serve as an attachment site for metathoracic leg muscles.

The DK ear comprises a deep longitudinal groove demarcated at its anterior end by two prominent cuticular knobs (Figs. 6.1, 6.12D). In a moderately small mantis (*Ameles heldreichi*, male body length 25.3 ± 1.1 mm, mean ± s.e., $n = 6$) the groove is 0.62 ± 0.02 mm long; in a large mantis (*Hierodula membranacea*, male body length 82.2 ± 1.3 mm, $n = 20$) the length is 1.29 ± 0.06 mm (Yager, 1990). Each groove wall contains a teardrop-shaped tympanum bounded ventrally by a ridge of cuticle and dorsally by the furcasternum (Fig. 6.2). The tympana face each other and are separated by only 100 to 200 μm. Because its walls are not parallel and the furcal pits lie in its floor, the groove actually forms a long, thin chamber of very complex shape and bioacoustic characteristics.

The tympana are not typical of most insect ears in that they are relatively thick (15 to 20 μm) and quite stiff (Yager and Hoy, 1987; Yager,

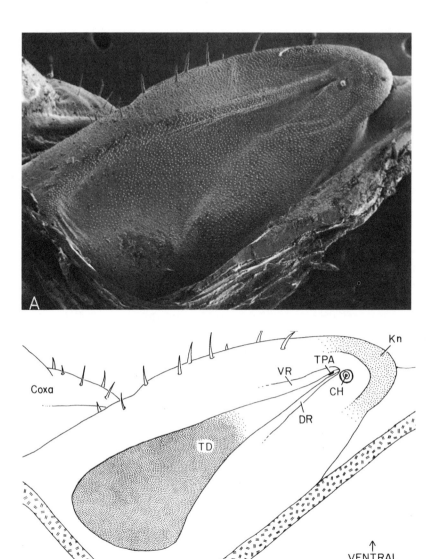

Fig. 6.2 A midsagittal scanning electron micrograph of the right groove wall of the DK ear of *Mantis religiosa* (Mantidae, Mantinae). The cuticular knob (Kn) marks the anterior end of the groove. The furcasternum (Fs) forms the groove floor, and the sternal apophyses (SA) originate midway along the groove. The tympanum is a teardrop-shaped depression (TD) in the center of the wall, partly bounded by a ventral ridge (VR) and a dorsal ridge (DR). The attachment site of the tympanal organ (TPA) is at the extreme anterior end of the tympanum. An unusual, conical sensillum (CH) of unknown function is also found on the groove wall. Scale bar: 200 μm. From Yager and Hoy (1987). Used with permission.

1996a). Pushing with a minuten pin does not deform the tympanum, but rather displaces it. A very large tracheal sac adheres tightly to the inner surface of the tympanum (Fig. 6.1). Such air spaces are a key feature of virtually all tympanate ears including our own and are necessary for adequate sensitivity (Michelsen and Nocke, 1974). We currently have no information about the nature of tympanal movements in response to sound.

The mechanical vibration induced by sound is transduced to neural signals by a tympanal organ attached to each tympanum (Figs. 6.1, 6.2). The attachment site is not near the center of the tympanum as in most other insects (Michelsen and Larsen, 1985; Robert et al., 1994), but rather at the extreme anterior end where the tympanum is narrowest (again raising the question of exactly how the tympanum moves). Three ligaments anchor the tympanal organ in place: one is the attachment to the tympanum and two are fixed to the ventral cuticle laterally on the metathorax (Figs. 6.1, 6.13). Each tympanal organ contains 35 to 45 scolopophorous sensilla in three groups, one group associated with each ligament. A sensillum comprises one bipolar neuron and one scolopale (scolopale cell with cap and rods) plus a number of supporting cells, the standard composition for insect auditory transduction (Yack and Fullard, 1990; Yack and Roots, 1992). All of the scolopales are oriented away from the neuron somata in the tympanal organ, so those in the lateral ligaments are approximately 180° out of line with those in the tympanal ligament. Vibration of the tympanum alternately stretches and relaxes the ligaments/tympanal organ, thus stimulating the sensilla.

Signals produced by the bipolar neurons of the tympanal organ travel to the metathoracic ganglion along the axons in Nerve 7 (the tympanal nerve; Figs. 6.1, 6.13). Besides the auditory afferents, this 35 μm diam nerve carries information from sensory hairs on the ventral cuticle. The auditory afferents synapse with interneurons in several discrete regions in the metathoracic ganglion as well as in the first three abdominal ganglia (which are fused with the metathoracic ganglion). With the exception of an occasional isolated fiber, the afferent terminations are entirely ipsilateral. Other tympanate insects for which the information is available (crickets, locusts, cicadas, and moths) show a consistent pattern of auditory afferent terminations in the anterior ring tract of the ganglion (Römer et al., 1988; Boyan, 1993). Mantids, however, have few, if any, fibers going to that region (Yager and Hoy, 1987). We do not know the exact significance of this major difference, but it suggests that processing of auditory information in the mantis central nervous system (CNS) may be quite different from that of other insects.

Function

We have studied the functioning of the mantis auditory system by both extracellular and intracellular neural recording in the thorax (Yager, 1989; Yager and Hoy, 1989; Yager, 1990; Yager and Triblehorn, 1995; Yager, 1996b). Our results tell us about the combined effects of the bioacoustics of the peripheral auditory system and the early stages of CNS processing.

The most common form for the physiological audiogram of DK-eared mantids is shown in Figure 6.3 (see also Figs. 6.5, 6.9, 6.11, and 6.12). Hearing is most acute at ultrasonic frequencies, generally between 25 and 50 kHz, and these mantids have little or no sensitivity below 10 kHz. The lowest thresholds we have encountered were 40 to 45 dB re: 20 μPA (= dB SPL [sound pressure level]) in the African mantis *Polyspilota aeruginosa*, while most species are in the 50 to 60 dB SPL range, and a few are less sensitive at 65 to 75 dB SPL (Yager, unpublished observations; Yager and Triblehorn, 1995). This threshold range is typical for many tympanate insects, although a few are more sensitive (Michelsen and Larsen, 1985; Hoy, 1992; Robert and Hoy, 1995). Neither our physiological results, nor the anatomy of the tympanal organ, nor behavioral testing suggests that mantids can discriminate different frequencies—they are tone deaf.

A particularly striking physiological characteristic of mantis hearing is the speed with which signals travel from the ear toward the brain. In *Mantis religiosa*, for instance, ascending signals reach the neck as little as 8 to 10 ms after the

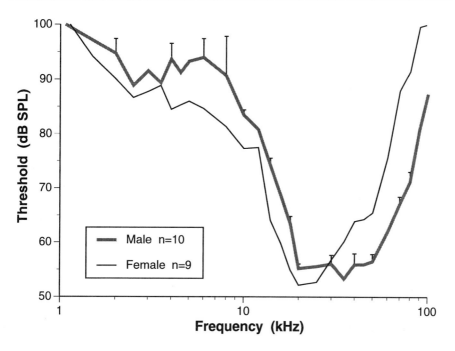

Fig. 6.3 Extracellular physiological tuning curves recorded from the ascending connectives of a typical DK-eared mantis, *Sphodromantis viridis* (Mantidae, Mantinae). The slight sexual dimorphism with males sensitive to higher frequencies than the females is common among DK-eared species. Standard deviation bars are shown for the males and are comparable for the females.

sound reaches the ear (Yager and Hoy, 1989). Clearly there can be very little analysis of the auditory information taking place in the thorax. This very short latency also suggests specializations for high conduction velocities, which we have found in two locations: (1) Electron microscopy shows that the auditory axons in the tympanal nerve have unusually large diameters for afferents (some >10μm) and a connective tissue sheath around each axon that may speed conduction much as myelin does in vertebrates (Yager and Scaffidi, 1993; Yager, forthcoming); and (2) the auditory interneuron that dominates the extracellular record has an axon diameter of 15 to 20 μm and correspondingly high conduction velocities of 4 to 5 m/s (Yager, 1989). Thus, at least part of the mantis's auditory system seems designed to serve a behavioral function requiring very fast reaction times.

We have only just begun to study the auditory interneurons involved in mantis hearing (Yager, 1989). Many neurons in the metathoracic ganglion of *M. religiosa* receive weak auditory input (only postsynaptic potentials), but we have found eight mirror-image pairs that respond strongly with bursts of action potentials (Fig. 6.4). We have detailed physiological information on one of these, named MR-501-T3. Two aspects of its behavior are especially salient to the present discussion. First, this is the neuron mentioned above that dominates the extracellular record and has a high conduction velocity. The very short latency also suggests that MR-501-T3 receives its input directly from the afferents rather than through intermediate cells. Second, both the right and left MR-501-T3 appear to receive excitatory input from both the left and right tympanal nerves, precluding a directional response.

Does the Mantis Have One Ear or Two?

The mantis has two tympana, two tympanal organs, and two separate nerves carrying auditory information to the CNS. As will be discussed be-

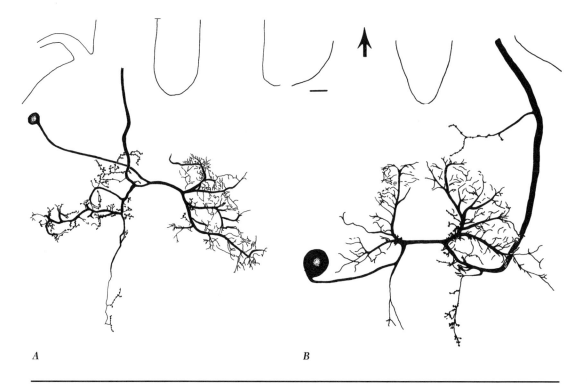

Fig. 6.4 Two examples of metathoracic interneurons in *M. religiosa* that receive strong auditory input. Both occur as mirror-image, left-right pairs. The cells were filled intracellularly with Lucifer yellow after physiological characterization. Both neurons are highly symmetrical with large axons traveling toward the brain. The neuron in *B* is MR-501-T3, the cell whose activity dominates the extracellular record in the ascending connectives. The arrow points anteriorly. Scale bar: 50 μm. The drawing of MR-501-T3 is modified from Yager and Hoy (1989) and used with permission.

low, the midline groove forms during nymphal development from two more laterally situated structures. Is it really appropriate, then, to say that these insects have a single ear—that the mantis is an auditory "cyclops"? This is more than a semantic issue, since a major function of auditory systems in animals is determining the direction to a sound source—a function that requires two ears (Lewis, 1983; Michelsen and Larsen, 1985).

The structure of the mantis DK ear with two tympana facing each other in a deep groove and separated by less than 200 μm ensures that the acoustic environment the two halves of the ear experience is identical, since the distance between the tympana is only a small fraction of a wavelength at frequencies below 100 kHz (Michelsen and Nocke, 1974). In addition, we have physiological evidence indicating that the groove itself—presumably by an acoustic property of the very complex shape of the chamber—confers an almost twofold increase (5 dB) in sensitivity to the ear (Yager and Hoy, 1989). Thus, acoustically, the ear is singular.

Physiological experiments show no hint of directional capability in the mantis auditory system (Yager and Hoy, 1989; Yager, 1996b). We presented identical stimuli from 90° to the right or left while simultaneously recording from both the left and right connective; the responses on the two sides of the CNS were always identical. It is especially telling that even cutting one tympanal nerve, thus providing maximally lateralized input to the CNS, failed to give lateralized neural responses. This means that the auditory system is actually, "antidirectional": while information from the left and right tympanal organs is initially segregated, it is all mixed together at the first processing site in the CNS. Hence, physiologically, the ear is singular.

Finally, none of a range of behavioral tests has shown any indication of directional competence (Yager and May, 1990; Yager et al., 1990; Yager, 1996b). For instance, the direction of yaw by tethered mantids in response to ultrasonic pulses is random. The turn by a mantis in free flight triggered by ultrasound is equally likely to be toward or away from the sound source. Thus, functionally as well as anatomically and physiologically, the mantis does truly appear to be an auditory cyclops.

Ontogeny of the DK Ear

As befits a hemimetabolous insect, mantis nymphs generally resemble adults (Yager, 1996a). There are, however, three notable differences relevant to the question of whether or not nymphs can hear. First, nymphs are much smaller than adults. For *Hierodula membranacea*, first instars are about 1/10 as long and 1/800 as heavy as an adult. Thus, nymphs will clearly have both different predators and prey. Second, nymphs do not reproduce, so courtship and mating behavior will not be present. Third, nymphs do not have wings and are not vulnerable to aerial predators such as bats. So, do mantis nymphs hear?

We have studied nymphal development of the DK ear in *H. membranacea* by rearing large numbers of nymphs individually and at each stadium, assessing both auditory competence and ear morphology (Yager, 1996a). In this species, males have eight and females nine nymphal instars. Extracellular recordings from the connectives have provided audiograms for nymphs of different ages (Fig. 6.5). The first hint of auditory sensitivity appears at the fifth stadium, but the first consistent responses are not present until the sixth. Thereafter, sensitivity increases at each molt with the largest increase at the molt to adult. Eighth- and ninth-stadium nymphs have auditory thresholds below 80 dB SPL. The tuning of the auditory system does not change dramatically during this period. Based on these results, nymphal development can be divided into an early earless period followed by gradual acquisition hearing.

By purely anatomical criteria, we divide the early earless period into two stages. First, the peripheral neural structures serving audition develop in the embryo and are all in place at the time of hatching. This means that the number of sensilla and their structure matches the adult and that the three ligaments are in the mature relationship to each other. Tympanal nerve recordings are not yet available from small nymphs, however, so we can not be sure that the organ is functional at the earliest stages. The second stage begins at the third stadium when the groove begins to form by a complex infolding (Fig. 6.6). The structures that later become the tympana are initially situated

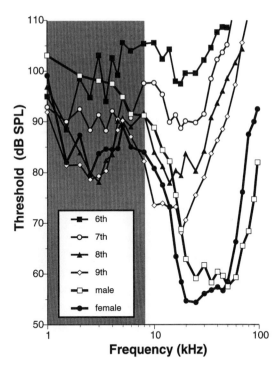

Fig. 6.5 Tuning curves from *Hierodula membranacea* (Mantidae, Mantinae) nymphs of different instars compared to adult males and females. The thresholds were obtained from extracellular neurophysiological recordings of the ascending connectives. Each curve is the mean from >12 animals. Fifth-stadium nymphs occasionally showed responses at >100 dB SPL. Frequencies in the grayed region are not detected by the metathoracic ear. From Yager (1996a) and used with permission.

more laterally and, as the groove forms, move toward the midline. The basic structure of the groove is in place by the sixth stadium.

The gradual acquisition of hearing begins at the sixth stadium and corresponds to two important anatomical changes. First, the tympanum begins increasing in size faster than the overall growth of the nymph. In particular, it broadens dramatically into the adult teardrop shape. Second, a large tracheal sac progressively forms behind the tympanum and becomes adherent to its inner surface. These two changes are both clearly linked to an increase in sensitivity: a greater surface area to transduce pressure change into motion and better impedance-matching across the membrane.

Whereas the transition from earless to eared in mantis nymphs has clear anatomical correlates, there are other less obvious, but equally important, changes that must take place as well. Most important of these is the formation of appropriate connections within the CNS. One possibility is that the neurons of the tympanal sensilla do not form connections in the CNS until the fifth instar when we see the first indications of auditory activity in the extracellular record. The auditory interneurons may also be immature in the young nymphs. This is the pattern locusts show in their auditory system development (Boyan, 1983). An intriguing alternative possibility is that the future tympanal organ is operational (as suggested by the anatomy) and makes appropriate synapses in the CNS of even the youngest nymphs, but serves a sensory function other than audition. If this were the case, the emerging auditory sensitivity might arise solely from the transformation of the periphery (although changes in the CNS could not be completely precluded). Paralleling the modality change we might also expect to see a transformation in function from behaviors relevant to nymphal ecology to others serving adult needs. Neurophysiological recordings from young nymphs will be a necessary first step to sort out these possibilities.

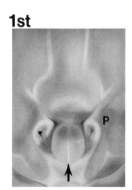
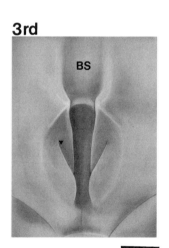
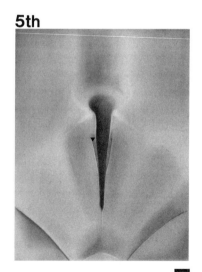

Fig. 6.6 External morphology of the DK ear in *H. membranacea* during the first half of nymphal development (first, third, and fifth instars). The groove walls are initially lateral, but fold inward to form the deep midline groove by the fifth instar. The black triangles mark the anterior end of the developing tympanum and the attachment site of the tympanal organ. The arrow points to the furcasternum. BS - basisternum, P - preepisternum. Anterior is toward the top. Scale bars: first and third instars, 50 μm; fifth instar, 100 μm. Modified from Yager (1996a) and used with permission.

Acoustic Behaviors

The discovery of the mantis ear came about in the absence of any behavioral evidence of hearing, and our current knowledge of acoustic behavior in these insects is incomplete, at best. In other tympanate insects, intraspecific communication is often the most evident function of hearing (Bailey, 1991; Fullard and Yack, 1993). Advertisement calls attract potential mates, and courtship monologues or dialogs precede copulation and may mediate mate choice. Agonistic signals accompany disputes over territory ownership or a mate. Also important in many insects is interspecific acoustic communication. Many insects, including some mantids (Wood-Mason, 1898; Williams, 1904; Stäger, 1928; Varley, 1939; Edmunds, 1972; Edmunds, 1976), produce defensive sounds as a means for startling or warning away predators. This is one-way communication, and many of the insects producing the sounds have no ears. Interspecific communication—in its broadest sense—may also take the form of "eavesdropping." Insectivorous bats that hunt using ultrasonic sonar are the dominant nocturnal predators for many insects (Fenton, 1985; Fullard, 1985; Hoy, 1992). The strength of the selection pressure bats exert is reflected in the number of insects that have evolved ultrasound-sensitive tympanate ears for defense (Spangler, 1988; Hoy, 1992; Yager and Spangler, 1995): crickets, bush crickets, locusts, lacewings, tiger beetles, and many moths. In these cases, hearing provides an early warning system that allows the insect to take appropriate evasive measures and often escape capture.

Intraspecific Acoustic Communication in Mantids?

At present, there is no evidence to support the notion that mantids communicate with one another acoustically. As noted above, several species produce defensive sounds. Particularly striking are *Gongylus*, which rubs the femur against the tegmen (Wood-Mason, 1898; Williams, 1904), and *Choeradodis*, which apparently stridulates on an expanded pronotal shield that acts as a "sounding board" (Robinson, 1969a). However, no one has reported sound production in a social context in any mantis.

Taking into account the paucity of information on mantis social behavior under natural conditions, it seems premature to rule out this possible function for audition. Certainly, the meeting of a slender, largely defenseless male with a much larger, perpetually hungry female would seem to afford ample reason for courtship communication. Since it appears that courtship and mating in many species occur at night (Edmunds, 1975; Robinson and Robinson, 1979), the signaling would most likely be olfactory and/or tactile and/or acoustic.

Bat Evasion

The major biological source of ultrasound in the environment is echolocating bats using sonar to locate and capture flying insect prey. In the more conventional view that mantids are diurnal creatures, they would have no opportunity to hear hunting bats. This view, however, fails to acknowledge the other, nocturnal side of mantis life (their Dark Side . . .). Numerous published accounts (and the experiences of entomologists in the tropics) establish that mantids actively fly at night. Edmunds (1986), for instance, visited lights for 15 minutes every other night for 36 months in Ghana and captured 1771 mantids of twenty-six species. Cumming (1996 and personal communication) reports collecting over one hundred *Miomantis* sp. in a single night at a black light in South Africa. I have collected forty *M. religiosa* at a streetlight in one evening in upstate New York. In all of these cases, males dominate the collections. In part, this is because females of many mantis species do not fly (females in 35% of mantis genera have reduced wings [Yager, 1990]), but it is also consistent with the hypothesis that males are flying at night to locate pheromone plumes emitted by sexually receptive females (Edmunds, 1975; Robinson and Robinson, 1979; Edmunds, 1988). Whereas we know that bats find mantids palatable and have no difficulty capturing them in flight (Yager et al., 1990), information about the actual level of aerial predation in the field has proved very difficult to obtain.

Our physiological data also lend support to the

idea of bat evasion as a function of mantis hearing (Yager and Hoy, 1989; Yager, 1996b). The most sensitive hearing range for mantids, 25 to 50 kHz (Fig. 6.3), corresponds exactly to the echolocation frequencies bats most commonly use. The very short physiological latencies suggest an auditory system adapted for the rapid behavioral response that would be necessary to survive a bat attack.

Tethered Flight Responses

A mantis suspended in a gentle wind by a stick glued to its pronotum flies readily for 10 minutes to 2 hours, depending on the species (Yager and May, 1990). The normal flight posture is streamlined with the body horizontal and all legs tucked tightly against the body (Fig. 6.7A).

A tethered, flying mantis responds to ultrasonic pulses with a short-latency, complex behavior (Fig. 6.7B; Yager and May, 1990). It completely extends its prothoracic legs (but not the other legs, distinguishing this from a nonspecific startle response), rolls its head to the side, changes the wing beat rate and excursion, and curls the abdomen upward by as much as 90°. Prothoracic leg movement begins on average 66 ms after stimulus onset, and the full response develops within 170 ms. Whereas the wing excursion increases, the wing beat frequency actually decreases by 5 to 10%. The strength of the response is generally proportional to the strength of the stimulus, but there is considerable variability. Behavioral tuning exactly matches the physiological audiograms. We could detect no indication that this behavior is directional relative to stimulus location; in fact, all of the components except the head roll are symmetrical.

With two exceptions (*M. religiosa* and *Tenodera aridifolia sinensis*), all of the more than forty hearing species we have tested in tethered flight show strong ultrasound-triggered behavior comprising the components described above (some species add flight cessation; Yager, unpublished data). However, not all species weight the components equally (Yager and Triblehorn, 1995; Triblehorn and Yager, forthcoming). Mantids like *Creobroter* and *Parasphendale* almost always perform the full behavior, but many other species emphasize one of the components such as the abdomen dorsiflexion or the arm extension. One subset of mantids that includes *Taumantis* and *Miomantis* are unique in using unpredictable combinations of the various components; for example, four consecutive trials might be arm extension and flight cessation, abdomen flexion only, no response, and arm extension and abdomen flexion (Yager and Triblehorn, 1995). The ultrasound-triggered behavior of this group is maximally "evitable" in the sense of Roeder (1975) and thus minimally predictable.

Behavior in Free Flight

Stationary flight on the end of a tether can establish the basic nature of the ultrasound-triggered behavior, but it provides little insight into the changes in flight path the behavior might cause. We have studied mantids in free flight using strobe light photography, and we have also staged encounters in the field between mantids and wild, hunting bats (Yager et al., 1990).

Batlike ultrasound elicits dramatic changes in mantids' flight path (Fig. 6.8). Normally, mantids fly in relatively straight paths at speeds of approximately 2 m/s. Following ultrasound stimulation by 150 to 200 ms, the flight path angles slightly upward, and the speed decreases. The animal then rolls steeply to one side into a spiral power dive during which flight speed almost doubles. The behavior is graded so that at lower stimulus intensities (the behavioral threshold is 60 dB SPL) there is simply a turn and/or a shallow dive. Mantids respond to typical bat cries (100–110 dB SPL at 10 cm) at distances up to 10 m; the most dramatic spiral dives occur at distances of 4 m or less. The mantis is equally likely to turn toward or away from the source of the ultrasound. Since bats can not generally detect their prey until it is within approximately 5 m (for 19 mm spheres [Kick, 1982]) and given the mantis's short latency to response, the prey would seem to have a good chance of evading the predator. In fact, calculations taking into account relative flight speeds, auditory thresholds, and latencies indicate that the mantis will have more than ample time to escape as long as the bat cries are above 90 dB SPL (measured at 10 cm), a source intensi-

Fig. 6.7 *A, Parasphendale agrionina* (Mantidae, Mantinae) in undisturbed tethered flight. A gentle wind stream was blowing at the animal's head. Note the streamlined posture, with legs tightly tucked to the body and the abdomen straight. The body length is approximately 45 mm. *B,* The same mantis approximately 200 ms after the onset of a train of 40 kHz pulses. Wing excursion increases and wing beat frequency decreases in addition to the movements of the head, prothoracic legs, and abdomen. From Yager and May (1990) and used with permission.

ty typical of most aerial insectivores. The advantage shifts to the bat, however, if it uses echolocation cries outside the 25 to 50 kHz band or at SPLs below 90 dB SPL.

We tested these predictions by releasing mantids into the airspace cruised by hunting bats (*Lasiurus borealis* and *Lasiurus cinereus*) at a field site in western Ontario (Yager et al., 1990). In over two hundred trials, we individually released mantids of two species: *Parasphendale agrionina*, whose audiogram matched the cries of the local bats, and the similarly sized *Miomantis paykullii*, whose most sensitive hearing is at 80 to 100 kHz (Yager and Triblehorn, 1995), well out of the frequency range used by the local bats. A group of two to six observers monitored the bat-mantis interactions. In five of the eleven attacks observed, the mantis executed clear evasive maneuvers—dives and turns or steep spiral dives—and in every case escaped capture (all were *P. agrionina*). The man-

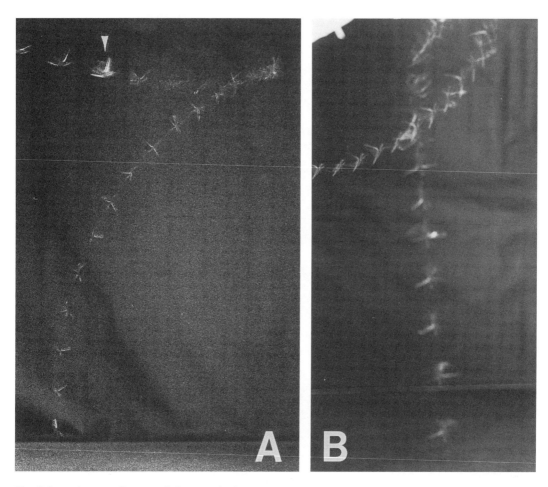

Fig. 6.8 Stroboscope-illuminated photographs (inter-flash intervals: 30 ms in *A*; 22 ms in *B*) of the aerial (free-flight) response of *P. agrionina* to ultrasonic pulses. The bright image at the arrow in *A* marks the onset of the ultrasound; this is out of the frame in *B*. Spiral power dives like these occurred close to the sound source; lower sound intensities yielded correspondingly weaker responses. From Yager et al. (1990) and used with permission.

tids in the six other attacks (two *P. agrionina* and four *M. paykullii*) showed no evasive maneuvers, and five were captured.

Summary

It seems clear that some mantids can use their ultrasonic hearing as part of an evasion system designed to protect them from echolocating, insectivorous bats. Thus, this independently evolved auditory system shows striking convergence in function with several other independently evolved insect auditory systems: noctuoid moths, green lacewings, crickets, and locusts (Yager and Hoy, 1989; Hoy, 1992). This finding also highlights the error in thinking of mantids as strictly diurnal creatures—clearly the selection pressure exerted by nocturnal, aerial predators has substantially influenced mantis evolution.

It would be premature, however, to assume that predator evasion is the only function of hearing in mantids. As the following section details, mantids show a range of auditory capabilities, and it would not be surprising to find other behavioral functions.

Patterns of Audition in the Mantodea

While a small and uniform taxon by insect standards, the 1800 to 2000 known mantis species nonetheless show a considerable range of body form and ecological style (Giglio-Tos, 1927; Beier, 1968). Adult body length extends from approximately 10 mm (*Haldwania liliputana* and several of the Iridopteryginae) to 150 mm (*Macrodanuria elongata*). Habitus ranges from the almost thread-like (Schizocephalinae) to the flattened (many of the Liturgusinae) to the rotund (*Eremiaphila*, among others). Some are winged, some wingless. Some are cursorial, some arboreal. Many are sit-and-wait predators, while others hunt more actively. Habitats range from desert to cloud forest. This diversity raises the question of whether there might be a corresponding array of auditory systems in the suborder.

We have approached this question using two strategies. First, we have studied physiologically and behaviorally approximately seventy mantis species that we have had in our colony over the last fifteen years. Second, the hard cuticular structure of the ear makes it relatively easy to examine in pinned specimens, and we have gleaned detailed anatomical data on 330 of the approximately 400 mantis genera from museum collections (primarily in London and Philadelphia).

Mantids are not uniform in auditory structure and function. In fact, four distinct patterns of auditory structure and function have emerged from these studies.

Ultrasonic Hearing with the DK Ear

It is important to add to the discussion above that several variations on this basic pattern have surfaced in the physiological survey (as yet, they have no recognizable anatomical correlates). Most are relatively minor: small differences in the sharpness of tuning, in the range of best frequencies, or in absolute sensitivity.

One taxonomically disparate group of DK-eared mantids, however, shows an unusual and striking suite of characters (Yager and Triblehorn, 1995). The dominant feature is an increase in most sensitive frequency to 60 to 120 kHz instead of the usual 25 to 50 kHz (Fig. 6.9). At 60 to 70 dB SPL, absolute sensitivity for some of these species is somewhat lower than in other DK-eared mantids. This is also the group of species mentioned above that shows the maximally evitable behavioral responses.

No obvious phylogenetic pattern links the species showing ultra–high-frequency hearing. The species (*Taumantis ehrmanii*, *Miomantis paykullii*, *M. ehrenbergi*, *Empusa fasciata*, and *Ameles heldreichi*) all have close relatives with typical hearing and behavior. In fact, *Litaneutria minor*—in size and general ecology a New World twin (and tribe-mate) to *A. heldreichi*—hears best at 30 to 50 kHz (Yager and Triblehorn, 1995). A more tenable hypothesis is that evolution of this DK variant was ecologically driven and independently arose in several taxa. All of the species with ultra–high-frequency hearing are sympatric with bats in the exclusively Old World families Rhinolophidae and Hipposideridae that use echolocation calls above 80 kHz. These bats are prevalent aerial insectivores and could potentially exert a strong selective pressure on mantids. Recent behavioral observations in South Africa have shown *Miomantis natalica* successfully evading attack by *Rhinolophus clivosus* echolocating at 85 kHz (Cumming, 1996, and personal communication). This scenario nonetheless leaves unanswered the question of what specific behaviors make the mantids with atypical DK hearing especially vulnerable to predation by rhinolophid and hipposiderid bats.

Primitively Earless Mantids

No members of four families of mantids—Chaeteessidae, Mantoididae, Metallyticidae, and Eremiaphilidae—show any anatomical trace of a metathoracic ear (Fig. 6.10). There is no groove, and the furcasternum, which forms the floor of the groove in DK-eared animals, is exposed and oriented dorso-ventrally instead of longitudinally. The areas homologous to the groove walls lie lateral to the furcasternum facing posteriorly, and the only indication of a "tympanum" is a very narrow, vertical slit. Of these taxa, only *Eremiaphila brunneri* has been tested physiologically: it showed no auditory sensitivity at any frequency.

The argument that these animals display the

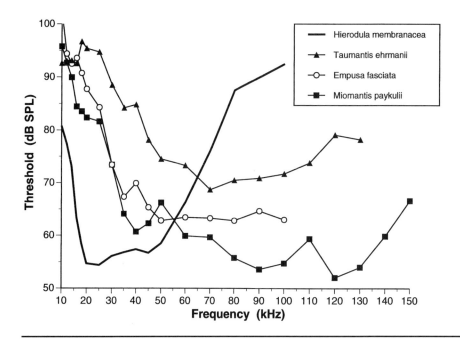

Fig. 6.9 Tuning curve of a typical DK-eared mantis (*H. membranacea*) contrasted with curves from three DK-eared mantids with exceptionally high-frequency hearing. *T. ehrmani* and *M. paykullii* belong to the same tribe (Mantidae, Mantinae, Miomantini) while *E. fasciata* is in a different family (Empusidae, Empusinae). Thresholds were obtained from extracellular recordings of the ascending connectives; curves are means from >10 animals.

primitive form of the ventral metathorax rests on three observations. First, the ventral metathorax is largely indistinguishable from the serially homologous structures of the mesothorax, which vary little in external morphology across the entire suborder (Fig. 6.10; Yager, unpublished observations). Second, the adult metathorax in these mantids remains essentially unchanged from that of first instars (compare Fig. 6.6 with Fig. 6.10; Yager, 1996a). Third, the relevant anatomy in these mantids matches closely that found in cockroaches, the best outgroup for comparison (Thorne and Carpenter, 1992; Yager and Scaffidi, 1993). A strong caveat is in order here, however: These arguments are most compelling for the families from the New World tropics, Chaeteessidae and Mantoididae. The highly specialized body form of the Eremiaphilidae, for instance, may very well be a response to the geologically recent desertification of northern Africa and the Middle East (Cox and Moore, 1993), and the ear may have been lost during this period (especially since they have also lost flight). The Metallyticidae (limited to the Philippines, Borneo, and nearby islands) are very incompletely studied, with few specimens in museum collections. The most parsimonious biogeographic argument might well limit the truly primitive "ear" to the Chaeteessidae and Mantoididae.

Both Chaeteessa and Mantoida are small animals (<25 mm) with very long wings (Giglio-Tos, 1921; Beier, 1968; personal observations). The several species of Mantoida are apparently not common. Natural history reports remark on their exceptional flight ability, and they have been captured at light traps at night (Beebe et al., 1952; Crane, 1952). Thus, these mantids are probably at risk for bat predation.

Secondary Loss of Ears

Taking the classification scheme of Beier (1968) as an approximation to mantis phylogeny, hearing has independently been lost 15 to 20 times in the suborder. By far the most frequent circumstance for this loss is auditory sexual dimorphism (Yager, 1990).

Some mantis species are sexually dimorphic predominantly in body size, the most spectacular cases occurring in the Asian Hymenopodinae (*Theopropus* and *Hymenopus*, for instance) where males may be only 40 to 50% as long as females (Beier, 1934; personal observations). However, the more frequent and general pattern of dimorphism comprises a suite of characters: females have much greater body mass and sometimes length, reduced wings, reduced ocelli, and shortened antennae. The degree of dimorphism varies among species but is evident in approximately 35% of the genera studied in museum collections. There are also marked behavioral differences unrelated to mating between the sexes, as has been noted by Edmunds (1972, 1976) for defensive displays.

In every dimorphic species of mantis that we have examined physiologically, the males have ultrasonic hearing with normal sensitivity, but the females have markedly reduced or absent hearing (Fig. 6.11). This parallels anatomical differences in the "ears": males have the usual DK structure while females have a variant (called DNK, *d*eep groove, *n*o *k*nob) in which the cuticular knobs are strongly reduced or absent, the groove is splayed open, and the area of the tympanum may be reduced. Anatomical data from the museum studies replicate this pattern precisely.

Our database indicates that an especially compelling correlation exists between auditory structure/function and wing length (Yager, 1990). Anatomically, females with the shortest wings display the most deviant ear anatomy (both wing length and ear form are consistent not only within a species, but also—with only a very few exceptions—within a genus). Physiologically, females with shortened wings also have reduced auditory sensitivity. Structurally and functionally, this relationship between wings and ears is graded—the shorter the wings, the lower the auditory sensitivity (Fig. 6.11, bottom). Even large, heavy females that do not show any attempt to fly in behavioral tests have fully functional DK ears as long as they have long wings. ("Long wings" normally means that they extend at least to the tip of the abdomen. However, in groups with markedly elongated bodies such as the Angelinae, the Schizocephalinae, or some of the Vatinae, fully developed, functional wings may be relatively shorter, and these animals have DK ears.)

To separate the effects of sexual dimorphism in general from a specific correlation with wing length, we would want to evaluate brachypterous or apterous males. In contrast with females, males with reduced wings are very unusual in the Mantodea. We have thoroughly studied one such case, *Yersiniops solitarium* (Yager, 1990). In this species, both sexes are apterous, and both totally lack auditory sensitivity. The ventral metathorax is not anatomically dimorphic. The meta-

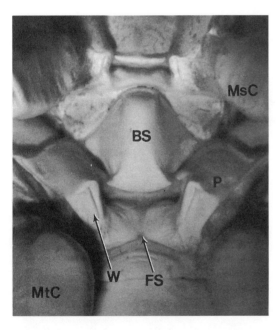

Fig. 6.10 The caudal, ventral metathorax of adult *Eremiaphila brunneri* (Eremiaphilidae) showing striking similarity to the same region of first-instar nymphs in a DK-eared species (compare to Fig. 6.6). The area between the metathoracic coxae (MtC) and caudal to the basisternum (BS) contains the furcasternum (FS) and two laterally situated walls (W) with vertical slits. These walls are homologous to the deep-groove walls in hearing mantids and the slits to the tympana. Similar anatomy is visible in the mesothoracic segment between the mesothoracic coxae (MsC). P - preepisternum.

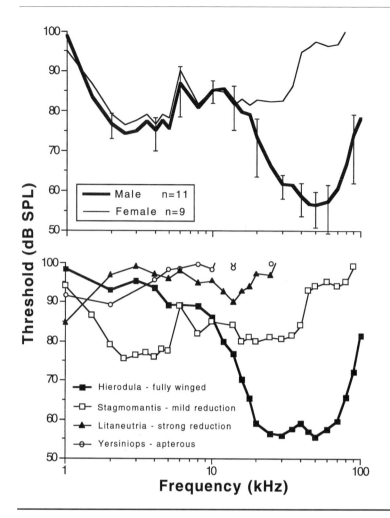

Fig. 6.11 Reduction of hearing in females of sexually dimorphic mantis species. *Top*, *Stagmomantis carolina* (Mantidae, Mantinae) males and females have very different audiograms above 20 kHz (sensitivity below 10 kHz derives from mesothoracic structures, not the DK ear). The males have very long wings; females are mildly brachypterous. Thresholds are from extracellular physiology. *Bottom*, The degree of hearing reduction in females correlates with reduction in wing length. *S. carolina* and *L. minor* (Mantidae, Amelinae) males are fully winged and have sensitive hearing. *Y. solitarium* (Mantidae, Amelinae) males are apterous and have hearing identical to the females. Curves are means from >8 animals. Top graph modified from Yager (1990) and used with permission.

thoracic anatomy of other short-winged males in museum collections (*Apteromantis, Geomantis, Bolivaria, Opsomantis*) matches that of females with comparable wing length (Yager, 1990, and unpublished observations).

Mantids with Two Ears

An auditory system with two ears—and the ability to determine sound direction that they confer—is the norm in the animal world, and for good reason: knowing the direction to a sound source is crucial for successful social interactions like courtship and for surviving in a world full of noisy dangers, not the least of which is the approach of a predator (Lewis, 1983; Michelsen and Larsen, 1985). Even the very recently discovered "single" ear of some tachinid flies is functionally two in its clever and unique mechanism for sound directionality (Miles et al., 1995). It might come as little surprise, then, that one group of mantis has evolved two ears instead of the usual single ear (Yager, 1996b). Contrary to expectation, however, the two ears have nothing at all to do with locating a sound source. Instead, it is the mantis's way of dealing with the problem of being tone deaf. Both ears are located in the ventral midline—but in different body segments. Neither has any directional capability, but they listen in two nonoverlapping frequency ranges. Thus, while most mantids are auditory cyclops, these are auditory bicyclops.

In contrast to the V-shaped audiogram of the typical DK-eared mantis, the audiograms of several hymenopodid mantids (*Creobroter gemmatus, C. pictipennis, Pseudocreobotra wahlbergi,* and *Hestiasula brunneriana*) have a W shape with frequency ranges of equal sensitivity at 2 to 4 kHz and 25 to 50 kHz (Fig. 6.12A; Yager, 1996b). Hearing in the intermediate range of 8 to 15 kHz is minimal. The low-frequency hearing of *H. brunneriana* is the most sensitive hearing we have ever recorded in any mantis, with thresholds as low as 25 dB SPL at 3 kHz. Using simple ablations to isolate body segments neurally, it became clear that the source of the low-frequency sensitivity is exclusively in the mesothorax, while the ultrasonic sensitivity arises from the metathorax (Fig. 6.12B).

The auditory organ in the mesothorax is serially homologous to the metathoracic ear (Fig. 6.12D). The latter is a conventional DK ear in both anatomy and physiology. The groove of the mesothoracic ear is much more open and, in fact, resembles the DK ear at mid-nymphal development. Although there is a tympanum in the form of a very broad groove, differentiation from the rest of the groove wall is not especially clear. The entire wall is thin, membranous cuticle that is quite "floppy." A tympanal organ comprising 35 to 45 scolopophorous sensilla attaches at the anterior end of each tympanum, and the axons travel in Nerve 7 to the mesothoracic ganglion.

The serially homologous peripheral auditory system poses the question of serial homology of the auditory CNS. Although we have not as yet done the critical intracellular recordings, substantial extracellular evidence indicates that there are, indeed, serially homologous sets of interneurons. For instance, mesothoracic auditory interneurons matching the physiological characteristics of metathoracic interneurons travel in the same portions of the ascending connective and have the same axon size relationships (judged by spike height [Yager, 1996b]). Silver-intensified cobalt backfills reveal a cell in the mesothoracic ganglion identical in its three-dimensional anatomy to MR-501-T3. From the functional standpoint, it is important to note that there is no physiological evidence of interaction between the two ears. At least below the brain, these mantids have two entirely separate auditory systems.

Rather than a single ear that hears a range of frequencies, these hymenopodid mantids have evolved two ears that are frequency "specialists." These mantids can—and apparently do—discriminate acoustic information based on its frequency and respond appropriately. In the case of the metathoracic ear, ultrasound triggers the suite of evasive behaviors described above, and the ear clearly functions in bat avoidance. How these mantids use their low-frequency hearing continues to be a frustrating puzzle. Despite extensive attempts to elicit sound-triggered behavior in social, hunting, and threatening contexts, we remain without any substantive clues.

Evolution of Hearing in the Mantodea

Insects in general display a remarkable capacity for evolving auditory systems in the unlikeliest of places: on wings, on legs, on mouth parts, under their "chin," to say nothing of more mundane abdominal and thoracic sites. In fact, tympanate hearing has evolved independently no fewer than fifteen times among insects (Michelsen and Larsen, 1985; Fullard and Yack, 1993; Robert et al., 1994). Even within this group, the mantis ear is an oddity and, as such, may offer an opportunity to learn something of how sensory systems evolve. A suite of at least three changes must converge to form an auditory system: peripheral structures capable of transducing sound must appear, components of the CNS must be created or co-opted to process the incoming audi-

110 Hearing and Vision

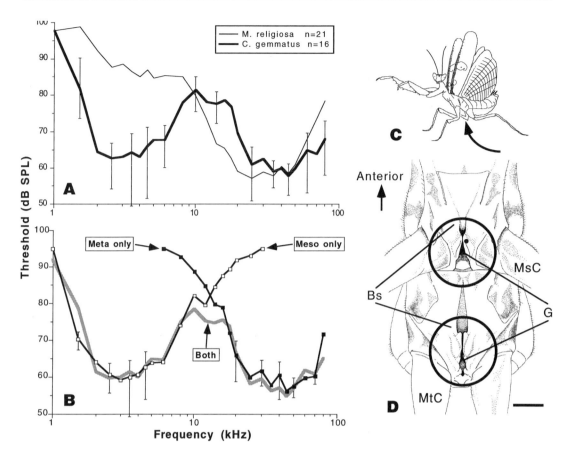

Fig. 6.12 An auditory bicyclops, *Creobroter* spp. (Hymenopodidae, Hymenopodinae). *A*, Extracellular physiological audiograms contrast the V-shaped curve of a typical DK-eared mantis, *M. religiosa*, with the W-shaped curve of *C. gemmatus*. There is no significant auditory sexual dimorphism in these species, and data from both sexes are included. *B*, Data proving that two separate auditory organs contribute to the W-shaped audiogram in *C. gemmatus*. The control audiogram (both) was recorded from the ascending connectives in the neck as in *A* above. The other two curves were recorded from neurally isolated thoracic segments. The auditory organ in the mesothorax transduces only frequencies below 10 kHz, and the one in the metathorax only frequencies above 20 kHz. The three audiograms were recorded in rapid succession in each of 13 animals. *C*, The arrow indicates the location of the ears in a female *Creobroter* performing a defensive display. Typical body length is 35 mm. *D*, Ventral mesothorax and metathorax of *Creobroter* with the two ears circled. The ears lie between the mesothoracic coxae (MsC) and metathoracic coxae (MtC) just caudal to the basisternum (Bs). Both comprise a deep groove (G), but the walls of the mesothoracic groove are flared outward. The black circle in the mesoear marks the attachment site of the tympanal organ at the anterior end of the tympanum. Scale bar: 1 mm. Modified from Yager (1996b) and used with permission.

tory information, and a combination of CNS commands and peripheral effectors must coordinate to produce an appropriate behavior. Since it is unlikely that on an evolutionary scale anything ever appears completely *de novo*, the focus of this discussion is really on the history of mantis audition.

An Auditory Phylogeny

Fully acknowledging the disadvantage imposed by the absence of a contemporary phylogeny for the Mantodea, it is nonetheless worthwhile to formulate an hypothesis for the general evolutionary history of audition in the suborder. As an initial assumption, I am trusting that Beier's (1968) taxonomic scheme substantially reflects monophyletic groupings—but would also argue conversely that the distribution of auditory patterns may contribute to a sensible rearrangement of taxa that eventually prove polyphyletic.

The phylogenetic hypothesis derives directly from the discussions above of the four patterns of audition in the Mantodea, with one additional assumption: that there was only one innovation of the DK ear (Yager, 1989; Yager, 1992). (While this is reasonable, parsimonious, and most probable, the alternative possibility that the ventral thoracic midline is a "hot spot" for ear evolution [Yager and Spangler, 1995] in mantids leading to multiple independent appearances cannot be entirely dismissed.) The underlying data are primarily anatomical, but physiological audiograms support all of the general patterns; for example, DK morphology means good auditory sensitivity, and all other anatomical forms show reduced or absent hearing.

The hypothesis proposes that the Chaeteessidae and Mantoididae represent primitively earless lineages; the Eremiaphilidae and Metallyticidae may also be, but, alternatively, they may show extreme secondary loss of their ears. All other extant mantids derive from eared ancestors. In many cases, predominantly in auditory sexual dimorphism, the ear is secondarily reduced or lost entirely. In one group of mantids, the Hymenopodinae + Acromantinae (we have not yet tested the third hymenopodid subfamily, the Oxypilinae), a second, independent auditory system evolved.

This straightforward-sounding scheme is not without difficulties and puzzles, a sampling of which include: (1) The Old World Acromantinae (Hymenopodidae) are all DK eared, but their New World subfamily mates are earless; (2) both the Oligonychinae and the Thespinae have a ventral metathorax like Chaeteessa, despite being placed in the generally DK-eared family Mantidae; (3) both sexes of all Photininae examined (also in the Mantidae) have highly aberrant "ears" that look like an extreme form of the DNK morphology. These are the sorts of inconsistencies that may hint at important biogeographic patterns in mantis ear evolution or, alternatively, may flag particular taxa for careful taxonomic and/or phylogenetic re-examination.

Origins of the Mantis Auditory System

The fossil record of mantids can most generously be described as impoverished, and is no help in studying the evolution of the DK ear. A small number of mantids have been preserved in amber, but, besides also being geologically young, these are almost exclusively nymphs (G. Poinar, personal communication). The oldest fossil mantids come from the lower Cretaceous of Asia and are 100 to 145 million years old (Sharov, 1962; Gratshev and Zherikhin, 1993). However, these fossils comprise wing pieces and body parts that provide no information about the ventral metathorax. We are left with a comparative strategy for studying the history of the mantis ear and fortunately have an almost perfect outgroup for comparison. Cockroaches and mantids share many morphological characters, and their very close phylogenetic affinity is well accepted (Thorne and Carpenter, 1992). One character they do NOT share, however, is tympanate hearing. We have tested neurophysiologically twelve cockroach species from four of the five families (eleven of the twenty subfamilies; McKittrick, 1964) and find no auditory sensitivity at any frequency above 5 kHz (Yager and Scaffidi, 1993). Responsiveness below 5 kHz derives from the subgenual organs in the tibiae, exquisitely sensitive substrate vibration detectors that may also be used to detect airborne sound (Shaw, 1994).

Do cockroaches have structures in their metathorax homologous to the mantis's peripheral auditory system? Do cockroaches have a "tympanal" nerve? Does the interneuron MR-501-T3 exist in the cockroach? The answer to all of these questions is unequivocally yes (Yager and Scaffidi, 1993; Yager, 1996a). The external anatomy of the caudal mesothorax and metathorax of adult *Periplaneta americana* are essentially identical to each other, to the same region in first-stadium mantis nymphs, and to the caudal mesothorax of all adult mantids. The furcasternum is oriented dorso-ventrally (rather than longitudinally in the DK ear), and lateral to it are two regions of membranous cuticle containing a slitlike depression.

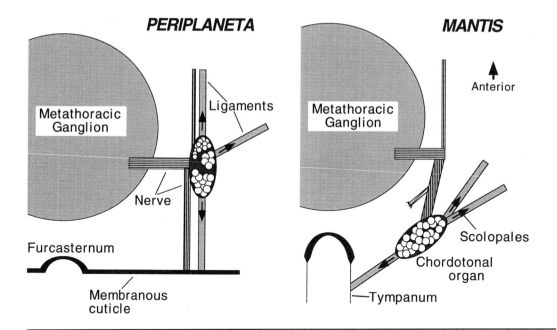

Fig. 6.13 Schematic drawings comparing the structure of the mantis DK ear with the homologous structures in the cockroach *Periplaneta americana*. Nerve 7 of the cockroach (= the mantis tympanal nerve) has an anterior and posterior branch and a chordotonal component (= the tympanal organ in mantids). The structures in the two animals are almost identical except for the simple change in geometry caused by the infolding that forms the deep groove with the furcasternum as its floor. This is reflected clearly in the orientation of the three ligaments in each animal.

In all of these cases, Nerve 7 has a chordotonal component with three ligaments, one of which is attached to the dorsal end of the slitlike depression near the furcasternum. The chordotonal organ always contains thirty-five to forty-five scolopophorous sensilla. As Figure 6.13 shows, the difference in Nerve 7 between an adult cockroach and an adult mantis is one of simple geometry (Yager and Scaffidi, 1993). We are currently mapping the afferent terminations of the Nerve 7 chordotonal neurons in the metathoracic ganglion of *P. americana*; at least at the superficial level they do not differ from the mantis. Within the metathoracic ganglion, our cobalt backfills show a neuron that has the distinctive anatomy of MR-501-T3, and Ritzmann and colleagues (Ritzmann et al., 1991) have recorded from and stained intracellularly what appears to be exactly the same cell.

Given all these similarities, how do we account for the auditory competence of the mantis compared to the cockroach? The most compelling insight comes from comparing the peripheral anatomy of the caudal metathorax in the first-instar mantis nymph (= the adult cockroach; Fig. 6.6) with the adult DK-eared mantis (Figs. 6.1, 6.12D; Yager and Scaffidi, 1993; Yager, 1996a). The nymph and the cockroach lack two essential features of a tympanate ear that the adult possesses—the broad, thin tympanum and the impedance-matching air sac—although all three have the same neural components. A second useful comparison looks at species with and without mesothoracic hearing: the anatomical differences boil down to the presence of a large tracheal sac behind the membranous cuticle in the auditory bicyclops (Yager, 1996b, and forthcoming). In other words, these comparisons suggest that the difference between earless and eared is simply a matter of building an appropriate peripheral transducing mechanism.

Although useful as a starting point, this view

is clearly incomplete. The transformation from earless to eared also entails behavioral changes, which, in turn, imply some degree of rewiring in the CNS. But how much? Experiments selectively ablating different components of Nerve 7 in *P. americana* show that the nerve is bifunctional (Yager and Tola, 1994). The nonchordotonal afferents innervate socketed hairs that monitor leg position and motion. The chordotonal afferents respond to vibration; as yet we do not know the normal route of stimulation (for example, substrate, body, or airborne). Pollack and his colleagues (Pollack et al., 1994) have implicated Nerve 7 (presumably the chordotonal component) of *P. americana* in a vibration-triggered escape response that may function in parallel with the cercal escape system. Thus, the behavioral function of the primitive "auditory system" may be much the same as for present mantis ear—vibration-triggered escape from predators. What has changed is the context and the exact nature of the escape behaviors.

Unanswered Questions

The subject of unanswered questions is an appropriate, if ironic, topic with which to end the chapter since there really are relatively few answered questions about mantis audition. A truly multidisciplinary understanding of how and why mantids hear requires far more data on mantis natural history and phylogeny than are currently available. This frustrating dearth of information—seemingly inexplicable given the inherent, almost universal fascination these creatures engender—plagues all of us working with mantids.

I will simply draw attention to three key questions regarding mantis audition that should command special attention:

• Do mantids use intraspecific acoustic communication? If so, it is most likely to occur in courtship—one more compelling reason to focus careful attention on this aspect of mantis natural history. We particularly need to look at the hymenopodids to try to solve the mystery of their low-frequency hearing.

• In both proximate and ultimate senses, what explains the graded correlation of wing length and auditory sensitivity? The central issue is the loss of auditory sensitivity. Does the loss in these females imply that an auditory system is costly to maintain and that the evolutionary cost/benefit analysis for these animals has swung in a very different direction from that of other mantids? Or is ear loss secondary to some more fundamental process? We are currently exploring the hypothesis that females in highly dimorphic species are neotonous, retaining a suite of juvenile characters that includes immature wings and ears (Haron and Yager, 1996). If true, it may help us understand how heterochrony can play a role in the evolution of insect sensory systems (Gould, 1977).

• How has the CNS of the mantis altered to link input from a "new" sensory modality to the context-specific neural system controlling a "new" behavior? Despite the complex suite of coordinated movements mantids perform in response to ultrasound, I suspect that exploration in the CNS will demonstrate that the modifications have actually been relatively small. After all, neither the sensory modality (vibration detection) nor the resulting behavior (an aerial deimatic display?) has really appeared *de novo*. The evolution of the mantis auditory nervous system has, in all probability, proceeded very economically using existing resources (interneurons and motor neurons) and control frameworks (basic neural circuits). The innovations—no less potent for their simplicity—will be in the way the circuits are linked together, in the nature and strength of connections between neurons.

ACKNOWLEDGMENTS

My thanks to Drs. Dan Otte and Don Azuma at the Academy of Natural Sciences in Philadelphia, to Dr. David Ragge and Mrs. Judith Marshall at the Natural History Museum in London, and to Dr. David Nickle at the Smithsonian Institution for their help and hospitality. J. Triblehorn and A. Cook generously allowed me to cite unpublished data. This review was written with support from NIDCD Grant DC01382.

7. Binocular Vision and Distance Estimation

Karl Kral

For insects, the ability to estimate distances and movements in space can be a matter of life and death. For decades, research on the visual systems of insects has been devoted to the question of whether insects are capable of spatial perception and, if so, on what processes or structures their spatial perception is based. For instance, do insects, like humans, need a number of spatial cues in order to see in spatial dimensions, or does a single factor suffice? Because insect eyes are immobile, oculomotor factors such as convergent eye movements and accommodation (via a lens) can be excluded from the outset. Quite another matter, however, is the use of stereoscopic depth perception (distance calculations based on disparities between the two retinal images) and motion parallax (distance calculations based on the change of visual angle due to movement of the animal's own body). Both of these capacities could play a decisive role in insect spatial vision. In this context it should, of course, be noted that spatial vision in insects is much simpler than in humans: For insects, spatial orientation probably is based on only a few critical stimulus parameters, such as the presence of a dark shape (a potential predator, prey, or mate) or of a luminance edge running in a particular direction (a potential perch site).

A second question in which my colleagues and I are interested is whether the visual system of insects is capable of spatial perception from the time that a mantis hatches or whether visual experience is needed for the capacity to develop either in general or in particular cases. For instance, does either the perceived size of an object or an innate ability to maintain size constancy allow an insect to judge the distance to an object because it knows the relationship between visual angle and distance? Many animals are innately capable of perceiving a looming object because the shrinking or expanding retinal image is interpreted as an increase or decrease in object distance, respectively. Could this also be the case in insects? Selective deprivation experiments immediately after hatching and during the early posthatching phase could help to clarify these questions. Such studies, however, can produce fruitful results only when a generally suitable experimental animal is available. The praying mantis is just such an animal.

Mantids have, for a number of reasons, proved to be extremely suitable subjects for work aimed at clarifying distance measurement and depth perception in insects. They are rather large insects that are (relatively) easily bred and kept under laboratory conditions, and they have a very sophisticated visual system that is integrated with an equally sophisticated mechanism for capturing prey. Work in behavioral physiology, neuroanatomy, and neurophysiology has shown that mantids have all of the prerequisites for localizing prey with stereoscopic mechanisms: a highly acute, slightly inward-pointing frontal portion of the compound eyes (the fovea); an extensive binocular visual field; and neurons that receive input from both eyes and/or are movement sensitive (Rossel, 1979, 1983, 1986; Collett, 1987; Prete and Mahaffey, 1993; Prete et al., 1993; Mathis and Rossel, 1993; Prete and McLean, 1996; for a review see Schwind, 1989; for neurons responding to moving objects, see Young, 1991, 64–81; Rind and Simmons, 1992; Simmons and Rind, 1992). In addition, there is evidence that mantids rely not only on stereoscopic factors for spatial vision, but also on movement factors, such as movement parallax (Horridge, 1986; Poteser and Kral, 1995).

Binocular Determination of Prey Distance

In his now classic behavioral studies on *Mantis religiosa* Zänkert (1939) was the first to show that the distance to an object of prey could be correctly estimated only if both eyes were used. His research notes indicate that *M. religiosa* will notice a meal worm at distances of up to 30 cm. Within this range, the mantis stares at the prey with an attentiveness that is most pronounced when it is within 5 cm. If the prey is as close as 2 to 3 cm, a predatory strike is elicited. However, when a monocular animal tries to fixate a meal worm, it often twists its head around the upper and long axis, which turns its sighted eye downward. In Zänkert's words: "Mealworm recognized at a distance of 10 cm. Mantis creeps toward it, assumes strike position at 3 cm, strikes, misses. Constantly tries to fix the prey with the intact left eye. Turns head. Approaches to within 1–0.5 cm."

In the late 1960s, Maldonado and his co-workers did an extensive series of studies on how mantids capture their prey. In one of their first experiments, they painted over the left eye of *Coptopteryx viridis*, imbedded each mantid into a plaster block, and placed the block into a holder in the center of a cylindrical arena. A fly, mounted on a magnet, was moved around the inner wall of the cylinder from right to left. As soon as the fly entered the lateral part of its visual field, the mantid quickly turned its head toward the fly, but it struck only when the prey was within the visual field of both eyes. Maldonado et al. found that mantids with an occluded eye struck considerably less often than normally sighted mantids. Strikes were also performed only when the fly was very close to the normal eye, and the strikes that were emitted were atypical. That is, the prey was often caught with only one foreleg, was hit in one way or another by different parts of the foreleg, and often apparently was caught more or less by chance. These monocular mantids also tried constantly to remove the paint from their eye. The cleaning reflexes occurred even when the prey was right in front of the mantid's head.

One of my former students analyzed this compensatory behavior more closely by using video recordings of *Tenodera aridifolia sinensis* that had been subjected to monocular occlusion for several days (Jakobs, 1993). With one eye occluded, *T. a. sinensis* behave, generally, as do other mantids: They make exaggerated head movements that turn the intact eye toward the prey while simultaneously rotating their head toward the side with the occluded eye. As with *C. viridis*, *T. a. sinensis* struck only when prey was brought very close, and in some cases the mouth parts may have even touched the prey, resulting in a clumsy, unaimed strike. Like Maldonado and Levin (1967), we found that monocular mantids constantly use their forelegs to try to remove the lacquer from the frontal eye region. If they succeed in doing so, they capture prey as well as normally sighted animals even if the rest of the eye remains covered.

Returning to Zänkert (1939), one finds that he, too, studied prey capture by mantids with partially occluded eyes. He found that if the dorsal parts of the eye were painted over, strike success was not affected. When the medio-frontal parts of both eyes were occluded, however, the strikes were much less accurate. This was the first concrete indication of the functional significance of the frontal eye region in localization of prey. Maldonado and Barros-Pita (1970) continued their work on mantids with partially occluded eyes by using precise optical measurements to eliminate certain carefully defined parts of the eyes of *Stagmatoptera biocellata*. Using the same experimental set-up described above, they, too, confirmed that the frontal parts of both eyes are absolutely necessary for distance estimation.

The Role of the Acute Visual Zones in Binocular Distance Determination

The Acute Visual Zone

The first systematic studies of the frontal areas of the mantid's compound eyes were done by Barros-Pita and Maldonado (1970). They created a map of the ommatidia by stretching nylon mesh over one eye and using a goniometer to move a pseudopupil over the eye's surface. (The pseudopupil is an illusory pupil or black spot in a compound eye that is located over the ommatidia from which no light is being reflected back toward the observer.) Their data suggested that when certain regions of the eye are painted over, the outlines of a visually fixated object are no longer visible. Beyond object recognition, however, these critical retinal areas are important for determining object position in the visual field via the processing of binocular cues.

In 1979, Samuel Rossel determined the precise dimensions of the ommatidial angles in a mantid eye using the pseudopupil method and electrophysiological recordings from the photoreceptors. His most important finding was that the horizontal and vertical interommatidial angle ($\Delta\phi$; the angle enclosed by the optical axis of two neighboring ommatidia) is distinctly smaller in the fovea than in the rest of the eye (0.6° versus 2°; Fig. 7.1). His measurements also showed that in the acute zone, the facet diameter is 50 to 70 μm, almost twice as large as in the rest of the eye, and the ommatidia are almost twice as long as elsewhere. The latter is due to the fact that, in the acute zone, there is a considerably greater radius of curvature than in the rest of the eye due to a local flattening of the superficial curvature. This causes the ommatidial axes at the edge of the center of the acute zone to be slanted so that they are no longer perpendicular to the surface of the eye. In turn, this slants the optical axes relative to the anatomical axis of the eye, so that the frontal eye region appears to be more curved anatomically than it actually is optically. Hence, Rossel's optical determination of interommatidial angle is preferable to an anatomical determination (also see Horridge and Duelli, 1979). Further, the rhabdomes of the ommatidia have smaller diameters (1.5 μm) in the acute zone and are considerably longer than in the rest of the eye, and the crystallin cones are elongated where the ommatidial angles are smallest (Fig. 7.2).

Rossel also showed that over the entire eye, under light adapted conditions, there is a generally constant 1:1 relationship between the interommatidial angle ($\Delta\phi$) and acceptance angle ($\Delta\rho$; the spatial extent of the light-collecting area of an ommatidium; Fig. 7.1, inset). This is surprising given the heterogeneity of the two angles. Rossel found that the acceptance angle ($\Delta\rho$) is 0.7° in the acute zone and increases toward the periphery to 2° (Fig. 7.1). In twilight and darkness, however, $\Delta\rho$ increases in the frontal region to about 2°.[1]

The anatomical and optical characteristics of the mantid's eye indicate that the acute zone is capable of very high spatial resolution compared to the rest of the eye (Horridge and Duelli, 1979; Rossel, 1979; Horridge, 1980; for an overview, see Young, 1991). This leads to the question of

---

1. According to Snyder (1977), $\Delta\rho$ is produced by the connection of the two functions $\Delta\rho_{lens}$ and $\Delta\rho_{rhabdome}$. The term $\Delta\rho_{lens}$ is the half-width of the intensity distribution, which is the result of diffraction caused by the tissue surrounding the lens of the individual ommatidium. For an eye working at its diffraction limit, $\Delta\rho_{lens} = \lambda/D$, where λ is the wavelength of light in a vacuum and D is the diameter of the facet. The term $\Delta\rho_{rhabdome}$ is the half-width of the rhabdome opening function and is expressed as the angular projection of the rhabdome diameter (d_r) out into space ($\Delta\rho_r = d_r/f$ insofar as no additional light is collected by the crystallin cone). The term f is the distance between tip of the rhabdome and the node and is related to the thickness of the lens and the length of the crystallin cone. The curves in Figure 7.2, which are based on Rossel's (1979) angle measurements, show how close his calculations for the acceptance angles are to those based on Snyder's theoretical calculations, $\Delta\rho^2 = (\lambda/D)^2 + (d_r/f^2)$. Here, Rossel comes to the conclusions that the large facets of the ommatidia in the acute zone are the prerequisite for the small acceptance angle ($\Delta\rho$) because $\Delta\rho$ in the acute zone is smaller than the limit set by diffraction for ommatidia in the rest of the eye. Further, $\Delta\rho$ in the acute zone is limited by diffraction ($\lambda/D > d_r/f$), whereas the limiting factor for the rest of the eye is the opening function of the rhabdome ($d_r/f > \lambda/D$) (cf. data in Fig. 7.2).

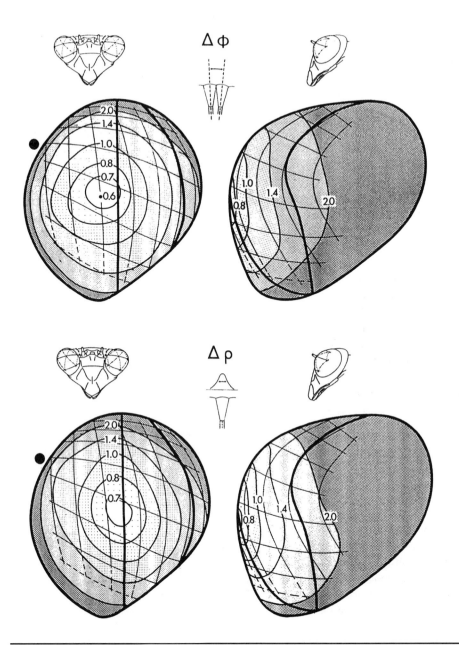

Fig. 7.1 The upper two diagrams indicate the distribution of the interommatidial angles ($\Delta\phi$) over the left compound eye of *Tenodera australasiae*. The acute zone is located frontally and slightly toward the inner edge of the eye. The $\Delta\phi$, beginning at 0.6°, increases continuously and almost symmetrically to values of 2.5° in the outermost portions of the eye. The $\Delta\phi$ values along the concentric isolines are the same. The thick line running vertically through the frontal part of the left eye borders the binocular visual field. In the lateral part of the left eye, the thick line covers the ommatidia which look in the same direction as those of the inner edge of the right eye. The lower two diagrams indicate the distribution of the physiological opening angle ($\Delta\rho$) over the left eye of *T. australasiae*. The $\Delta\rho$ angles were measured in light-adapted mantids and vary from 0.7° in the acute zone to 2.5° in the peripheral areas. From Rossel, 1979, and used with the kind permission of Springer-Verlag.

whether this high spatial resolution is possible without a considerable loss of contrast.[2] The decrease in the ommatidial angle increases the number of ommatidia (number of receptors) per spatial angle unit, improving detail resolution. At the same time, however, the related decrease in facet surface area and/or diameter and decrease in rhabdome size reduces the number of photons captured per ommatidium, which reduces the local array's ability to resolve contrast. The facets in the acute zone are greatly enlarged due to local flattening of the eye's surface, and, at the same time, the rhabdome opening angle ($\Delta\rho_{rhabdome}$) is small and this would seem to permit a kind of compromise (see Snyder et al., 1977). Any loss of contrast in the proximal stimulus might be recovered at the neuronal level through lateral inhibition. On the other hand, any gain in spatial resolution by the acute zone ommatidia could be lost at the neuronal level through convergence.

Determinations of light sensitivity throughout the mantid eye (Rossel, 1979) suggest that the high spatial resolution capacity of the acute zone is present only under light-adapted conditions. When dark adapted, the photoreceptor cells in and out of the acute zone react with about the same sensitivity to a point of light. This indicates that light is less efficiently taken up in the acute zone, although the larger facets could capture considerably more light. But due to their relatively small acceptance angles ($\Delta\rho$), the photoreceptors in the acute zone react less sensitively than the rest of the eye to a large, diffuse light source. In bright daylight, however, the photoreceptors in the acute zone are considerably more sensitive to a point of light than those in the rest of the eye.

---

2. The diffraction criterion is determined by the equation $r = 1.22 \lambda^2 f/A$, where r is the radius of the central airy disk and A is the ommatidial aperture. The contrast-perception criterion is described by the equation $v = A/\lambda$, where v indicates spatial frequency of the visual stimulus. These two factors are in competition with each other (see, e.g., Young, 1991, 47–57). The formulas show that there is good resolution when the aperture A is as large as possible, or the focal length f is small, but an aperture (A) that is as small as possible provides optimal contrast perception.

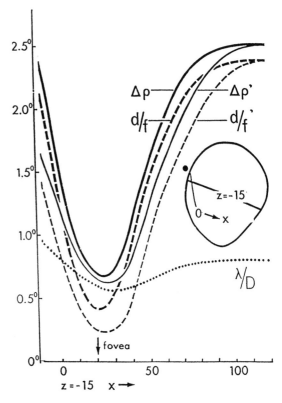

Fig. 7.2 The distribution of both the calculated and the measured physiological acceptance angle ($\Delta\rho$) along the facet row $z = -15$ in *T. australasiae*, which passes through the acute zone. The value $\Delta\rho'$ was calculated from the anatomical components d/f' and λ/D. The value d/f is the derivative of the rhabdome opening function from the experimentally determined $\Delta\rho$ values and λ/D. From Rossel, 1979, and used with the kind permission of Springer-Verlag.

Rossel (1979) was also able to show that the spectral sensitivity of photoreceptor cells is the same throughout the eye. The main maximum sensitivity is at a wavelength (λ) of 500 to 520 nm (green), and there is a small auxiliary maximum at $\lambda = 370$ nm (ultraviolet).

With regard to the foveal specialization of the mantid's compound eyes, some interesting comparative studies by Kirmse and Kirmse (1985) should be mentioned. They found that related to the total surface area of the eye, the fovea made up some 29% in adult *T. a. sinensis*, 37% in adult *M. religiosa*, and 30% in larval stages of *Sphodromantis lineola* (unfortunately, there are no data on the adult stage of *S. lineola*). Interestingly, male mantids, too, have a relatively large acute zone which, perhaps, is related to the males' flying ability.

Binocular Distance Measurement via Triangulation: The Convergence Theory

The facts that mantids can successfully measure distance to prey only with binocular vision, and that to do so requires input from the acute zones, inevitably lead to questions regarding the mechanisms behind it all. An early attempt to explain these mechanisms involved the application of a long-standing theory of depth perception—the theory of convergence—to mantids. The reasoning behind the theory is this: Each acute zone is entirely within the binocular visual field, and the sight lines emanating from the acute zone ommatidia of the left and right eyes converge in the extended midsagittal plane of the mantid's head. The points of convergence define the area both within which prey can be precisely localized and in which it is captured most successfully. Because the interommatidial angles ($\Delta\phi$) in the acute zone are small compared to the rest of the eye, the rectangular fields defined by the intersection of the sight lines are rather small. Further, with increasing distance to the eye, these fields become elongated. As no additional resolution is possible within the rectangles, depth perception should be best where the rectangles are smallest (i.e., close to the mantid and within the area of convergence of the acute zone sight lines [Barros-Pita and Maldonado, 1970]). According to this theory, the acute zones allow the highest resolution of absolute distance to an object of prey. Interestingly, this type of convergence theory was also considered by Zänkert (1939). On the basis of his mantids' visual fixation behavior, Zänkert said that "strikes are only elicited when adjacent intersections of symmetrical sight lines extending [to] the elongated midsagittal plane of the mantid head touch the object of prey. The points of intersection are between 2 and 4 cm for adult *M. religiosa*, i.e., in the optimal striking range of the forelimbs."

Barros-Pita and Maldonado (1970) also hold the opinion that precise measurement of distance to prey must involve a mechanism of triangulation. They base their opinion on the fact that the number of ommatidia per unit area is larger in the acute zone than elsewhere in the eye while, on the other hand, the radius of curvature in the rest of the eye is larger than in the acute zone. In turn, a smaller ocular radius in the acute zone means greater binocular triangulation density and thus greater sensitivity for small changes in distance. The authors do, however, mention that some values for the acute zone, especially at its edges, do not quite fit their conclusion that a small eye radius and greater density of ommatidia are unique characteristics of the acute zone.

Horridge (1977) noted that precise fixation of the prey in the midline would suffice to stimulate two corresponding ommatidia and thus to determine its distance. This sort of distance determination would have to be based on a point-to-point interaction between the eyes. One wonders whether an insect brain could really be loaded down with the enormous number of neural connections that would be necessary for triangulation. Horridge (1977) believes, however, that precise fixation could obviate the need for complicated neural connections between the eyes, though he does not explain how this might work. Certain distances could be related to certain trigger zones made up of groups of corresponding ommatidia. This arrangement would require fewer neurons (Via, 1977) but would be a rigid one. The influence of environmental movements (in the background and by the mantis and its prey) on this system also remains an open question. Perhaps what speaks most strongly against the convergence theory is simply the variability of both the mantid's behavior while capturing prey (e.g., Prete and Hamilton, this volume, chap. 10) and the behavior of the prey itself.

Our video recordings of *M. religiosa* capturing prey in natural surroundings show that prey is both stalked and successfully captured not only when it is positioned within the midsagittal plane (and in the range of the acute zone ommatidia) but also when it is considerably outside of this area, in the dorsal or ventral binocular visual field. This means that the number of ommatidia involved in distance measurement can vary dramatically from situation to situation. The predation behavior that we saw in *M. religiosa* in the field is very much in agreement with laboratory studies on *Sphodromantis lineola* by Prete and his colleagues. For instance, Prete (1993) found that strikes can occur when a small moving visual stimulus is as much as 24.5° to the right, to the left, or below visual field center and, less infrequently, when it is 24.5° above visual field center.

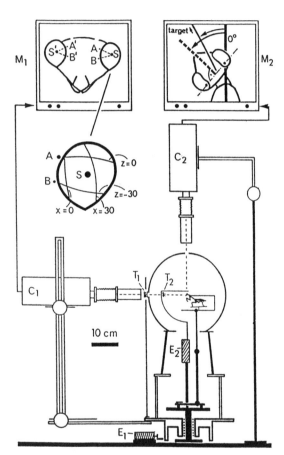

Fig. 7.3 The experimental set-up designed to determine the saccade-sensitive regions in the eye of a praying mantis. The mantid is in the middle of the glass globe. C_1, C_2 - video cameras; E_1 - the motor that drives the globe, the camera C_1, and the target T_1; E_2 - the motor that drives the target T_2; M - video monitors; S,S'- positions of the pseudopupils; z, x - facet rows. From Rossel, 1980, and used with the kind permission of Springer-Verlag.

The significance of the acute zones for the estimation of prey distance should not, however, be underestimated, as we shall see in the following section.

Binocular Distance Measurement via Retinal Image Disparity

By means of photographic and optoelectronic analyses, Lea and Mueller (1977) found that mantids always try to bring the retinal image of a potential prey object onto the acute zone of both eyes by making saccadic (jerky) and/or smooth head movements. Rossel (1980) used the pseudopupil method to determine the stimulus intensity necessary to elicit a saccadic head movement and in what retinal areas an image would elicit such a movement (Fig. 7.3). Saccades, which could reach peak velocities of about 560°/second, were unlikely to occur when the image was in the acute zone, but Rossel found that there is a ring around the acute zone in which saccades are easily elicited (Fig. 7.4). He also found that if an object moves against a homogeneous background, it is tracked with smooth head movements over a long distance so long as the object moves at low speed (8°/second). If object speed increases (35°/second), the head lags behind the leading edge, and the mantid has to perform periodic saccades to "pull" the object back into its acute zone. If the background is patterned (heterogeneous), however, head movements will be primarily saccadic, and the higher the background contrast, the more likely will be the appearance of saccades. However, observations in natural (structured) surroundings indicate that even just before the mantid strikes, saccadic head movements predominate, which strongly suggests that the pre-strike retinal image falls in the saccade-sensitive zone (i.e., outside of the acute zone). This would have to mean that stabilization of the object image in the acute zone is not an absolute prerequisite for a successful strike. The orthodox convergence theory, which is restricted to objects in the midsagittal plane and which suggests a relatively rigid mechanism of distance measurement (Baldus, 1926; Barros-Pita and Maldonado, 1970; Horridge, 1977), seems insupportable in view of the data on saccade sensitivity in mantids.

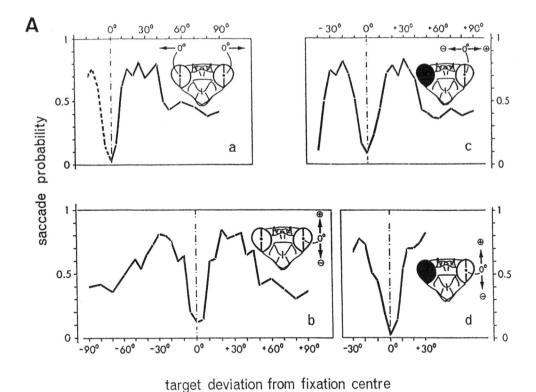
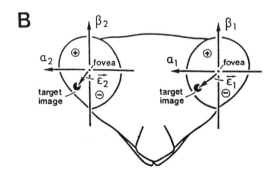

Fig. 7.4 A, Graphs indicating the probability that a saccade will occur when the image of a visual stimulus is in various areas of the retinae. The four graphs present data on saccade sensitivity in the anterior-lateral (a, c) and dorsoventral (b, d) directions in a binocular (a, b) and a monocular (c, d) *Tenodera australasiae*. The deviation of the object from fixation center (0°) is indicated on the abscissa, and the saccade sensitivity is indicated on the ordinate. B, A frontal projection of a co-ordinate system projected onto the eye of *T. australasiae*. The zero point is in the acute zone. The vectors ε_1, ε_2 define the position of the target image on the eye surface. From Rossel, 1980, and used with the kind permission of Springer-Verlag.

Rossel (1980) tried to explain binocular distance measurement that is not limited to the mid-sagittal plane of the head with a relatively simple principle: He argued that the saccade sensitivity of the areas outside the acute zone is related to the (saccade-free) acute zone itself (see Fig. 7.4). He assumed that each eye has the necessary neural equipment to relate a retinal image position to the center of the acute zone. Figure 7.4 shows the projection of a coordinate system whose origin is in the acute zone on the surface of the eye. The retinal positions are determined by the vectors ε_1 and ε_2. Under the assumption that the coordinate systems in both eyes are the same, Rossel (1980, 278) concluded that "The cooperation between the two eyes during fixation, stalking and distance measurement is limited to a comparison of angular coordinates that define the vector position in each of the two eyes."

Three years later, Rossel published additional data that support the existence of stereoscopic vision in mantids. He based these experiments on the concept that when a mantid sees an object in three-dimensional space, the positions of the retinal images can be defined in relation to the acute zones. In other words, for each eye the direction (orientation) of an object in space is determined by the angle (α) between the sight lines through the acute zones and the sight lines through the object (Fig. 7.5A). The binocular disparity is then the difference between the angles α_{left} and α_{right}; the binocular eccentricity is the average of the sum of the two angles ($[\alpha_{left} + \alpha_{right}]/2$). At greater object distances (i.e., beyond the strike zone), the difference between these two angles becomes negligibly small, and the disparity value becomes constant. In 1983, Rossel had the ingenious idea of mounting prisms in front of the eyes of mantids. The effect of the prisms on retinal disparity is demonstrated in Figure 7.5B. The prisms deflect the sight lines to such an extent that the animal—insofar as it actually uses disparity to measure distances—would have to perceive the object as being at a much lesser distance. The disparity for the corresponding distances increases with increasing prism strength. The results of the prism experiments show a clear correlation between prism strength and those object distances that elicit a strike: the mantids strike at prey when the disparity of the retinal images reaches certain distinct values.

Subsequently, Rossel executed an additional series of experiments which examined how binocular eccentricity codes the horizontal deviation of an object from the median plane. In these experiments, each mantid's head was glued to its prothorax to prevent it from fixing an eccentric object with its acute zones. Under this condition, Rossel found that binocular depth perception occurs within almost the entire binocular visual field. But beyond an eccentricity of 20°, strike frequency diminishes dramatically. In a paper published three years later (1986), Rossel reconsidered the saccadic head movements that bring the object's image into the area of highest acuity. He demonstrated that in monocular mantids there is a close correlation between the amplitude of a saccade and the angle (α) that is defined by the sight lines through the acute zone and the object. He then presented binocular mantids with an object in their immediate vicinity, so that the angular difference ($\alpha_{left} - \alpha_{right}$) was as great as possible. Under these conditions, the saccadic head movements placed the object neither in the right nor in the left acute zone but, rather, in between them. This result is also indicative of binocular interactions between the two eyes.

Rossel postulated the following mechanism to explain how both eyes could work together without requiring the extremely complicated neuronal network that the convergence theory would ultimately require. In Rossel's words:

> Each optic lobe selects from the detailed retinal image a few key characteristics that would indicate potential prey. High impulse rates in corresponding characteristic detectors mean that the trigger characteristic is close by. Up to that point, the monocular data transmission could run in parallel paths. Using an X/Y coordinate system whose zero point is in the acute zone, active characteristic detectors code the average retinal position of the object image. This would require X/Y position-detecting neurons whose impulse rates would vary with the angular values of the X/Y coordinates. As the X/Y coordinates can be both

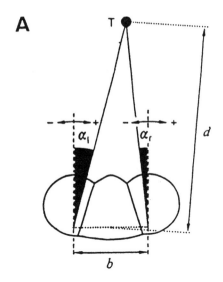 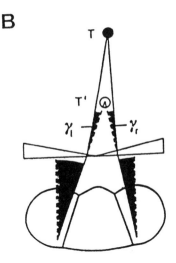

Fig. 7.5 Schematic representation of the principle of binocular disparity in mantids. *A,* When the mantid fixates some target, T, the position of the target relative to each eye is defined by the angle α between the sight lines from the acute zone and the sight line that leads to the target. The value α_l is the angle that occurs in the left eye, and α_r is the angle in the right eye. The binocular disparity is the difference in these two angles ($\alpha_l - \alpha_r$). The interocular distance (*b*) is the horizontal distance between the two acute zones. The distance to the target (*d*) is defined as the distance between object center and that point on the mantid's head that is exactly between the optical centers of the radii of curvature of the acute zones. *B,* Diffraction of the sight lines through two prisms. For the mantid, the target appears to be at T' rather than at T. The angles γ_l and γ_r are the degrees of angular divergence created by the left and right prisms, respectively. The strength of the prism is the sum of the two angles ($\gamma_l + \gamma_r$). From Rossel, 1983, and used with the kind permission of Macmillan Magazines Ltd.

positive and negative, only four neurons would be necessary in each optic lobe to specify every retinal object position precisely. The direction of the object could then be calculated from the average of the corresponding coordinate signals, and the distance from their difference. (Rossel, 1986, 278)

Here it should be noted that with such a minimal model, Rossel probably only wanted to show the possibility that a complex binocular image-matching procedure such as has been realized in higher mammals need not exist in mantids. This hypothesis does, however, have one liability: If there were a number of objects of prey in the visual field, the mantid would seem to be at a loss. How, after all, could each eye know for itself whether it is viewing the same object as is the other eye? To account for this possible dilemma, Rossel (1986) added the idea that perhaps each optic lobe selects only the most attractive object, that is, the one that moves most or is fastest (and, consequently, creates the highest impulse rate), and suppresses further transmission of all the others. The strongest signal could then overcome competition through lateral inhibition, as is common in biological systems. Recently, Rossel (1996) was able to identify the tricks that the mantid uses to efficiently solve the problem of having a number of objects in its visual field. If a praying mantis is shown two prey targets simultaneously, its visual system is still able to match pairs of retinal images that correspond to the same target in space. The matching process requires, however, the targets to be separated by at least 20°. Binocular target separation is therefore based on a very rough angular scale. Thus, the praying mantis seems not to have the correspondence problem, because potential false matches are associated with disparities that are too large to be detected by the visual system.

It seems almost unbelievable that at the end of the last century, Sigmund Exner (1891) postulated the principle—applicable also to insects—of stereopsis based on binocular disparity and that this was almost ninety years before anyone else actually took up the subject. One wonders if people were reluctant to believe in such an idea. Theoretically we should assume that since the horizontal distance between insect eyes is very small, for both optical and physiological reasons, stereopsis should be possible for insects only when an object is within a few millimeters to a few centimeters. We humans have eyes that lie relatively far apart, with interpupil distances averaging 6 cm, a large binocular visual field, and small angular separations between photoreceptors. These factors endow us with stereoscopic vision to a distance of about 25 m. In actual fact, the stereoscopic area for praying mantids is probably also somewhat greater than the calculations suggest, especially when movement is involved (Maldonado et al., 1970; Collett, 1987). Rossel found that mantids view a target with several ommatidia and hence might somehow interpolate the position of the retinal image. What he meant by this is that moving prey constantly activates new ommatidia, thus permitting the position and distance of the prey to be updated constantly. Theoretically, this would provide the mantid with a more accurate signal than is to be expected from the interommatidial angle alone.

Binocular Distance Measurement Immediately before and during Walking, Jumping, or Flying

Use of Peering Movements to Measure Distances

If we watch how grasshoppers or mantids move in their natural environment or in the laboratory, we notice that they perform translatory side-to-side movements of the head or body by alternate-

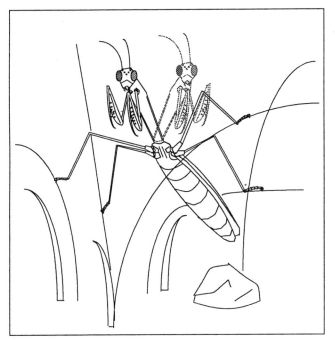
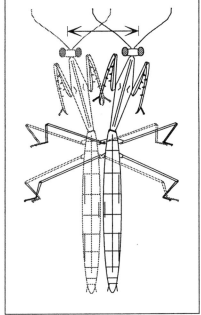

Fig. 7.6 Drawings based on snapshots of a *Mantis religiosa* in the field. The mantid is making side-to-side peering movements presumably to determine the distance to an object. Drawing courtesy of M. Poteser.

ly extending and bending their mesothoracic and metathoracic (walking) legs (Fig. 7.6). During these movements, the animal turns its head in the direction opposite to each side-to-side movement, so that the head and eyes continue to be directed forward. The distance that the horizontal movement covers and the average speed of the movement can vary considerably. Additional sideward displacement of the prothorax can considerably increase peering amplitude. It is precisely this flexibility of the prothorax that gives peering movements such a variety of forms. Mantids can turn toward an object just by moving the prothorax, without changing the position of the abdomen. The resulting peering movement is then extremely complex as far as the head-thorax-abdomen positions are concerned. In any case, however, the positions of the two pairs of walking legs remain unchanged during the peering motion.

When there are salient stationary objects in the animal's vicinity, especially contrasting vertical borders, peering movements tend to be directed toward the objects, suggesting that the peering movements could have something to do with spatial orientation and/or distance measurement. Zänkert (1939) assumed that this object-oriented peering served to gather information on the characteristics of an object. This idea should not be rejected out of hand.

Horridge (1986) was the first to look more closely at object-oriented peering movements in mantids: he studied peering behavior in *Archimantis latistyla, Tenodera australasiae,* and *Orthodera ministralis* in the field as it occurs during ambulation from one branch to the next and in the laboratory as it occurs when mantids attempt to reach an object with the legs. Even though Horridge's observations strongly suggest that peering is involved in distance measurement, they are still not direct proof given the fact that a clear and quantifiable peering-related behavior pattern was not described.

My colleagues and I characterized the sort of input-output relationship required to uncover the function of peering in the characteristic peering-jump behavior of young mantid nymphs. This was the first indication of a possible involvement of peering in the localization of and distance measurement to stationary objects by mantids (Walcher and Kral, 1994; Poteser and Kral, 1995). We placed young nymphs of *T. sinensis* on an off-center island surrounded by a round arena that had black bars painted on a white inner wall. When they were inclined to flee, the nymphs fixated on the vertical edges of the bars and very clearly peered at them. In this manner, they surveyed all the edges included in the pattern. After their surroundings had been examined completely, each decided to make a fleeing jump after one or two additional peering movements. The jump was always aimed at one of the vertical edges of the pattern and ended with a precise landing on that vertical edge. Figure 7.7A, B is taken from a video recording and shows the reconstructed sequence of movements during the comparison of two patterns at different distances that finally ended with a jump aimed at the nearer object. Only rarely did a mantid miss and land in the water. Similar arena experiments with nymphs of the South African *Polyspilota* sp. also showed that when mantids have stationary objects at different distances to peer at, they jump to the nearest one. Compared to locusts (Wallace, 1959) or crickets (Goulet et al., 1981) in similar circumstances, the mantid's ability to differentiate among distances was very good.

The distance to stationary objects can be measured relatively (i.e., with successive distance comparisons) or absolutely. A distance-measurement mechanism that works successively, however, requires that the animal be able to retain information on distance over a short period of time in order to compare the individual measurements. This means that the animal must be able to perform a content-related memory task. Apparently, a monocular mantid can perform the relative measurement task, but the absolute distance measurement always requires two fully intact eyes. This is suggested by the fact that mantids will not jump when one eye is nonfunctional or when one or both eyes are only partially functional. The reason is that determining jump distance requires information from the front, sides, and acute zones of the eyes.

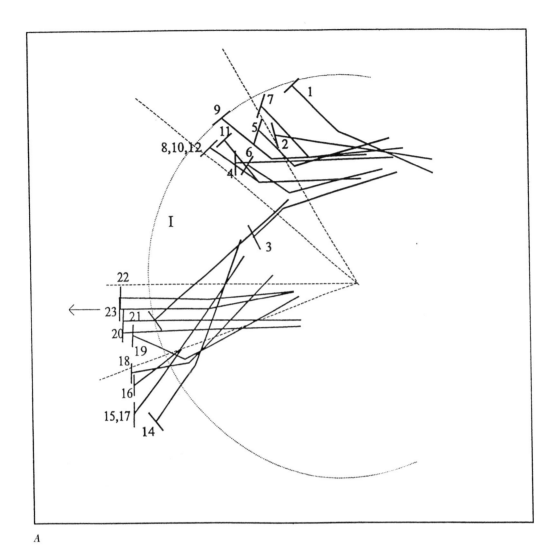

A

Fig. 7.7 *Above and right.* Sequential movements of a young *Tenodera aridifolia sinensis* nymph traced from video recordings made while it was peering at two rectangular objects at different distances. The mantid chose the closer target, jumped, and landed on its vertical edge. A, Each of these stick figures is made up of three lines which indicate the median axis of the head (short line) and the long axis of the prothorax (center line), and the long axis of the pterothorax-abdomen. The curved line is the edge of the platform. In this example, the mantid turns in succession to two objects whose edge positions are indicated by the dashed lines (the lowermost pair indicate the edges of the nearer object). The mantid first tried to get as close as possible to the more distant object (movements 1–9). This is followed by peering movements (8–12). During movements 2–4, the nearer object was already within the mantid's visual field, and it turned toward the nearer object immediately after peering at the more distant one. Again, the mantid attempted to get as close to the object as possible and leaned farther over the edge of the island that it did when viewing the more distant object. Movements 14–18 and 20–23 were peering movements directed toward the object's edges. Two distinct peering techniques could be distinguished. In movements 14–18, translation of the head was primarily due to a movement of the prothorax-abdomen joint. Movements 19–23 were caused by flexions and extensions of the legs. The jump (arrow) occurred immediately after movement 23. B, An illustration of one movement sequence. From Poteser and Kral, 1995, and used with kind permission from Company of Biologists Ltd.

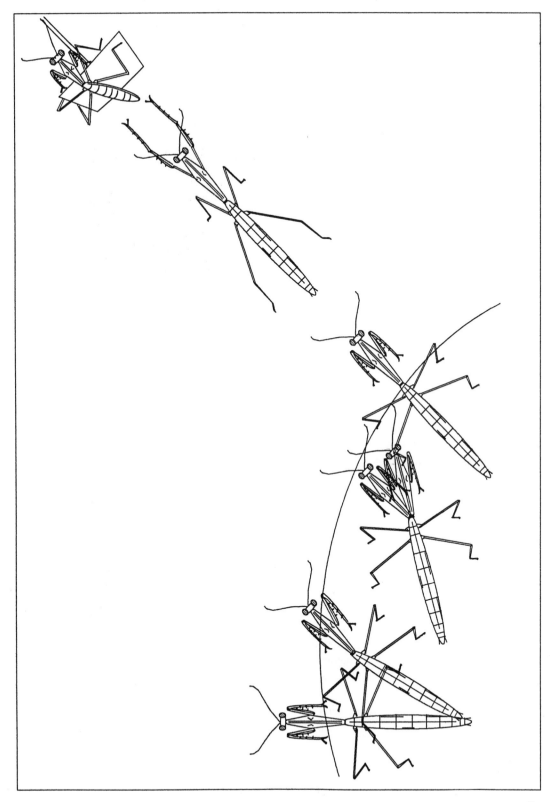

B

Collett (1978) demonstrated that, in locusts, the role of input from the lateral eye regions controls head position during peering: During side-to-side peering, there should be no image movement in the lateral eye region, if the head is held stable in the yaw plane. Experiments in which the lateral retinae were occluded confirmed this by showing that visual input from the lateral visual fields does, indeed, have the expected effect on peering behavior. Hence, it seems to be the case that the "central peering generator" is under visual control. (See analogous findings on moths in the next section.)

The peering-jump behavior of young mantids clearly indicates that object-related peering must be involved in distance measurement, but the fact that it seems to be so does not constitute proof. What could be done to test whether peering movements are directly involved in distance measurement and are not just an ancillary behavior, for instance, as a preliminary to flight? This question can be answered by using a relatively simple movement trick to fool a peering mantid. Wallace (1959) and more recently Sobel (1990) played this trick on the locusts *Schistocera gregaria* and *S. americana*: When an object is moved from side to side in antiphase to the insect's peering movements, locusts underestimate the distance to the object. When the object is moved in the same direction as the peering movements, however, distance to the object is overestimated. This is because, under normal conditions, the nearer the object, the greater the retinal image movement. So, when an experimenter moves an object against the animal's own movement, the retinal image shift is increased. When the object is moved with the animal's movements, the image shift is decreased. Sobel (1990) used the take-off speed of the locusts as a measure of this misestimation. This was the first demonstration that insects, too, can process retinal image movement—generated in this case by peering movements—to measure distance to an object.

We played essentially the same trick on mantids (Poteser and Kral, 1995). Mantids were offered an attractive target (an upright black square) in front of a white background either within the optimal jumping range (but outside

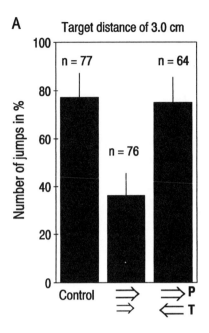

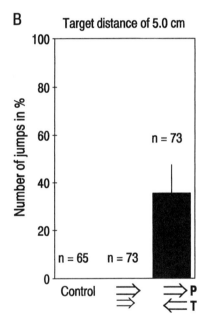

Fig. 7.8 Illustration of the willingness of *Tenodera aridifolia sinensis* nymphs to jump after object-directed peering when the object was stationary (control), when the object was moved in the same direction as the peering movements (matched arrows), or when the object was moved in antiphase to the peering movements (reversed arrows), and when the objects were (*A*) within the optimal jump range and (*B*) beyond the optimal jump range. From Poteser and Kral, 1995, and used with kind permission from Company of Biologists Ltd.

the optimal stereoscopic visual field) or outside the jumping range. As we expected, the animals jumped at the target only under the first set of conditions. If the object was then moved in phase with the peering movements but at a slightly slower speed, the tendency to jump decreased by more than half, and the animals over-jumped. If the object was moved in antiphase to the peering direction but at the same speed as peering, normal jumping tendency was retained, but the jumps fell short (Fig. 7.8A). With their relatively long legs, the mantids usually just managed to reach the edge of the object but usually could not hold on and fell into the water. As expected, the animals did not jump at an object offered outside their jumping range, either when it remained stationary or when it was moved in phase with the peering movements. When a distant object was moved in antiphase to the peering movements, however, the mantids could usually be induced to jump; willingness to jump was almost 40% in this case. In these instances, the jumps were much too short, and the animals landed in the water without reaching the object (Fig. 7.8B). The latter findings agree with our field observations on *M. religiosa* nymphs and adults. These mantids often peered at salient twigs that were out of their reach. If a sudden breeze happened to move the twig in the counter-peering direction, the mantids would often try to reach for it with their forelegs.

It is our conviction that this movement trick has provided direct proof that, like locusts, praying mantids use retinal image movement as a distance measurement. (Prete et al. [1993, Experiment 4] also found this to be the case.) To use retinal image movement during peering as a measure of object distance, mantids can use any of three strategies: (1) With uniform peering movements, retinal image movement (amplitude and speed) will change consistently with object distance; (2) if peering movements change depending on the distance of the object, retinal image movement will be the same for every distance; and (3) if peering speed is held constant and peering amplitude changes depending on object distance, retinal image speed in conjunction with the mantid's own movements will determine object distance.

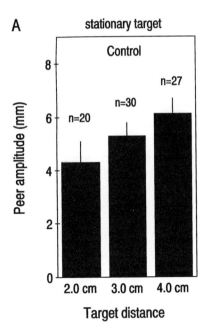

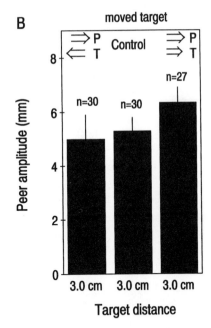

Fig. 7.9 Average peering amplitude measured immediately before a jump. *A*, Stationary objects at three distances, and (*B*) objects 3 cm away that moved in the same direction as the peering movements (matched arrows) or objects moved in antiphase to the peering movements (reversed arrows). After Poteser and Kral, 1995, and used with kind permission from Company of Biologists Ltd.

There is good experimental support for the last of these three strategies. There seems to be a distinct correlation between peering amplitude (but not peering speed) and an object's distance (Fig. 7.9A). On the basis of the following findings, one might go so far as to say that distance-dependent retinal image speed controls peering amplitude. That is, peering amplitude changes not only with actual, but also with apparent, object distance. Figure 7.9B shows this clearly: Object countermovement, which creates the illusion of nearness, leads to a decrease in peering amplitude, but object movement, which creates the illusion of distance, increases peering amplitude. So it seems that when mantids measure distances, they not only consider retinal image movement but also their own movement. The role of self-movement is apparent in the observation that mantids never jump at a moving object unless they, too, are in active motion (Poteser and Kral, 1995; Kral and Poteser, 1997).

We are currently pursuing experiments related to this latter hypothesis, and our preliminary findings confirm that a mantid's own movement is indeed included in distance measurement via retinal image shift. The underlying mechanism probably relies on a complex regulatory system that seems to involve Erich von Holst's reafference principle (von Holst, 1969, 135; Mittelstaedt, 1971), but we will only really understand the visually controlled peering mechanism when we are thoroughly familiar with the pertinent sensorimotor couplings and their hierarchical arrangements. Certainly, information from proprioceptive systems, such as from the neck hairs (Mittelstaedt, 1957; Liske and Mohren, 1984), will also have to be taken into account.

The findings described above allow us to assume that peering mantids use motion parallax to determine the distance to a stationary object. The principle of motion parallax, which Helmholtz discussed in the second half of the nineteenth century (1866), is fairly simple, and we are continuously confronted with it, usually unconsciously. When an observer moves, the retinal images of nearer objects shift farther than the images of more distant objects. The changes in image speed and/or distance moved (the parallaxes) indicate relative object distances. The distance between an object in the foreground and one in the background can be determined from the relationship between the parallaxes (Cornilleau-Pérès and Gielen, 1996). The question arises as to whether mantids, and insects in general, could use parallax differences in distance discrimination and figure-ground discrimination.

To answer this question, my student Michael Poteser did the following experiment. *Polyspilota* sp. nymphs were offered black rectangular objects in front of either a black-and-white–striped or an unstructured background. The contrast, brightness, and contours of the objects in the foreground and background were so similar that Michael himself had trouble telling them apart and could do so only when he moved his head back and forth. The mantids, too, seemed to be able to distinguish the foreground objects only when they performed peering movements. When they did, however, they were able to distinguish between foreground objects that were several centimeters away even when there was less than 3 mm between the objects. However, since the mantids always chose the nearest object, the experiment failed to demonstrate that *relative* motion parallax was being used: The mantid could simply have been recognizing the nearest object as the one with the largest or fastest retinal image shift. What we do not know is if, under different conditions, a mantid might choose to jump to a more attractive distant object rather than a less attractive nearby object. For instance, both in the laboratory and in the field, we were able to show that there is a distinct preference for an upright object among objects with different orientations. In its natural environment, it may be very important that a praying mantis moving from one perch to another recognize the best and most safely reachable target rather than the closest target.

In addition to side-to-side peering movements, mantids also perform other (less impressive) characteristic motions that may also be involved in visual orientation. These are considered in the following section.

Use of Back-and-Forth Rocking Movements to Provide Visual Information

In the field we often see flying *M. religiosa* males land on an exposed, upward-pointing twig or on the tips of blades of grass after which they usually turn and assume a hanging position. Then—and this is also true of females—they creep very slowly down into the protective grass. Calculations based on video recordings of this slow walking behavior show an average walking speed of 2 to 3 cm/s. During their downward trek, the mantids display forward-backward rocking movements, the forward component of which is approximately three times the amplitude of the backward component (e.g., 1.5 mm versus 0.5 mm; Fig. 7.10). Under field conditions, forward movements last between 120 and 500 ms. We noticed that the back-and-forth movements had greater amplitudes (>5 mm) and were faster (<200 ms) and that overall the mantids made faster progress the more strongly the perch moved in the wind. There are no side-to-side movements during the back-and-forth movement, but the mantid will stop and make peering movements from time to time, especially when new structures enter its visual field. Also noteworthy is the fact that while moving forward, it often waves its forelegs in the air as if it wants to check the distance to another perch. This behavior is reminiscent of the way birds such as pigeons use their outstretched legs to provide orientation when landing. In both of these cases, the legs are within the binocular visual field, and in

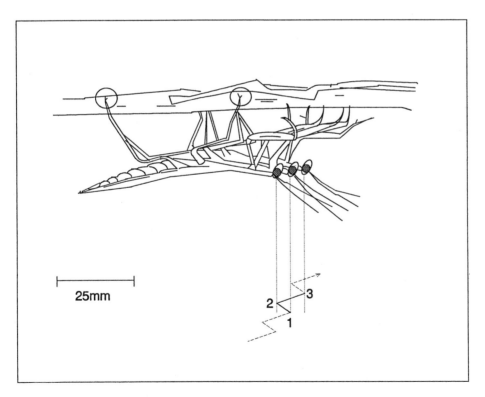

Fig. 7.10 Composite drawing based on video recordings made in a natural environment of a *Mantis religiosa* climbing along a twig while performing typical back-and-forth rocking movements. The forelegs are stretched forward in a characteristic way, as if the animal were using them as a measuring instrument.

the case of *M. religiosa*, the antennae are stretched forward (Kral, unpublished observations).

It seems that mantids use their back-and-forth rocking movements not only as camouflage in the form of wind mimicry or movement mimicry (Rupprecht, 1971; Edmunds and Brunner, this volume, chap. 13) or as a threatening gesture to scare off enemies like birds or sphegidae (Meissner, 1909), but also as a means of obtaining important visual information about climbing routes, to detect new perches, and to avoid obstacles. This may also be the case for rocking movements performed by other insects, such as saturiid moths (Collett, 1965) or stick insects (Pflüger, 1977). In the Automeris moths, side-to-side rocking movements seem to be under precise bimodal (visual and proprioceptive) control. Unfortunately, to my knowledge no further studies on this topic have been made on moths. In the case of stick insects, forward or side-to-side rocking movements may play a role in the localization of predators.

In our efforts to determine the functional significance of both peering and back-and-forth rocking movements, I believe that it must be borne in mind that these behaviors occur to very different degrees in the various mantid species. For example, *T. a. sinensis*, *Empusa fasciata*, and *Polyspilota* sp. make very vigorous peering movements, often in series, at the end of a locomotor sequence. In contrast, *M. religiosa* usually makes rather simple peering movements that are nearly integrated into the locomotor movement sequence, and some mantids apparently do not peer at all. This makes one wonder about the visual factors that may have contributed to the evolution of these visually controlled movement sequences derived, perhaps, from locomotor movement patterns.

Visual Information for Flight and Landing from the Mantid's Own Movement

Mantids are not, generally speaking, strong flyers (but see Brackenbury, this volume, chap. 9). Degeneration up to a complete loss of the wings occurs rather commonly in these sedentary insects, and females are more likely to be affected (Yager, this volume, chap. 6). Both sexes of the species discussed in this chapter have well-developed wings, but only the males fly. Due to their heavy abdomen, the females can only flutter briefly, but they do use their wings for threatening gestures (Edmunds and Brunner, this volume, chap. 13).

During our field work in the Slovenian Karst we saw that *M. religiosa* males flushed from grass about 0.5 m high would fly for a few to more than 20 m, but only in strong sunlight and with an air temperature over 30°C. Their fluttering flight, characterized by a wing beat frequency of about 20 Hz, lasted several seconds and could reach heights as much as several meters above the grass tips. These escape flights could be straight, curved, or looped so that the mantids ended at their starting point.

Video recordings showed that when *M. religiosa* males sit upright on a blade of grass, they take off either by turning to the side and bending forward or by retaining an upright posture and taking off backward. In the latter case, they make a loop after taking off, so they are first in a forward-slanting and then in a horizontal flight posture. Prior to landing, they reduce their speed and return to a vertical orientation, so that they land head up. The preferred landing site in wind-blown grass is a vertical perch such as a blade of grass or a stem. Interestingly, these mantids always land with assurance, even if the perch is moving in the breeze. Presumably, both the landing and the flight are visually controlled by movement-induced retinal image flow.

Opposite page

Fig. 7.11 Series of drawings based on a video recording of a flying male of *Mantis religiosa* in the field just before landing on a blade of grass. Drawing courtesy of M. Poteser.

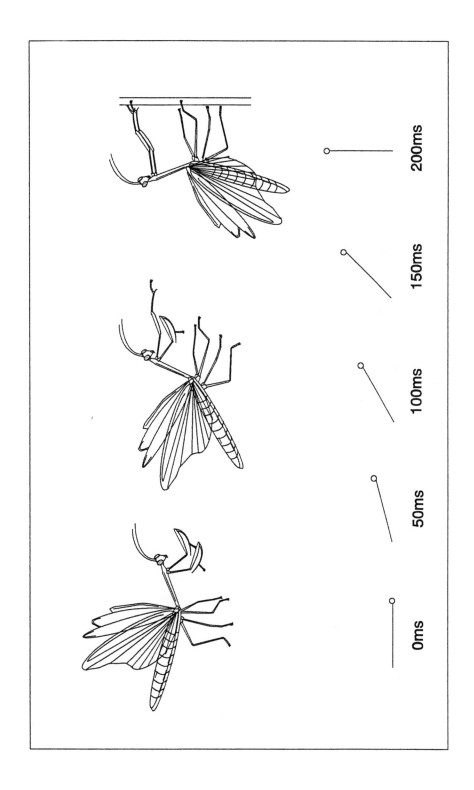

Quantitative studies of free-flying mantids in the field are difficult because the observer cannot predict either the course of flight or the landing site. In *M. religiosa* males, the beginning of the change from a horizontal flight posture to a distinctly erect prelanding body posture was seen to be related to the first localization of the landing site (Fig. 7.11). During the 120 to 200 ms between beginning of the discrete landing reaction and the landing per se, a number of optical parameters could be assessed by the mantid, including the properties of the landing site such as its distance, the retinal speed of its image, and the increase in the size of its image. Any of these cues could elicit the landing reaction. The change to an erect posture and simultaneous reduction of flight speed that precedes the landing cause a downward motion of the retinal flow field that is counteracted to some degree by a downward movement of the head. It is conceivable that these postural changes could go so far as to eliminate expansion of the approaching site's image so that a more or less static retinal image develops just prior to landing. Any lateral movements of the head and/or the body of the mantid during the approach could produce motion parallaxes that could play a decisive role in perch localization and obstacle avoidance. Our video recordings indicate that the landing reaction begins at a distance at which the landing site is already in the binocular visual field (i.e., at just a few centimeters). Further video analyses will be necessary to determine just how the mantis manages these landing maneuvers so well.

The Influence of Development and Experience on Binocular Distance Measurement

Rilling et al. (1959) compared normal mantids raised on flies with mantids that were only hand fed to demonstrate that, as adults, the former were more successful at capturing prey than the latter. This experiment led the authors to formulate the reiterative learning hypothesis, which assumes that after every molt, the correct recognition of ideal strike distance has to be relearned due to changes in the animal's morphology such as longer forelegs and larger eyes (Przibram, 1930). Obviously, this requires experience.

In contrast, Maldonado et al. (1974) offered a hypothesis based on the optical triangulation mechanism (as explained above) and believed that, after every molt, nymphs of *Stagmatoptera biocellata* automatically adapted their estimation of prey distance to their new morphological situation. Here, no experience is necessary. As the authors themselves put it, "This would mean that the same ommatidium or groups of ommatidia, connected to the same integrative area of the central nervous system through an unchangeable wiring (but that is not true), tell the brain an object is at maximum catching distance, irrespective of the instars and the different body dimensions" (Maldonado et al., 1974, 603). In support of this hypothesis, Balderrama and Maldonado (1973) showed that mantids can estimate the distance to prey equally well in all nymphal stages. Their work also showed, however, that young nymphs can capture prey that is well beyond the length of the striking forelegs but older juveniles and adults cannot (but see Prete and Hamilton, this volume, chap. 10). According to the authors, in contrast to older mantids, the younger nymphs could lunge much farther (relatively speaking) during the strike and so increase the functional range of the raptorial forelegs. Of course, this implies that in terms of prey localization and estimation of prey distance, the visual "measuring area" is less limited in young mantids than in older ones. However, Rossel et al. (personal communication) were able recently to find just the opposite in a large series of tests on free-hanging nymphs and adult animals. He found that adult females could, indeed, lunge quite far forward during the strike and capture prey at a considerable distance (see also Prete and Cleal, 1996). The possibility cannot, however, be excluded here that, in adults, the greatest strike frequency correlates with relatively short distances.

Thus, there appear to be some indications that in the course of postembryonic development, which can last for several months and cover six to ten nymphal stages, changes occur in prey capture behavior and by implication in depth perception as well. This developmental issue was tackled in my laboratory by Köck et al. (1993), who studied prey capture behavior in both binocular and

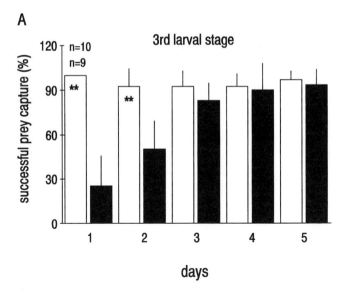

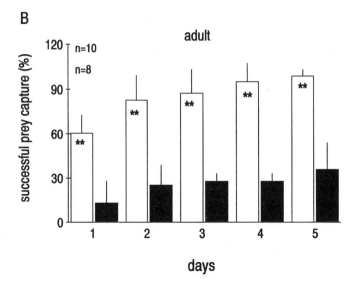

Fig. 7.12 Depiction of the average prey-catching efficiency of binocular (white bars) and monocular (black bars) (*A*) third larval instar and (*B*) adult *Tenodera aridifolia sinensis* as a function of the number of days spent in their respective instars. Significance levels (t-test): ** = $p \leq .005$. Modified from Köck et al., 1993, and used with kind permission from Elsevier Science Ltd.

monocular *T. a. sinensis* throughout their nymphal development and adulthood. Behavioral observations revealed that, in all age groups, monocular animals were at a disadvantage in prey capturing compared to binocular controls for the first few days after visual deprivation. Specifically, they let prey approach more closely prior to the strike and consequently overestimated prey distance, which reduced the functional strike range by more than 50% and decreased strike efficiency by as much as 80%. Young mantids, however, were able to compensate completely for monocular deprivation in just a few days, after which strike distances were the same as for binocular controls. Older nymphs and adult monocular animals were not able to adapt, however, and continued to overestimate strike distance (Fig. 7.12A, B). Hence, the latter absolutely require use of both eyes for correct estimation of distances.

Considering the findings for *T. a. sinensis* and *S. biocellata* (Balderrama and Maldonado, 1973), my colleagues and I believe that the use of binocular cues for prey distance determination is critical in early life but not absolutely essential. In later life, only the binocular mechanism is available for distance measurement. This assumption is supported by the fact that some binocular characteristics only appear or are optimized in the course of postembryonic development (Köck, 1992; Köck et al., 1993; Leitinger, 1994; Leitinger et al., 1994). For instance, newly hatched *T. a. sinensis* were found to have a horizontal binocular overlap of 48.5°, but adults have an overlap of 73.6°, meaning that the horizontal limit of the binocular visual field (i.e., the region for depth perception) increases by more than half as the mantid matures. Further, in the second nymphal instar, scanning electron microscopy reveals no frontal specialization on the surface of the eye in the form of distinctly enlarged facets. This characteristic does not appear until the third instar and becomes more pronounced in the course of development (see Table 7.1). Similarly, the original spherical shape of the eye in *T. a. sinensis* evolves progressively into a heterogeneous form with a flattened frontal area. Longitudinal sections through the entire eye show that, in the second nymphal instar, the ommatidia in the frontal region have already become about one-third longer than in the lateral region. This is made possible by a concave indentation in the basal membrane, but the difference in length only becomes visibly greater with increasing age, almost doubling by the time the adult stage has been reached (Leitinger, 1994). Thus in early nymphal stages the interommatidial angle ($\Delta\phi$), calculated from the facet diameter (D) and the radius (R) of the eye ($\Delta\phi = D/R$), seems to be both fairly constant and relatively large. Only in the course of postembryonic development does the frontal interommatidial angle decrease. Leitinger (1994) found that in the second nymphal stage the interommatidial angle was 2.4° in the frontal region and 2.5° in the lateral region of the eye. These values were 2.1° and 3.0° in the fourth nymphal stage, but 0.6° and 2.0° in the adult stage. In contrast, Sabine Martin, a former student of S. Rossel (personal communication) used the especially large mantid species *Hierodula mebranacea* to show that, when the divergence angles of the ommatidia are

Table 7.1 Postembryonic development of facet pattern in the compound eye of *Tenodera aridifolia sinensis*

Age	Dorsal (μm)	Frontal (μm)	Frontal/medial (μm)	Lateral (μm)	Ventral (μm)
2d	12.0 ± 0.0[a]	20.0 ± 0.0	20.0 ± 0.0	12.0 ± 0.0	14.0 ± 0.0
3d	14.1 ± 0.1	31.2 ± 0.8	21.2 ± 0.8	14.0 ± 0.0	15.1 ± 0.8
4th	14.2 ± 0.2	31.2 ± 2.0	21.0 ± 0.0	17.0 ± 3.0	14.0 ± 0.0
5th	18.0 ± 1.6	34.7 ± 3.4	23.3 ± 2.5	19.3 ± 0.9	19.0 ± 1.0
7th	23.3 ± 2.4	50.0 ± 0.0	31.2 ± 2.2	25.0 ± 0.0	23.7 ± 2.2
Adult	31.2 ± 2.2	70.0 ± 0.0	38.7 ± 2.2	36.2 ± 2.2	32.5 ± 2.5

Source: From Köck et al., 1993, with kind permission from Elsevier Science Ltd.
[a] means with standard errors; $n = 5$.

used as an indicator, the acute zone appears to be completely developed from hatching onward (the divergence angle seems to be only slightly larger than in adults).

Our results, however, lead us to another, more far-reaching, question: Is the postembryonic development of binocular characteristics such as an acute zone, a horizontal binocular visual field, and the underlying mechanisms for depth perception completely preprogrammed, or are these characteristics dependent upon visual experience? If the latter were the case, one would assume, given the ontogenetic behavioral findings on *T. a. sinensis* (Köck et al., 1993; Walcher and Kral, 1994), that the critical developmental stages may cover the first four to five of the seven or eight nymphal instars. We know that, in some insects, operations such as monocular light deprivation performed during the first days after emergence can lead to severe irreversible changes in visually controlled behavior (e.g., fly visual pattern discrimination [Mimura, 1986, 1987]; fly visually guided choice behavior [Hirsch et al., 1990]). Further, it was recently shown in crickets that dark rearing during the postembryonic developmental phase has an effect on eye development, leading to an increase in the number of ommatidia, a decrease in interommatidial angles (e.g., frontal $\Delta\phi = 2.47°$ versus $4.04°$; ventral $1.16°$ versus $2.54°$), and changes in the visual field (Deruntz et al., 1994). This surprising work on crickets shows that, in insects, visual experience might be essential for the development of normal binocular vision. One might well expect that what is true for crickets is also true for praying mantids.

Assuming that behavioral adjustments are necessary due to the migration of the ommatidia during development, Mathis et al. (1992) studied *Hierodula mebranacea* by depriving one eye of light throughout the entire postembryonic period and then examining visual fixation and binocular distance estimation in the adults. After removal of the lacquer from the surface of the deprived eye, the experimental mantids were found to behave the same when capturing prey as binocular controls. In another manipulation, the distance to the prey object was optically changed with prisms

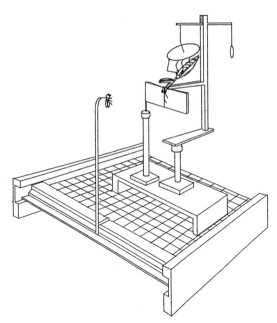

Fig. 7.13 Schematic diagram of the set-up used by Samuel Rossel to demonstrate binocular stereopsis in mantids. As was the case with Mittelstaedt's work (1957), a praying mantis (*Tenodera australasiae* or *Mantis religiosa*) is fixed by the prothorax but otherwise has relative freedom of movement. The insect is offered a live blowfly at a certain distance (about 200 mm) and in a certain position. As soon as the mantid visually fixates the target, the motor of an *x/y* recorder is used to move the target slowly in the direction of the mantis until it strikes. A pair of prisms are mounted in front of the mantid's head such that their thicker (base) surfaces face laterally. By increasing the retinal disparity without changing the image size, the prisms cause the mantid to misestimate the distance to the target. The capture distance can be read from a video monitor and compared to the interocular distance to determine the binocular disparity. Modified from Rossel, 1983, and used with the kind permission of Macmillian Magazines Ltd.

to exclude a possible monocular distance-measuring mechanism (Fig. 7.13). Based on this experiment, the authors concluded that binocular spatial vision in this mantis is by no means dependent on experience. The results do not, however, exclude the possibility that longer-term deprivation might have some effect.

Finally, I would like to mention that my colleagues and I have also done some research on *Mantispa styriaca*, an insect that bears a striking resemblance to the praying mantis but belongs to the lacewing family (Neuroptera; Fig. 7.14). The adults are 6 to 12 mm long and, unlike mantids, are holometabolous, developing from an egg via a free, and then a parasitic, nymphal stage and a pupal stage (Brauer, 1869; Kral, 1989).

Adult *M. styriaca* demonstrate prey capture behavior like that of a young mantid nymph: They are sedentary ambush predators that turn and fixate potential prey with both eyes and then strike with both forelegs. This being the case, we might assume that *M. styriaca* uses a visually controlled mechanism like that of the praying mantis for capturing prey. We explored this possibility by measuring the binocular visual field (Fig. 7.15) and the ommatidial parameters of *M. styriaca* and by conducting several histological studies (Eggenreich and Kral, 1990; Kral, 1990; Kral et al., 1990; Mayer and Kral, 1993). Surprisingly, *Mantispa* showed good agreement in many respects with the situation in young mantid nymphs. For instance, we found an approximately equal horizontal overlap of 40 to 50° and about the same diameter of the facets set in semispherical eyes so that there was no recognizable frontal zone. We also found an effective ommatidial acceptance angle ($\Delta\rho$) of approximately 2° under light adaptation and an interommatidial angle ($\Delta\phi$) between 1.8 and 2.3°. In contrast to mantids, however, longitudinal sections through the eye showed that it is not an apposition eye (Horridge, 1980) but rather a superposition eye (i.e., with a small superposition pupil) that is adapted to daylight vision (Kral, 1990; Kral et al., 1990).

This short excursion into the mantispids shows that the ability to estimate distances accurately enough for successful prey capture is not dependent on the existence of a specially developed acute zone in the frontal eye regions. However, like young mantid nymphs, *Mantispa* captures prey that creeps by rather slowly whereas adult mantids like *T. a. sinensis* are capable of snatching an extremely fast-flying blowfly or hornet from the air. In the latter case, an unsuccessful strike could prove fatal for the predator!

Directions for Future Research

If the praying mantids do indeed use disparities between the two retinal images in assessing object distance, then there ought to be evidence of it in the brain in terms of visually driven neurons that are stimulated by both eyes. If binocular neurons can be identified (Mathis and Rossel, 1993), then their receptive fields and the shape and orientation of their receptive fields must be determined. The next question that would have to be answered is how the differences in retinal images are recognized by binocular neurons and how the retinal disparities are used to control

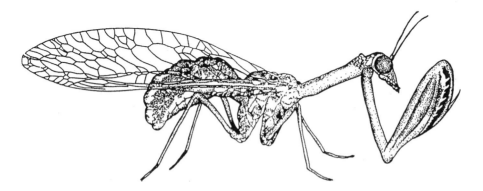

Fig. 7.14 Drawing of a female *Mantispa styriaca* (Neuroptera) caught in a pine grove on the southern tip of the Istrian peninsula. Drawing courtesy of W. Draxler; from Kral et al., 1990, and used with the kind permission of Gustav Fischer Verlag.

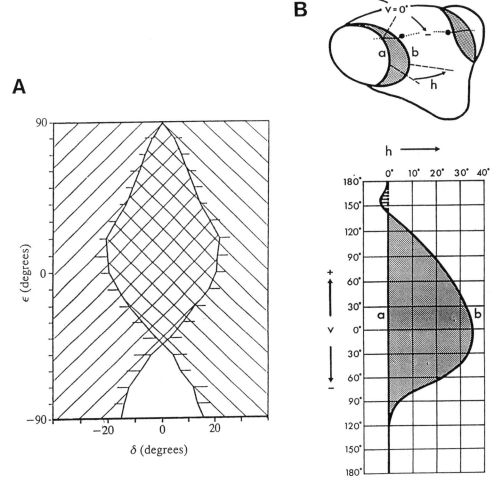

Fig. 7.15 Graphic indication of the binocular overlap in the visual field of (A) *Mantispa styriaca* and (B) *Mantis religiosa*. The visual field was measured with a pseudopupil in both cases. The cross-hatching in A and the strip stripes in B indicate the locations of the binocular areas of each of the visual field. The symbol (abscissa in A) is the angle between the optical axis of a marginal ommatidium and the vertical symmetry plane. The symbol (ordinate in A) is the angle of the same axis to the horizontal symmetry plane. The equator is = 0°. In B, the shaded area in the graph indicates the inward-pointing ommatidia of the right eye. Doubling this area reveals the binocular field of view, which has a maximum horizontal extension of about 70°. Abbreviations: h - horizontal; v - vertical; a - angular deviation between the optical axis of the forward-looking ommatidium; b - the optical axis of the ommatidium at the inner edge of the eye. A is from Eggenreich and Kral, 1990, and is used with the kind permission of Company of Biologists Ltd.; B is from Rossel, 1979, and is used with the kind permission of Springer-Verlag.

movement of the forelegs. Experimental findings could then be used to create a biologically relevant mathematical model of the underlying neuronal network. A clear and quantifiable model of this sort could contribute greatly to a better understanding of the binocular mechanisms of prey localization and of stereopsis in general.

There are also a number of interesting questions to be asked concerning stereopsis itself: Did stereoscopic vision develop in the course of evolution only in predatory insects? Is it limited to mantids? Is the acute zone a prerequisite? How does a visual system with stereopsis differ from one that cannot process stereoscopic information? Is binocular coupling the special trademark of stereoscopic vision?

For praying mantids, motion parallaxes are probably an important, if not the most important, source of information on spatial depth perception. This, too, could be a major research area. The first step would be to identify visual neurons that are clearly related to peering movement and that react selectively to motion to the right or left and/or up and down. If the topography and stimulus-response behavior of these neurons were known, then the neural network could be identified that receives information on motion parallax, processes it, and passes it on to the motor neurons. Here again, the logical goal would ultimately be a biologically relevant model of the underlying network.

One additional potentially rich area for research revolves around the role of multisensory integration in the control of mantid behavior, in particular the connections between the visual and proprioceptive systems, which cooperate to determine the position of the eyes in space. Learning, too, may affect such sensory integration. This may also be an important factor in learning behavior. There is still a great deal of work to be done in these areas, and a truly satisfactory understanding of these questions can be achieved only through close cooperation among specific working groups.

ACKNOWLEDGMENTS

I thank Reinhard Ehrmann (Entomological Institute, Goch, Germany) for classifying mantid species and Eugenia Lamont for translating the text into English. I am also grateful to the colleagues in our laboratory for valuable and stimulating comments. This book chapter was supported by Austrian Science Foundation Grant P09510-BIO.

8. Prey Recognition

Frederick R. Prete

When I took my first graduate school course in sensation and perception in the mid-1980s, I learned that anuran amphibians (i.e., frogs and toads) had "fly detectors" in their retinae. As it was explained to me, these amphibians had such small brains that they simply did not have the necessary neural machinery to do any complex processing of visual information. Consequently, their retinal ganglion cells were "wired up" in such a way that they automatically responded to any small, moving black spot—that is, the putative silhouette of a fly. In turn, the responses of these "fly detector" retinal ganglion cells automatically led to a reflexive flick of the amphibian's tongue. "Doesn't it make perfect sense," my professor concluded, "that nature would have equipped these creatures with a mechanism whereby they respond immediately and reflexively to their prey?" Well, it did make sense to me. Later, however, I found out that the story wasn't true.

The lesson that I had been taught about anurans and their fly-detecting retinae was, by the time that I heard it, long outdated but still widely believed. Originally, the idea was based on several well-known articles, written in the 1950s, that argued this: "[In frog's], 'on-off' units [class R3 retinal ganglion cells] seem to possess the whole of the discriminatory mechanism needed to account for this rather *simple* behavior [of prey capture], . . . and it is difficult to avoid the conclusion that the 'on-off' units are matched to this stimulus image (image of a fly at 2 in. distance) and act as 'fly detectors'" (Barlow [1953, 86], cited in Ewert [1987, 356], emphasis added; also see Lettvin et al., 1959, 1961; Maturana et al., 1960). The so-called higher visual centers in the frog's brain were thought to be involved only in the transformation of the incoming visual information from the "fly detectors" into the appropriate motor output.

What makes the fly-detector model so seductive, of course, is that it is clear, simple, accessible to the nonspecialist, easy to remember, and powerful—it can explain the whole of prey recognition in an uncomplicated sentence or two. Ironically, however, it is quite inconsistent with the real-world behaviors of anurans. That is, anurans do not live exclusively on flies or fly-sized prey. They eat a variety of items that simply cannot all be described as fly-sized dots at a distance of two inches. For instance, J-. P. Ewert's elegant review of his own work on the neuroethology of prey recognition in toads includes a striking photograph of the toad *Bufo bufo spinosus* prepared to snap at a very large (and clearly nonflylike) earthworm (Ewert, 1987, 338; for analogous work on frogs, see, e.g., Ingle, 1976). In reality, for a variety of anurans, the defining characteristics of prey (versus nonprey) are based on "a schema [rather than an invariant set of stimulus parameters] which includes various prey objects within the limits of the defining spatiotemporal features" (Ewert, 1989, 59). In other words, within certain size limits, if a moving object is compact (like a small square or disc) *or* is elongated parallel to its direction of movement (a so-called worm configuration), anurans recognize it as prey. If, however, the object is too large *or* it is elongated perpendicular to its direction of movement (an "antiworm" configuration), anurans do not recognize the object as prey. This means, of course, that the same elongated object (e.g., a 5 × 20 mm rectangle) will be recognized as prey if it is moving parallel to its long axis, but ignored if it is rotated 90° (Ewert, 1987, 1989).

So, what does any of this have to do with mantids? Well, in my opinion, the history of the study of prey recognition by mantids parallels precisely the history of the study of prey recognition by anuran amphibians. It is frequently argued that mantids recognize small flylike objects as prey and invariably release a predatory strike when such an object moves to the appropriate position in the center of the mantid's visual field (see Prete and Wolfe, 1992; Prete, 1995, for reviews of this issue). The popular wisdom has been that, like toads of old, mantids simply do not have the neural machinery to do anything more complicated than automatically release a predatory strike when a fly-sized dot moves into just the right spot in their visual field.

Like the original anuran story, this explanation of mantid prey capture is clear, simple, and seductive but, unfortunately, inconsistent with the actual predatory behavior of mantids. To make matters worse, the idea that a predatory strike is automatically released when a small spot enters the appropriate area of a mantid's visual field has led to considerable confusion in the literature between the terms *prey recognition*, on the one hand, and *distance estimation, visual orientation,* and even *predatory strike,* on the other. For instance, papers purporting to explore prey recognition sometimes only describe experiments that demonstrate at what distances small black dots will elicit predatory strikes, and some assume that simply orienting toward a stimulus means that the stimulus has been recognized as a potential meal. And, sometimes, the predatory strike is assumed to be a necessary and sufficient indicator that prey has been recognized—as if there are no modulating factors that might influence the mantid's behavior.

As an aside, I think that it is interesting that the confusion between prey recognition and prey capture has its complement in the amphibian literature: It was once argued that the prey-capture motor patterns of anurans were actually essential components of prey recognition by these creatures (e.g., Grüsser and Grüsser-Cornehls, 1970; Grüsser-Cornehls, 1976, cited in Ewert, 1987)!

What Does *Prey Recognition* Mean, Anyway?

If I were to ask you what I should do first if I wanted to study prey recognition in an animal about which I knew little or nothing, what would you say? I think that I am safe in guessing that the first thing would be, "Find out what the animal eats!" Certainly, you cannot explain prey recognition until you know what is being recognized as prey, In other words, what group of items constitute the perceptual category "prey" for the animal in question. Once the category is known, it is a relatively straightforward (although not a trivial) task to test specific hypotheses regarding the stimulus parameters that define prey from the point of view of the predator. So, then, what *do* mantids eat?

It has been known for millennia in Asia and (at least) for centuries in Europe that mantids are opportunistic predators that prey primarily but not exclusively on arthropods and their larvae (Prete and Wolfe, 1992). Certainly, the best source of quantitative data on the creatures that mantids normally eat is the ecological literature, especially that of Hurd and his colleagues (e.g., Bartley, 1983; Fagan and Hurd, 1994; Hurd and Eisenberg, 1984a, b, 1990; Moran and Hurd, 1994; also see Barrows, 1984; Hadden, 1927; Paradise and Stamp, 1993; Rau and Rau, 1913; Hurd, this volume, chap. 3). Interestingly, however, there is also a rich but often ignored body of natural history writings and some anecdotal reports that are consistent with, and complementary to, the ecological literature. All together, this information teaches us several important things about mantid prey recognition. First, mantids are *generalized* predators. Second, under natural conditions, the particular mix of prey items that any mantid eats is limited only by the rates at which prey items are encountered and successfully captured (Bartley, 1983). Third, the mix of prey items encountered by a mantid could include arthropods representing dozens of families (e.g., Bartley, 1983), including creatures as diverse as beetles, bees, and wasps; crickets and grasshoppers; caterpillars; and butterflies. In addition to these relatively smaller creatures, however, mantids may also capture very large prey such as same-sized conspecifics and even newts, lizards, frogs, small birds, small turtles, and mice (e.g., Edmunds, 1972; Johnson, 1976; Nickle, 1981; Ridpath, 1977; Tulk, 1844; for reviews see Kevan, 1985; Prete and Wolfe, 1992).

Now think for a moment about the variety of visual images created by these prey items. Given mantids' eclectic tastes, it seems impossible to me that they are identifying prey by some simple matching-to-template procedure (e.g., "the object looks like a fly at 20 mm"). Certainly, butterflies, grasshoppers, other mantids, and mice do not look like flies—in fact, it is hard to imagine at first just what they do have in common. So, the idea, now somewhat widely held, that prey recognition and prey capture by mantids simply involve the release of a strike when a fly-sized spot appears in just the right areas of a mantid's visual field seems untenable.

Two Early Experiments

In spite of the facts that people have known for at least a thousand years that mantids will capture and eat same-sized conspecifics and that published reports of mantids eating small vertebrates began appearing in the mid-nineteenth century, the first and, perhaps, most influential systematic study of mantid prey recognition did not take into consideration either of these pieces of information. In fact, publication of this frequently cited paper (Rilling, Mittelstaedt, and Roeder, 1959) was, apparently, the consequence of the junior author's persistent eagerness to demonstrate the effectiveness of the visual stimuli that she had designed (Prete and Wolfe, 1992). The paper lures of which Rilling was so proud consisted of a variety of small ovals or rectangles, some with legs or wings affixed, that were dangled, swung, and/or twirled in front of tethered *Parastagmatoptera unipunctata* and *Hierodulae* sp.

There are two intriguing aspects to Rilling's paper. The first is that Rilling constructed no stimuli that were other than flylike. That is, all were small ovals or rectangles with legs or wings affixed, and they were dangled in the air. None resembled or moved like other mantids, or caterpillars, or butterflies, or any other elongated creatures that might cross a mantid's path. Evidently, the underlying assumption, and the inevitable conclusion to which Rilling came, was that the releasing stimulus for mantid prey capture is a flylike stimulus. Specifically, Rilling and her colleagues concluded that mantids respond preferentially to small (0.3–0.8 cm^2) bug-shaped prey that move with "jerks and stops and starts" (p. 172).

The second fascinating aspect of Rilling's study is that, despite its many methodological problems (e.g., stimulus movements were uncontrolled and the mantids' prothoraxes were immobilized), the paper became authoritative (Prete and Wolfe, 1992). In fact, the study became so influential that it has been cited in all of the subsequent work on mantid prey recognition of which I am aware to the exclusion of a more recent study that demonstrated that mantids strike most at *elongated* stimuli as much as twice the surface area of Rilling's most potent stimulus!

In the more recent study, Holling (1964) constructed a series of elongated black paper stimuli for which "the width was 1/3 the length, and the sizes ranged from 8.5 mm long (about the length of a housefly) to 69 mm long (about the length of a mantid)" (p. 343). These dummies were affixed to a thin wire and rotated within the striking range of untethered *Hierodula crassa*. Holling was able to elicit the most predatory strikes (i.e., on about 73% of trials) with stimuli that were 8 × 24 mm, but, as his Figure 5 reveals (p. 344), mantids did, in fact, strike almost as frequently (on more than 65% of trials) at stimuli that were approximately 4 × 12 and 6 × 18 mm. Strike rate dropped precipitously as stimulus size decreased or increased beyond this range. This study paints quite a different picture of mantid prey recognition than does Rilling's: Holling's results suggest that mantids recognize rather large, elongated stimuli as prey.

Unfortunately, and to the best of my knowledge, Holling's results were neither integrated with nor cited in the studies of mantid prey recognition published between 1964 and 1990, despite the fact that Holling's results were complementary (not contradictory) to Rilling's.[1] A contributing factor to this omission may have been the fact that Holling's data were embedded in a paper that dealt more with theoretical issues in ecology than with mantid prey-catching behavior per se. So, with the notable exception of studies in the field of ecology (e.g., Bartley, 1983), work published during these twenty-five years generally assumed that flies or fly-sized objects were the only stimuli that needed to be considered when studying mantid vision, mantid prey recognition, mantid prey localization, and the like. This unfortunate oversight has contributed to most of the confusions noted in the beginning of this section. That is, the frequent repetition of Rilling's results in the Introduction or Discussion sections of papers dealing with virtually any aspect of mantid behavior, in conjunction with the consistent use of flies as stimuli in all types of experiments with mantids, led to the now deeply entrenched belief that mantid prey recognition is tantamount to discovering how mantids visually localize small, individual (i.e., flylike) spots.

Rilling and Holling Reconciled

In 1990, two articles appeared that seemed to reconcile the Rilling and Holling data. Coincidentally, both researchers presented mantids with three-dimensional black stimuli that moved along the ground as would prey in the wild. For the sake of argument, I want to begin by explaining the second of these two articles. Iwasaki (1990) presented sixth- and seventh-instar *Tenodera aridifolia sinensis* (Sauss.) with various stimuli that traveled around the perimeter of a circular stage on which a nymph was free to move. The array of stimuli that Iwasaki used were three-dimensional rectangles with square faces measuring 2 × 2 to 8 × 8 mm and lengths that were one, two, four, or eight times greater than the width. In essence, this was a series of compact or elongated stimuli that moved around the arena parallel to their longest axis. (Remember that the elongated stimulus configurations are what Ewert [1987] called *worm* stimuli.) Based on a comparison of the residual variances of several least-square analyses, Iwasaki concluded that the best predictor of whether or not a stimulus would elicit a predatory strike was its volume (also see Iwasaki, 1991).

Iwasaki's statistical analyses notwithstanding, I do not think that this is the most plausible interpretation of his data: There is no known mechanism by which a mantis could compute the volume of such an object—especially a dark black object moving at 45 cm/s and seen from a continuously changing perspective. Given what we know about the limits of resolution of the nymph's visual system (Kral, this volume, chap. 7), I am sure that the stimuli were perceived solely in terms of the lengths and widths of their silhouettes.

Interestingly, I also presented mantids with an array of three-dimensional, rectangular, black balsa wood stimuli in my first experiment, although I was unaware of Iwasaki's work at the time (his papers did not appear in print until after I had completed my experiments). Perhaps, we both took the same cue from Holling (1964). My stimuli had 3 to 30 mm square faces and lengths that ranged from 3 to 100 mm (Prete, 1990), and my protocol differed from Iwasaki's in several ways. First, I tethered *Sphodromantis lineola* (Burmeister) by affixing a small (5 × 13 × 12 mm high) balsa wood block to the dorsal midline of their pterothorax (between wing joints). During experiments, a mantid was suspended by its tether within a circular white arena and given a hollow styrofoam ball which it reflexively held in its mesothoracic and metathoracic (walking) legs. Such an arrangement gives the mantis the illusion that it is unencumbered and standing on something solid. When held like this, mantids be-

---

1. Holling's data have been used, however, to support various hypotheses about the relationship between the putative "ideal" size (meaning radius) of prey and the morphology of the mantid's raptorial foreleg (e.g., Bartley, 1983, Loxton and Nicholls, 1979).

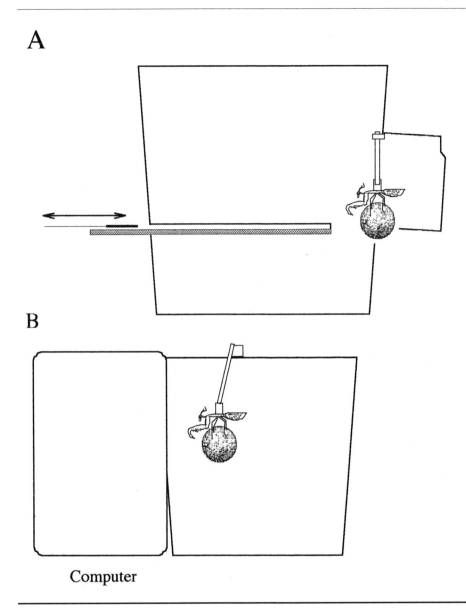

Fig. 8.1 Schematics of the two types of set-ups in which mantids were presented with various types of visual stimuli. *A*, Mantids watched mechanically driven stimuli that slid along the arena floor powered by a variable-speed mechanical arm. *B*, Mantids watched computer-generated stimuli move around a black-and-white computer screen. Reprinted with from Prete and McLean, 1996, with the kind permission of S. Karger.

have normally, displaying their entire repertoire of behaviors, including walking (stepping movements simply rotate the styrofoam ball). I watched the mantids via an overhead video camera. Figure 8.1A shows one version of this set-up.

My experiment also differed from Iwasaki's in that, like Holling, I impaled the stimuli on a wire and rotated them as I moved them along the arena floor either directly toward, or orthogonal to (and directly under the head of), the mantid. From the mantid's vantage point, 1 mm at ground level subtended approximately 1° of visual angle. To add to the lures' attractiveness, I borrowed a trick used by Rilling: I affixed a short length of black rubber band to the middle of one side of the wooden blocks, which fluttered as the blocks rotated (Fig. 8.2).

I learned several things about mantid prey recognition from this (embarrassingly) simplistic experiment, all of which have been confirmed in subsequent, better-controlled efforts. First, I learned that visual tracking behavior—which is often used as an indication that a mantis recognizes an object as prey—actually has little to do with prey recognition per se. Just like you and I do, a mantis will orient to and visually track a variety of stimuli, many of which will never be identified as a potential meal. (By the way, Iwasaki found this, too.) Conversely, mantids will often strike at objects without first making any perceptible tracking movements. Second, I learned that mantids display two distinct appetitive behaviors that indicate that they have classified a stimulus as prey: The first, of course, is the characteristic rapid grasping movement of the raptorial forelegs, or the strike. The second, which I termed an *approach*, I defined as two or more stepping movements toward a stimulus (i.e., steps that caused the ball to rotate away from the stimulus). So, on any given trial, a stimulus that was recognized as prey could elicit one approach and/or one or more strikes. The third thing that I learned was that in this set-up, stimuli moving orthogonally are poor releasers of predatory behavior compared to approaching stimuli.

The data that I collected with approaching stimuli are presented in Figure 8.2A. This graph represents approximately seven thousand trials delivered to twenty-four mantids. As in all of my graphs, the data points are average response rates for adult females (response rate equals the number of strikes or the number of strikes plus approaches divided by the number of trials for a given stimulus), and the vertical bars indicate one standard error.

First, note that if I plot the number of strikes per trial against stimulus volume, as did Iwasaki (filled circles in inset, B), the data give us no more information than does the simpler representation of plotting the strike rate against the surface area of the largest side of the stimulus (crosses in inset B). This makes sense, of course, because the surface areas of a side are linearly related to the volumes of the stimuli. More important, however, plotting the data in these terms hides a key point about mantid prey recognition. That is, graph A in this figure demonstrates that neither volume nor surface area per se is the critical stimulus parameter for prey identification. When I averaged response rates across all stimuli of a given width (thick line) or length (thin line), I discovered that width and length were treated quite differently by the mantids. The most predatory behavior was elicited by stimuli that were 3 to 35 mm long (open circles, thin line; remember that 1 mm subtends approximately 1°); when stimuli were longer than 35 mm, their appeal steadily declined. On the other hand, stimulus width defined a much smaller envelope of acceptable dimensions (filled circles, thick line). In this case, stimuli elicited high levels of predatory behavior if they were 3 to 20 mm wide (with a peak at 10 mm) but elicited virtually no predatory behavior if they were more than 20 mm wide.

As soon as I had plotted these data, I realized two things. First, Rilling's (1959) and Holling's (1964) results were not incompatible. Each, however, reported only one-half of the answer to the question of what mantids recognize as prey. That is, my mantids were identifying two types of stimuli as prey: small, compact stimuli (like Rilling's) and elongated stimuli moving parallel to their long axes (like Holling's). My second realization was that my mantids were reacting just like the amphibian predators that I spoke about earlier. They appeared to be differentiating between worm and antiworm stimulus configurations (Ewert, 1987; Ingle, 1983; Schürg-Pfeiffer and Ewert, 1981; for similar preferences in salamanders see Roth, 1987). I confirmed this hunch by presenting a group of *S. lineola* with an array of black, flat (two-dimensional) stimuli that consisted of a series of squares ranging from 1.5 × 1.5 mm to 30 × 30 mm, and a series of rectangles, one edge of which measured 1.5 to 30 mm and the other edge of which measured 1.5 to 114 mm. Once again, the mantids were tethered in a white arena, but this time the stimuli did not rotate; they simply slid along the arena floor toward the mantis and then withdrew (Prete, 1992a, b).

The key data from these experiments are presented in Figure 8.3. These graphs represent a total of several thousand trials given to approximately thirty mantids. Graphs A–D show the average response rates to three types of stimuli: square stimuli (the first point in each graph), worm stimuli (stimuli moving parallel to their long axes; the heavy line), and antiworm stimuli (stimuli moving perpendicular to their long axes; the dashed line). In these trials, the arena floor was white. Graphs F–H show the average responses to the same stimuli when they moved against a patterned floor that was designed to mimic the heterogeneous background against which prey might normally be seen (Fig. 8.3E). All told, the data in these seven graphs confirm

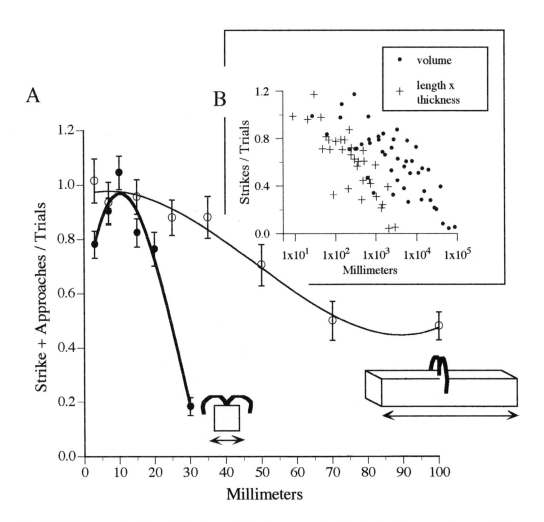

Fig. 8.2 Response rates of female *Sphodromantis lineola* to a number of three-dimensional black balsa wood stimuli that rotated as they moved toward them across the floor of an arena like that depicted in Figure 8.1A. Each stimulus had a short length of rubber band affixed to one side to increase its attractiveness. *A*, The mean response rates for stimulus thicknesses averaged across lengths (heavy line) and the mean response rates for stimulus lengths averaged across thicknesses (thin line). Note that thickness (dimension of the edge orthogonal to the direction of movement) defines a very different envelope of acceptable sizes than does length (dimension of the axis parallel to the direction of movement). *B*, The same data plotted as Iwasaki's (1990) data were for the same type of stimuli. *A* is redrawn from Prete, 1990, with the kind permission of S. Karger.

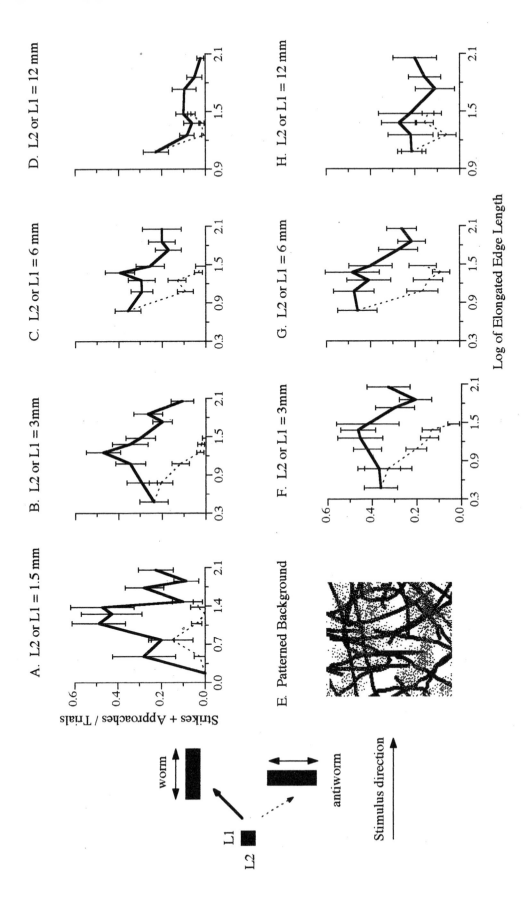

that the mantids were responding appetitively based primarily on stimulus configuration (worm versus antiworm), not on stimulus size. Specifically, the mantids responded the most to squares when the edge lengths were 3 to 6 mm (approximately 3–6°; the first points in graphs B, C and F, G). However, when it came to elongated stimuli (all of the data points except the first ones), response rates were high only if the stimuli were in the worm configuration, and their leading edge was less than 12 mm (the heavy lines in graphs A–C and F–G).

I would like to reiterate the point that these mantids did not respond to the stimuli based on size per se. You will note that in each graph each point on the heavy line represents the average response rate to a worm stimulus of a given size and is positioned directly over a point on the dashed line that represents the average response rate to an antiworm with exactly the same surface area. Clearly, stimulus geometry in relation to direction of movement was the critical factor under these experimental conditions.

Subsequent experiments revealed that this worm versus antiworm preference is not unique to *S. lineola*. We know that female *Tenodera aridifolia sinensis* display the same worm versus antiworm preference even when the stimuli are quite cryptic. For instance, in Figure 8.4 you will see two graphs that show the average response rates to square, worm, and antiworm stimuli that were patterned to match the arena floor. We all know

Facing page

Fig. 8.3 *A–D*, The response rates to two-dimensional black stimuli that were square (the first point in each graph), elongated parallel to their direction of movement (worms; the heavy lines), or elongated perpendicular to their direction of movement (antiworms; the dashed lines). These stimuli moved against a white background. *E*, Patterned background. *F–H*, The response rates to the same stimuli moving against the patterned background. Mantids responded appetitively to small squares and worms but they did not recognize antiworms as prey. Redrawn from Prete, 1992a, b, with the kind permission of S. Karger.

that mantids face the problem of having to recognize prey that is less obvious than a black stimulus moving against a white background, and I wanted to be sure that the worm preference could be seen under more realistic conditions; if the preference failed to show up using cryptic stimuli, I would have doubted its ethological relevance. As it turned out, however, the worm versus antiworm preference is quite robust. As the graphs indicate, both *S. lineola* and *T. a. sinensis* (including four wild-caught specimens) displayed the worm versus antiworm preference even to these stimuli (Prete and McLean, 1996). Incidentally, you will also note that in this experiment, female *T. a. sinensis* responded to slightly smaller stimuli than did *S. lineola*. We are currently exploring this and some other species differences.

Pursuing the solution to the prey recognition question became even more exciting when my students and I discovered that stimulus configuration was only the first in a series of stimulus parameters that shaped the choice behavior of our subjects. Our next discovery was that the direction of stimulus-background contrast also made a difference—that is, a dark stimulus on a lighter background is perceived as prey by *S. lineola* but the reverse is not. (It is interesting to note here that there are also differences in the appetitive responses of anuran amphibians to darker versus lighter stimuli [e.g., Ewert et al., 1982].) In Figure 8.5A you will see that when mantids were presented with black or white, 6 × 24 mm stimuli delivered against a patterned background, only the black worms were recognized as prey even though the mean luminance contrast was actually greater between the white stimuli and the background than it was between the black stimuli and the background (Prete, 1992b).

This preference for relatively darker versus lighter stimuli is as robust as is the mantid's worm versus antiworm preference: My students and I have never been able to get a mantid to strike at a stimulus brighter than the background against which it moves even when the mantids were completely dark adapted. This was somewhat surprising to me given that movement-sensitive visual interneurons in grasshoppers and cockroaches will respond to dark stimuli moving

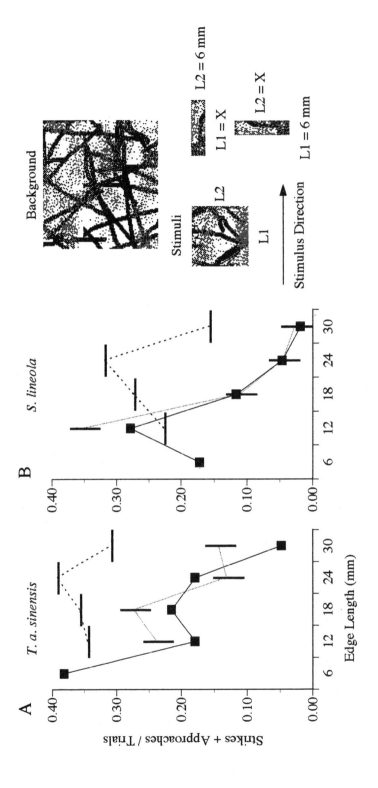

against a brighter background when they are light adapted (e.g., during the day) and to the same stimuli with the contrasts reversed when they are dark adapted (e.g., during the night; Edwards, 1982a, b; O'Shea and Rowell, 1976; Rowell and M. O'Shea, 1976a; Rowell and O'Shea, 1976; Rowell et al., 1977). This does not mean, of course, that mantids don't see these relatively brighter stimuli. What it means is that bright stimuli *mean* something different to mantids because they are processed differently than dark stimuli. For instance, we know that small bright stimuli are recognized as droplets of water by both *S. lineola* and *T. a. sinensis*, and, consequently, they elicit a behavioral response sequence appropriate for drinking rather than prey catching (Prete et al., 1992).

Incidentally, this preference for relatively darker visual stimuli is consistent with data collected by Bowdish and Bultman (1993) in their investigations of the role of visual cues in food avoidance learning by *T. a. sinensis*. After exposure to toxic milkweed bugs, their mantids were presented with stimuli that were black, orange, or black and orange. The authors found that under all conditions, mantids struck most frequently at solid black (i.e., the relatively darkest) visual stimuli.

As one might imagine, our mantid's worm versus antiworm preferences were due in part to the fact that mantids seem to attend preferentially to the leading edge of a moving stimulus—once again, a preference seen in anuran amphibians (e.g., Burghagen and Ewert, 1982). For instance, Figure 8.5B shows the average response rates of *S. lineola* to a series of black 6 × 24 mm worm stimuli that had either a white or a black antiworm stimulus affixed to their undersides at the trailing edge, in the middle, or at the leading edge. Worm stimuli with a white antiworm attached were presented against a patterned background (as shown), and the stimuli with a black antiworm attached were presented against a white background (Prete, 1992b). Within background conditions, antiworms placed at the leading edges of the worm stimuli depressed appetitive behaviors compared to the other stimulus configurations.

We saw a similar effect when we superimposed a 6 × 6 mm small black square on white worms and antiworms and presented the compound stimuli against a patterned background (Fig. 8.5C; Prete, unpublished data). Here again, an antiworm configuration—that is, a broad leading edge—suppressed responses to the small black square.

All told, then, I think that these early experiments make Rilling's and Holling's data understandable. Clearly, the mantids with which I have worked do not respond appetitively to visual stimuli based simply on size: Stimulus configuration and relative brightness are also critically important. Further, I am convinced that my mantids were not engaging in any sophisticated computational tasks such as figuring out stimulus volume. What I think that they were doing, however, is much more elegant: They were recognizing prey by means of a visual system that seems to be separating objects into "prey" and "nonprey" categories by weighing several stimulus parameters. Even more interestingly, my mantids seemed to be defining the category "prey" in terms of an algorithm very similar to that used by amphibian predators such as frogs, toads, and salamanders!

Refining the Algorithm

Having learned that at least two species of mantid recognize both compact and wormlike stimuli as prey (as long as the leading edge is not too big), I set out to determine if there were additional stimulus parameters that played a role in circumscribing the category "prey." The first (and perhaps the most obvious) parameter that occurred to me was stimulus speed. Up to this point in time

Facing page

Fig. 8.4 The mean response rates of two species of mantid to patterned, cryptic two-dimensional stimuli that moved across a similarly patterned background. Although there were species differences in the sizes of the smallest stimuli that were preferred, in general, worm configurations were preferred over antiworms of the same size. Redrawn from Prete and McLean, 1996, with the kind permission of S. Karger.

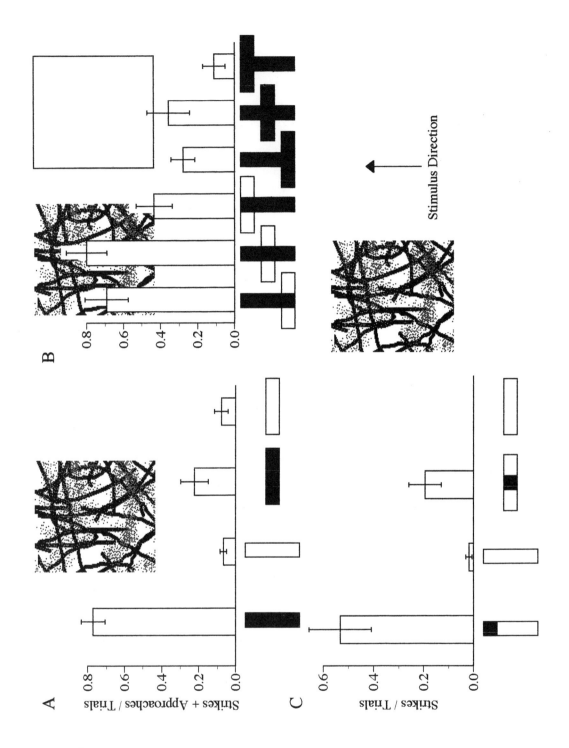

(as far as I knew), there were no published data on the effects of stimulus speed on mantid prey recognition. Quite to the contrary, in fact, it was the norm for people to jiggle, shake, vibrate, swing, twirl, or rotate stimuli in uncontrollable ways in order to elicit prey capture from these bugs—I had done the same at times.

I developed two protocols to make my stimulus presentations more precise. In the first, a variable speed mechanical arm slid a stimulus that was impaled on a long, thin wire across the floor of a circular arena like the one that I described earlier (Fig. 8.1A). In the second protocol, stimuli were presented via computer animations. In this case, the stimuli were shown on an 11.7 x 17.5 cm black-and-white computer monitor inserted into a window cut into the side of a white arena (Fig. 8.1B). The computer had a vertical refresh rate of 60 Hz and 512 x 342 bit-mapped resolution (pixel size = 0.34 x 0.34 mm). Computer animations had the benefit of allowing me to create stimuli that would have been impossible otherwise.

Results of the first experiments on stimulus speed were so dramatic that I was taken aback (Prete et al., 1993). My students and I presented *S. lineola* with a series of mechanically driven black stimuli that moved over a white or a patterned arena floor, at speeds of 18.5, 26.5, or 34.3 cm/s. We also varied the angle of approach that the stimuli took. However, in this experiment, neither the background pattern nor the angles made any difference. The results, summarized over a total of about sixteen thousand trials, appear in Figure 8.6A–C. Each graph shows the average strike rate elicited by a 6 x 6 mm square (first data point in each graph), a series of worm stimuli that were 6 mm wide and 12 to 60 mm long (heavy line), and a corresponding series of antiworm stimuli (dashed line). Clearly, the long-overlooked parameter of stimulus speed plays a critical role in prey recognition, at least for *S. lineola*.

When we watch an image move across a movie screen, its movement is called *apparent* because it is an illusion created by the rapid, sequential appearance of a series of still images. Such, too, is the case with computer animations. As expected, we discovered that mantids respond to the apparent movement of computer-animated stimuli just like they do to "real" stimuli (Prete and Mahaffey, 1993b). The results of one of our first experiments appear in Figure 8.6D. Here, mantids were presented with black square stimuli with edge lengths that subtended 12 to 47°. On any given trial, a square appeared in the center of the computer screen (indicated by the open square in the inset), moved along a predetermined path (indicated by the line in the inset—the line was not actually visible during the experiment), and then the stimulus either disappeared (when it reached the point indicated by the solid square in the inset) or repeated the path (depending on the experiment). We called this particular path Erratic Path 1. In this experiment, apparent speed was manipulated by running the animation at 2, 10, or 20 frames per second (fps), while holding constant the distance moved by the stimulus from frame to frame (10°), the trial length (48 seconds), and the viewing distance (23 mm). At these frame rates, the apparent stimulus speeds were 20, 82, and 120°/second. As you can see, the effect of apparent speed is not subtle: It is a critical factor in prey identification. We have since confirmed these results in a number of experiments including some in which we have immobilized the head and prothorax in order to precisely control the retinal position of the stimuli.

Facing page

Fig. 8.5 *A*, The average response rates to 6 x 24 mm black-and-white stimuli that moved against a patterned background. Only relatively darker worm configurations reliably elicited appetitive behaviors. Lighter stimuli and antiworms were not recognized as prey. *B*, Demonstration that mantids attend preferentially to the leading edge of a moving stimulus. Within each background condition (i.e., left three bars and right three bars), an antiworm placed at the leading edge of a worm stimulus depressed the response rate to the worm more than an antiworm placed at the middle or trailing edge of the worm. *C*, Further evidence that the broad leading edge of a white antiworm will suppress response rates to a preferred 6 x 6 mm black square. *B* is redrawn from Prete, 1992b, with the kind permission of S. Karger.

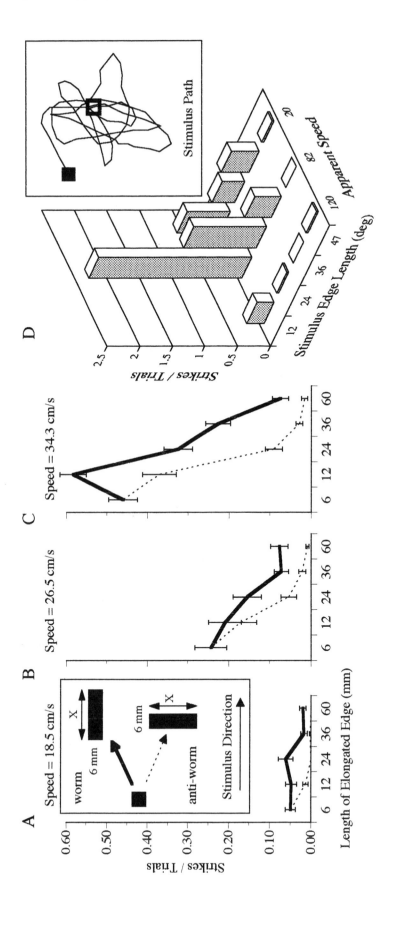

The results are always virtually the same (e.g., Prete and Mahaffey, 1993b; Fig. 8.4).

Theoretically, the speed of a potential prey item could inform a mantid about either the object's identity (e.g., what species it is) or its distance. It is unlikely that speed is used for the former, however, given the mantid's eclectic tastes. They eat slow-moving prey and fast-moving prey, even when the latter are moving slowly (e.g., a walking fly). So, speed is always confounded with prey type. However, speed could be used as a cue to object distance (e.g., see Kral, this volume, chap. 7). We know that mantids do recognize the distance between themselves and a potential prey item. For instance, they may approach (walk up to or stalk) prey before striking at it, or depending upon the distance to be spanned, the strike may be accompanied by a lunge of varying magnitude (e.g., Cleal and Prete, 1996; Prete and Cleal, 1996; Prete et al., 1990, 1992b; Prete and Hamilton, this volume, chap. 10).

However, because insects (including mantids) have immobile eyes with fixed-focus optics, the distance of an object cannot be inferred by ocular convergence as it can be by many vertebrates (Srinivasan, 1992). So, instead, some insects extract the relative distances of objects from the relative speeds of the retinal images of those objects (e.g., Collett and Patterson, 1991; Collett, 1978; Srinivasan et al., 1989; Srinivasan, 1992; Sobel, 1990). Put simply, if absolute object speed is held constant, or if the insect moves in relation to stationary objects, the retinal images of nearer objects will move faster than will the images of more distant objects (Kral, this volume, chap. 7). You will recognize this phenomenon if you recall looking out of the window of a moving car. The telephone poles at the side of the road, for instance, move across your field of vision quite quickly, but buildings in the distance move by very slowly. For the mantis, then, an image that subtends a visual angle small enough to be considered prey (say, 10°) but is moving quite slowly probably represents a very large distant object (such as a bird flying by). However, an image that subtends the same 10° and is moving quickly probably represents a small nearby object (and, perhaps a potential meal).

If this reasoning is correct and mantids perceive slower-moving images as more distant, then objects that are otherwise recognized as prey but are moving slowly should elicit primarily approaching behavior, the purpose of which would be to bring the object to within the range of a strike. On the other hand, faster-moving preylike images—presumably perceived as being closer and, consequently, already within the range of a strike—should elicit only strikes. When my students and I tested these hypotheses, that is precisely what we found. Mantids emitted virtually no strikes but did try to approach mechanically driven stimuli moving at only 12 cm/s, but they emitted only strikes to the very same stimuli moving at 36 cm/s (Fig. 8.7A, B). Intermediate speeds (data not shown) elicited intermediate levels of each of the behaviors (Prete et al., 1993).

Our interpretation also explains an interesting phenomena that we noticed when we first presented mantids with computer-generated black squares moving along an erratic path. When mantids viewed the stimuli from 23 mm away, all that they emitted were strikes (Fig. 8.7C). When the same stimuli were viewed from 45 mm away, however, they elicited virtually no strikes but did elicit some approaching behavior (Fig. 8.7D). So, it seems that in this experiment, too, when the mantids were farther away from the computer screen and, consequently, the stimuli appeared to be moving more slowly, they were seen as being too far away to strike at, but when the stimuli

Facing page

Fig. 8.6 A–C, The average response rates to a 6 x 6 mm black square and a series of worm (heavy line) and antiworm (dashed line) stimuli for which one edge length remained 6 mm. Each graph shows the data for a different stimulus speed. D, The response rates to a series of computer-animated square stimuli that moved along an "erratic" path (inset) at three apparent speeds. In both experiments, stimulus speed was a critically important stimulus parameter for prey recognition. Redrawn from Prete and Mahaffey, 1993, with the kind permission of Cambridge University Press.

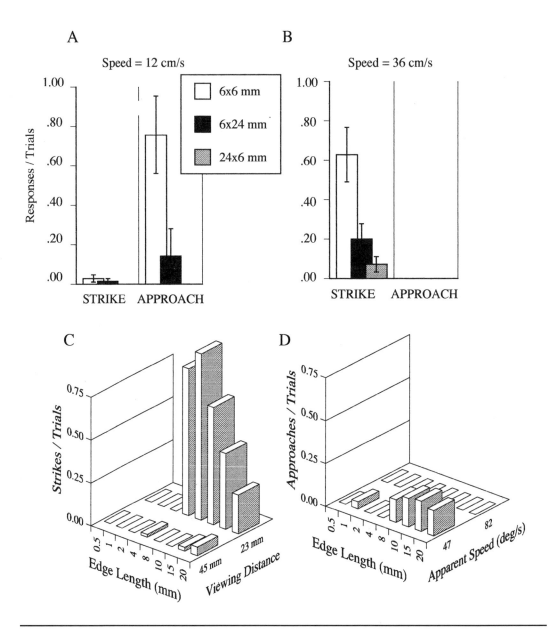

Fig. 8.7 *A, B,* Demonstration of differences in responses depending upon whether a stimulus is moving slowly or quickly. The former elicit primarily approaching behavior because they are presumably perceived to be farther away. On the other hand, faster-moving stimuli elicit primarily strikes, because they are perceived to be nearer. A similar phenomenon can be demonstrated with computer-animated squares moving along an erratic path. *C,* When viewed from a distance of 23 mm, the stimuli elicit a high level of striking, but *D,* when viewed from 45 mm away, they elicit approaching behavior. The critical difference between these two conditions is apparent stimulus speed (82 versus 47°/second, respectively). *A, B* are redrawn from Prete et al., 1993, with the kind permission of S. Karger. *C, D* are redrawn from Prete and Mahaffey, 1993, with the kind permission of Cambridge University Press.

were closer and appeared to be moving more quickly, they elicited strikes.

Dr. Kral has contributed a very interesting chapter to this book in which he discusses the mantid's use of retinal image motion to judge the distances of nonprey objects in the environment (and it is of note that all of our findings agree with his). As was Dr. Kral, I was curious as to whether I could trick a mantis into mistaking the apparent distance of a stimulus by moving the background against which the stimulus moved. For instance, I wanted to know if I could make a slow stimulus appear to move more quickly by moving the background in the direction opposite that of the stimulus. If I could do so, that would mean that my mantids were comparing stimulus and background speeds. My students and I attempted to do this trick by presenting mantids with computer-generated cartoon crickets that looked like they were walking across the computer screen at 5°/second. The crickets were presented against one of three backgrounds, a white one; one with very small, regularly spaced black dots on it; and one with a random pattern of small black squares that made the background look speckled. The latter two backgrounds were either stationary or moved in the opposite direction to that which the cricket was walking. I reasoned that if mantids compared the movements of the background and the crickets, the crickets would appear to be moving faster when the background moved, and they should elicit high rates of striking. However, if the mantids only take into account the apparent speed of the cricket, then background movement should either have no effect or should inhibit prey-catching behavior.

Apparently, mantids process the images of preylike stimuli differently than they do the types of stimuli in which Dr. Kral is interested. That is, we could not trick our mantids.[2] Although our mantids visually tracked all of the cartoon crickets, the moving backgrounds inhibited striking behavior. So, in the case of prey recognition, mantids seem only to attend to the apparent speed of the small-field stimulus (the potential prey object) when making judgments about whether to strike. They apparently do not compare the speed of the small-field stimulus with that of the background (Prete and Mahaffey, 1993b). The fact that coherent large-field (i.e., background) motion suppresses appetitive behaviors to small-field motion is usually explained as being a product of the fact that insects interpret large-field motion as self-motion and small-field motion as object motion (e.g., Rowell, 1971; Olberg, 1981; Egelhaaf, 1985a–c; Egelhaaf et al., 1988; Ibbotson and Goodman, 1990; also Kral, this volume, chap. 7). It seems that mantids act like other insects in this regard.

All of my research indicates that mantids recognize an object as prey if it falls within an envelope defined by a number of stimulus parameters. So far we have seen that those parameters include (1) overall size (about 5–30° for a compact stimulus moving erratically), (2) geometry in relationship to direction of movement (worm, not antiworm), (3) length of the leading edge (about 3–30° for stimuli moving linearly), (4) relative contrast (darker than the background), and (5) apparent speed (about 35–85°/second). There are a few more parameters that must be added into this mix, but I did not come to them until I stumbled over a more subtle question.

Up until this point in my research, I had taken for granted that I knew what a mantis defined as an "object"—some solid, clearly visible image with a distinct border, I thought. That is, I had simply assumed (as had others) that questions regarding the perception of anything more subtle than a black rectangle were appropriate only when dealing with "higher" animals. I was wrong.

When you or I see a cat walking behind a picket fence, the entire cat is not visible to us. Despite this, however, we do not interpret what we see as several furry "slices." We perceive the synchronously moving images as being portions of a single object, and the cat is perceived as being whole.

---

2. I must add here that the fact that these two types of visual stimuli are processed differently demonstrates the complexity of this insect's visual system. Even more exciting is the fact that both of these stimulus classes are handled differently than the visual cues that represent nonprey ingestants including water to mantids (Prete et al., 1992).

In this example, our perception of the intact cat is only in part a product of having learned what a cat looks like. The truth of the matter is that our propensity to see the synchronously moving portions as a single, intact object is mostly a product of a perceptual capacity that is ubiquitous among sighted organisms (insects included). This propensity causes us to perceive a pattern of synchronously moving small shapes as a single unified object, even if the pattern is moving against a patterned background against which it would be undetectable if motionless (e.g., Cutting et al., 1988; Juelez, 1971; Prazdny, 1985). It was a surprise to me (though it should not have been) when I discovered that *S. lineola* also has this perceptual propensity!

The key experiment on this phenomenon used square, computer-animated stimuli that consisted of randomly arranged small black rectangles (Fig. 8.8A). When these were stationary and superimposed on a similarly patterned background (shown to the right of the graph in Fig. 8.8A), the stimuli were invisible to us (and presumably to the mantids, since they ignored them). However, as soon as the stimuli began to move, they became clearly discernible both to us and, based on their responses, to the mantids (Fig. 8.8A; Prete and Mahaffey, 1993b). What is interesting about these results is that the individual component rectangles that made up the stimuli were each too small to elicit appetitive response on their own. So, the mantids must have perceived the synchronized components as a single, intact object.

The results of this experiment changed the way I looked at the issue of mantid prey recognition. After this experiment, I realized that there was something amiss in the common wisdom about how mantids work, so to speak. Much of the literature that I had read argued either explicitly or implicitly that mantids have neither the ability nor the need to solve complex visual discrimination tasks like the one faced in the previous experiment. For instance, it has been suggested that when faced with a simpler task than the previous one—choosing between two simultaneously appearing, preylike stimuli—the mantid's visual system must somehow "automatically" attend to only one of the stimuli to avoid an irreconcilable confusion (e.g., Rossel et al., 1992). But what if the mantis *was* faced with two equally desirable potential targets? What would one predict? Based on the experiment that I just described, I predicted that a pair of equally desirable stimuli might, in fact, constitute what ethologists call a *super sign stimulus*, a stimulus that is unusually potent by virtue of the fact that it exaggerates some particularly attractive stimulus parameter. While it may be true that in nature it is unlikely that two equally attractive prey objects appear simultaneously in a mantid's visual field and behave precisely the same way, it does not follow that the mantid's visual system is designed to automatically pick the single most attractive target or perceive two targets as one (e.g., by "fusing" the images as some have suggested). To the contrary, I tend to think that it is precisely because the simultaneous appearance of two equally appealing targets is so unlikely that mantids could not have evolved a strategy for dealing with them. I reasoned that if two equally appealing images appeared at the same time, their affects should be additive: Together, they should elicit a very high rate of striking, and that is precisely what I found.

Figure 8.8B shows the results of an experiment in which mantids with immobilized heads and prothoraxes watched a pair of computer-generated stimuli that moved in a narrow (100-frame-long) erratic path. The inset shows the positions of a pair in animation frames 2, 26, 55, 62, and 91 in reference to the center of the mantis's visual field (denoted by the cross). In this experiment, pairs of square stimuli were aligned vertically and were symmetrical to the midline in frame one of each trial, but the horizontal distance between the squares varied from 36 to 74° across trials. Now, note the scale of the ordinate in the graph: It is one-third that of the other graphs in Figure 8.8. That is, rather than posing a perceptual difficulty for the mantids, the pair were actually more potent releasers of predatory behavior than was a single square. I should also point out here that, generally speaking, *S. lineola* respond *less* to stimuli when their heads and prothoraxes are immobilized as in this experiment. So, the very high rate of striking to the paired

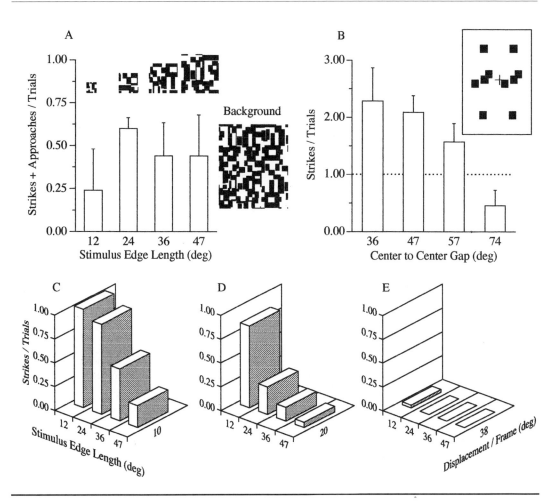

Fig. 8.8 *A*, The response rates elicited by the square, computer-generated stimuli (positioned over each bar) when they moved in an erratic path against a similarly patterned background. Mantids responded as if the synchronously moving subthreshold components that make up each stimulus represented a single unified object. *B*, The response rates elicited by a pair of synchronized computer-generated stimuli that moved in a narrow erratic path in front of mantids whose heads and prothoraxes were immobilized. The inset shows the positions of a pair of stimuli in several frames of the animation. The cross which represents the center of the visual field was not visible on the screen. The pair of stimuli elicited a higher rate of responding than would just one of the squares under similar circumstances. *C–E*, Demonstration that whether or not a mantis perceives a series of discrete retinal images as a single moving object depends upon the distance between the successive images. *C, D,* When the displacement of the images is 20° or less, they are perceived as a single object. *E,* When the displacements are 38°, the images are ignored. *A–C* are redrawn with permission from Prete and Mahaffey, 1993, with the kind permission of Cambridge University Press.

squares is even more dramatic than it would otherwise be (Prete and Mahaffey, 1993b; also see Prete and McLean, 1996).

I hope that, by now, you see that what makes a particular visual stimulus attractive to a mantid is a rather complicated affair based on the weighing of a variety of stimulus parameters. The latter two experiments have added a sixth parameter to the list: The synchronized movement of a sufficient number of small-field stimuli (even a constellation of subthreshold stimuli) will be perceived as a single, potential prey object as long as a sufficient number of other parameters are met. Now, at first, this might seem like a rather arcane point. Who cares, you might be thinking, whether a few disconnected dark spots moving together at the right speed (especially like those in Fig. 8.8A) are perceived as prey by mantids. What does that have to do with real life? Recall the taste aversion experiment to which I referred earlier (Bowdish and Bultman, 1993). In this experiment mantids learned to avoid milkweed bugs after several distasteful encounters with them. Milkweed bugs (like many other bugs) are multicolored: They are orange and black. Hence, when viewed against a dark background the orange markings will appear lighter than the background and (for reasons that I have explained) will not contribute to the mantid's perception of the bug as prey. In fact, Bowdish and Bultman found that completely orange stimuli were not very attractive to their mantids. On the other hand, the black portions of the bug will likely be darker than the background against which it is viewed and, consequently, will contribute to the mantid's perception of it as prey. So, in real life, identifying a discontinuous but synchronously moving pattern of dark shapes may be precisely the capacity needed to recognize a cryptic or multicolored bug.

The results of the previous two experiments demonstrate that the mantid's visual system sums visual inputs over space and responds to them as if they represented a single object (as long as the total retinal area stimulated at any one time is not too large). The limits of this capacity were revealed in an experiment in which the frame-to-frame displacement of square computer-animated stimuli was increased incrementally. We found that a series of still images, in which each image was displaced 20° or less from its predecessor, created the illusion of a single moving object for *S. lineola* whereas images displaced more than 20° did not (Fig. 8.8C–E). This phenomenon is, of course, also experienced by people. If two stimuli are flashed in different places on a screen, one immediately after the other, it appears as if the first image "jumps" to the second position if the distance between the two images is not too great. So, once again, I found that the fundamental rules governing object recognition by *S. lineola* are the same as those used by other organisms, and this is not surprising. Recognizing objects in one's environment is an essential task for any organism whether it is a cockroach, a mantid, a toad, or a person. Quite frankly, it would be more surprising if mantids were different.

The spatial and temporal proximity of successive retinal images is the seventh stimulus parameter that affects prey recognition for *S. lineola*.

Up to this point, then, how might one sum up the fundamental principle underlying mantid prey recognition? A clue to the answer lies in the leftmost column of Figure 8.8A. Note that when this 12° × 12° *patterned* stimulus moved against a matching background it elicited a relatively low level of responding compared to a 12° × 12° *solid* black stimulus moving against a white background (leftmost column in Fig. 8.8C) and only a small fraction of the response rate elicited by *a pair* of 12° × 12° black squares moving against a white background (leftmost column in Fig. 8.8B). What changed across these stimulus conditions was the overall retinal area, or number of sampling units, that were subject to luminance decrements at any given time during a trial. That is, the 12° × 12° stimulus pictured in Figure 8.8A darkens the receptive fields of only a subset of the ommatidia that are sampling the 12° × 12° area of the visual field in which the stimulus is located at any given moment. A solid black stimulus of the same size darkens all of the receptive fields within that same 12° × 12° area (Fig. 8.8C), and two 12° × 12° stimuli (Fig. 8.8B) will darken the receptive fields of even more sampling units. Hence, I would argue that within the boundaries of the stimulus parameters already defined (overall size, geometry in relationship to direction of movement, length of the leading edge, relative contrast, apparent speed, and the constraints necessary for summation over space and time), the probability that an object will be perceived as prey increases in proportion to the number of contiguous (or at least very closely spaced) sampling units that experience a luminance decrement. In support of this hypothesis, consider the following experiment, the results of which appear in Figure 8.9.

For this experiment, I created an array of nine 25° × 25° black computer-generated stimuli that varied systematically along two dimensions: the density of black pixels around their borders (level B, for border) and the density of black pixels that made up their internal spatial detail (level D, for detail). The stimuli are pictured in Figure 8.9B. I began with a stimulus that mantids would track but which elicited no predatory behavior (D1B1, lowermost in 8.9B). (This beginning point is of particular importance in that it underscores the fact that visual tracking and prey recognition are not the same thing.) To this stimulus I added 93 or 186 black pixels that were concentrated around the border (levels B2 and B3, respectively) and/or that were randomly distributed within the interior of the stimulus (levels D2 and D3, respectively). In the figure, the specific constellation of additional pixels that created each level of each dimension appear above and to either side of the stimulus array, and the total number of pixels added is indicated under each stimulus.

The response rates of the mantids to the stimuli are presented in Figure 8.9A, C. You will note that the columns in these graphs are arranged in the same order as are the stimuli in Figure 8.9B. In addition, the average rate at which these mantids tracked or struck at a solid black 25° × 25° square is indicated above each corresponding graph. The first point to be made is that mantids tracked each of the nine stimuli at moderately high rates, so we know that they saw all of them. But, even though all of the stimuli were tracked, not all of them were recognized as prey. As you can see in Figure 8.9C, increasing the number of black pixels anywhere in the stimulus increased the probability that the mantids would emit a strike. For instance, each of the stimuli in the center row of the array (D3B1, D2B2, and D1B3) have the same number of pixels added to them (186 more than D1B1) and elicited the same rate of striking despite the fact that the pixels were distributed differently in each stimulus. You will also note that stimuli with progressively more pixels elicited progressively more strikes and that a solid black stimulus elicited the most strikes of all.

Prey Recognition and Visual Space

In the preceding chapter Dr. Kral has reviewed what we know about the optics of the mantid's compound eye. As one might expect, in those species of mantid studied, there is a forward- and slightly inward-looking area of high acuity that has come to be called the *fovea* because it is analogous in many ways to the vertebrate fovea (Horridge and Duelli, 1979; Rossel, 1979, 1980; see Young, 1989, for a lucid synopsis). In the mantid's fovea, the separation between the optical axes of adjacent sampling units, or ommatidia—that is, the interommatidial angles—are smaller than they are in other parts of the eye, and their acceptance angles are smaller. Together, these two properties create an area of high resolution in the lower center of the mantid's visual field. As a point of comparison, the mantid's fovea is approximately an order of magnitude poorer in absolute resolving power than is a human's peripheral retina and two orders of magnitude poorer than is the human fovea (Young, 1989).

The fact that more ommatidia are sampling the lower center of the mantid's visual field does more than allow the mantid to resolve more spatial detail. It also provides more temporal resolution. That is, as an object moves through this area, luminance changes created by the edges of an object will be sampled at a higher rate than if the object were moving in the periphery. This being the case, and given the data depicted in Figure 8.9, one would predict that if all else is equal, a potential prey item moving through the lower center of the visual field would be more likely to elicit predatory behavior than an object moving in the periphery because more sampling units will be affected per unit of time, and this is precisely what occurs.

Figure 8.10A shows the average strike rates emitted by a group of mantids that watched a black 12° × 12° square move in a straight line either up, down, or horizontally through their visual field. (The arrows indicate the stimulus directions.) Prior to this experiment, each mantid's head and prothorax were immobilized so that the path of the stimulus could be precisely positioned relative to visual field center (Prete, 1993). You

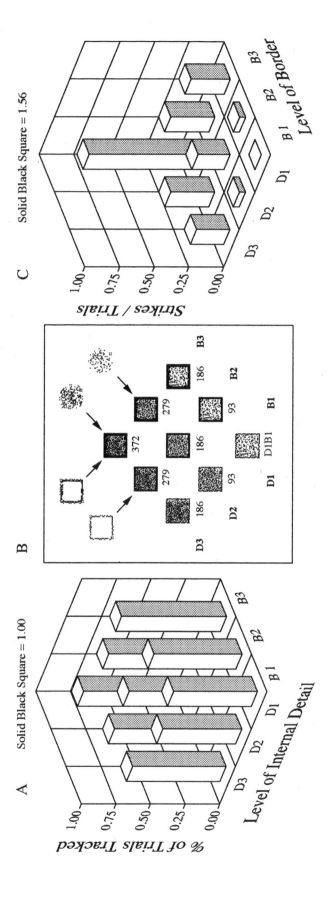

will note that the vertically moving stimuli elicited the highest rates of striking when they passed through visual field center and that horizontally moving stimuli elicited the highest rates when they passed through, or 24° below, visual field center.

Interestingly, you will also note that the direction that the stimulus moved made a difference. Stimuli moving downward elicited a higher strike rate than did stimuli moving upward. I did not anticipate that stimulus direction would make much of a difference but it does, and the effect is quite robust. We found the same direction preferences when we passed a black worm stimulus through visual field center (Fig. 8.10B; Prete, 1993). This was quite an interesting phenomenon when my student Bob Mahaffey and I first discovered it. It suggests, of course, that the fovea is exquisitely sensitive to both stimulus speed and stimulus direction. Unfortunately, I do not, as yet, know the underlying mechanisms.

The fact that the fovea is exquisitely sensitive to moving objects does not mean that an object appearing *only* in the fovea will elicit a strike, even if the object is preylike in other respects. The overall distance moved also plays a role in whether the object will elicit appetitive behavior. For instance, if a worm stimulus appears 24° from visual field center, moves through visual field center, and then disappears when its trailing edge is 24° past the center, it is not very likely that a mantis will strike at it, irrespective of the direction that it is traveling (Fig. 8.10C–E, condition 2). Likewise, the same stimulus will do poorly if it begins as much as 58° from visual field (i.e., in the periphery) and disappears when its leading edge reaches the center of the visual field (Fig. 8.10C–E, condition 1). In contrast, downward and horizontally moving stimuli elicit their highest rates of responding when they begin in the periphery and do not disappear until their *trailing* edge is 24° *past* visual field center (Fig. 8.10C–E, condition 3).

These experiments give us the final three stimulus parameters that shape mantid prey recognition: All else being equal, an object is more likely to be perceived as prey if it appears in or just below visual field center, if it is moving downward over the retinae, and if it travels a sufficient distance over the retinae.

Some Interesting "Anomalous" Data

Up to this point, I have presented a relatively parsimonious explanation of how mantids (at least *S. lineola* and *T. a. sinensis*) recognize an object as prey. However, as is always the case, the original data (both published and unpublished) from which this synopsis was created reveal a number of fascinating details, subtleties, and anomalies that remain to be explored. One, about which I have written only briefly, is the fact that there is considerable variability in the response rates of mantids, both between individuals and, over time, within the same individual (Prete et al., 1993). This fact is often lost in the graphs that depict average rates of responding, especially in cases where only the "best" or highest responders are used. I have always resisted the temptation to use only high responders, a decision to which the characteristically large error bars in my graphs attest. The size of the standard error bars reveal a point that I often fail to make explicit: Mantid prey recognition is by no means a simple, or invariant, reflex. It is a complex behavior modulated by all of the neuronal and physiological intricacies and subtleties that underlie comparable behaviors in other animals both large and small.

Facing page

Fig. 8.9 *A, C,* The response rates to the nine computer-generated stimuli depicted in *B.* The stimuli differ in the number of black pixels that they contain (noted under each stimulus) and the location of the pixels. Pixels were either concentrated around the border, denoted by levels of "B," or dispersed within the center, denoted by levels of "D." The position of the columns in the graphs corresponds to the positions of the stimuli in *B.* Although all stimuli were tracked, the rate of striking was simply a product of the number of black pixels in each stimulus irrespective of where the pixels were located. Redrawn from Prete and McLean, 1996, with the kind permission of S. Karger.

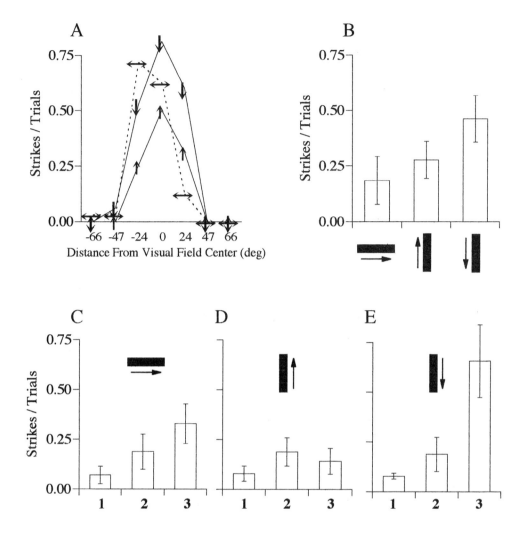

1. The stimulus began 58 deg from visual field center and disappeared when its leading edge reached center.
2. The stimulus began 24 deg from visual field center and disappeared when its trailing edge was 24 deg past center.
3. The stimulus began when its leading edge was 58 deg from visual field center and disappeared when its trailing edge was 24 deg past center.

Fig. 8.10 *A*, Demonstration that the highest strike rates are elicited by a computer-generated square stimulus when it passes through or slightly below visual field center. This area of the visual field corresponds to the mantid's fovea. The arrows indicate stimulus direction *B*. The preference for downward-moving stimuli seen in *A* is also evident when mantids are presented with computer-generated worm stimuli. *C, D,* The overall distance that a preferred stimulus moves through the visual field influences whether or not it will elicit appetitive behaviors. Redrawn from Prete, 1993, with the kind permission of Cambridge University Press.

J. H. Fabré (inadvertently) made this point when he described his observation that the same stimulus could at one moment elicit a defensive response and then, at the very next moment, elicit a predatory strike. Of course, he envisioned the underling motivational states differently than do I. Allow me to finish the thought that I began in my introduction to this book:

> The wing-covers open, and are thrust obliquely aside . . . parallel screens of transparent gauze, forming a pyramidal prominence which dominates the back; the end of the abdomen curls upwards. . . . The murderous forelimbs . . . open to their full extent, forming a cross with the body. . . . The object of this mimicry is evident; the Mantis wishes to terrorize its powerful prey, to paralyse it with fright; for if not demoralized with fear the quarry might prove too dangerous. . . . [However], with the smaller crickets . . . the Mantis rarely employs her means of intimidation; she merely seizes the heedless passer-by as she lies in wait. (Fabré, 1912, 74–75)

We know absolutely nothing about the reasons that mantids' responses to visual stimuli vary the way that they do except that both associative and nonassociative learning can play a modulating role in some cases (e.g., Bowdish and Bultman, 1993; Prete et al., 1993). We do know, however, that for many insects, the sensitivity of the sensory receptors that are involved in the initiation of feeding behaviors is affected by hunger level (Bernays, 1985; Bernays and Chapman, 1974; Bernays and Simpson, 1982; Long and Murdock, 1983), and it may also be the case that both visual sensitivity and/or chemoreceptors on the antennae and/or mouthparts of mantids are affected by hunger level, too. For instance, prey will elicit both tracking and approaching behavior from unrestrained mantids from much greater distances when the mantids are food deprived for 7 to 8 days than they will when the mantids are food deprived for only 2 to 3 days (Prete et al., 1993). In addition, there are plenty of anecdotal reports that hungry mantids will attack unusually large prey (Tulk, 1844; Johnson, 1976; Ridpath, 1977; Nickle, 1981; for reviews see Kevan, 1985; Prete and Wolfe, 1992). However, despite the ease with which hunger level can be manipulated, recent experiments have considered this variable only in passing (e.g., Prete et al., 1993), and earlier experiments were too poorly controlled, in my opinion, to offer much reliable information. I would venture to guess, however, that visual sensitivity (i.e., the threshold for the proximate eliciting stimulus for mantid predatory behavior) is modulated by a number of motivational (i.e., physiological) parameters just like taste receptor sensitivity (i.e., the threshold for the proximate eliciting stimulus for grazing behavior) is modulated in insects such as flies and grasshoppers (e.g., Bernays, 1985; Bernays and Chapman, 1974; Bernays and Simpson, 1982; Dethier, 1969; Long and Murdock, 1983).

A second set of interesting "anomalous" data is that pertaining to nonpredatory ingestive behaviors. It has been long thought that mantids are strictly visually guided carnivores. In fact, it was originally proposed that mantids be placed in a genus separate from the phasmids (walking sticks and leaf insects) with which they had been grouped because, unlike the latter, "the *Mantes* . . . confine themselves entirely to [live] food taken from the animal kingdom" (Lichtenstein [1797], 1802, 5). However, when sufficiently deprived, at least two species of mantid will eat fruit which, apparently, they identify by olfaction and locate both olfactorily and visually (Harmer and Shipley, 1922; Prete et al., 1992). We do not know if the antennal drumming that sometimes precedes prey capture conveys mechanosensory and/or olfactory cues to mantids, but it is a possibility. It is almost certainly the case that olfaction and vision work in tandem when mantids identify, localize, approach, and begin eating nonprey ingestants such as fruit (Prete et al., 1992), but we know nothing about the underlying mechanisms of this behavior.

If a novice were to peruse the literature on mantid predatory behavior, I think that he or she would come away with the belief that the mantid's entire visual system is dedicated to the identification and localization of moving prey. That was certainly my initial impression. Quite to the contrary, however, the identification of moving prey is only one task (albeit an important one) that the mantid's visual system must perform. Another interesting task, for instance, is the identification of droplets of water for drinking. In contrast to prey items that are identified, in part, by virtue of the fact that they appear as small luminance decrements in the visual field, water droplets are identified as small luminance increments (or bright spots) in the visual field (Prete et al., 1992). This parceling out of small-field luminance decrements and increments into visual pathways that subserve two distinct behavioral repertoires seems to be absolute. As I noted earlier, although we have tried many times, my students and I have never been able to get a mantis to respond to a small-field luminance increment (a bright stimulus on a dark background) as if it were prey, even when the mantids were unequivocally dark adapted. Once again, of course, we know nothing about the neural or physiological mechanisms underlying any aspect of drinking behavior in mantids, although we do know that there are interesting species differences (Prete et al., 1992).

One More Surprising Result

Just when I thought that I had the entire story of mantid prey identification wrapped up, one of my students noticed an interesting phenomenon. When computer-generated stimuli were moved horizontally across the mantid's visual field, worm/antiworm preferences switched; that is, mantids struck more frequently at horizontally moving antiworms than at horizontally moving worms (Prete, 1993). At first, the result seemed at odds with all of our previous data until I thought about how mantid prey recognition must have evolved.

It is fairly well accepted that mantids and cockroaches (the Blattodea) shared a common ancestor as recently as thirty million years ago (see Roy, this volume, chap. 2, and Prete and McLean, 1995, for references on mantis systematics and evolution). Now, let me make two speculative (but plausible) assumptions. The first is that the common ancestor was an omnivore like a cockroach. So, what needs to be explained is how predatory behavior and prey recognition evolved in mantids (rather than how the cockroach's omnivory

evolved had the common ancestor been a predator). My second assumption is that the common ancestor had a visual system that was sensitive to, and which mediated a behavioral preference for, vertical (versus horizontal) stripes. This preference for vertical stripes is both robust and ubiquitous among extant insects: It is vertical luminance edges that insects tend to approach, orient to, and/or alight upon. Such a preference can be seen in, for instance, backswimmers (Schwind, 1978), flies (Harris et al., 1993), dragonflies (Olberg, 1981), ladybirds (Collett, 1988), grasshoppers (Collett and Patterson, 1991; Sobel, 1990), and mantids (Kral, this volume, chap. 7).

Consider this evolutionary scenario: As the common, roachlike ancestor began relying more heavily on live prey, there was considerable selective pressure for it to recognize small, compact objects and worm configurations as potential food items. The reason, of course, is that this is precisely what the insects moving in its environment would look like. So, my hypothetical common ancestor would be at a selective advantage if it responded with predatory behaviors to such stimuli. I can imagine a visual-motor system evolving in which (1) some descending, movement-sensitive visual interneurons became preferentially responsive to small compact stimuli and stimuli with relatively small leading edges and (2) the movement-sensitive interneurons interfaced with the motor neurons responsible for appetitive behaviors such as approaching and grabbing at the prey. Perhaps some visual interneurons that were originally sensitive to vertical luminance edges, and that played a role in running or jumping toward a dark edge (e.g., Kral, this volume, chap. 7), were co-opted into a prey detector system. Under these conditions, the common ancestor could have come to respond appetitively to the preylike stimulus configurations described in this chapter, while retaining some of the original preferences for vertical luminance edges. The latter preference may have been retained because there could be no selective pressure against it. That is, there are no naturally occurring stimuli that are short vertical stripes moving horizontally across the retinae. Evolutionary pressure to lose the horizontally moving–antiworm preference could only come, for instance, from poisonous beetles walking upright on their hindlegs. But even if such an upright-walking beetle did exist, it would be unlikely that it could make a *horizontal* excursion through a mantid's visual field given that mantids usually perch parallel to the long axes of stems or branches; there is no "stage left" or "stage right" from which such a beetle could enter the visual field. On the other hand, when so perched, a normal beetle walking along a stem will look like a worm stimulus moving upward (ventral to dorsal across the retinae) if it is walking away, or as a worm stimulus moving downward (dorsal to ventral across the retinae) if it is walking toward the mantis.

Obviously, the final questions are these: Is there a visual subsystem in any insect related to the mantids that might represent the type of subsystem from which the mantid's prey detection system might have evolved? And, if such a system exists, is there any evidence that it exists in mantids? The answer to both of these questions is yes. The subsystem is the orthopteroid descending contralateral movement detector system (the DCMD).

A Preliminary—and Admittedly Speculative—Model of the Mechanism Underlying Prey Recognition

The DCMD is a best described as a visual subsystem that mediates emergency (e.g., escape) responses by shunting visual information regarding small-field movement to the appropriate motor neurons of the thoracic ganglia. Although the DCMD is known to exist in a number of orthopteroid insects, it has been studied most extensively in grasshoppers (Barker, 1993; Rowell, 1971; Burrows and Rowell, 1973; O'Shea and Rowell, 1976; Rowell and O'Shea, 1976a, b; Rowell et al., 1977; Rind, 1984; Osario, 1987a, b).

In diurnal insects, such as grasshoppers (and presumably mantids), retinotopically organized information is processed sequentially by a series of three neuropils that lie behind each eye and together make up the optic lobe: the lamina, the medulla, and the lobula (Fig. 8.11A). Retinula cells in each of the many thousands of ommatidia that make up the grasshopper's eye synapse with large monopolar cells (LMC) in the first of these neuropils, the lamina. In the light-adapted state (i.e., during the day), information regarding luminance decrements is passed from the medulla to a giant movement detector cell (LGMD) in the third neuropil, the lobula. In turn, the LGMD responds with a rapidly habituating burst of impulses to any relatively small, dark moving object; it is, however, insensitive to large-field (panoramic) movement (O'Shea and Rowell, 1976). The LGMD synapses onto a large, descending interneuron, the descending contralateral movement detector cell (DCMD), which follows the responses of the LGMD on a one-to-one basis. Hence, the LGMD and the DCMD can be thought of as a single functional unit. The DCMD ultimately synapses on premotor and motor neurons that control the escape jump. A phasic lateral inhibitory network (LIN) between the incoming afferents is situated (most probably) in the proximal medulla (Rowell et al., 1977). The incoming afferent neurons also give rise to feed-forward inhibitory pathways that bypass the LIN and directly inhibit the LGMD (Fig. 8.11B; Rowell et al., 1977).

At first, it may seem odd that I have introduced the idea that a DCMD-like subsystem could play a role in mantid prey recognition, but I think that I can make the case. First, this system seems to be ubiquitous among orthopteroid insects (of which the mantis is one). It even exists, apparently, in the mantid's closest relative, the cockroach (e.g., Edwards, 1982a, b). Hence, it is plausible that such a system, if it exists in mantids, is derived from the hypothetical common ancestor about which I speculated earlier. Second, the DCMD system subserves what I would call emergency responses like the grasshopper's escape jump (or the predatory strike of the mantis?) that must be performed quickly when the appropriate visual stimulus occurs.

The third reason that I think that a mantid DCMD-like subsystem could subserve prey recognition is that there are several key aspects of the DCMD system's functional organization that are consistent with prey recognition by mantids. In general, the DCMD is broadly tuned. It is not

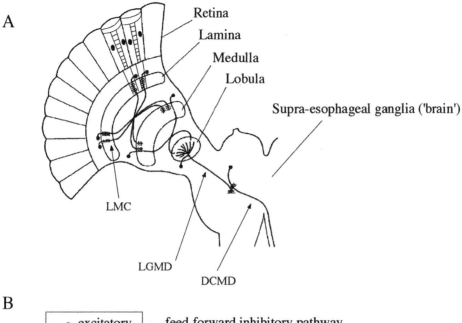
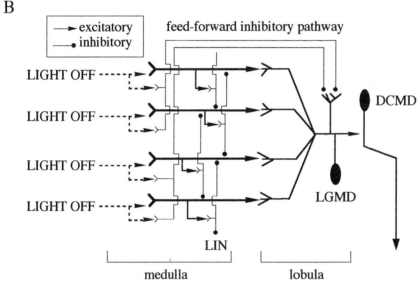

Fig. 8.11 *A*, Schematic of an idealized visual system of a diurnal insect (such as a grasshopper or mantid), which shows the relationship between the incoming parallel afferent channels (from the retina to the medulla) and the LGMD-DCMD subsystem. *B*, Detail of the hypothesized relationship between the structures pictured in *A*, and the two inhibitory subsystems that affect the response characteristics of the DCMD. *A* is redrawn from Young, 1991, with the kind permission of Cambridge University Press. *B* is redrawn from Rowell et al., 1977, with the kind permission of the Company of Biologists, Ltd.

organized such that it responds to a very specific or narrow range of stimuli. Likewise, mantids respond appetitively to a range of stimuli which are defined by roughly the same parameters that define the stimuli to which the DCMD system responds. In specific, the feed-forward inhibitory pathway and the phasic lateral inhibitory network (LIN) between incoming afferent channels biases the system in favor of small moving stimuli that are in the same size range as the compact stimuli to which mantids respond. For instance, panoramic (i.e., background) movement inhibits the DCMD via the feed-forward pathway, and (as I explained earlier) background movement also inhibits mantid responses to preylike visual stimuli. Further, a visual stimulus subtending 40° will elicit no activity from the DCMD. However, the DCMD's response will progressively increase as the size of the stimulus is incrementally reduced to 5° (O'Shea and Rowell, 1976; Rowell and O'Shea, 1976a, b; Rowell et al., 1977). This is the same size-response relationship seen in my mantids. But even more provocative is the finding that the appetitive response rate of intact *S. lineola* and the cellular response rate of the grasshopper DCMD are affected similarly by the very same visual stimuli. For instance, a large (nonpreferred) visual stimulus moving next to a small (preferred) stimulus inhibits the spike rate of the DCMD and the strike rate of *S. lineola* in inverse proportion to the distance between the two stimuli (Fig. 8.12; Rowell et al., 1977; O'Shea and Rowell, 1975; Prete and Mahaffey, 1993b; Prete and McLean, 1996).

The fourth reason that I think a DCMD-like subsystem could underpin mantid prey recognition is this: The most fundamental characteristic of the DCMD subsystem is that it responds to moving stimuli. This responsivity is a product of the spatial and temporal summation of a series of localized responses in the large dendritic fan of the LGMD caused by a succession of small-field luminance changes at the retina (Palka, 1967; Rowell and O'Shea, 1976a). Earlier, I tried to

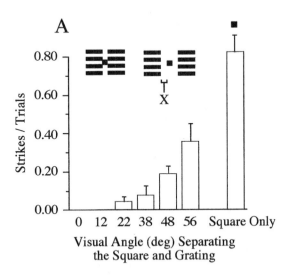
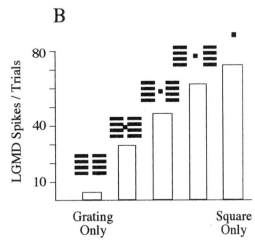

Fig. 8.12 *A*, Demonstration that the proximity of a large stimulus (in this case a pair of gratings) affects the rate at which mantids will strike at a preferred 12° × 12° square in inverse proportion to the distance between the gratings and the square. *B*, Demonstration that the response rate of the grasshopper DCMD to the same type of compound stimuli parallels the behavioral responses of the mantids. *A* is redrawn from Prete and McLean, 1996, with the kind permission of S. Karger. *B* is redrawn from O'Shea and Rowell, 1975, with the kind permission of Macmillan Magazines.

make the case that prey recognition in mantids also requires a succession of local luminance changes that meet certain spatial and temporal criteria such as the total retinal area over which luminance decreases, the time within which a the series of local decrements occur, and so on.

Finally, I think that a DCMD subsystem is sufficient to explain mantid prey recognition because it is inherently sensitive to the visual field location in which a small moving object appears (and remember that so is the mantid). The DCMD subsystem is location sensitive by virtue of the geometry of the LGMD's dendritic fan. Because of that geometry, incoming afferent channels that are presynaptic to it have different potencies depending upon their proximity to the LGMD's spike-initiating zone: "The MD system does not respond equally to identical stimuli at all points on the retina. Instead, there is a rather complex gradient of sensitivity in which responsiveness is greatest to stimuli presented to the center of the eye and declines most rapidly at the extreme edges of the visual field" (O'Shea and Rowell, 1976, 305–306).

Some Preliminary Evidence

If my speculations are on target (and stranger things have happened), I should be able to do two things. First, if I record extracellularly from one of the two cervical nerve cords of *S. lineola*, I should see a number of large action potentials (like those that characterize the DCMD) only when I present the contralateral eye with visual stimuli that the mantis recognizes as prey. (Remember that the DCMD sends its axon down the contralateral nerve cord.) Second, I should be able to construct an artificial neural network computer model (an ANN) of a DCMD-like system, and it should preferentially recognize the same preylike stimuli as does a real mantis. Regarding the former point, our preliminary electrophysiological data do, indeed, support my contention, and I will present some of those data later. The second point requires a bit of an explanation.

Dr. John Bonomo (currently in the Mathematics and Computer Sciences Department at St. Mary's College), one of his senior honors students (Richard Ward), and I collaborated on an ANN project that ultimately surprised us by its success. Briefly, the ANN model that we constructed included (1) an input layer consisting of a grid for which each "pixel" in the visual field corresponded to one functional receptor unit (or ommatidia) in the mantis, (2) a rectangular area in the center of the input grid that was more "sensitive" than the rest of the array (i.e., a "fovea"), (3) a lateral inhibitory filter that operated on the input grid as does the phasic lateral inhibitory network in grasshoppers, (4) an habituation system that modeled the response decrements of the ommatidia to repetitive stimulation, and (5) an output layer that mimicked the hypothesized mantid DCMD—that is, the output layer returned a "strike" or "no strike" message depending upon the particular stimulus "seen" by the input layer.

Now, what is so interesting about ANN systems is that they actually learn. After building one, you "train" it with a set of stimuli and tell it when it has made a correct or incorrect decision. Based on this feedback, it systematically readjusts its internal architecture (the analog of changing synaptic connections between neurons) by means of the algorithms programmed into it. We used a standard backpropagation algorithm (Wasserman, 1989).

Once trained, you "test" the ANN with novel stimuli (i.e., input data) and see if it has learned to behave as does the real, biological system. So, for instance, if you wanted a system that could recognize cars, you might train it to recognize a Ford Escort, a Chevy Corvette, and a Jeep. Once it has learned to recognize these cars reliably (based on your feedback), you can test it with cars that it has never seen to see if the system has learned to recognize "cars in general." Then, if your ANN is behaving properly, you can perform experiments on it to see what role the various component parts play in the overall functioning of the system. For instance, you can turn off lateral inhibition, an experiment that is impossible to do in a real animal. As you can imagine, such a tool can be especially informative if you want to know the necessary and sufficient conditions that lead to the evolution of a particular neural system, or the function of one component of a certain type of neural system.

Our ANNs or mantoids, as we called them, were trained with inputs that mimicked the computer-generated stimuli to which real mantids respond: squares of various sizes that moved in random erratic paths at various speeds. Those mantoids that learned to respond to the training stimuli as would a real mantid were later tested with squares of novel sizes, with worm and antiworm stimuli, and with their lateral inhibition or habituation systems turned off.

Figure 8.13 shows the results for a typical set of experiments with some real mantids and one of our eight ANNs (the one we called Mantoid 14-9). In graph A you will see the average number of strikes per trial emitted by seven real *S. lineola* when they watched three computer-generated square stimuli (5 × 5 mm, 10 × 10 mm, 20 × 20 mm; indicated by the relative size of the symbols) that moved in an erratic path at four speeds. As you can see, the mantids responded typically: Overall, they preferred the intermediate-sized square and emitted the most strikes when the square was moving at 105°/second.[3]

Graph B shows the response rates of a typical mantoid presented with test stimuli after training. (I should point out that there was mantoid-to-mantoid variability that mimicked the variability between real mantids.) The relative sizes of the symbols indicate the four stimulus sizes (1 × 1, 2 × 2, 3 × 3, and 4 × 4 pixels). Speed was represented to mantoids by the number of samplings that the neural network took before the "image" moved to the next location in the input matrix. For example, at speed 4 (the slowest), the image remained at one location for four samplings, at speed 3 it remained at one location for three samplings, and so on. So, speed 1 was the fastest. Mantoid 14-9 was trained only on the three smallest stimuli (indicated by filled symbols); it had never seen the largest stimulus prior to being tested. You will notice that, like a real mantid, it preferred only intermediate-sized squares moving at the second fastest speed.

We found this result interesting in and of itself. It was encouraging to know that our ANNs could be trained to mimic mantid prey recognition and that they would reject novel, nonpreylike stimuli. However, what was even more exciting was that even though the mantoids had never been trained with elongated stimuli, they consistently displayed a worm versus antiworm preference when we tested them with downward-moving stimuli (1 × 4 versus 4 × 1 pixels, Fig. 8.13E). And, as if this were not enough, the mantoids even displayed the "anomalous" switch in preference when worms and antiworms moved horizontally rather than vertically (e.g., Fig. 8.13H)! I surmise that the vertical-worm and horizontal-antiworm preferences in both mantoids and mantids are products of the elongated shape of the fovea in each (e.g., for mantids see Barros-Pita and Maldonado, 1970).

The remaining graphs in Figure 8.13 elucidate the roles played by the lateral inhibitory network and by receptor habituation in the mantoid (and, perhaps, mantid; C, D for square stimuli; F, G and I, J for worms and antiworms). In general, the loss of the lateral inhibitory network caused our mantoids to lose their ability to discriminate stimulus speed, size, and configuration. Graphs C, F, and I demonstrate how the loss affected Mantoid 14-9. The loss of receptor habituation caused the mantoids to lose their ability to discriminate stimulus speed and configuration, and their ability to discriminate size became aberrant. Again, graphs D, G, and J demonstrate how the loss affected this particular mantoid.

So, at the time of this writing, I had convinced

---

3. This speed preference is slightly higher than it was for the mantids whose response rates are depicted in Figure 8.6D. The apparent peak shift may be due to the fact that in the previously reported experiment, I did not test mantids with stimuli moving at 47 or 105°/second. Hence, 105°/second may be the true peak. On the other hand, the peak shift could be due to differences in motivation, age, genetic mix of the cohorts, ambient temperature, and so on. However, the most likely explanation is that the computers were different and, hence, there were subtle differences in the CRT refresh rates and the "smoothness" of the stimulus movement on the screen. In any event, the data in the two graphs are concordant. All of this underscores a point that I made earlier: We know nothing about the factors that modulate prey recognition over time either within or between mantids.

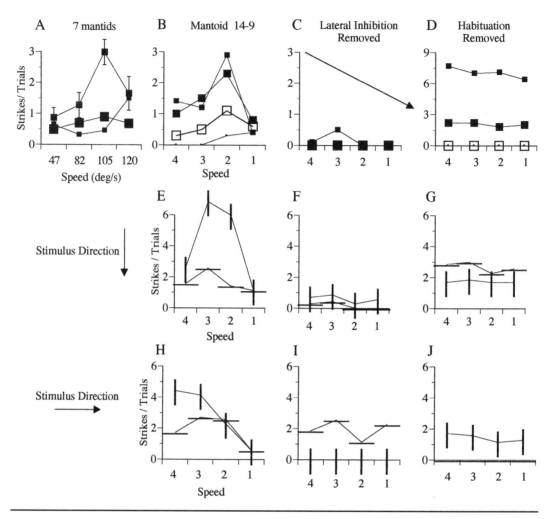

Fig. 8.13 *A*, Typical response rates of seven *S. lineola* to three square computer-generated stimuli moving along an erratic path at four speeds. *B*, The response patterns of an ANN (artificial neural network) that was modeled after the grasshopper DCMD, when it was presented with square stimuli moving at four speeds. (Speed 4 is the slowest.) In *A–D*, the ordinal relationship of the stimulus sizes is represented by the different-sized symbols. The ANN had not "seen" the largest stimulus (open square) before the test. Note that its behavior (as depicted in *B*) parallels that of real mantids. *E, H*, The response rates of Mantoid 14-9 to worm and antiworm stimuli that it had not "seen" before this test. Note, again, that its responses mirror those of real mantids. *C, F, I* and *D, G, J*, indicate what happened to the mantoid's behavior when lateral inhibition and habituation (respectively) were turned off.

myself (for better or for worse) that a DCMD-like visual subsystem is sufficient to account for the prey recognition abilities of mantids. I think that in part, at least, I am correct. Here's why.

Preliminary Electrophysiological Data

Two very skillful graduate students, Mark Gonka and Tim Laurie, have been working diligently in my lab to sort out the ways in which descending visual interneurons underpin prey recognition in several mantid species, including *S. lineola*. Our experiments have begun, of course, by doing extracellular recordings from both cervical nerve cords simultaneously while presenting monocular mantids with the same computer-animated visual stimuli that we have used in the behavioral experiments that I described above.[4] Our preliminary results are very promising and do, indeed, appear to support my hypotheses about a DCMD-like prey recognition system for *S. lineola*.

Of all of the stimulus parameters that define prey, we believe that (1) stimulus location in the visual field, (2) overall size of a compact stimulus (such as a square) or the length of the leading edge of an elongated stimulus (such as a rectangle), (3) stimulus speed, and (4) the direction of stimulus contrast relative to the background are fundamental. So, we decided to begin our electrophysiological investigations by seeing if we could find large, descending, contralaterally projecting, movement-sensitive visual interneurons that would respond differentially to changes in these stimulus parameters.[5]

Figure 8.14 displays a typical set of responses recorded from the contralateral and ipsilateral nerve cords (relative to the open eye) while the mantis watched a prey-sized (11°), computer-generated square stimulus move horizontally through its visual field, at 77°/second, at three elevations. Based on the behavioral data presented earlier (e.g., Fig 8.10A), you may have guessed that any interneurons responsible for prey recognition should respond most when a preylike stimulus moves through the center or lower center of the visual field because those are the locations from which it would elicit the most striking behavior. Traces A–C demonstrate the existence of what appears to be a large, contralaterally projecting (contralateral), descending, movement-sensitive interneuron that was preferentially responsive to movement through or below visual field center. In contrast, the largest descending ipsilaterally projecting (ipsilateral),

movement-sensitive interneuron(s) (traces D–F) showed no such location preference and appeared to be more broadly tuned, generalized movement detectors.

There are several specific points that should be made about these data. First, consider the ipsilateral traces (traces D–F). Each begins with an onset response, which followed the appearance of the stimulus with a latency of 100 to 168 ms. We think that these onset responses function to alert the mantid to the fact that something has changed in its visual field. Then as the stimulus moved from the periphery toward the center, the largest ipsilateral cell(s) began to fire in synchrony with the sequential (100 ms) displacements of the stimulus on the computer screen. These responses were relatively consistent until the square reached a point about 47° to the left of center. This was the edge of the right eye's visual field.

Compare the responses of the largest ipsilateral cells to those of the largest contralateral cell(s) (traces A–C). The first difference is that the contralateral onset responses were more modest. Second, the contralateral cells responded most vigorously when the stimulus moved from about 28° to the right to about 46° to the left of center and (as noted) when it was at or below (horizontal) visual field center. This response pattern suggests that these large movement-sensitive cell(s) are responsible for monitoring small-field movement in the same area of the visual field from which the stimulus would elicit the most predatory behavior.

Interestingly, we found similar results when a prey-sized stimulus moved vertically. For instance, in Figure 8.15 you will see that the largest contralateral cells responded vigorously to a downward-moving 11° × 11° square moving at 77°/second but, again, only when it was viewed by the inward-facing fovea (compare the Contra trace in Fig. 8.15A to that in Fig. 8.15B). On the other hand, the largest ipsilateral cells responded to the stimulus irrespective of its location (compare the Ipsi trace in Fig. 8.15A to that in Fig. 8.15B).

Of course, if the largest descending, contralateral interneurons are really playing the key role in prey recognition, their response properties should parallel the mantid's behavioral responses to the other fundamental stimulus parameters

---

4. The reason that we are recording from the cervical nerve cords is that descending visual interneurons must send their axons though the cervical cords to reach the thoracic motor neurons that control the lunge and the strike.

5. The following data were first presented at the 1997 Meeting of Midwest Neurobiologists in St. Paul, Minn., USA.

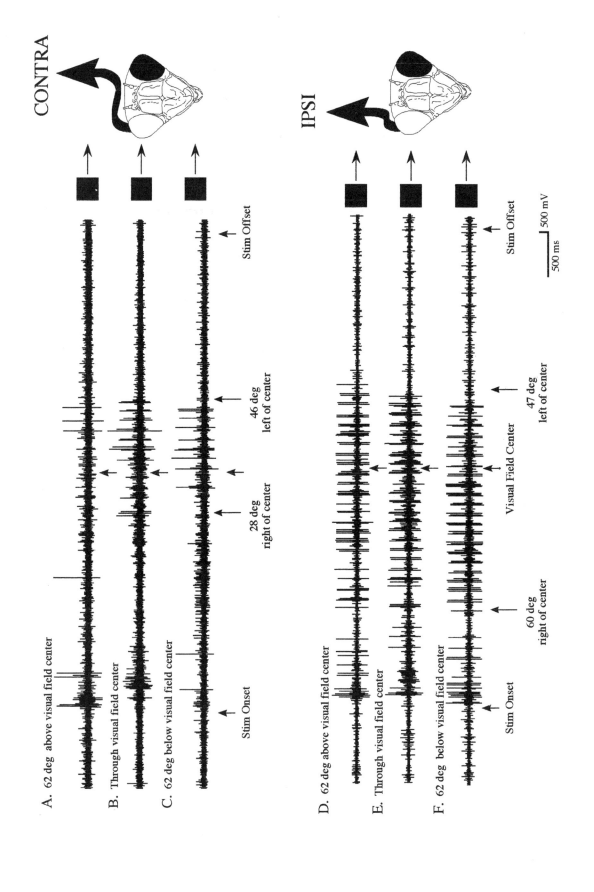

that define prey. For instance, the mantid's behavioral responses to preylike stimuli are sensitive to both stimulus size and speed, and we found that the responses of the largest, contralateral movement-sensitive cells parallel the mantid's behavioral responses. Figure 8.16 shows the responses of contralateral and ipsilateral interneurons to three different-sized squares (1, 11, and 77°), moving horizontally at four speeds (18, 77, 110, or 121°/second), through the center of the mantid's visual field.

Although this seems like a dizzying array of data, there are a few key points that emerge from it all. Each pair of traces contained within a rectangle (A, B, C, etc.) includes the simultaneous recordings from the contralateral and ipsilateral cervical nerve cords, and all of the pairs of traces are normalized such that the distance between stimulus onset and offset is the same along the horizontal axis.

Consider size first: The smallest square stimulus was below threshold for both the largest contralateral and ipsilateral movement-sensitive interneurons (graphs A–D). However, the prey-sized square elicited vigorous responses from both the largest movement-sensitive cells in each of the connectives (E–H). Finally, the 77° square, which is larger than preferred prey size, elicited no responses from the largest contralateral cells but did elicit responses from the largest ipsilateral cells. Once again, these data are consistent with our hypothesis that the largest, descending, contralateral movement-sensitive visual interneurons play the key role in prey identification and that the largest ipsilateral cells are involved in movement perception, generally.

Now consider the effects of stimulus speed by looking just at the changes in the responses to the 11° (prey-sized) square (E–H). When the square was moving at a slow to moderate rate (E and F), both the ipsilateral and contralateral cells responded vigorously to the stimulus and, as before, the contralateral cells responded to movement within a more restricted area of the visual field than did their ipsilateral counterparts. Further, if you look carefully at the traces in E, you will note that the spike rate of both the largest contralateral and ipsilateral cells increased when the stimulus moved through the visual field just left of center. (Look at the right end of each train of large spikes.) This increased spike frequency no doubt reflects the higher density of inward-facing (foveal) ommatidia.

As we have explained, behavioral responses to a prey-sized stimulus decline sharply when the stimulus moves too quickly (i.e., around 120°/second; Fig. 8.6). Graphs G and H in this figure show what happens at the cellular level when this occurs. We believe that there are two critical differences between the traces in G and H, and the corresponding traces in graphs E and F. First, there is a deterioration of the patterning of the responses of the largest contralateral cells, and, second, there is a dramatic increase in the frequency of intermediate-sized spikes in both the contralateral and ipsilateral traces. Our working hypothesis is that these two differences represent two different phenomenological states for the mantid. From the insect's point of view, the spike patterns in E and F represent the presence of a discrete, prey-sized object moving at a moderate speed in an otherwise stable visual environment. On the other hand, the less patterned and, overall, more "crowded" signals (especially in the contralateral connective) seen in H, represent quite a different perception to the mantid, in our opinion. This level of activity from so many movement-sensitive cells is most likely interpreted as a visual space in which there is a great deal of generalized and/or random movement, a per-

---

Facing page

Fig. 8.14 Typical extracellular recordings taken from the ipsilateral and contralateral nerve cord (relative to the open eye) of an immobilized female *S. lineola* while it watched an 11° × 11° square move through its visual field at three elevations. A–C, The contralateral recordings revealing large, descending visual interneurons that responded most when the preylike stimulus moved through the center or just below center of the visual field. D–F, In contrast, the ipsilateral recordings revealed interneurons that are tightly tuned to stimulus movement and which respond over a much wider area of the visual field.

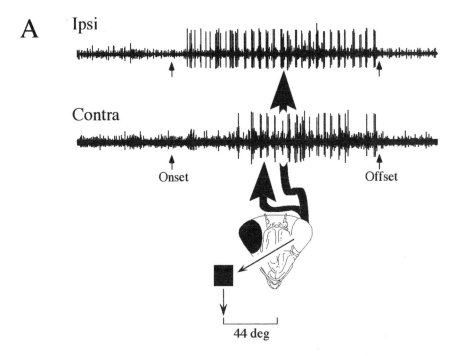

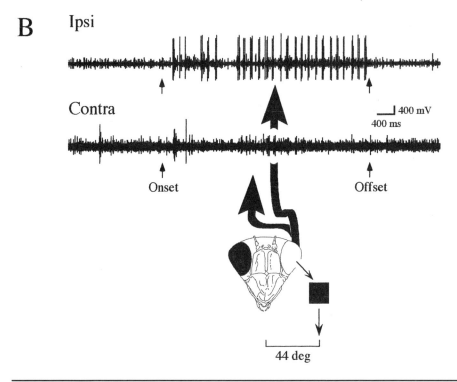

Fig. 8.15 Extracellular recordings done as described in the previous figure except that the mantid watched the stimulus move downward at various locations in the visual field. Large contralaterally projecting interneurons responded only when the stimulus moved through the area of the visual field sampled by the fovea (Contra in *A* versus *B*). On the other hand, ipsilaterally projecting cells received input from sampling units over a wide area of the visual field (Ipsi in *A* versus *B*).

ception quite different than that represented by the traces in F.

The last fundamental stimulus parameter is contrast to the background: *S. lineola* will strike at relatively darker but not at relatively lighter prey-sized stimuli. Hence, the interneurons responsible for prey identification should also be preferentially sensitive to luminance decrements rather than luminance increments. Figure 8.17 demonstrates that this is the case. Here you see the responses of the contralateral and ipsilateral interneurons to both the leading and the trailing edge of a long black bar as it moved vertically through the center of the visual field. The figure really needs little explanation. Although there are some differences in the response frequency of the largest ipsilateral cells depending upon whether their receptive fields are becoming darker or brighter, the more dramatic difference is in the responses of the largest contralateral cells. In a number of experiments, we have found that the latter respond to luminance decrements at a rate almost fifteen times greater than they do to luminance increments. In contrast, the largest ipsilateral movement-sensitive cells respond to luminance decrements at a rate only three times greater than that to luminance increments. All told, all of our data support the contention that it is the largest, descending, contralateral movement-sensitive cells that play the key role in prey identification and, perhaps, in triggering the predatory strike.

Directions for Future Research

Any study that explores the psychophysical, electrophysiological, or neuroanatomical aspects of mantid vision is an appropriate direction for future research, given the fact that virtually nothing is known about the neural underpinning of any aspect of this insect's visual system. My lab is currently vigorously pursuing the prey recognition issue and doing some comparative psychophysical and electrophysiological studies, but this is just the tip of the iceberg.

Mantids use vision for a variety of tasks other than prey recognition, and few of these have ever been systematically explored even with behavioral studies. For instance, we know that vision plays a role in locomotion and flight, in identifying and locating drinking water and nonprey ingestants, and (for males) in localizing and (perhaps) mounting a mate. However, we know little if anything about how vision is used in these behaviors. Further, we know nothing about the underlying neural systems, nothing about how the underlying neural systems are integrated, and nothing about the way that the visual system interfaces with motor systems. Finally, we know nothing about how mantid vision is modulated by motivation. Certainly, visual sensitivity is affected by hunger, thirst, and the reproductive status of the insect. We just have no idea of what way it is affected.

The mantis is a wonderful animal. My heart pounded the first time I saw one, and I am still enamored. The pursuit of any one of the experimental avenues that I have suggested will yield a richly rewarding scientific career. Of that, I am sure.

ACKNOWLEDGMENTS

The work reported here has been supported over the last several years by grants from Youngstown State University, Denison University, DePaul University, The State of Ohio Board of Regents, and The Edmond and Marianne Blaauw Ophthalmology Fund (through Sigma Xi). I would also like to thank the students at the three universities mentioned who have helped me collect data for most of these experiments. In particular, I would like to acknowledge my graduate student, Mr. Mark Gonka. While in my lab, he collected a voluminous amount of electrophysiological data on mantid descending visual interneurons and collaborated fully with me in writing the last section of this chapter.

Overleaf

Fig. 8.16 Pairs of graphs (e.g., A, B, C), each showing the responses of ipsilaterally and contralaterally projecting cells to (*A–D*) a 1° square, (*E–H*) an 11° square, or (*I–L*) a 77° square that moved horizontally through the center of the mantid's visual field at four speeds (18–121°/second). The two largest stimuli (second and third columns of graphs) elicited vigorous responses form ipsilateral cells, but only the prey-sized stimulus (second column) elicited responses from the contralaterally projecting interneurons.

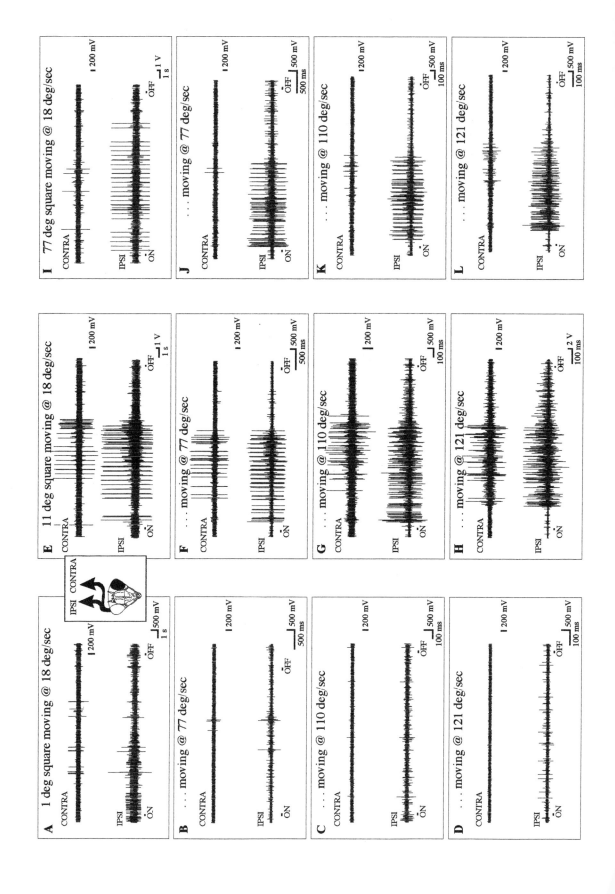

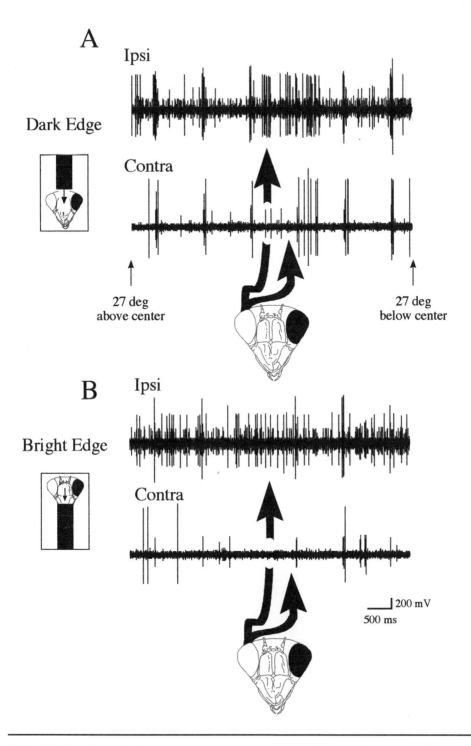

Fig. 8.17 *Above.* Extracellular recordings representing the responses of ipsilaterally and contralaterally projecting cells to (*A*) the leading or (*B*) the trailing edge of a long, computer-generated black bar that moved downward through the center of the mantid's visual field. The most dramatic difference between the effects of the luminance decrement (*A*) and increment (*B*) appear in the responses of the contralaterally projecting interneurons. Once again, these data are consistent with the idea that it is the contralaterally projecting movement-sensitive interneurons that are responsible for prey recognition.

Motor Behaviors

9. Flight and Wing Kinematics

John Brackenbury

Mantids are usually relatively large sedentary insects that prefer to lie in wait for their prey rather than actively hunting for it. These are not the ideal characteristics of a regularly flying insect and, not surprisingly, when mantids do indulge in flight over extended distances (i.e., greater than normal leaping range), it is usually at night when the risks of predation by birds (although not bats, as will be seen later) are minimal. These excursions are far more likely to be performed by the comparatively slender male; the female is disadvantaged by her swollen abdomen, particularly in the gravid state, which increases wing loading and impairs stability. Some species of mantid can readily be induced to fly from the hand, especially if released into their natural surroundings, and on such occasions their characteristically graceful, undulating flight—elsewhere referred to as "goldfinch flight" (Yager and May, 1990a)—presents a graphic demonstration of their vulnerability to aerial predators. In contrast, short bursts of flight activity are probably used on a much more regular basis, during flight-assisted leaping (Brackenbury, 1990, 1991b). The recruitment of the wings may be so transitory on these occasions as to defy detection except with the aid of high-speed photography.

As a corollary to their lifestyle, mantids are frequently cryptically colored, and the wings form an important part of the camouflage in both fully winged and brachypterous species. The forewings are especially important since, when folded, they form the largest exposed part of the body. The mottled tegmina of *Blepharopsis mendica* (Fig. 9.1A), a species found in North Africa and the Middle East, provide a good example of disruptive patterning. Close inspection reveals how the precise pattern of mottling is templated on the underlying distribution of wing veins. The desert-dwelling *Eremiaphila* (Fig. 9.1B) shows the kind of adaptation common to many desert insects: crypsis combined with flightlessness. In the southern and eastern Mediterranean countries *Eremiaphila* occurs naturally in a number of color variants ranging from slaty blue to sandy, each being optimally adapted to a local shade of coloring of the desert floor. The truncated wings enhance the molded, pebblelike contour of the body, and camouflage is so effective that the insect is able to scurry freely across open ground, running down its prey in the habit of a cicindelid beetle. The stop-start nature of its movements has earned *Eremiaphila* the soubriquet "tears of the desert" in the Negev desert in southern Israel. Brachyptery is not necessarily accompanied by loss of function of the flight muscles, at least to judge from the electromyographic activity recorded from flightless phasmids (Kutsch and Kittman, 1991). The retention of the flight motor generator means that even atrophied wings can be used in terrestrial display.

Advertisement or warning coloration is well exemplified in the brilliantly colored hindwings of some species of mantid, such as the European *Iris oratoria* shown in Figure 9.2. Significantly, the tegmina retain the drab coloration of the rest of

A   B

Fig. 9.1 *A*, Disruptive patterning on tegmina of *Blepharopsis mendica*. *B*, Crypsis in *Eremiaphila brunneri*; the legs carry disruptive banding and the truncated tegmina enhance the rounded, pebblelike contours of the body.

the body for cryptic purposes. Exposure of the striking eye spot of the hindwings occurs during a dramatic threat display where it forms part of a suite of behaviors including audible rustling of the wings, extension of the prehensile forelegs, and a very pointed staring at the source of aggression. As in *Blepharopsis*, the color pattern is intimately based on the underlying tracery of the wing veins. The ground color of the remigium is red surrounding cells of transparent wing membrane enclosed between the cross-veins. The vannus has a blue-black or deep violet eye at its base highlighted by concentric areas of brown spots. Even in flight it is likely that the mimicry continues to work: a predator in pursuit of a fleeing mantis, if located directly behind its prey, will be presented with the apparition of a pair of large owllike eyes blinking (along with the wing movement) at approximately 20 to 25 Hz. The violet coloration of the eye spots may be significant in that avian eyes are sensitive as far as the ultraviolet end of the light spectrum.

Flight and Behavior

One of the few existing studies on the ecobiology of mantid flight was prompted by the discovery that some species share with moths and green lacewings aerial bat avoidance behavior (Yager and May, 1990a, b; Yager, this volume, chap. 6). This trait is strictly flight-gated since grounded mantids fail to display it. The response is mediated by a single midline acoustic organ in the ventral metathorax tuned to frequencies of 20 to 60 kHz. In tethered flight other insects show steering responses to batlike pulses, including gryllids, tettigonids, acridids, and cicindelids. The mantid *Creobroter gemmatus* reacted to pulses of ultrasound with a short-latency display of foreleg extension and abdominal dorsiflexion. This is the typical behavior of mantids about to launch into flight and is also remarkably similar to the defense posture of mantids such as *Iris oratoria* under threat (Fig. 9.2). Yager and May (1990b) also drew a parallel between aerial bat evasive behavior and the primary defense to threat on the ground. The purpose of the latter response is understood to be enhancement of crypsis and prevention of detection (Robinson, 1969a; Edmunds, 1972, 1976; Edmunds and Brunner, this volume, chap. 13). The more intense "deimatic" display of the female involving, in the case of *Parasphendale agrionina* studied by Yager and May, exposure of the bright yellow bands on the truncated hindwings, is thought to be a secondary deterrent reaction following discovery by the predator (Maldonado, 1970). The ultrasound-triggered avoidance response may have its origins in earthbound defensive display, carrying the startle function into flight. According to Yager and May, this correspondence of behavioral components probably even extends to the increased forewing stroke angle observed during aerial startle, which is the counterpart of wing raising during terrestrial display.

The average flight speed of *P. agrionina* is 1.9 m/s, only one-half the cruising speed of a bat, so unpredictability in flight pathway is essential if a harried mantis is to evade capture. In free flight *P. agrionina* can resort to sharp turns and dives, and Yager and May (1990a) were able to identify three distinctive types at normal flight pattern. In the first, employed during stable runs over relatively long distances, the body is held at a shallow angle suitable for tracing out a long, straight path. The second, "goldfinchlike" pattern is characterized by sequences of slow climbing flight alternating with stretches of slow level flight and appears to represent a searching mode. The third technique is used in anticipation of landing: flight is slow, with the body tilted at a high angle and the forelegs outstretched. Bat sounds elicit a variety of escape maneuvers, including level

Fig. 9.2 Threat display of female *Iris oratoria*. The wings are raised, exposing the hindwing eye spot. Other components of the response include rubbing the hindwing remigium against the tegmina, abdominal dorsiflexion, and staring at the aggressor.

turns; diving turns with a strong roll component and accelerated wing beat; and steep spiral dives that just shave the ground. The impressive turning repertoire of flying mantids may stem in part from a mechanical device available only to long-legged insects with large, billowing hindwings: the ability to slew the hindlegs to the side making them interfere with the motion of the ipsilateral wing (Brackenbury, 1992).

Wing Folding and Structure of the Wing Base

The structure of insect wings reflects that they are a compromise between contradictory requirements: they need strength but must also be able to deform; they must perform as coherent aerofoils but also, in most cases, need to be folded against the body when not in use. Various strategically arranged creases in the wing fabric permit these conformational changes to be accomplished.

The main stiffening element of the mantid forewing is the thickened radial vein (Fig. 9.3). This, together with the other veins crowded toward the anterior surface of the wing, constitutes the leading edge spar. A concave flexion line, the claval furrow, crosses the wing base at an oblique angle, dividing the remigium in front from the clavus lying medially and behind. A second, convex crease, the jugal fold, runs roughly parallel to the claval furrow and separates the more heavily sclerotized distal segment of the clavus from the soft membranous jugum which forms the posterior part of the wing base. When the wing is retracted after use, inboard movement of the remigium forces the jugum to fold under along the jugal fold. The latter is a true fold line, as distinct from a flexion line, which is normally active during flight (Wootton, 1979, 1981, 1990, 1992). The hindwing (Fig. 9.4) possesses a much more complicated folded nature due to the presence of a greatly expanded, highly pleated vannus, which is in fact the principal lifting surface of the entire wing assembly. The vannus is supported by a system of anal veins radiating out from their connection to the third axillary sclerite in the wing base. The anal veins support the ridges of the pleats, smaller intercalary veins lying along the troughs. The vannus is divided from the stiffer anterior remigium by a convex fold line, the vannal fold. In the folded state, the vannus is reflected beneath the remigium, along the vannal fold, and the pleats are pressed flat against the notum (Fig. 9.5). Retraction of the wing against the body during folding is pivoted on the junction between the first and second axillary sclerites and the base of the leading edge; the underfolding of the vannus is facilitated by the continuity between the vannal fold and the convex fold lying between the proximal and distal median plates.

Folding is driven in part by the automatic recoil of elastic elements in the wing base, as soon as the flight muscles have ceased their activity, and partly as a result of the action of a flexor muscle inserting onto the third axillary sclerite. The sclerite is drawn upward and inward along an arc, pivoted on the dual attachment of the third axillary sclerite to the second axillary sclerite and the posterior wing process. As the remigium is drawn mesad, the vannus, or in the case of the forewing, the jugum, is folded beneath. The main details of this process are similar throughout the different orders of insects: a general description is given by Snodgrass (1935), and a detailed description of folding and the wing base in Heteroptera may be found in Betts (1986).

The kinematics of wing opening at flight initiation have been investigated during flight-assisted jumping in *Mantis religiosa*, *Iris oratoria*, and *Ameles spallanziana* (Brackenbury, 1990, 1991b). Separation of the wings from the notum is begun by a rapid dorsiflexion of the abdomen. At the same time, this movement serves to displace the center of gravity forward to a position more in line with the wing attachments to the body. This effect is further enhanced by extension of the long forelegs, in the manner of a swimmer preparing to dive into the water (Fig. 9.6). In males, the V-shaped attitude of the body is maintained at least for a short time after becoming airborne, but because of their swollen abdomen females often pitch backward and stall soon after take-off.

From rest, the forewings are drawn forward almost in the sagittal plane so that they begin the first downstroke from a position in which they lie back-to-back, leading edges directed outward and trailing edges in contact. Similarly, the fully elevated hindwings are arranged closely together, the remigia directed forward and outward, and the vanni in almost full contact along their dorsal surfaces. *A. spallanziana* and *I. oratoria* both hold this position for a brief fraction of a second (Fig. 9.7) before take-off, ensuring precise synchrony between extension of the hindlegs and the first downward sweep of the wings. It is tempting to view this careful timing of kinematic events as a device to achieve maximum impulse from the combined forces of the leg and the depressor muscles, and several species of Diptera exhibit a comparable synchrony between wing depression and leg extension at flight initiation (Trimarchi and Schneiderman, 1995). The flea-beetle *Chalcoides aurata* frequently uses the same device although the measured take-off momentum is no greater than on other occasions when leaping is wingless (Brackenbury and Wang, 1995). If the latter finding is representative, it suggests that critical timing of the action of the wings and legs might have more to do with directional control than power amplification. In support of this argument it is significant that flight in *Drosophila melanogaster* was appreciably more forward directed when launching was made with the aid of the wings than when the launch was wingless (Trimarchi and Schneiderman, 1995).

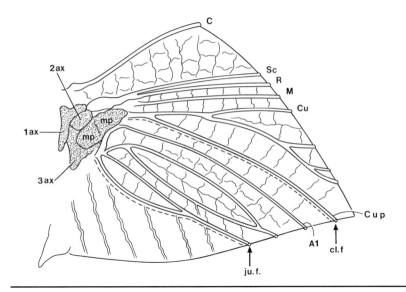

Fig. 9.3 Forewing axilla of *Sphodromantis*, dorsal view. 1ax, 2ax, 3ax - axillary sclerites; mp - median plates. Veins: C - costa; Sc - subcosta; R - radial; M - median; Cu - cubital; CuP - postcubital; A1 - first anal. Fold lines: ju.f - jugal fold; cl.f - claval furrow.

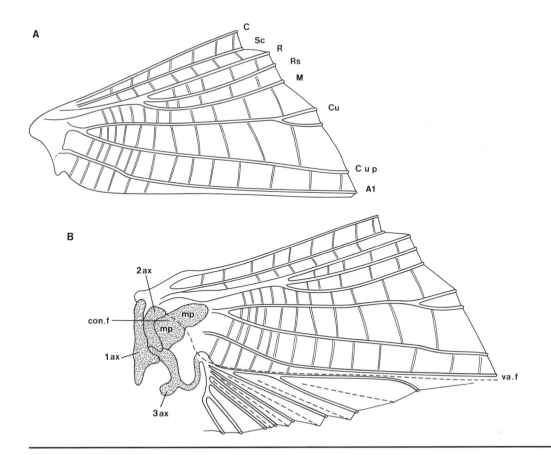

Fig. 9.4 Hindwing axilla of *Sphodromantis*, dorsal view. *A*, wing folded; *B*, wing extended. 1ax, 2ax, 3ax - axillary sclerites; mp - median plates. Veins: C - costa; Sc - subcosta; R - radial; M - median; Cu - cubital; CuP - postcubital; A1 - first anal; Rs - radial sector. Fold lines: con.f - convex fold line; va.f - vannal fold.

A.

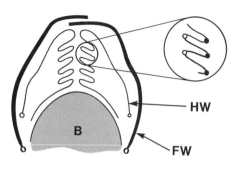

B.

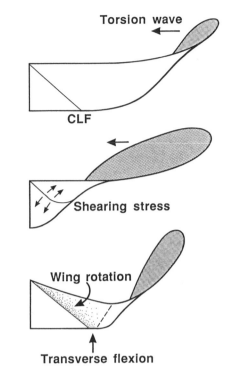

Fig. 9.5 *A*, Diagrammatic cross section through the abdominal region showing the folded wings. FW - forewing, HW - hindwing, B - body. The small circle denotes the leading edge of the wing. *B*, Action of the claval furrow, CLF, during forewing supination. *a*, A wave of torsional deformation spreads inward from the wingtip; *b*, In the absence of supination of the remigium along the claval furrow, shearing stresses would invade the wing base; *c*, Rotation of the remigium about the claval furrow produces a corrugation which restricts wing flexion to the outer part of the wing.

A

B

Flight Kinematics

Even to the naked eye, mantid wings appear to beat slowly and through rather large angles. Quantitative data on free-flight kinematics are lacking, but the wing beat frequency in tethered flight by *Parasphendale agrionina* was 21 to 25 Hz (Yager and May, 1990b). The forewing stroke angle was 60°, most of this being above the horizontal, but still a long way short of the condition in which the wings come together as at flight initiation in *I. oratoria* and *A. spallanziana*. The hindwings of *P. agrionina* beat through 129°, with 90° of this being above the horizontal, in agreement with qualitative observations on freely flying mantids, showing regular downstroke peeling of the hindwings (Brackenbury, 1991b). Both wing pairs undergo changes in shape and profile throughout the stroke which benefit aerodynamic performance but also buffer the stresses experienced within the wing fabric. After making contact at the top of the upstroke, the hindwing vanni peel apart from their leading to their trailing edges, imparting strong positive camber to the wings (Fig. 9.8B), which is further enhanced by the downward deflection of the remigium along the vannal fold. Peeling is a widely used technique for gaining additional lift during take-off and climbing flight, particularly in insects with expanded hindwing vanes. It has been recorded in locusts (Baker and Cooter, 1979), Heteroptera (Betts, 1986), butterflies, bush-crickets, and beetles (Betts and Wootton, 1988; Brodsky and Grodnitsky, 1986; Ellington, 1980; Brackenbury, 1991a), all in free-flight conditions, and also during tethered flight in stoneflies (Brodsky,

Facing page

Fig. 9.6 *A*, "Diving" posture of male *Mantis religiosa* during flight take-off. Both wings have reached the bottom of the downstroke and are about to supinate in preparation for the upstroke. *B*, Upstroke phase of wing beat: the forewing is flexed downward outboard of the claval furrow (white arrow), and the hindwings have developed conical camber. Photographed at an exposure of 1/20,000 second using a high-speed, laser-triggered photographic system. From Brackenbury and Dack, 1992.

A

B

Fig. 9.7 *A*, Male *Ameles spallanziana* poised with wings fully elevated in preparation for flight-assisted leap. *B*, Wing upstroke: the leading edge of the forewing is turned downward; the hindwings are approaching the top of the upstroke and their leading edges are about to come into contact or near-contact.

Fig. 9.8 *A*, Upstroke in *Iris oratoria*. The forewing tip shows pronounced ventral flexure; note the heavily thickened radial nerve, the main stiffener of the leading edge of the wing. *B*, Start of the downstroke; the hindwing vanni have almost completely separated at the end of the peeling phase.

1979a, b, 1981, 1982). The extra force generated results from the interaction between the wings which throws additional vorticity into the wake. Peeling belongs to the general category of "nonsteady" lift-generating mechanisms first identified in insects by Weis-Fogh (1973) and analyzed in detail by Ellington (1980, 1984a, b).

The forewing preserves a relatively flat profile during the downstroke, but around stroke reversal prominent longitudinal, torsional, and transverse flexures make their appearance in the wing surface. These are coincident with supination, which begins as a backward or nose-up rotation of the remigium about the claval furrow (Figs. 9.9, 9.10). The hindwing supinates a little in retard of the forewing, again beginning with rotation of the remigium along the vannal fold. The upstroking hindwing is characterized by a pronounced conical camber (Figs. 9.6B, 9.8A). The terminal stages of the upstroke consist of a wave of pronation, beginning at the leading edge and finally throwing the opposite vanni into contact.

Functions of the Flexion Lines

Controlled deformability is central to insect wing design (Wootton, 1990, 1992). The elasticity of the wing fabric, the distribution of the veins, and the incidence of fold and flexion lines all influence the response of the wing to dynamic loading during flight. Some changes in wing profile are purely passive and result from elastic-inertial interactions within the wing. The dipteran wing, for example, undergoes some active regulation of profile from the wing base (Wisser, 1987), but is primarily designed to permit automatic camber changes as the wing tip twists in response to aerodynamic forces acting behind the torsional axis (Ennos, 1988a, b, 1989). The degree of camber that develops is determined by the number and angle of divergence of the main veins from the leading edge spar. In insect wings possessing large deformable vanni and/or prominent flexion lines, such as those of Orthoptera, Plecoptera, Trichoptera, and Dictyoptera, relative movements of wing areas along specific flexion lines are more important in determining profile changes, and these may involve direct muscular action from the wing base.

Flight and Wing Kinematics   191

Fig. 9.9 *A, Iris oratoria,* start of upstroke. Note the torsional distortion of the forewing tip outboard of the claval furrow and downward deflection of the costal border. The latter appears to be associated with an upward bulging of the radial vein from the wing surface. In the hindwing, the remigium has already supinated, producing reversed or conical camber on the upper wing surface. *B, Ameles spallanziana.* Downward deflection of the forewing tips outboard of the clavus indicates that upstroke is just commencing. The movement of the hindwing lags slightly behind, positive camber on the vannus indicating that it is approaching the bottom of the downstroke.

The role of the most commonly occurring flexion lines, the claval furrow, and the median flexion line (Brodsky, 1985; Wootton, 1979, 1981, 1990, 1992) in regulating profile changes was worked out by Pfau (1977, 1978) and largely confirmed by Zarnack (1972, 1983) in the wing of *Locusta migratoria.* Jensen (1950) first described the so-called Z-profile adopted by the locust forewing during supination. The profile results from a levering down of the costal field along the median flexion line and a simultaneous levering up of the anal field along the claval furrow. The costal, radial, and anal fields thus form the consecutive limbs of the Z. Although these conformational changes bestow the beneficial

Fig. 9.10 *Mantis religiosa.* Both wing couples have just supinated at the start of the upstroke. Note ventral flexure of the forewing tips just outboard of the corrugation formed along the claval furrow and ventral flexure also in the hindwing tips.

effect of reducing lift and drag on the upstroke (Jensen, 1956; Weis-Fogh, 1973; Nachtigall, 1981a, b) their principal importance, as Wootton (1979) notes, lies in controlling the angle of attack and passive deformation of the distal and most aerodynamically effective part of the wing. According to Pfau (1977, 1978), the mechanics of Z-profile formation is controlled by elastic elements in the wing base, but active forces are responsible for profile changes accompanying pronation on the downstroke, namely contraction of the basalar and subalar muscles which pull down the costal and radial fields with respect to the anal field.

The behavior of the mantid forewing during flight is basically similar to that of the locust. The most obvious supinatory displacement is the backward rolling of the remigium about the claval furrow, but both *A. spallanziana* (Fig. 9.7B) and *I. oratoria* (Fig. 9.9A) show evidence of a dip in the leading edge corresponding to the costal limb of the Z-profile of the supinating locust wing. This second displacement appears to be less marked and, if so, falls broadly in line with the finding by Brodsky and Ivanov (1983) that the median flexion line is only poorly expressed and is therefore unlikely to be involved in Z-profile formation in Trichoptera, Neuroptera, Megaloptera, and Mecoptera. The present author also found little evidence of Z-profile formation in the freely flying bush-cricket *Tettigonia viridissima* (Brackenbury, 1990). These comparisons suggest that wings that are superficially alike in structure may not necessarily behave the same way in flight and, furthermore, tethered and free-flight conditions may not necessarily expose the wings to the same kinds of aerodynamic loading.

The Clavus

Flexion lines permit large-scale deformations of the wing, but in addition they can lead to the formation of strengthening elements which embody the principle of the corrugation. For example, the corrugation that forms along the claval furrow of the mantid forewing during supination (Figs. 9.9, 9.10) provides a stiffener for the wing base, isolating it from the torsional and flexural forces acting on the outboard section of the wing. At lower stroke reversal, inertia of the fast-moving wing tip, combined with reversal of movement of the wing base, produces a torsional wave that begins to travel inboard. Simultaneously, the remigium supinates along the claval furrow. The effect of the claval corrugation is to check the inward passage of the torsional wave (Fig. 9.5B): as Figure 9.10 shows, the flexural distortion of the wing stops abruptly at the marginal insertion of the claval furrow. The flexural behavior of the mantid forewing is similar to that described by Brodsky (1979a, b, 1981, 1982) in stoneflies. Brodsky suggests that the sudden unbending of the wing near the top of the upstroke, which he refers to as the *jerk*, provides extra vorticity and therefore lifting force in a manner resembling the classic *quick supination* or *flip* originally described by Weis-Fogh (1973).

When different insect groups are compared, a general correlation is seen to exist between claval morphology and the degree of passive torsion expressed along the wing. At one end of the spectrum, an extremely narrow wing base with a very poorly developed clavus, as is found in Diptera, permits generous twisting which, in this particular case, is beneficial since it also allows the passive development of pitch (Ennos, 1988a, b). The cockroach wing, with a short claval furrow, undergoes strong supinatory torsion, but the classic wedge-shaped clavus of Plecoptera and the elongated clavus of Trichoptera dampen torsion to a much greater extent (Brodsky and Ivanov, 1983). The mantid forewing also lies at this end of the spectrum. An interesting intermediate condition is found in the scorpion-fly *Panorpa germanica* (Ennos and Wootton, 1989); the hindwing base has only a small clavus allowing passive supinatory twisting; the forewing base, with a larger clavus, suppresses passive torsion, and the setting up of pitch in the distal wing is controlled by flexion along the claval furrow.

Hindwing Camber

Mechanical factors controlling the development of camber in umbrellalike hindwings, such as those of mantids and locusts, have been described by Wootton (1979). During the downstroke, the outer edges of the vannus become pulled taut by the lifting force on the wing, but the inner parts of the vannus initially remain pleated. This forces the veins to flex ventrally like sections of hoops, allowing the pleats to open out forming a positively cambered aerofoil (Fig.9.8B, 9.9B). Wootton likened the movements to the opening of an umbrella. The veins supporting the vannal pleats are of two different kinds: the anal vein occupying the ridge of each pleat, analogous to one spoke of an umbrella, is relatively stout with a high cross section at right angles to the wing plane. These veins impart strength to the pleat and also resist dorsoventral bending. Their greater susceptibility to lateral bending in the plane of the wing is compensated by the network of cross-veins. The trough veins of the pleats are slender, with elliptical cross sections oriented parallel to the wing plane: these bend easily in the dorso-ventral direction to allow camber formation.

Future Perspectives in Mantid Flight

The relative ease with which mantid flight can be monitored, particularly in free-flight conditions, makes this group of insects potentially ideal candidates for studies on the ethology of flight. In this respect, the work of Yager and May (1990a, b) on bat-mantid interactions provides a useful pointer to the future. Another clear avenue is the relationship between vision and aerial behavior. The visual capabilities of mantids are well documented, as a scan of the chapter headings in this book adequately testifies, and an analysis of skilled visuomotor tasks during aerial displays in Diptera (Collett and Land, 1975; Land and Collett, 1974) has already been carried out. The escape response of mantids, in the form of accurately directed, flight-assisted leaping, provides a convenient experimental paradigm of the visuomotor control of targeted locomotory behavior. It would then become realistic to attempt, in the same species and in essentially natural conditions, a comparison of visuomotor targeting in two skilled tasks: predatory strike and directed locomotion. Such studies could make an important contribution to the understanding of higher-order neural function in insects.

10. Prey Capture

Frederick R. Prete, Kristy Hamilton

We do not want to begin this chapter with a simple overview of the pertinent literature. Instead, we want to start with a discussion of behavior analysis in general. So, if you know anything about how mantids capture prey, we would like you to abandon what you know and rethink the subject with us. If, on the other hand, you do not know anything about the topic, so much the better.

Let's say that we came from another planet and that we were creatures with stubby, virtually immobile arms. After arriving on earth, we were immediately fascinated by the forelimb dexterity—especially the catching ability—that humans display. In fact, we were so fascinated that we got a job in the biology department at a local university just to study how you folks are able to do all of the reaching and grasping that you seem to do so effortlessly. Luckily, the department offered us some start-up funds with which we were able to hire a few undergraduates to serve as subjects.

Our first experiment went like this. We invited a subject—let's call him Bill—into the lab. Once there, we encased Bill's torso in a concrete vest—being careful that it didn't interfere with any of his arm movements—and we bolted the vest to the wall. After Bill was firmly anchored, we left him for a couple of days in order to pique his appetite.

When we returned to the lab, we brought along a bucket of apples, Bill's favorite fruit. One of us—let's say it was FRP—positioned himself about two meters to Bill's left and about one meter from the wall against which Bill was fastened, and then began tossing apples so that they flew past Bill at about chest height, from left to right. After each toss, FRP repositioned himself a little closer to or farther away from the wall and kept on until he had tossed about four thousand apples.

Surprisingly, after all of these trials, Bill tried to catch only about 500 apples, and he was successful on only 140 trials. Interestingly—but predictably—the proportion of trials on which Bill attempted to catch an apple was normally distributed around a mean Bill-to-apple distance. What we mean by that is Bill made most of his catching attempts when the apples were 50 cm away from the wall against which he was fastened, and he made progressively fewer attempts when the apples were closer or farther away. Specifically, Bill was most successful at catching apples when he grabbed at the ones that were about 40 cm from the wall.

Now, our question to you is this. The experiment has accurately described some aspects of Bill's catching behavior, but to what extent has it told us about human catching behavior in general? Based on our results could we make the claim that human catching behavior is quite stereotyped and that people are maximally motivated to catch things that are 50 cm away from an arbitrary point behind their head (in this case, the wall), or that the reason that humans most often catch apple-sized objects that are 50 cm away is because their fingers must form a precise angle around the object in order to catch it and they can only do so when the object is at this distance? We don't think that we could make these claims based on these data. A human critic might argue that reaching and grasping are not rigidly stereotyped behaviors in people, although the experimental set-up made it look as if they were. If Bill had been free to bend, to lean, to shift his weight from foot to foot, and so on, he would have displayed a much wider range of catching behaviors than he did. The fact that Bill tried to catch an apple on only about 16% of the trials and succeeded on only about 4% seems to indicate that the experimental set-up was not optimally designed for human apple catching. Further, the fact that Bill caught apples that were both closer *and* farther away than 50 cm suggests that it is the range of distances over which he was able to operate that best describes Bill's catching abilities (rather than the average distance at which he could catch an apple). The human might go on to argue that moving objects in the world appear at a variety of (unpredictable) distances. If people had evolved to catch things only at 50 cm away, they would be at quite a disadvantage compared to competitors who could catch things over a wider range of distances.

Finally, regarding our claim that humans must form specific angles between their fingers and thumb to successfully catch an apple, the critic might point out that if Bill most frequently tries to catch apples that are 50 cm away—because he can't reposition himself to catch things at other distances—it will appear that his hand-closing behavior is stereotyped because the motor patterns displayed under these circumstances are just a subset of all of the catching behaviors of which Bill is capable. As anybody who has ever played ball knows, when you are free to move around, your fingers and thumb can form a variety of angles around an object as it is caught: Grasping per se is not stereotyped. For instance, you might first touch the object with your fingertips and draw it into your palm, or it might strike the palm first, after which your fingers will close around it, or (like Bill) you may grasp it simultaneously with your finger tips and thumb. All of these options are possible, and all of them happen in humans allowed to move around freely.

Now, let's say that we take these criticisms seriously and change our experimental approach.

Instead of attaching Bill to a wall, Kristy has him stand on a tiny tabletop, say 45 cm square. Once he is balanced up there, she dangles apples in front of him to see if he can grab them. Interestingly, she notes once again that Bill's catching behavior looks stereotyped. But is it? The same critic might argue that if Bill is to keep his balance, he simply can't display the complete range of movements that he would display if he was, for instance, playing catch with you in the park. This experimental set-up, like the former, makes Bill's behavior look stereotyped although it is not. Only if you let Bill move around freely will you see how variable Bill's behavior is, how dexterous his arms really are, and what the relationship is between his arm movements and the movements of the rest of his body parts. Only then will you really learn for what abilities natural selection has shaped Bill's neuromuscular system. That is, natural selection chooses traits that confer advantages to organisms; it certainly would not have selected for stereotyped catching behavior unless things always appeared in precisely the same relative location. And we know that they do not.

We hope that these facetious scenarios have piqued your curiosity about mantid prey-catching behavior. Our intent was to place you in a position from which you would appreciate prey-catching behavior from our point of view and think seriously about how you'd go about analyzing it if you had to. Although you may not agree with us, we do hope that you empathize with us.

As you may have realized by now, these facetious scenarios mirror the key experiments in what are arguably the two most influential papers published on mantid prey capture behavior. We want to make it clear, however, that our intent is not to criticize the original experiments. They are elegant, insightful, and informative. What we do want to do, however, is initiate a serious reevaluation of the way that the results of these experiments have been interpreted. That is, over the years, the data derived from experimental set-ups in which mantids have been restrained or movement restricted have come to be thought of as exhaustive descriptions of what mantids do in the wild rather than descriptions of what they do under specific experimental situations.

If our facetious scenarios made any sense, then you already have some idea about our point of view regarding mantid predatory behavior. Contrary to popular opinion, we do not think that the motor patterns involved in prey capture are stereotyped although they are, of course, constrained biomechanically (as are all behaviors); you can, however, make them look stereotyped under certain circumstances. Neither do we think that the motor patterns involved in prey capture are best studied in restrained or movement-restricted preparations, although such preparations are appropriate for studying specific aspects of prey-catching behavior.

We are going to flesh out our position in several steps. First, we are going to explain just what motor patterns are under consideration. Next, we will comment on a few early, influential studies on mantid prey-catching behavior. Finally, we will present an analysis of the prey-catching behavior of free-ranging *Sphodromantis lineola* (Burmeister) based on data collected via high-speed video recordings.

The Motor Patterns

Praying mantids such as *S. lineola* are opportunistic, sedentary predators that feed primarily on adult and larval arthropods (Barrows, 1984; Bartley, 1983; Hurd and Eisenberg, 1984). However, they will also capture much larger prey—even small vertebrates—if sufficiently motivated (Tulk, 1844; Frank, 1930; Johnson, 1976; Ridpath, 1977; Kevan, 1985; Nickle, 1981; Prete and Wolfe, 1992). During the day, prey recognition is mediated by vision, and whether or not an object is recognized as prey is determined by the extent to which it meets a certain number of criteria, including its size, speed, configuration, orientation in relation to its direction of movement, contrast to the background, and the distance and direction that its image moves across the retinae. The probability that a mantid will display any predatory behavior(s) is proportional to the degree to which the threshold levels of these stimulus parameters are met or surpassed (for data and references, see Prete, this volume, chap. 8).

Predatory behavior per se is sometimes preceded by orientation and/or tracking movements of the head and/or prothorax—but neither of these behaviors indicate that an object is or will be recognized as prey (i.e., that predatory behavior will ensue). The purpose of these orienting movements is to move the image of the object onto the fovea of the compound eyes and into the binocular visual field (Lea and Mueller, 1977; Rossel, 1983, 1986; Kral, this volume, chap. 7).

Although generally an ambush predator, once prey is identified as such, mantids may pursue their target (Holling, 1966; Prete et al, 1993). Then, when prey is sufficiently close, the terminal act of prey capture is performed. Contrary to popular belief, prey capture consists of two distinct behavioral components which, unfortunately, are generally referred to as if they were one. The key component is a rapid extension and grasping movement made by the raptorial forelegs. This is the strike. The other component is a displacement of the mantid's body effected by the mesothoracic (middle) and metathoracic (hind) legs. This is the lunge. A lunge does not always accompany a strike. Whether or not a lunge occurs depends upon how far away the prey is perceived to be.

A Few Early Studies

The mantid's dramatic method of prey capture and the comparative ease with which the behaviors can be elicited have made it an interesting model for the study of what was thought to be a somewhat stereotyped response to a relatively circumscribed set of visual stimuli (e.g., Barros-Pita and Maldonado, 1970; Corrette, 1980, 1990; Levin and Maldonado, 1970; Maldonado and Barros-Pita, 1970; Maldonado, Benko, and Isern, 1970; Maldonado and Levin, 1967; Maldonado, Levin, and Barros-Pita, 1967; Mittelstaedt, 1957, 1962; but see Copeland, 1979, for a different point of view). Interestingly, beliefs as to the precision of mantid vision have actually been derived from behavioral studies of the predatory strike: The idea is that the mantis *must* be able to precisely localize prey *because* its strike so stereotyped—that is, if prey could not be precisely localized, the strike would miss its target. In fact, the claim that the mantid has the ability for "perfect localization [of prey] for a successful hit" appears at the very beginning of the first article that we want to review.

In this article, the authors explain that they are interested in "the distance estimation problem" because it is "an indispensable indication for a successful strike" (Maldonado, Levin, and Barros-Pita, 1967, 238). These statements suggest an underlying assumption that mantid predatory behavior is made up of a series of discrete, stereotyped components: The mantis precisely localizes its prey, (if necessary) moves into the right position, and then executes a stereotyped strike. In fact, over the last several decades a number of investigators have argued that prey capture is so stereotyped that the forelegs must form precisely the right angle around the prey and that prey must initially contact the midpoint of the femur if a strike is to be successful. One has even argued that the strike is so stereotyped, and mantids must move into precisely the right position prior to it because they have not evolved the ability to steer the strike to the right or the left (Corrette, 1980, 1990). In a related argument, some have claimed that prey capture is optimally triggered by a flylike stimulus (Rilling et al., 1959) that appears at the optimum distance from the mantid. The general argument has even been extended to include the claim that the mantid's raptorial forelegs have evolved to be optimally designed to capture and hold the optimum fly-sized prey when just the right femur-tibia angle is formed around it during the strike (e.g., Holling, 1964; Loxton and Nicholls, 1979). Do you see the problem here (as explained, for instance, in Gould, 1979)? In these arguments, the emphasis is on finding the way in which mantids are ideally adapted to capture the ideal prey under the ideal conditions. The problem is that the arguments do not lead to models that are consistent with what unrestrained mantids do under normal conditions.

Although a few (now unfamiliar) studies preceded it, probably the most frequently cited, and arguably the most influential, early study of mantid predatory behavior was published by Maldonado, Levin, and Barros-Pita in 1967. In a series of ingenious and innovative experiments this team studied the strike of *Coptopteryx viridis* (G.-Tos). As noted above, the explicit purposes of this study were "to determine the optimum mantid-prey distance for hitting," and the "relationship between these maximum distances and the length of the forelegs." The latter apparently contributed to the investigators' decision to describe distances in terms of percentage of foreleg length (%FL) rather than in millimeters. This rather unintuitive metric was made more enigmatic because in this and some subsequent papers distance to the prey (which is divided by foreleg length to yield %FL) was measured from the anterior end of the prothorax—that is, an arbitrary point behind the mantid's head (recall the experiment with Bill)—and foreleg length (FL) was considered to be only that portion of the foreleg extending beyond the anterior end of the prothorax. Hence, mantid-prey distance was always overestimated, and foreleg length was always underestimated.

In the experiment, dead flies were moved mechanically around the inner surface of a cylindrical arena in which a mantid was tethered. In the first set of experiments, mantids were embedded in a plaster block, reminiscent of Bill in our opening fantasy. In other experiments, the mantid's feet were permanently affixed to a plate with tacky wax, and, in a final series of experiments, untethered mantids were confined within a small arena. (By the way, throughout this chapter, we refer to the first type of preparation as immobilized, and the latter two types as movement restricted.)

Although this and several subsequent studies by this lab generated some very important data about how mantids capture prey, the experimental set-ups constrained the mantids in precisely the same way that Bill's behavior was constrained. For instance, both the response and success rates for mantids embedded in plaster blocks were as low as were Bill's for about the same number of trials, and the mantids' predatory behavior seemed as limited as was Bill's. Not surprisingly, the investigators found that there was a positive correlation between the length of a mantid's foreleg and the mantid-prey distance at which it successfully captured prey. They also found that immobilized mantids strike most frequently when prey was at 70 to 80%FL, but that they are most successful at capturing prey at slightly shorter distances.

Interestingly, subsequent misinterpretations of these data led to two still widely held but erroneous beliefs about the predatory strike. The first has to do with the fact that mantids struck relatively less frequently at prey that was only 40 to 50%FL away. (Maldonado and his team did not test at prey distances shorter than 7–8 mm.) Although the original investigators understood that this phenomenon required "further analysis," their data came to be understood to mean that mantids either will not or *cannot* capture prey when it is this close or closer.

The second misconception has to do with what the authors called "an outstanding observation: the maximum motivation class—70–80%[FL] does not agree with the that of maximum efficiency—40–50%[FL]" (p. 245). The way the authors handle this inconsistency is perfectly reasonable given their experimental set-up. However, the relationship between the authors' explanation and their set-up was, apparently, forgotten by subsequent investigators and led, therefore, to an odd set of beliefs. The original authors' explanation began with a convincing drawing (Maldonado et al., 1967, Fig. 7) of what happens when an immobilized mantid tries to capture an object that is slightly out of its reach. The first frame of the diagram depicts a mantid striking at a sphere that is so far away that it is touched only by the tibia. Because the mantid is immobilized, of course, the hypothetical sphere is propelled out of its grasp. This would be equivalent to our subject Bill trying to catch an apple that was thrown just out of his reach. You can imagine that, if he reached for it and just managed to touch it with his finger tips, he would not be able to grab it.

What Maldonado and his colleagues argued next was that when a mantid is sufficiently close to a prey object, the tibia and femur close symmetrically and simultaneously around the object so that the prey is trapped between the femoral and tibial spines. In order for this to occur properly, the mantid must make some initial postural adjustments and execute a precise series of movements during the strike itself. These behaviors were revealed to the investigators by mantids whose feet were tacked down or who were restricted to, but otherwise untethered within, a circular arena. When presented with the same mechanically driven flies, the authors found that their mantids initially made some preparatory movements that placed them in what they called the "optimal" position for striking—this included precise positioning of the prothorax (e.g., p. 252). Then, during the strike, the mantids lunged a distance that was on average "about 35% of the 'foreleg' extension" (p. 255). You will note that such a lunge precisely makes up the difference between the two distances that had been dubbed the "maximum motivation class—70–80%[FL]" and the class "of maximum efficiency—30–40%[FL]."

Interestingly, not only did the investigators find that the lunge spanned just the right distance, they also found that the angles assumed by the forelegs prior to and during the strike represented the optimum positions necessary in order to allow the femora and tibiae to simultaneously grasp the prey. In their words, the "lunges tended to place the animal at a distance at which the grasping mechanism worked in optimum conditions" (p. 256). So, their conclusions were that the maximum motivation distance to strike lies between 70 and 80% of the "foreleg length . . . optimum hitting is achieved at . . . 30–40% of 'foreleg' extension . . . [and] the gap between these to distances . . . is filled by a decisive dynamic component . . . the lunge. Computation previous to the strike has to take this lunge into account in order to hit successfully" (p. 256). Now, all told, these concluding remarks were insightful and reasonable, given (as I have said) the constraints of the experimental set-up. The historical and scientific issue at hand is the degree to which these conclusions can be generalized to unrestrained *C. viridis* or, more important, to mantids in general.

The reason that we have spent so much time discussing this study is that the authors' conclusions—but not the context in which they were voiced—have become entrenched and have slowed the recognition of the plasticity of mantid predatory behavior (but, once again, see Copeland [1979] for a refreshingly different but, unfortunately, too long ignored point of view).

It is not surprising that, by using virtually the

same methodology, Balderrama and Maldonado (1973) found that the strike and lunge of *Stagmatoptera biocellata* also appeared quite stereotyped. Unfortunately, they, too, expressed the relative distances of the two behavioral components as percentages of foreleg length (the static component, or the strike, covered 60%FL, and the dynamic component, or the lunge, was 20%FL). However, comparison of the results of these experiments to other studies is a bit difficult in that some details of the methods were not given in the paper, and the mantis-prey angle (the angle between the long axis of the mantid's prothorax and the prey) remained fixed at an arbitrary 40°. The investigators concluded that mantids can precisely estimate the location of prey at all instars and that strikes do not occur within a wide range of distances as one would expect if the behavior's success was based on what they termed "random probabilities" (Balderrama and Maldonado, 1973, 334). From this, we assume that they mean simply that the strike is rather stereotyped, that is, that the strike is not a broad sweep of the general area of the visual field in which the prey is perceived to be located. On the contrary, however, we think that it is just that.

In contrast to claims that the strike is invariant and stereotyped, there were a few studies that argued that it is not only variable, but could be modified by learning. Unfortunately, these papers have all but disappeared from the field's collective memory. For instance, Rilling et al. (1959) and Gelperin (1968) claimed that learning can modify prey capture efficiency and prey selection in *Parastagmatoptera unipunctata* and an unspecified species of Hierodulae (Rilling et al., 1959) and in *Paratenodera sinensis* and *Mantis religiosa* (Gelperin, 1968). Although Maldonado (1972) could not duplicate Gelperin's results, he did find that prey-catching behavior in *Stagmatoptera biocellata* could be modified by learning (Maldonado and Tablante, 1975). In addition, in an extremely valuable but (unfortunately) relatively uninfluential paper, Copeland and Carlson (1979) found that the lunge of *Tenodera aridifolia sinensis* (Sauss.) was quite variable in both distance and direction. Estimating lunge distance from the displacement of a styrofoam ball that tethered mantids held in their middle legs and hindlegs, Copeland and Carlson reported lunges ranging from 6.6 (± 1.77 s.d.) mm to 23.3 (± 1.18 s.d.) mm when stimuli were positioned at the mantid's eye level. Even more important, they found that a tethered *T. sinensis* would direct its lunges appropriately even toward stimuli that were presented at positions as much as 48 mm above or 49 mm below its head, from 13 to 30 mm in front it, or as much as 45° to the side of its head. These data paint a different picture of the mantid's method of capturing prey than do earlier studies. It is too bad that they did not have more of an impact (see Prete and Wolfe [1992] for an explanation of why they did not).

A Helpful Coincidence

When each of us saw a praying mantis for the first time, we were struck by the flexibility and variability of its predatory behavior. They can catch prey when it is above their head or below their feet. They can catch prey that is on the ground directly below their head and prey that is more than an "arm's length" away. Sometimes they lunge, sometimes they do not. And, interestingly enough, they can catch prey that is far to their right or to their left. They are extraordinary beasts.

Our early observations taught us one important thing about mantids: They never read the scientific literature. They are remarkably unlike the relatively stereotyped creatures described in most of the articles that we read. The closest approximations to a mantid's real-world behavior were those in parts of Jonathan Copeland's doctoral dissertation (Copeland, 1975; also see Copeland and Carlson, 1979). As mentioned earlier, Copeland found that the lunge of *T. a. sinensis* was, in fact, quite variable. Following Copeland's lead, Prete decided that his first experiment should be an analysis of the strike of completely unencumbered *T. a. sinensis*. That first manuscript was submitted for publication early in 1989. Sadly, the paper was rejected on 2 May of that year. Among the criticisms in one of the reviews was the suggestion that the author read an obscure doctoral dissertation completed at the University of Oregon in 1980. The dissertation was obscure in the sense that it was not cited in the mainstream literature on mantids and its author had never published a paper based on it. The dissertation was "Motor Control of Prey Capture in the Praying Mantis, *Tenodera aridifolia sinensis*," (Corrette, 1980). A particularly interesting point about Corrette's data was that his movement-restricted *T. a. sinensis* initiated strikes and successfully captured prey over considerably smaller areas than did Prete's freely moving mantids. However, it was very hard to tell precisely where those areas were because the data were presented in a scatter plot in which prey distance (expressed as %FL) was plotted against relative prey angle (in degrees). So, to help readers compare the two data sets, Prete converted Corrette's scatter plot into a simpler graphic and added it to the manuscript.

Interestingly, while the revised manuscript was out for review, an article appeared in the *Journal of Experimental Biology* in which Corrette summarized the results of his dissertation of a decade earlier (Corrette, 1990). Happily, in that paper he had changed the way in which he depicted mantid-prey positions: Rather than using the original scatter plot, he, too, used a pair of polygons (compare Prete et al., 1990, Fig. 1, with Corrette, 1990, Fig. 4B.). This coincidence allowed comparison between the behavior of movement-restricted and freely moving *T. a. sinensis* as shown in Figure 10.1. In the graph, relative prey positions in the midsagittal plane at strike initiation (A) and at the hit (B) are plotted in relationship to the mantid's eye, and the axes simply represent distance in terms of %FL. Corrette's data are represented by the superimposed polygons.

The Data on *T. a. sinensis:* Making a Long Story Short

Based on a remarkably detailed analysis of how his mantids captured prey, Corrette (1990) came to a number of general conclusions, many of which were not applicable to the behavior of unrestrained mantids. Although it is unnecessary to analyze all of the conclusions, we want to mention several that have become more or less widely held. The first is that the strike is made up of "several stereotyped components which together [form] a single movement sequence of all six legs. . . . That is, the strike and lunge form a single attack movement sequence which is adjusted . . . by the same underlying mechanism " (Corrette, 1990, 147, 174). Although this is generally believed to be the case, the fact of the matter is that a lunge does *not* always accompany a strike; hence, prey capture is not a single movement sequence of all six legs. The strike and lunge are separate behaviors that can, when necessary, be coordinated. The appearance that they are one stems from the fact that if you entice mantids to strike at prey only when it is some distance away a lunge will always accompany a strike and it will appear that the two are inextricably bound.

The second conclusion is that "the attack is only steered within the median plane of the prothorax. . . . Lateral steering has presumably not evolved, since it would require a complex neuromuscular control beyond the capabilities of the mantid nervous system" (p. 176). In actuality, unrestrained mantids can easily steer their strike to one side or the other (see below). The third widely held belief is that "the movements of mantid prey capture have a single goal: to bring the spined region of the femora into contact with the prey. . . . The exact mechanism for grasping prey sets limits on the final position of the femora. . . . The largest spines on the femur are necessary for accurate capture of the prey, and the grasping mechanism . . . implies that an exact angular orientation of the spined surface of the femora is important for optimal function" (p. 176). Again, this is not necessarily the case, as we will explain in a moment. In actuality, the femoral spines do two things: First, they make it more likely that prey will be snagged and held by the foreleg and, second, the movable spines serve as tactile sensilla providing sensory feedback letting the mantid "know" that it has something in its grasp (Prete, 1990a; Prete and Cleal, 1996).

Basically, mantids are no different than other bugs. Other bugs can steer to the left and to the right, and so can mantids; other bugs have spines that are tactile sensilla, and so do mantids; other bugs drink water, and so do mantids; other bugs can reach, walk, and jump, and so can mantids. Natural selection did not create an ideal, single-

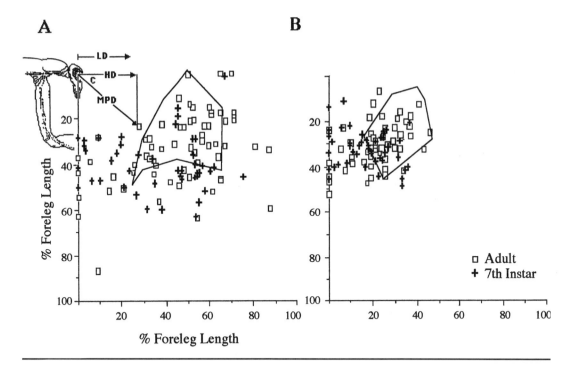

Fig. 10.1 The relative positions (indicated by rectangles and crosses) of prey at (A) strike initiation and (B) the Hit (the time of initial foreleg-prey contact) for free-ranging *Tenodera aridifolia sinensis*. The polygons represent the same information for *T. a. sinensis* that were restricted to a small platform (from Corrette, 1990). Based on these data, free-ranging mantids displayed considerably more variability in several key measurements of prey capture ability, including MPD (mantis-prey distance) and HD (horizontal distance measured along the axis of the prothorax) over which prey could be caught, LD (lunge distance measured along HD during the strike), and angle C (the angle formed by the vectors HD and MPD at strike initiation). Distances are expressed as percentage of foreleg length (coxa + femur + tibia; tarsus excluded). Reprinted from *J. Insect Physiol.*, vol. 36, F. R. Prete, C. A. Klimek, and S. P. Grossman, "The predatory strike of the praying mantis, *Tenodera aridifolia sinensis* (Sauss.)," 561–65, 1990, with the kind permission of Elsevier Science Ltd., The Boulevard, Langford Lane Kidlington, OX5 1GB, UK.

purpose automaton that we now call *mantis*. It created an insect in which the basic mechanisms and structures that existed in the putative roach-like ancestor were modified over time such that mantids are able to be sufficiently successful predators to maintain themselves within their respective niches. Mantids are just as intricate, just as complex, and just as sophisticated as are other insects. In fact, they share many of the characteristics of the other orthopteroids, which includes their closest relative, the cockroach.

Unencumbered Mantids Catching Crickets

If you were interested in our ability to catch moving objects—that is, if you were interested in the mechanisms by which one of us could integrate visual information and target-directed motor output—you would not do to us what we did to Bill in the opening scenarios, would you? Instead, you'd probably film one of us catching apples that were thrown to as many relative locations as possible. In fact, you'd probably try your best to be sure that our catching behavior was tested over its entire range. Well, that is just what we did to examine the prey-catching behavior of adult, female *S. lineola*.

The results that we are going to present were gathered by presenting free-ranging mantids with adult crickets impaled on a thin stainless steel wire. The mantid's prey-catching behavior was recorded with a high-speed video system (200 fps; 20 &s shutter speed) in the lab of Roy Ritzmann (Department of Biology at Case Western Reserve University). The mantids were taped standing on one of two large flat stages or, sometimes, on a raised platform. One of the stages was glass under which a mirror was mounted so that when the center of the camera lens was placed level with the stage, both side and ventral views could be filmed simultaneously.

Prey Capture Zones in the Midsagittal Plane

Most early studies of mantid predatory behavior were based on the insect's ability to capture prey presented in the midsagittal plane, and we followed suit. The measurements used to analyze the movements of these mantids are diagrammed in Figure 10.2, although the figure may seem a bit overwhelming. We will try our best to explain what happened without having to use this rather enigmatic jargon.

The first thing that we learned from these data was that *S. lineola* actually have two basic positions from which they will strike at prey in the midsagittal plane. They assume one position when prey is relatively high (more than 35° off of the substrate) and another when it is relatively low (less than 35° off of the substrate; Fig. 10.3A). We think that these are really two distinct postures (and not two halves of a continuum) because, when we compared the positions of various body parts prior to high and low strikes, all but one pair of the measurement distributions were significantly different (Fig. 10.3B–I). You can see how different these two are by comparing the typical high and low postures depicted in Figure 10.4. (By the way, these data are consistent with data collected on free-ranging *T. a sinensis*; Prete et al., 1990, 1993).

We want to reiterate the point that both free-ranging *T. a. sinensis* and *S. lineola* can capture prey that is quite close to, and even directly below, their heads. This is an important point because, as noted earlier, it is often claimed that prey capture at close range is impossible for any of the following reasons: (1) mantids are biomechanically constrained from doing so, (2) a lunge must always accompany a strike, or (3) prey capture at close range is a result of a reflexive grasping movement elicited by being bumped by the prey. Our data do not support any of these claims (also see Prete et al., 1990).

When we learned that initial strike postures fall into these two classes—either high or low—it dawned on us why previous investigators had seen the strike as so stereotyped. If you are familiar with the classic studies on mantid predatory behavior, you will recognize the fact that low strikes were rarely (if ever) included in the analyses. This is important, as you will see, because low strikes are much more variable than are high strikes (although even high strikes are not as invariant as has been claimed).

Based on a number of regression analyses, we can summarize the differences in initial high and low strike postures like this: Prior to high strikes, mantids tilt their head upward and raise their prothorax proportionate to the elevation of the prey. In addition, their abdomen pitches downward, and they straighten their middle legs in direct relation to the degree to which their head and prothorax are tilted upward (Fig. 10.4A). In contrast, prior to low strikes mantids position themselves increasingly closer to the substrate and shift their body farther forward over their mesothoracic tarsi (middle feet) as the elevation of the prey decreases (Fig. 10.4B).

Overall, these data taught us that when orienting to prey, *S. lineola* make postural adjustments that accomplish two goals: One is to assume a position that will allow it to reach the prey given the mechanical constraints of its body. The second is to position the prey in the lower half of its visual field. Regarding the latter point, we found only one strike in which the entire image of the prey was above visual field center, and only one other strike in which the center of the prey image was more than 2° above visual field center. These results agree with the fact that stimuli moving 0 to 25° below visual field center elicit significantly more predatory strikes than the same stimuli moving 0 to 25° above center (Prete, this volume, chap. 8).

In spite of the relatively large range of positions from which we were able to elicit strikes in these experiments, neither of us would claim that

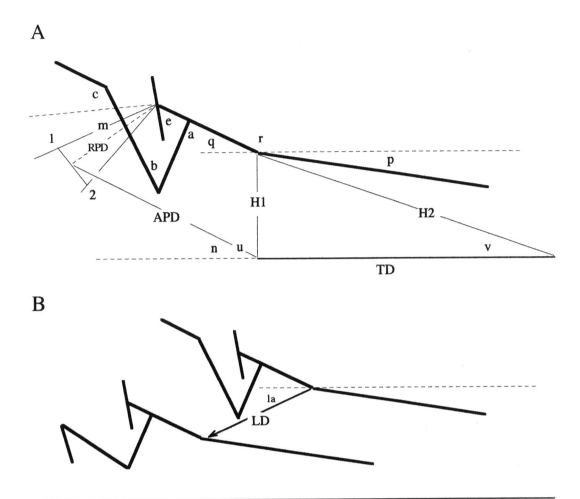

Fig. 10.2 The measurements used to describe the positions of mantids and their prey at various times during the strike. In all cases, lowercase letters represent angles and uppercase letters represent distances. Most of the measurements are self-explanatory; however, a few should be clarified. The pitch of the prothorax and abdomen from horizontal (dashed line) is indicated by angles q and p, respectively, and angle r is that formed between the long axes of the prothorax and abdomen. Following the convention of Corrette (1980, 1990), the relative positions of the mesothoracic and metathoracic legs are measured indirectly via three line segments (H1, H2, and TD) that form a triangle, the vertices of which are the tips of the ipsilateral tarsi and the articulation point between the prothorax and pterothorax. Rostral-caudal movements in the sagittal plane are reflected in the changes in the interior and exterior angles formed, respectively, between H1 and H2 and horizontal (angles u and v, respectively). Extensions of the mesothoracic and metathoracic legs are indicated by changes in the lengths of H1 and H2, respectively. The position of the prey in relation to the mantis was measured from two points of view. The "relative" prey distance was determined by constructing two lines that defined the visual angle subtended by the body of the prey, and measuring the length of the line segment (dashed line RPD) that bisected the visual angle and extended to a perpendicular line segment tangential to the outline of the prey (line segment 1-2). "Relative" prey angle (angle m) represents the position of the center of the plane upon which the image of the prey is projected (line segment 1-2) relative to the center of the mantis's visual field; angle m was measured between RPD and a line perpendicular to the long axis of the mantis's head. "Absolute" prey distance (APD) is the distance between the tip of the mesothoracic tarsus and the intersection of RPD with the projection plane of the prey. Absolute prey angle (angle n) was measured between APD and horizontal. Reprinted from Prete and Cleal, 1996, with the kind permission of S. Karger.

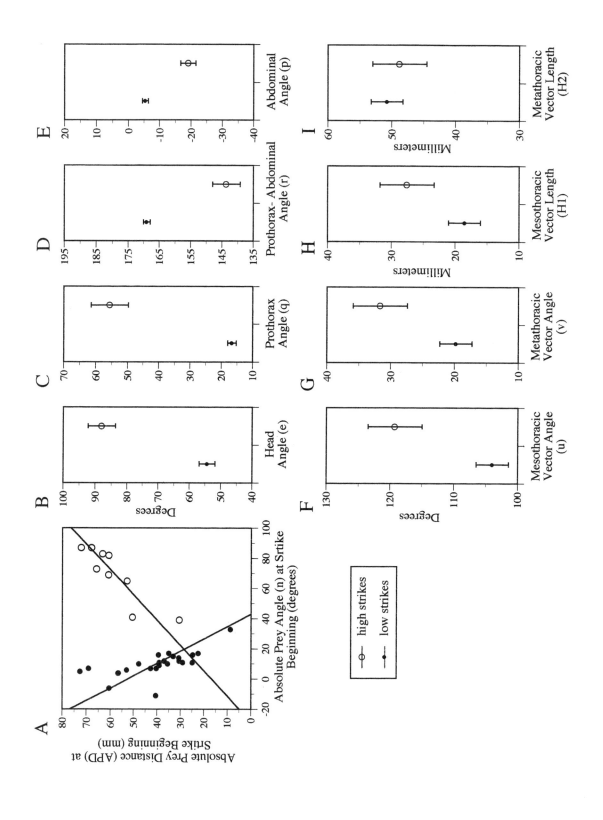

we have mapped the entire midsagittal capture zone of *S. lineola*. For one thing, high strikes are more difficult to elicit than are low strikes, so it may be that *S. lineola* will, if sufficiently motivated, emit high strikes (and, for that matter, low strikes) over an even larger range than we saw. Further, we are pretty sure that prey capture zones vary between sexes and species.

Lunges in the Midsagittal Plane

Depending upon the position of the prey relative to the mantid, the strike may be accompanied by a displacement of the body effected by the middle legs and hindlegs. This displacement is called the *lunge*. Its name notwithstanding, however, the lunge is not a simple linear movement "directed ... along the long axis of the prothorax" which functions to place "the base of the forelegs in the position that the compound eyes had at the beginning of the lunge" as has been claimed (Corrette, 1990, 163, 177). When a lunge occurs, even if within the midsagittal plane, it is quite variable and, frankly, quite fascinating.

Figure 10.5 shows tracings of several strikes that illustrate the variability in body displacement that can occur between the beginning of the strike and what has come to be called the *hit*. The hit is the point in time at which the foreleg(s) first contacts the prey. Figure 10.5A depicts a low strike in which there was no measurable body displacement, only the neck extended. Figure 10.5B depicts a low strike in which there was a

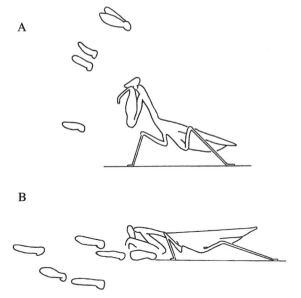

Fig. 10.4 Tracings from video frames that indicate representative prey positions, normalized to the mesothoracic tarsus, from which prey elicited (*A*) high or (*B*) low strikes. In each figure, the mantis was drawn in an arbitrary pose characteristic of the strike beginning. Please note in *B* that the rightmost cricket was *not* in contact with the mantis prior to the strike in which it was captured. Reprinted from Prete and Cleal, 1996, with the kind permission of S. Karger.

slight downward and forward displacement of the body. Figure 10.5C depicts a low strike in which there was a dramatic forward and downward displacement accompanying the strike—the lunge distance was 29 mm. This lunge is particularly interesting in that in order to reach the prey, the mantid's rear tarsi (feet) actually released their grip on the substrate and slid forward 30 mm. When the mantis pulled the cricket off of the wire, the tarsi re-grasped the platform, and the mantis pulled herself backward.

Finally, Figure 10.5D depicts two high strikes, each of which began from the same initial tarsi position. In the lower of the two strikes, body movement was both upward and forward. In the higher strike, the movement was primarily upward. In both of these cases, prey were at virtually the same initial distance from the mantis (28 and 27 mm, respectively), the retinal images of the prey overlapped considerably (31–17° and

---

Facing page

Fig. 10.3 Graphs indicating the differences in prey positions and in the positions of several key body parts between high and low strikes. *A*, Absolute prey distances (APD) plotted as a function of absolute prey angle (n). Strikes fell neatly into two categories—low strikes in which angle n > 30° and high strikes in which angle n < 30°. *B–H*, The validity of this division is supported by the significant difference between all of the key measures of mantis body position at the beginning of high versus low strikes except metathoracic vector length (H2). Reprinted from Prete and Cleal, 1996, with the kind permission of S. Karger.

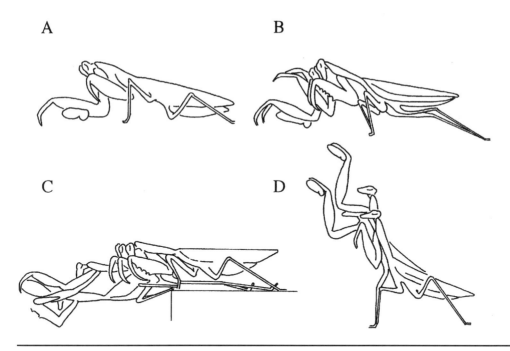

Fig. 10.5 Tracings from video frames of several strikes that illustrate the variability in body displacement (i.e., the lunge) that can occur between the beginning of the strike and the hit. *A*, A low strike in which there was no measurable lunge. *B*, A low strike in which there was a slight downward and forward lunge. *C*, A low strike in which there was a dramatic forward and downward lunge during which the metathoracic tarsi released (or lost) their grip and slid forward. *D*, Two high strikes emitted by the same mantis, each of which began from the same tarsi position. In the higher of the two, body displacement was primarily upward. In the lower of the two, the body displacement was a primarily forward. Reprinted from Prete and Cleal, 1996, with the kind permission of S. Karger.

20–14°, respectively, relative to visual field center), the initial pitch of the abdomen was the same, and the legs were in the same initial positions at the beginning of each lunge. Yet, in spite of the nearly identical initial postures, the lunge movements were very different. In the lower strike, the body displacement was produced by several discrete movements in which the prothorax was tilted back without a change in abdominal angle, the body moved forward, and the rear legs lengthened. In the higher strike the prothorax tilted back and, at the same time, the abdomen lowered so that there was only a 2° net change in the prothoracic-abdominal angle, and the middle legs extended considerably more than did the rear legs. In essence, in this strike, the prothorax and abdomen pitched backward as a unit as the mantid moved forward and up. Hence, the net displacement of the head and prothorax was primarily upward.

We wanted to give you the details of these two lunges to make two points. The first is that lunges are not stereotyped: They can be quite variable, even when the mantid begins from virtually the same initial body position. The second point (which follows from the first) is that the specific motor program that directs an accurate prey capture sequence must take into account a variety of proprioceptive inputs that denote the relative positions of the mantid's body parts in order to accurately move the animal toward its target. For instance, in the last two strikes, the only differences between the initial postures were in head angle, prothoracic-abdominal angle, and prothorax angle. So the proprioceptive cues associated with the relative positions of these body parts must have been responsible for the differences in the two lunges.

Similarly, Figure 10.5A and C depict two lunges that began from similar body positions. Once again, however, the lunges were considerably different. In the Figure 10.5A lunge, there was virtually no change in any measurement of body position between the strike beginning and the hit. In the Figure 10.5C lunge, however, the head raised, the abdomen raised as the prothorax lowered, and the middle and rear legs extended. The differential cues associated with these two dramatically different lunges were most likely both interoceptive and exteroceptive, including proprioceptive information about body position and information regarding prey distance detected by differences in retinal image movements (binocular disparity cues are ineffective at distances greater than 25 mm [Srinivasan, 1992; Prete and Mahaffey, 1993b]).

Surprisingly, lunge distances in the high strikes (range = 6–11 mm) were not correlated with prey distance or with prey elevation in our study. The reason is that it takes mantids longer to emit a high strike than a low strike because high strikes are preceded by a more protracted sequence of orienting behaviors that include elevating the body by extending the middle legs. With the body already elevated, the remaining distance to be covered by the lunge is comparatively small (about 15–30 mm). Further, the maximum prey distance at which a successful high strike can occur is limited by the length of the middle legs (which are already quite extended at the beginning of these strikes; note Fig. 10.5D). In contrast, we found that low strikes were initiated from a much wider range of relative prey distances (16–60 mm), and that in these strikes the distances spanned by the lunge were correlated with prey distances.

Without delving into the minutia, we found a number of correlations between the changes in body position over the course of the strike and the prey position at the beginning of the strike. Put simply, for high strikes, as prey distance and/or elevation increased, mantids pitched farther backward over the course of the strike and extended their middle legs farther. These changes are characteristic of upward-directed lunges. For low strikes, as prey distance increased, mantids pitched and leaned farther forward, and the rear legs extended farther over the course of the strike. These changes are characteristic of forward-directed lunges.

Foreleg Angles

There has been considerable theoretical concern over the angle formed by the forelegs as the prey is being grasped. The idea that a precise femur-tibia angle is required to assure a successful capture has gained considerable acceptance despite the fact that the idea is based on the behavior of movement-restricted preparations. Interestingly, when you look at the graphs in Figure 10.6, you will understand why people came to the idea. These graphs depict the prothorax-coxa, coxal-femur, and femur-tibia angles (a, b, c, respectively) of one foreleg (the one closest to the video camera) at the strike beginning (SB) and four subsequent points in the strike (midpoint = MP; hit = H; beginning of tibia closure = TCb; and end of tibia closure = TCe) for high (A–E) and low strikes (F–J). The range of elapsed times from the strike beginning (time = 0; graphs A, F) to each subsequent point, and the average elapsed times (ET) between points (± standard deviations) are indicated above each graph.

You will note that in high strikes—which make up virtually all of the strikes analyzed in earlier studies—foreleg positions are relatively consistent over the course of the strike. But remember that high strikes occur over a much smaller range of relative prey positions than do low strikes. So, one would expect them to be relatively less variable. On the other hand, look at the foreleg angles during low strikes (F–J). Not very stereotypical looking, are they?

Although our results, generally speaking, are in agreement with other people's analyses of foreleg movements during the strike, there are a few interesting differences that we would like to point out. First, although several studies have claimed that initial foreleg-prey contact must be with the ventral, spined surface of the femur and/or that the femur and tibia must form a precise angle around the prey at the moment of the

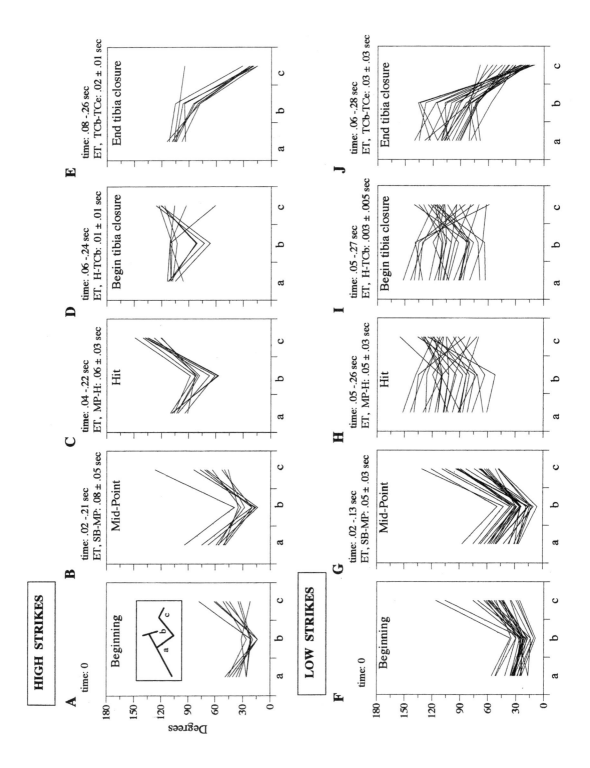

hit (e.g., Loxton and Nicholls, 1979; Maldonado et al., 1967; Corrette, 1990; Rossel, 1991), our data do not support these claims. Prey contact actually occurs at a variety of points along the entire length of the femur and tibia, even in successful strikes. (This is also the case for *T. a. sinensis* [Prete, 1990a; Prete et al., 1990].) Second, we found considerable variability in the femur-tibia angles (Fig. 10.6C) at the hit, especially in low strikes (Fig. 10.6H). For instance, in high strikes, femur-tibia angle at the hit ranged from 120 to 149°, but in low strikes it ranged from 70 to 155°. Overall, the range in positions that the foreleg assumed prior to the hit demonstrates, once again, that, overall, the strike is not as precisely orchestrated as high strikes make it appear to be.

Lunge Velocity

Interestingly, in low strikes we found that lunge velocity was correlated with prey distance; when prey was farther away, mantids lunged faster (0–295 mm/s). So, the elapsed time to the hit (ET.Hit) was relatively constant across prey distances (interquartile range = 0.056 second). Interestingly, Corrette found that lunge velocity of his movement-restricted *T. a. sinensis* was correlated with initial prey distance but in the opposite direction (Corrette, 1990). That is, when prey was farther away, the lunge was slower (44–130 mm/s). Although this result may seem unintuitive, it is really quite reasonable. The difference is due to the differences between the initial postures, and subsequent movement patterns of high and low strikes. Low strikes—in which the body displacement is primarily forward—are powered by extensions of the middle and hind femora, and the hind tibiae. Hence, the mantid can generate considerable force in a low strike. However, you will remember that in studies like Corrette's, low strikes were not considered, and high strikes—in which the body displacement is primarily upward—are powered only by extensions of the middle legs. Further, high strikes begin with the middle legs already more extended than do low strikes, and the mantid cannot generate the same forces from this position as it can at the beginning of a low strike. And the farther away the prey is, the more extended are the middle legs at the strike beginning so the more mechanically disadvantaged is the mantid.

The Strike after the Hit

Traditionally, analyses of strikes have been done only up to the movement of the hit or the "theoretical hit" (i.e., the point at which the forelegs are at the angle that they "should" be when the hit occurs [e.g., Corrette, 1990]). There is some good reasoning behind this approach in that resistance from impaled prey perturbs the natural movements of the mantid's legs. Unfortunately, this practice has left unanswered several questions about the prey capture sequence. We were are able to address several of these questions with data from twelve accurate strikes (five high and seven low) in which mantids missed the prey because it passed between their raptorial forelegs. In Figure 10.7, we have plotted the positions of the prothorax-abdominal joint every 10 ms (every other video frame) from strike beginning (SB) to the frame in which body movement stopped (STOP) for eight representative strikes. Frames in which the tibia began to close (TC.b) and stopped closing (TC.e) are marked with arrows, and for each strike, the frame in which prey contact would have been made (Hit) is indicated by an open circle.

Facing page

Fig. 10.6 The prothorax-coxa, coxal-femur, and femur-tibia angles (a, b, c, respectively; inset graph *A*) of one foreleg at the beginning (SB: graphs *A*, *F*), and each of four subsequent points in the strike (midpoint: graphs *B*, *G*; hit: graphs *C*, *H*; beginning of tibia closure: graphs *D*, *I*; and end of tibia closure: graphs *E*, *J*). Graphs *A–E* are high strikes and graphs *F–J* are low strikes. The range of elapsed times from the strike beginning (time = 0: graphs *A*, *F*), and the average elapsed times between points (ET ± s.d.) are indicated above each graph. Note especially the high degree of variability in joint position in low strikes (*F–J*) compared to high strikes. Reprinted from Prete and Cleal, 1996, with the kind permission of S. Karger.

As you can see, in all cases, mantids continued to move upward or forward for another 10–40 ms after the point at which they would have hit the prey before their movement changed directions. In addition, tibial closure occurred in spite of the fact that prey was missed. Together, these results demonstrate that initial contact with the prey is not necessary to stop the lunge or to trigger tibial flexion, as has been claimed. In fact, we now have data based on electrophysiological recordings from the pertinent foreleg muscles that confirm that the sequence of movements normally made by the foreleg segments during the strike do not require contact with the prey for

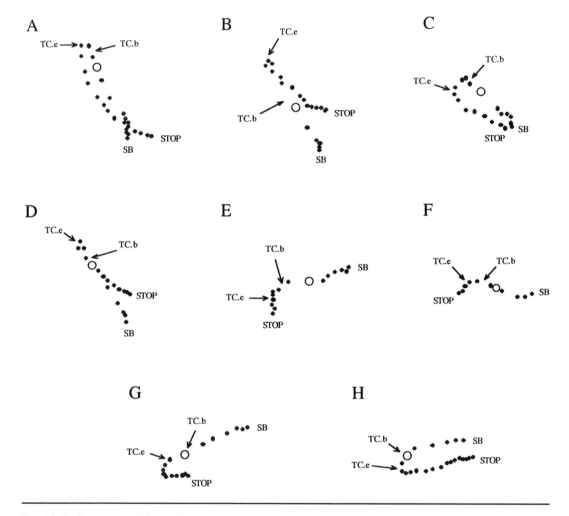

Fig. 10.7 The position of the prothorax-abdominal joint (represented by a dot) every 10 ms from the strike beginning (SB) to the video frame in which body movement stopped (STOP) for eight representative strikes (A–H). Each of these strikes was accurate, but prey passed between the raptorial forelegs; hence, the movements of the mantids after the frame in which the prey would have been hit (open circle) remained unimpeded. Frames in which the tibia began to close (TC.b) and stopped closing (TC.e) are marked with arrows. Note that, in all cases, mantids overshot their target; lunges continued for 2–8 frames (10–40 ms) beyond the point at which the hit would have occurred before the movement changed directions. In all cases tibial closure occurred in spite of the fact that prey was missed. These data demonstrate that contact with the prey is not necessary to stop the lunge or trigger tibial flexion. Reprinted from Prete and Cleal, 1996, with the kind permission of S. Karger.

their initiation or completion (Prete and Bullaro, unpublished data).

The Prothoracic Tibial Flexion Reflex

The fact that contact with the femoral spines is not necessary to ensure a successful strike or to initiate tibial closure prompted us to examine just what role proprioceptive feedback from the moveable spines might play in prey capture. We did this by first determining what tactile stimuli elicits reflexive tibial closure (the prothoracic tibial flexion reflex [PTFR], Copeland, 1975; Copeland and Carlson, 1977). Then, we compared the behaviors of femora on which the spines were immobilized (covered with wax) with intact femora. Prete had already done these experiments with *T. a. sinensis* (Prete, 1990a), and we were very interested to see if the *S. lineola* responded in the same way. It turns out that they do.

The frequencies with which the tibial flexion reflex was elicited by tactile stimuli are presented in Table 10.1. The response rates fell neatly into either high or low categories (rates = 52–73% or 0–13%, respectively). Put simply, stimuli were either strong releasers of the tibial flexion reflex or they were not.

Stroking the unwaxed femoral spines reliably elicited the tibial flexion reflex when the tarsus was off of the substrate, but it did not when the tarsus was resting on the substrate. And, as expected, the reflex was blocked when the spines were immobilized with wax. Interestingly, quickly pulling the tibia open also reliably elicited tibial flexion irrespective of whether the spines were immobilized. Presumably, this type of stimulus mimics the impact of prey against the tibia during prey capture. Unlike *T. a. sinensis*, random tapping of the ventral femur elicited the reflex in about half of the trials for *S. lineola*, but like *T. a. sinensis*, flexing only the largest femoral spine rarely elicited the reflex. In short, both flexion of the femoral spines when the tarsus was off of the substrate and sudden pressure against the tibia were sufficient to elicit reflexive tibial closure.

We also compared data collected from videotapes of five successful strikes, of 158 minutes of eating behavior by mantids with one foreleg waxed, and of eleven successful strikes by free-ranging mantids with both forelegs waxed. What we found, simply, was that mantids with immobilized femoral spines still could capture prey successfully. However, they had a tendency to release the prey after it was caught. That is, without the appropriate feedback from the movable femoral spines, the mantids acted as if they did not know that they had something in their grasp.

What We've Learned about Strikes in the Midsagittal Plane

The predatory behaviors of mantids have been analyzed with a variety of techniques, from simple observation to high-speed videography, and under a variety of conditions, from virtually immobilized to free ranging, and each study has taught us a little more about these insects. However, it is particularly interesting to us that the degree of freedom afforded the mantids in each study appears to be positively correlated with the degree of behavioral plasticity that is ultimately attributed to them.

We know now that when faced with prey in their midsagittal plane, free-ranging *S. lineola* attack it with one of two distinct behavioral routines depending upon the height of the prey in relationship to the surface on which the mantid is standing. We called these high and low strikes. In each of the routines, a characteristic initial posture is followed by a different constellation of movements which includes a rapid grasp by the raptorial forelegs and, if the prey is sufficiently distant, a displacement of the body upward and/or forward.

The facts that the initial postures between high and low strikes differ significantly and that the image of the prey need not be positioned precisely in the visual field prior to the strike (especially in low strikes) support our original idea that proprioceptive cues coding the relative positions of the mantid's body parts are critical in the central programming of the prey capture sequence. Our findings suggest that prothoracic angle and (to a lesser degree) head angle and the degree to which the middle legs are extended provide the critical proprioceptive cues. One can see that if the mantid consistently places the image of a potential target in the medial-ventral portion of the visual

Table 10.1 Frequencies with which the prothoracic tibial flexion reflex is elicited by tactile stimuli

Stimulus	Trials	Responses
Stroke unwaxed femoral spines:		
Tarsus off substrate	140	73%
Tarsus on substrate	100	7%
Stroke waxed femoral spines:		
Tarsus off substrate	140	5%
Tarsus on substrate	140	0%
Flex largest femoral spine, all spines unwaxed:		
Tarsus off substrate	160	13%
Pull tibia, spines unwaxed	140	62%
Pull tibia, spines waxed	140	56%
Tap ventral femur, spines unwaxed	140	52%
Tap ventral femur, spines waxed	140	3%
Stroke dorsal femur, spines unwaxed	140	7%
Stroke dorsal femur, spines waxed	140	3%

field (as we found that they do), then, together, the positions of the head, prothorax, and middle legs would be very good indicators of the relative elevation of the prey in the midsagittal plane.

By what means, then, might the mantid determine the distance to the target? Well, there are two possibilities. First, we know that at short distances (less than 25 mm [Srinivasan, 1992]) mantids use binocular disparity (Rossel, 1983, 1986). Like many other insects, however, they also use retinal image speed as a cue to object distance (Prete et al., 1993). Together, proprioceptive and visual cues probably provide the mantid with sufficient information to capture prey successfully anywhere within the limits of its reach.

A Fantasy

Let's assume that as mantids evolved from their putative roachlike ancestors, their initial predatory behavior involved running after and, perhaps, lunging at potential prey. Under these conditions, it would be critical that the insect's CNS be able to compute the target's position in relation to its thorax. This would be so because the degree to which the insect should turn, the distance that it should run, and the degree to which the target is out of reach each represent measurements that must be converted into leg movements and, of course, the legs are attached to the thorax. With an immobile head and thorax (which the roachlike ancestor surely had) these calculations could be accomplished by considering visual cues only. That is, if the position of the visual field relative to the legs does not change (because the head cannot move), then cues that localize the object within the visual field would be sufficient to indicate the position of the prey relative to the legs. Orienting so that the target's image falls in the ventral-medial visual field would ensure that the target is on the ground and straight ahead. CNS programming of the attack sequence would then be relatively simple:

Pursue the prey while keeping its image in the lower center of the visual field, then lunge at it or grasp at it depending upon its distance (determined by retinal image size, apparent object speed, and/or binocular disparity). This is similar to the type of simple system that Corrette (1990) thought existed in modern mantids. The problem with his model is, of course, that modern mantids have mobile heads and prothoraxes.

As mantids evolved greater head and prothorax mobility, the location of an object's image in the visual field became a progressively less reliable indicator of its position relative to the legs. Greater head mobility means that the mantid's CNS has to be able to ascertain both the position of the object in the visual field and the position of the visual field in relationship to the rest of the body.

Of course, as mantids evolved they not only gained head and prothorax mobility but also prothorax length, which separates the prothoracic legs from the middle legs and hindlegs. Based on the data presented here, we hypothesized that because of the elongation of the prothorax, mantids must process information on prey position in what might be called *pterothorax-centered* space (i.e., centered around the portion of the thorax to which the middle legs and hindlegs are joined). This hypothesis makes several predictions, the first two of which are supported by the data that we have already presented. First, if the mantid's CNS calculates the position of prey based on pterothorax-centered space, then the precise prey location relative to the forelegs at the hit and the angle of the forelegs at the hit should vary between strikes (because the forelegs are not always in precisely the same location relative to the middle legs and hindlegs prior to initiating prey capture; also see Prete et al., 1990). Second, when free-ranging mantids lunge they should overshoot their targets somewhat because the attack sequence is programmed based on the positions of the middle legs and hindlegs rather than the prothoracic legs (note Fig. 10.7). Third, mantids should be able to successfully capture prey that is not aligned with their pterothorax-abdominal axis.

Prior to our work, the third hypothesis had never been explicitly tested (but see Prete et al., 1993, Fig. 8). However, cockroaches (which are the mantid's closest living relative) have the ability to turn accurately away from puffs of air directed at them from a variety of angles (e.g., Burdohan and Comer, 1990; Stierle et al., 1994; Comer et al., 1994; Nye and Ritzmann, 1992), and we saw no reason to rule out the possibility that mantids either could have evolved the ability to turn accurately or could have retained this ability if it was present in their common ancestor. Of course, if mantids have the ability to turn toward prey as they strike, it would mean that they could integrate visual and proprioceptive information in terms of pterothorax-centered space. This would be the only way that their CNS could program prey capture sequences that involved a turn to one side or the other. And it also means something else: Mantids are remarkably more complicated than they have been given credit for.

Steering In the Horizontal Plane
Our Initial Thoughts

One of the things that we found particularly intriguing about the mantid's ability to steer its lunge in the horizontal plane (besides the fact that so many people said that they couldn't do it) was the fact that, as far as we could tell, the movements of the mantid's middle legs and hindlegs during the lunge seemed so similar to those made by the cockroach *Periplaneta americana* during its escape maneuvers (e.g., Camhi et al., 1978; Camhi and Levy, 1988; Nye and Ritzmann, 1992). This was especially intriguing given that mantids and roaches are so closely related. To see just how similar these two insects are in this regard, we analyzed a series of strikes in the same way that Nye and Ritzmann (1992) had analyzed the escape turns of the cockroach. Here is what we found.

As before, we have included the measurements used to analyze the mantid's movements in Figure 10.8. Although we will continue to avoid the use of jargon, these abbreviations will be useful in understanding some of the figures.

The Data

To facilitate our analyses, we divided the strikes into three groups: strikes at targets that were on

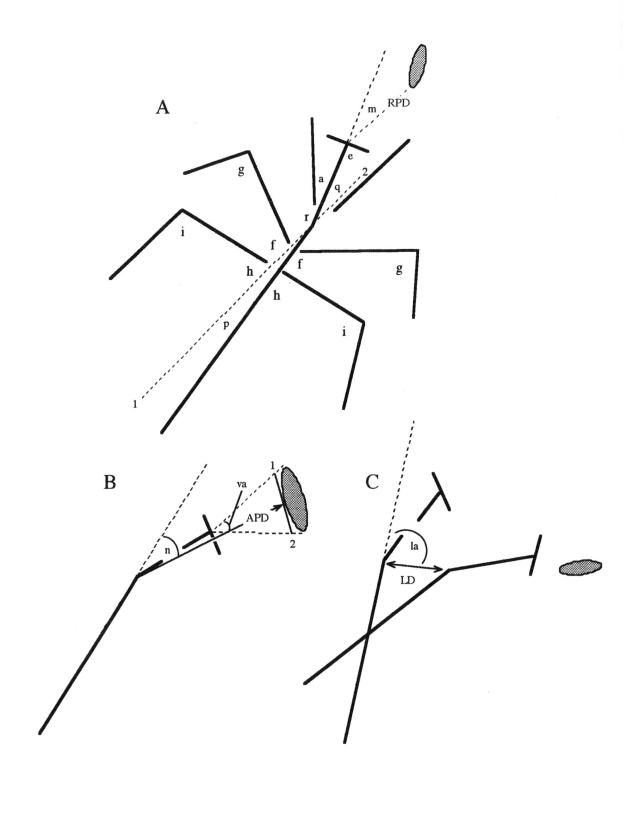

the ground in front of the mantid (Zone 2), and strikes at targets that were on the ground and more than 10° to the left or to the right of the mantid's abdomen axis (Zones 1 and 3, respectively).

The first thing that we found was that *S. lineola* can, indeed, capture prey that is not aligned with its abdomen axis. Figure 10.9A presents histograms of the number of strikes in which prey was situated at various relative and absolute prey angles. (Remember that Fig. 10.8 has all of the measurement abbreviations.) You can see that prey need not be precisely centered in the visual field (angle m) or be aligned with the abdomen axis (angle n) at the beginning of a strike.

On the far left of Figure 10.9A, there is a data point representing one interesting strike for which the initial relative prey angle was -122°. In this case, the prey was only 43° from the abdomen axis at the outset, but, because the mantis was actually facing away from the prey, the relative prey angle (m) was quite large. But, even excluding this unusual case, the centers of the prey images were still as much as 58° from visual field center and as far as 69° from the abdomen axis when the capture sequence began. These data agree with previously published reports on other free-ranging mantids (Holling, 1966; Prete et al., 1990, 1993).

Prior to initiating prey capture, mantids may orient their head and/or prothorax toward the prey. The assumption, of course, is that the mantid positions itself so that the image of the prey falls in the center of the visual field and on the most sensitive areas of the retinae. Presumably, this is important for prey identification (Prete, 1993) and/or localization (Rossel, 1991). However, if either identification or localization depended exclusively on precisely positioning the prey in the visual field, one would not predict the variability in relative prey positions seen in our data. Further, if accurate localization required precise positioning of the prey, it would require the mantid to align (or at least position consistently) the axes of its abdomen and prothorax and position its head consistently across capture attempts. Consistent orientation would let the insect know where it stands in relation to its visual field because the prothorax, head, and relative prey angles would vary little across capture at-

---

Facing page

Fig. 10.8 The measurements used to describe the positions of mantids and their prey at various times during the strike. In all cases, lowercase letters represent angles and uppercase letters represent distances. The diagrams represent ventral views and the prefixes R and L indicate the animal's right and left, respectively. Again, most of the measurements are self-explanatory, but a few need clarification. The position of the long axes of the prothoracic foreleg coxae in relation to the long axis of the prothorax (foreleg angle) is indicated by angle a. The long, dotted line segment 1, 2 indicates the long axis of the pterothorax-abdomen (abdomen axis) at the strike beginning (SB, the video frame 5 ms prior to the onset of continuous movement that culminated in the strike). Angles p and q indicate the angles of the abdomen and prothorax axes, respectively, from line segment 1, 2. The abdomen angle was set to zero at SB. Angle r is the angle formed between the long axes of the prothorax and abdomen. We used the change in the angle formed between each femur and the long axis of the pterothorax (femur-thorax angle) as a reliable, indirect measure of the change in coxa-femur angle. The mesothoracic femur-thorax angles are designated angles f, and the metathoracic femur-thorax angles are designated angles h. Prey position was measured from two points of view (A, B). The relative prey distance (RPD) was determined by constructing two lines that defined the visual angle (va shown in B) subtended by the body of the prey and measuring the length of the line segment that bisected the visual angle and extended to a perpendicular line segment tangential to the outline of the prey (dashed line segment RPD shown in A). Relative prey angle (m) is the angle formed between RPD and the long axis of the prothorax. Absolute prey distance (APD; shown in B) is the distance between the prothorax-pterothorax joint and the intersection of RPD with the projection plane of the prey. Absolute prey angle (angle n) is the angle formed between APD and the abdomen axis. Reprinted from Cleal and Prete, 1996, with the kind permission of S. Karger.

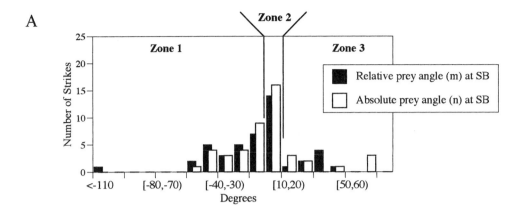

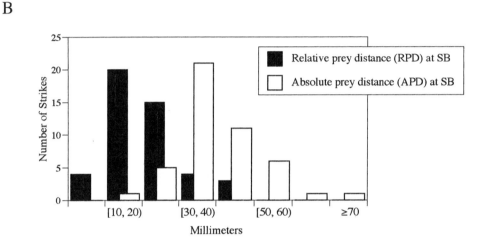

Fig. 10.9 These frequency histograms indicate the positions of prey in the horizontal plane at the beginning of predatory strikes. *A,* The number of strikes in which prey was situated at various relative (m) and absolute prey angles (n). Zone 1 corresponds to the animal's left, zone 3 to its right. *B,* The frequency with which mantids struck at prey at various relative (RPD) and absolute (APD) prey distances. Reprinted from Cleal and Prete, 1996, with the kind permission of S. Karger.

tempts. Quite to the contrary, however, we found that these three angles varied considerably across trials and that absolute prey angle was strongly correlated with both initial head and prothorax angle, even if we included the peculiar strike with the extreme relative prey angle. This suggests that knowing the position of the prey in the visual field is insufficient to precisely localize it. Instead, it appears that information regarding the relative angles of the head and prothorax must be integrated with visual information in order to steer the attack sequence accurately. We will return to this point later.

The simplest and, ironically, most important data that we collected are graphed in Figure 10.10. These histograms demonstrate that when prey is to the right or to the left, mantids both reach with their forelegs and steer their body toward the prey as the attack progresses. The graphs really need little explanation. Columns on either side of the zero mark indicate that a particular angle either increased or decreased. You can tell that the mantids systematically steered when they struck at prey to the left or right by the fact that the columns for Zones 1 and 3 are on opposite sides of the graphs and the columns for Zone 2 cluster in the center.

We spent considerable time carefully analyzing the changes in the middle legs and hindleg angles that corresponded to the changes in body position that are represented in the histograms in the previous figure, and all of the changes were just what we expected (Cleal and Prete, 1996). The simplest way to understand these changes is to look at Figure 10.11. These stick figures were traced directly from individual video images of the strike beginning (dashed lines) and hit (solid lines) for six characteristic capture sequences. In the instances depicted in drawings A–D prey was in Zones 1 or 3, and in E and F the prey was in Zone 2. The shaded shapes are silhouettes of the prey. You will notice that when mantids attacked in Zones 1 or 3 their bodies pivoted dramatically in the direction of the prey. In general, if prey was in Zones 1 or 3, the rear leg joints and the middle femur-coxa joint contralateral to the prey tended to extend, and those joints ipsilateral to the prey either extended less or stayed the same. On the other hand, if the mantid struck within Zone 2 (E, F) all of the leg joints tended to extend.

We also did a careful analysis of the relationship between the changes in the middle leg and hindleg joints and the distances that the mantids lunged. These data are presented in Figure 10.12. The graphs are scatter plots of the changes in the leg angles plotted against lunge distances.

As expected, when mantids attacked in Zones 1 or 3, the middle tibiae (A) flexed, the hind tibiae (C) extended, and the hind femora (D) extended to a degree proportional to the absolute value of the distance lunged. In contrast, you will note that the changes in the middle femur-thorax angles (B) had a somewhat different relationship to lunge distance than did the changes in the other three angles. In this case, the changes in angle were not related to the absolute distance lunged, but rather to the distance and direction that the mantids turned during the lunge. These data strongly suggest that the hindlegs generate the power that moves the mantid forward during the lunge and the middle legs act to steer the lunge.

When mantids attacked in Zone 2 (E–H), leg movements were what you would expect when a lunge is directed forward rather than sideways. Again, the middle tibiae tended to flex, and the other three leg angles extended in proportion to the distance lunged.

Mantids, Cockroaches, Evolution, and the Ability to Turn

The behaviors and by implication the underlying neural organization of praying mantids are often considered to be so specialized or unique that they bear little relationship to those of other, even closely related, insects. However, a number of behavioral studies done in our lab and an increasing amount of neuroanatomical data (Nesbitt, 1941; Bacon, 1980; Grey and Mill, 1985; Boyan and Ball, 1986; Kerry and Mill, 1987; Liske et al., 1989) have revealed striking similarities between mantids and other orthopteroid insects. This is good. The more people who come to realize that mantids are, in fact, basically like other insects, the less likely it is that they will fall victim to the odd and sometimes implausible notions that have surfaced from time to time (Prete and Wolfe, 1992; Prete, 1995).

Interestingly, we found a number of similarities between the attack-associated leg movements of our mantids and the escape-associated leg movements of the cockroach *P. americana*. We think that this is interesting, of course, because of the close phylogenetic relationship between mantids and roaches.

The roach's escape-related leg movements have been divided into three distinct phases: the turn phase, the transition phase (from turning to running), and the run phase (Nye and Ritzmann, 1992; Camhi and Levi, 1988; Camhi et al., 1978). We think that the movements associated with the lunge of *S. lineola* correspond to the turn phase of the roach's escape. To understand the relationship between our data and the analogous roach data, two conceptual bridges must be established. The first has to do with the relative position of the eliciting stimulus. Roach escape turns are elicited by puffs of air aimed at the an-

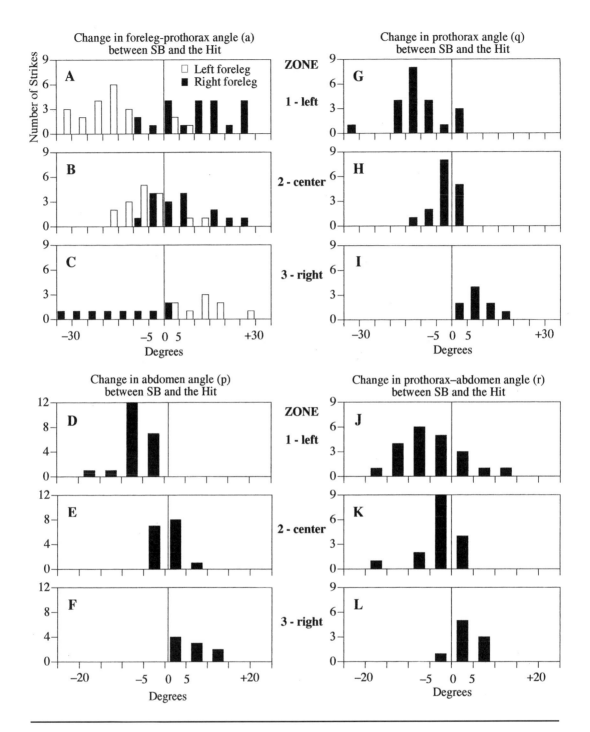

Fig. 10.10 Frequency histograms indicating the changes in four measures of body position between the strike beginning and the hit (time of initial foreleg-prey contact) when prey was in each of the three zones as defined in Figure 10.9. A–C, The changes in foreleg-prothorax angles (a) for the right (black columns) and left forelegs (open columns). G–I, D–F, The changes in prothorax (q) and abdomen angles (p), respectively. J–L, The changes in the prothorax-abdomen angle (r) between strike beginning and the hit. Reprinted from Cleal and Prete, 1996, with the kind permission of S. Karger.

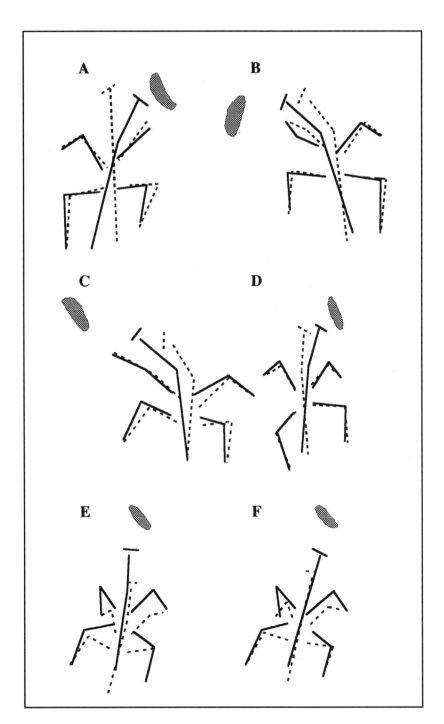

Fig. 10.11 Composite tracings from video frames showing the positions of several mantids at the beginning of a strike (dashed lines) and at the Hit (solid lines). The shaded object in each drawing is the prey at the strike beginning. Contrary to general belief, you will note that *S. lineola* can steer the strike in the horizontal plane. Reprinted from Cleal and Prete, 1996, with the kind permission of S. Karger.

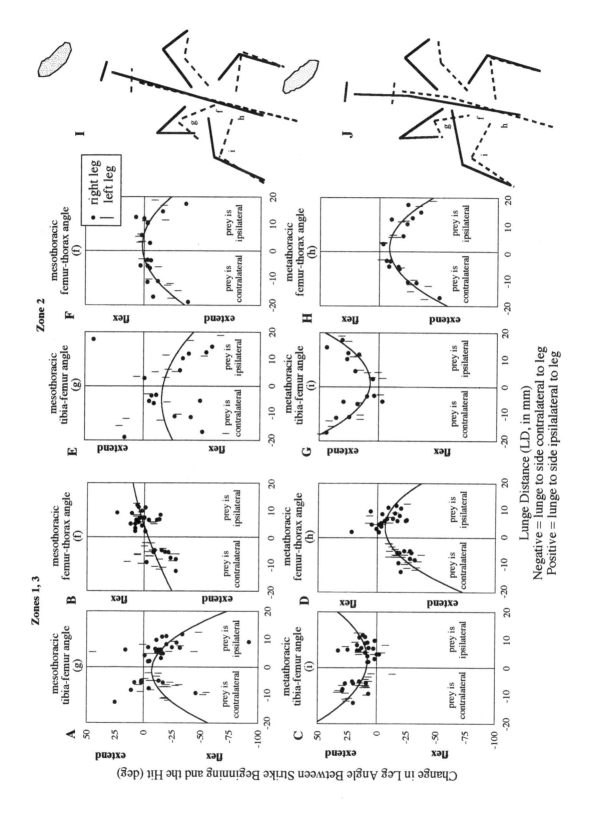

tennalike cerci on the rear end of the roach's abdomen (Nye and Ritzmann, 1992). If you think of the roach as being oriented such that its head is pointed at 12 o'clock, air puffs could come from locations corresponding to, for instance, 2 o'clock to 5 o'clock. These would elicit responses that turn the roach away from the stimulus. So, a puff from 5 o'clock would elicit a small turn to the left. A puff from 3 o'clock would elicit a larger turn to the left. In contrast, the mantid is faced with just the opposite task. It must move toward the eliciting stimulus. One might think of these two situations as if the roach was being pushed and the mantid pulled by stimuli that lie at opposite ends of a vector. For instance, a roach turning away from an air puff from 4 o'clock is analogous to—and should elicit similar leg movements in—a mantid turning toward a stimulus at 10 o'clock. In general, then, a puff of air that deviates by some number of degrees clockwise (or counterclockwise) from directly behind the roach is analogous to a prey item that deviates clockwise (or counterclockwise) by the same number of degrees from directly in front of the mantid. Second, because the roach turns away from and the mantid turns toward the eliciting stimulus, roach legs that are ipsilateral to the stimulus should behave as do mantid legs that are contralateral to the stimulus and vice versa.

In the roach study to which we want to compare our data, Nye and Ritzmann (1992) found that in tethered roaches, the ipsilateral middle coxa-femur joint extended irrespective of wind direction and that the contralateral middle coxa-femur joint extended when puffs deviated 45° from directly behind the roach but extended less or flexed as the puff moved anteriorly. This is precisely analogous to the movement patterns we saw in our mantids: In our insects, the contralateral middle femur-thorax angles extended in proportion to absolute prey angle. In contrast, the ipsilateral middle femur-thorax angle remained the same or extended just slightly when prey was close to the abdomen axis but flexed as the prey moved posteriorly. Nye and Ritzmann also found that the ipsilateral and contralateral rear coxa-femur joints changed their movements in relationship to the degree of the roach's turn. We found a similar patten in *S. lineola*: As prey moved posteriorly, the contralateral rear femur-thorax angle continued to extend, but the ipsilateral rear femur-thorax angle extended progressively less and then tended to flex when turns were largest.

Nye and Ritzmann (1992) classified three types of cockroach turns: In Type 1 turns, the movement was primarily forward—all leg joints extended. This is analogous to mantid attacks in Zone 2. In roach Type 2 turns, the rear coxa-femur joint contralateral to the stimulus flexed causing the contralateral rear leg to be pulled forward. This type of turn is analogous to those depicted in Figure 10.11A–D, in which the ipsilateral femora moved forward and the contralateral femora moved backward in relationship to the mantid's pterothorax. In both free-ranging *P. americana* and *S. lineola*, a Type 2 turn results in a large turn with less forward movement than a Type 1 turn (compare Fig. 10.11A–D with 10.11E, F). In roach Type 3 turns, both rear coxa-femur joints flex, pulling the cockroach backward as it turns. We did not see such turns in our mantids. These would be elicited, theoretically, by prey that is at absolute prey angles of ±90° or more. Of course, the fact that we did not see these types of turns does not mean that they do not occur.

Finally, we found that when our mantids turned to the side there were systematic correlations between the changes in leg joint angles that were similar to those seen in roaches (Cleal and Prete, 1996). These correlations suggest a coordi-

---

Facing page

Fig. 10.12 A–H, Changes in the positions of the mesothoracic and metathoracic legs plotted as a function of lunge distance. I–J, Two schematics of typical forward-directed strikes, for reference. Specific leg angles are indicated above each graph, whether or not the change in leg angle represents flexion or extension is indicated next to the ordinate, and whether or not the data refers to the ipsilateral or contralateral leg (in relation to prey position) is indicated in the lower left and right portion of each graph. Reprinted from Cleal and Prete, 1996, with the kind permission of S. Karger.

nated pattern of turning-related movements by *S. lineola* similar to those seen in the escape turns of *P. americana*.

We think that these data raise interesting questions regarding the evolution of the CNS control circuitry in *S. lineola* and, perhaps, in mantids in general. The escape turn of the roach requires that the insect's CNS process a complex array of sensory information so that its escape is directed appropriately (Burdohan and Comer, 1990; Stierle et al., 1994; Comer et al., 1994). For mantids, however, the situation is a bit more complex in that directional information corresponding to the relative location of the prey in relationship to the pterothorax (the point around which the mantid turns) must be supplied by proprioceptive inputs that code the position of the prothorax in both the horizontal and vertical planes (Prete and Cleal, 1996). Further, like other animals with mobile heads or eyes, mantids cannot localize prey based only on its position in the visual field. They must also consider the position of the visual field in relationship to the rest of the body. This is quite a different situation from that faced by the roach, for which the relative position of the cerci to the legs does not change.

All of our data on the relationship between prey position and the subsequent changes in the middle leg and hindleg angles suggest that information regarding prey position is integrated by the mantid's CNS in terms of pterothorax-centered space. This makes sense given the extent to which the highly mobile prothorax and head can displace the visual field relative to the leg joints that must direct the mantid toward its target (e.g., Prete et al., 1993; Prete and Cleal, 1996). It follows, then, that proprioceptive information regarding the positions of the mantid's head and prothorax also codes the position of the visual field in relationship to the pterothorax.

Finally, we would like to point out that the neural systems controlling the mantid's prey capture behavior must be integrated with at least two other systems. The first is the visual subsystem that recognizes an object as potential prey (Prete, this volume, chap. 8). That is, in order for a stimulus to elicit a predatory attack it must satisfy a sufficient number of a complex array of stimulus parameters, including stimulus size, speed, contrast, and geometry in relation to direction of movement (e.g., Prete, 1991, 1992a, b, 1993; Prete and Mahaffey, 1993b; Prete et al., 1993; Prete and McLean, 1996). The second system with which the attack control systems must be integrated is one that provides information about hunger level (e.g., Krombholz, 1977). This information probably acts on the visual system, modulating the system's sensitivity to certain visual stimulus parameters (i.e., by altering the releasing strength of the proximal eliciting stimulus for attack behavior), just as such feedback acts on the sensitivity of chemoreceptors in other insects (i.e., by altering the releasing strength of the proximal eliciting stimuli for eating behavior) (e.g., Dethier, 1969, 1987; Blaney and Duckett, 1975; Bernays and Simpson, 1982; Bernays, 1985). We think that these systems and their interrelationships provide a particularly fertile area for investigation by those interested in the evolution and functional organization of the invertebrate central nervous system.

Directions for Future Research

If our primary interest was the study of motor systems, and we wanted to study mantids, we would begin by adapting the work done on the roach escape response (especially that done by the Comer lab at the University of Illinois at Chicago), and on sensory-motor integration in the locust to the hunting behavior of the mantis (see Burrows [1996] for virtually all of the information available on locusts). We think that our data provide convincing evidence that at least some mantids display predatory behaviors, the complexity of which has been greatly underestimated. Examination of the subtleties and/or neural underpinnings of any aspect of mantid predatory behavior will keep a new investigator busy for his or her entire career. Our only other suggestion would be not to read any of the scientific literature until you have acquired a mantid and spent some time carefully watching how it behaves under natural conditions. It will be your best teacher.

ACKNOWLEDGMENTS

We would like to thank Drs. Roy Ritzmann and Joanne Westin for allowing F. R. P. to visit their laboratory and use their video equipment. This research was supported in part by a grant-in-aid of research from the Edmond and Marianne Blaauw Ophthalmology Fund of the National Academy of Sciences through Sigma Xi, the Scientific Research Society, by the Department of Psychology and the Committee on Biopsychology at the University of Chicago, by a Research Challenge Grant from the State of Ohio Board of Regents to Youngstown State University, and by Denison University, Granville, Ohio, where some of the data analysis was done.

11. The Hierarchical Organization of Mantid Behavior

Eckehard Liske

The bizarre appearance of the praying mantis as well as its special prey capture and courtship behaviors have been the cause of many legends and myths. At the beginning of this century legends still attributed magical powers to the female due, in part, to its supposed habit of engaging in sexual cannibalism, a behavior which actively fed the human imagination (Davis and Liske, 1985; Prete and Wolfe, 1992).

Since about the eighteenth century, the praying mantids (order Mantodea) have been a subject of scientific research. Comprehensive observations and descriptions regarding the biology of the mantids had already been done by Rösel von Rosenhof (1749, 1769) and J. H. Fabré (1897, in Guggenheim and Portmann, 1987). Kenneth Roeder (1963), one of the pioneers in the field of neuroethology, conducted comprehensive studies on the neural mechanisms underlying mantid behavior from approximately 1935 to 1963.

Roeder's experiments demonstrated that the segmental ganglia of the ventral nerve cord are responsible for sensorimotor coordination of local muscle activity, while the protocerebrum (the anteriormost brain division, sometimes referred to as the *brain*) serves a coordinating and integrating role during the execution of complex behavior. This idea that neurons located in the thoracic and abdominal ganglia of an insect are controlled by specific commands generated by "higher" neural systems has been confirmed by Huber (1965a, b) for the complex singing behavior of the field cricket. By means of specific lesions which he made in the cricket's brain, Huber demonstrated that the loss of the endogenic activities of the mushroom and central bodies of the protocerebrum abolishes species-specific song. Further, the results from a number of electrical brain stimulation experiments in a variety of freely moving animals support the generalization of the concept of a hierarchical organization within vertebrate and invertebrate central nervous systems (cat [Hess, 1954, 1957], cricket [Huber, 1952, 1955, 1960b; Otto, 1971], chicken [v. Holst and v. Saint Paul, 1960]).

In their natural environment animals are exposed to a multitude of stimuli, each of which may trigger quite a different reaction. However, organisms typically perform one behavior at a time, even though the eliciting stimuli for many behaviors may be present in their environment simultaneously. Such *singleness of action* has been described as the cornerstone of coordinated movement. Singleness of action is achieved by structuring independent behavioral acts into a priority sequence that governs choices. Such a priority sequence has been termed a *behavioral hierarchy*. In the event of a conflict situation—in which the eliciting stimuli for several mutually exclusive behaviors are present simultaneously—a behavioral hierarchy assures that an adaptive choice will be made. Such behavioral hierarchies are found in a variety of invertebrates (cricket [Huber, 1955, 1970], beetle [Simon, 1966], honeybee [Pflumm, 1968, 1969], crab [Kravitz, 1986]) and in all vertebrates examined thus far (Eibl-Eibesfeldt, 1972).

Aside from the disposition to respond to certain external stimuli, the parameters of some behaviors (such as their frequency or intensity) may depend on the coincident internal conditions of the organism (excepting, of course, in the case of some reflexes). Furthermore, these internal physiological conditions are subject to fluctuations, including circadian periodicity, which will cause them to interact in complex ways with a behavioral hierarchy. This means that, at different times, different behavioral patterns may be preferred.

The priority of one behavior over another can be assessed experimentally, making it possible to draw conclusions regarding a given priority sequence. Such an approach reveals which responses an animal will typically perform in a conflict situation. The concept of a hierarchy of behavioral patterns allows for a number of testable hypotheses about the possible underlying neural mechanisms regulating and controlling the hierarchically organized behaviors. However, it must be kept in mind that these mechanisms are, in turn, modulated by other neural and hormonal control mechanisms (Davis, 1976). Due to this complexity, and more important, to the system's plasticity, it is not surprising that the physiological mechanisms are still partially unclear.

The Goal

Praying mantids have a broad repertoire of behaviors that encompasses everything from simple reflex circles to complicated behavioral patterns. In this chapter, we would like to present the results of several studies in which we explored for the first time whether the behaviors of mantids are hierarchically organized. In examining this question, we studied males and females separately, when they were both hungry and satiated, and under conditions in which they were faced with various conflict situations. The behaviors with which we were primarily concerned were food acquisition, grooming, defense, and reproduction.

Animals and Protocols

Our studies were done on lab-reared adult *Tenodera aridifolia sinensis* (Saussure), with some supplemental experiments done on *Sphodromantis lineola* (Burmeister). We used standard rearing methods for both species (i.e., Liske and Davis, 1987). As noted, our experiments were conducted with both satiated and hungry mantids. Satiated mantids were fed house crickets *ad libitum;* "hungry" mantids were food deprived for 5 to 7 days prior to the experiments.

Dependent Measures
Feeding Behavior

Over the course of a single light-dark cycle, the amount of prey eaten (in the form of 3- to 4-week-old *A. domestica*) was recorded for each experimental animal (five males, eight females) every 3 hours (from 0900 until 2100). Partially eaten prey on the cage bottom was considered "devoured" if at least 50% of the body had been eaten. Every 3 hours the mantid's cages were cleaned and refilled with fifteen crickets. The ages of these mantids were between 0 (= adult molt) and 35 days.

Optokinetic Head Movements

Mantids were fixed with the vertical axis of their head at the center of a vertically oriented rotating drum (24 cm diam x 26 cm high) and presented with large visual stimuli (Fig. 11.1). The drum was lined with vertical, equally spaced black and white stripes, the widths of which could be varied in steps from 11.25 to 360° of visual angle. The stripe pattern was illuminated diffusely by an overhead halogen lamp—the mantid, however, was shielded from direct light exposure (for a detailed description, see Liske, 1977).

Turns of the mantid's head around the vertical axis were continuously registered by an electronic position-sensitive measuring system (Sandeman, 1968). In order to achieve this, the experimental animal was equipped with a fine wire glued dorsally between the two compound eyes and connected to a high-frequency sinusoidal

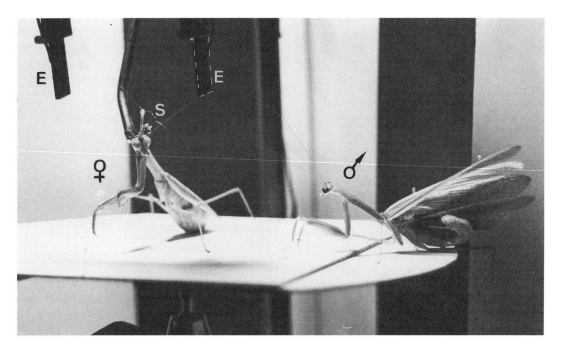

Fig. 11.1 The experimental set-up for triggering optokinetic reactions. *Left,* A female mantis is fixed in the center of a rotating drum lined with black and white stripes. Head movements are continually registered by means of an electronic position-sensitive measuring system (S - transmitter, E - receiver plates). *Right,* A courting male is approaching the female with abdominal "weaving movements" (a courtship behavior).

generator (20 kHz, 15 V) that functioned as a transmitter (labeled S on Fig. 11.1). The metallic probe did not contribute an additional moment of inertia and did not inhibit the mobility of the head. A small metal plate which functioned as a receiver was mounted at the sides of the mantid's head (labeled E on Fig. 11.1). Turns of the head to the right were converted into a positive, and turns to the left into a negative direct current signal, where the degree of the turn was proportional to the voltage amplitude (for additional details see Liske and Mohren, 1984).

During experiments, the drum rotated at a constant angular velocity (from 3.5 to 120°/second), 2 minutes in one direction and then 2 minutes in the other direction with a 5-minute break in between. This was repeated three to seven times for each mantid. The electrical output signals of the head-movement–measuring system and the switching pulses for the start and stop of the drum movement were saved on audio tape.

Because male mantids were consistently quite a bit more "restless" than females, only females were used for the quantitative evaluation.

Saccadic Head Movements

"Jerky," directed head movements, so-called fixation saccades, were triggered when a small visual stimulus was positioned within a mantid's visual field. This type of head movement often precedes prey capture.

The target (in this case, a cricket) was glued to a small vertical stick that was connected to a rotating plate. Beginning 90° to the right of the mantid, the cricket was moved "jerkily" in thirty steps through the horizontal visual field so that after 12 to 18 seconds the target had reached a point at 90° on the opposite side. The prey was located about 3 to 4 cm away from and even with the center of the mantid's head.

This experiment was done with the mantids in a semicylindrical experimental arena (30 cm diam × 33 cm high) which was lined with white paper. Again, the arena was diffusely illuminated with a halogen lamp in an otherwise dark room. The target was moved two to three times in succession (from right to left and back), and the number of saccades per 180 turns was used as the measure of the mantid's reaction.

Eye-Grooming Behavior

Eye-grooming behavior was elicited by applying one droplet of distilled water (1–10 µl) onto the dorso-anterior and lateral sector of one compound eye (for additional details, see Zack, 1978). If the mantid did not show a grooming reaction within 4 minutes after stimulus application, the droplet was carefully removed from the eye with the micropipette. After 4 minutes, the other eye was stimulated with a droplet of the same size. For each droplet size, the stimulation program was repeated three to seven times (depending on the cooperation of the mantids). As we have pointed out, male mantids are quite restless, which often made it difficult to apply the droplet precisely onto the eye.

Grooming behavior in tethered mantids was electronically registered with the position-sensitive transducer described above. The continual course of head movements during grooming caused unique changes in the electric signal which identified the individual grooming phases (Fig. 11.2). The grooming behavior of freely moving mantids was recorded with a video system at 60 fps (Panasonic Recorder NV-8030; Tamron Zoom Lens 1:1.8; 12.5–75 mm).

The terminology used by Zack (1978) for the *S. lineola* as well as his evaluation criteria were used to analyze the behavior of *T. sinensis*. However, a decisive and characteristic parameter of grooming behavior that was not considered by Zack is the mantid's reaction time. This is defined as the time interval between the application of the droplet and the beginning of the grooming reaction, which is characterized by the lifting of the prothoracic leg simultaneous with a rolling movement of the head (Fig. 11.2A, B).

In female mantids ($n = 2–4$) the head-grooming behavior was studied in detail, and the following parameters were quantitatively evaluated: (1) reaction time, (2) mean duration of one grooming cycle, (3) mean number of consecutive grooming cycles, and (4) mean duration of a grooming episode (a grooming episode consisting of n grooming cycles).

Sexual Behavior

To assess sexual behavior, a sexually mature male and female were placed facing each other, 60 cm apart, in a semicylindrical arena, lined with white paper (for details, see Liske and Davis, 1987). In some cases ($n = 15$) the animals were separated after a mean copulation time of 48 minutes (± 4.3 SEM); otherwise the observation period was terminated after the male actively abandoned the female.

Immediately after placing the mantids into the arena the behavior of both animals was videotaped. A total of forty matings (female $n = 40$, male $n = 31$) were evaluated.

Defense Behavior

When threatened, mantids show behavioral patterns of flight and deterrence (see Brunner and Edmunds, this volume, chap. 13). Neither nymphs nor adult mantids show a single, stereotyped defensive reaction; rather they have a flexible repertoire that depends on the particulars of the situation (Crane, 1952, Maldonado, 1970). In our experiments, mantids were placed into the semicylindrical arena, and their behaviors were either photographed or videotaped for later evaluation.

Electrophysiology: The Electroretinogram

When activated, the summed electrical potentials of receptors and higher-order neurons can be recorded extracellularly from the compound eyes and ocelli. This record is the electroretinogram, or ERG. To do these recordings, anesthetized animals were fixed to a plastic block with wax, and the head, prothorax, antennae, and mandibles were immobilized. The recording electrodes (100 µm chlorinated silver and platinum wires) were implanted in certain sectors of the compound eye or in the ocelli (with the point approximately 0.2 mm beneath the cornea), and a reference electrode was placed in the prothorax. Each man-

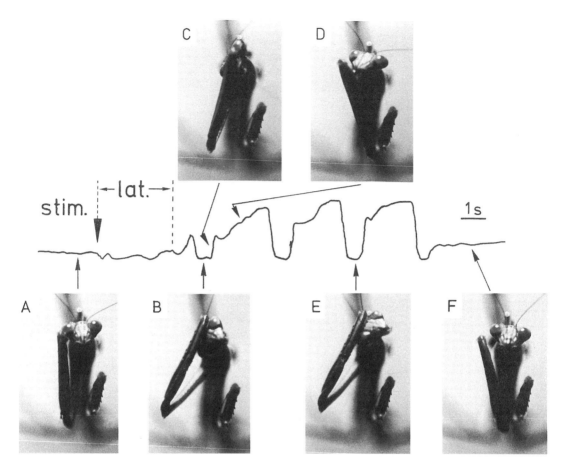

Fig. 11.2 Registration of head movement during the eye-grooming behavior. The kinetic flow of the head movements causes characteristic electrical signal changes in the position-sensitive measuring system (middle track). The signal changes can be assigned to individual grooming phases A–F. The diagram shows a grooming episode made up of three consecutive grooming cycles. One grooming cycle consists of the sequence B–D. A, F, Resting position; B, E, turning the head and lifting the ipsilateral prothoracic leg; C, the "wiping process"; D, cleaning the femur brush. Lat. - reaction time; stim. - application of stimulus (the stimulus was a 4 μl water droplet applied to the right compound eye).

tid was then placed in a Faraday cage lined with black paper with its head at a distance of 3 mm from the tip of a light source.

During the first test series, the eyes were stimulated with the white light of a halogen lamp operated with stabilized direct current (Osram Xenophot, 12 V, 75 W; cold light, Messrs. Schott). The intensities used were reduced and calculated by neutral glass filters of known transmission. By means of an electronically controllable compression lock (Prontor Magnet E/40) located in the ray path, the pulse duration was set for 1 second with a light energy of 1.7 W/m² (linear sector of the characteristic curve). Dark-adapted mantids were exposed to one light flash per hour. The reaction that we measured was the change in ERG amplitude.

In a second test series, mantids were exposed to colored light stimuli, and the spectral sensitivity of the dark-adapted eyes were compared to those of light-adapted eyes. For the comparisons, measurements were conducted in the morning (0830 to 0930), afternoon (1200 to 1600), and evening (2100 to 2200). For these experiments a

DIL interference filter (400–644 nm) was used. Neutral glass filters were used for lowering light intensity.[1]

In order to check the possible effects of circadian rhythms, simultaneous recordings from different sectors of the compound eyes and the median ocellus were done over periods of 72 to 96 hours. Here, the compression lock in the ray path was opened for 1 second once each hour, and the reactions were recorded. During the experiment the light energy at the eye was measured with a compensated, calibrated thermocolumn (Kipp and Zonen).

Control of the compression lock and data analysis were done on-line by means of a personal computer.[2]

Results

Feeding Behavior

Liske and Davis (1984) have reported on the eating behavior of satiated *T. sinensis* from 0 to 35 adult days old (i.e., days after the final molt). For females, the mean number of prey eaten increased from day 0 until a maximum was reached at days 16–17 ($x = 8.3 \pm 1$ prey/day).

---

[1]. The spectral sensitivity is defined as the reciprocal value of the light flow (quantum/second) with a varying wavelength, triggering a standard reaction (here: 1 mV "ON" amplitude). The spectral sensitivity can be determined from the intensity dependency (reaction degree-relation to the light flow) of the reaction for different wavelengths. Since in this case only the shape and position on the wavelength scale are of the interest, the relative spectral sensitivity was determined and the number of quanta (RQZ) was calculated for varying spectral lights.

[2]. Besides calculating the mean (x), and the standard error of the mean (SEM), we tested pairs of means for significant differences with two-tailed t-tests (Sachs 1974). The Kolmogoroff-Smirnoff test was used for checking the spread of two independent samples. Correlation and regression analyses answered questions regarding the degree of the functional correlation between sample values (PC-Software, Program STSC Statgraphics). Chi-squared tests were used to test for differences between observed and expected frequencies. Statistics were considered significant when $p = .05$.

After 4 weeks, the number of prey eaten decreased to two to four per day. In contrast, for the first 14 adult days, the males' intake was not variable and averaged just 1.6 (± 0.5) crickets per day. Older males ate less than one cricket per day.

Is the feeding behavior subject to a circadian rhythm? In order to answer this question the number of crickets eaten per day was analyzed separately for each experimental animal. One can clearly see that mantids of both sexes ate prey throughout the day (Fig. 11.3A), and there were no statistically significant time-of-day effects (Fig. 11.3B).

Within 3 hours of the start of the experiments (0900 or 2100), and independent of the time of day, food-deprived, 3- to 4-week-old adult females ($n = 7$) ate as many as fifteen crickets. Satiated females ($n = 7$), however, ate no prey (with one exception) in the first 24 hours after the beginning of the observation period. On the other hand, hungry adult males ($n = 7$) showed no differences in feeding behavior from satiated males ($n = 6$): Males older than 14 days ate on average less than one cricket per day (Fig. 11.3A).

The often overlooked fact that mantids can capture prey at night (i.e., in complete darkness) has also been demonstrated in laboratory studies done by Hurd et al. (1989). These observations were conducted under low infrared illumination, which is not visible to mantids. If prey accidentally touched the prothoracic legs of females, a strike was triggered. Prey captured like this were eaten immediately.

Sexual Behavior

In spite of the enormous interest in the reproductive biology of mantids (for references, see Davis and Liske, 1985; Maxwell, this volume, chap. 5), previous studies using behavioral/physiological methods have not been quantitatively evaluated (e.g., Roeder, 1935, 1963). Liske and Davis (1984, 1987) presented the first quantitative analysis for *T. a. sinensis*, and, given the importance of these results, the critical findings are summarized below.

Satiated Mantids

Immediately after placing potential sexual partners together in a brightly lit arena, target-direct-

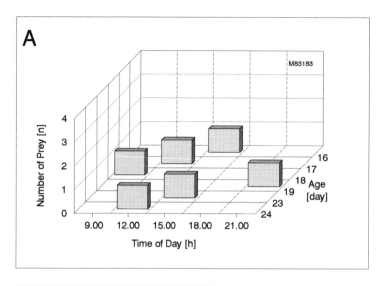

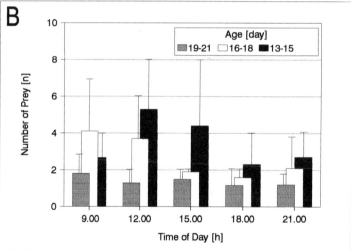

Fig. 11.3 The eating behavior of the Chinese praying mantis when fed *ad libitum*. A, "Nutritional histograms" of a male mantis. The mean number (± SEM) of prey eaten is presented as a function of time of day and the age of the mantid. B, The mean (± SEM) number of prey eaten is shown as a function of time of day for mantids in several age groups (*n* = 8 females) with the error of the mean.

ed saccadic head movements by both insects are observed, which suggests mutual visual fixation. Slowly and with vigorous antennal movements, the male approaches the female. Approaches can be described in terms of one of three different strategies: Strategy 1 was seen in fourteen (64%) of twenty-two matings. In these cases the male approached the motionless female frontally (41%) or from the side (23%). The mean duration of the approach (*t* = time from placing the mantids together until the male leaped onto the female) was 33 (± 24 SEM) minutes or 58 (± 22 SEM) minutes, respectively. Strategy 2 was observed in 18% of the matings. Here, the male approached the female from behind with a mean approach time of 45 (± 24 SEM) minutes. In Strategy 3, which was observed in 18% of the matings, the female approached the motionless male from the front with an approach time of 29 (± 24 SEM) minutes. Strategy 3 was observed if— and only if— the male was sexually immature (i.e., younger than 10–14 adult days).

We found that, when the distance between the two mantids reaches 30 to 40 cm, the male mantis begins a series of characteristic, complex abdominal movements, the so-called abdominal flexion display (or AFD). The display consists of two components, low-amplitude pumping movements of the outstretched abdomen in the dorsoventral direction (i.e., "pumping") and left-right movements with the abdomen curled dorsally (i.e., "weaving"). As the courtship progresses, the intensity (or excursion amplitudes) and repetition frequency (f_{AFD}) of these movements increases. For instance, a few minutes after the beginning of the courtship, we calculated the f_{AFD} to be 0.05/second, and 80 minutes thereafter it was 0.4/second (and frequency is correlated with time: $r = .87, p = .01$).

The abdominal flexion display was observed most often (86% of the time) when the male approached the female frontally and in 25% of the cases in which either strategies 2 or 3 were used.

We found that the male mounted the female from a mean distance of 7.8 (± 2.5 SEM) cm (range, 2–14 cm, $n = 23$) with a "flying leap," after which it assumed a characteristic position on the female. In 55% of the cases, the male landed on the back of his partner facing the rear and had to reposition himself. A few seconds to minutes thereafter, the tip of the male's abdomen curled in a characteristic manner, and the typical precopulatory abdominal search movements, the so-called S-bendings, began (Roeder, 1935; also see Liske, 1991a, b). These complex, cyclical abdominal movements also consist of two components. The first is a lowering and coincident twisting of the abdomen around its longitudinal axis so that the terminal segments are curled into an S shape. The second component is a superimposition of rhythmic, abdominal "searching" movements on the first component accompanied by a palpitation of the female's abdomen by the male's cerci (Davis and Liske, 1988). Simultaneously, the male's antennae "lash" about the female's head. As a rule, after several minutes the male's abdominal searching movements lead his genitalia to the female's bursa copulatrix.

For the duration of the copulation (1–3 hours) the male's abdomen remains twisted around its longitudinal axis; the abdominal tip is curled by approximately 180° and mechanically anchored by means of the phallomere within the female's bursa copulatrix. During this time the spermatophore is passed (Fig. 11.4A–C). The end of copulation begins with a separation of the abdomens and is followed by the male's active abandonment of the female. The end of copulation is independent of the time of day at which sexual behavior begins.

Hungry Mantids

The course of sexual behavior deviated from that described above if the female was hungry. In six out of seven cases studied, hungry females either attacked the approaching male (i.e., prey capture behavior was triggered) or displayed (in one case) the defensive "deimatic reaction" (Maldonado, 1970; Edmunds and Brunner, this volume, chap. 13). In two-thirds of these cases, however, the males were still able to perform the flying leap onto the female and copulate. In one case that we observed the male fled, and in another case the male was captured and devoured by the female prior to any signs of mating behavior. Interestingly, we found no differences in the sexual behavior of hungry versus satiated males.

Adult, sexually immature males showed only traces of the behavioral components of the courtship patterns. Apparently, however, they were not able to perform the flying leap onto the female.

Eye-Grooming Behavior

The eye-grooming behavior of male and female *T. a. sinensis* consists of rhythmically coordinated head and prothoracic leg movements and corresponds basically to that of *Sphodromantis lineola* (Zack, 1978).

After a reaction period—that is, the time from

Fig. 11.4 *A,* During copulation (1–3 h) the terminal segments of the male abdomen remain bent by 180°. *B, C,* After completion of copulation the spermatophore may still be visible at the female's genital opening.

placing the droplet of water on the eye to the beginning of the reaction—the prothoracic leg is lifted and the head is turned about its longitudinal axis, causing the "femur brush" of the leg (a hairy area on the inner side of the femur) to touch the surface of the eye (Fig. 11.2B, F). This movement is what Zack (1978) referred to as FE1. The femoral brush is then wiped over the surface of the eye by a simultaneous turn of the head and forward movement of the leg (Fig. 11.2C; Zack's FE2). The eye-grooming cycle is completed with the "cleaning" of the femur brush with the mandibles (Fig. 11.2D; Zack's FE3). Several such grooming cycles may be performed consecutively. We call a series of cycles a *grooming episode*.

Zack (1978) observed only females, did not consider the mantid's nutritional state, and did not analyze reaction time. Although we were able to examine the latter variables, we also found that quantitative measurements of the head/eye behavior were possible only in females.

Satiated Mantids

When mantids were placed in front of a plain white background, we found that the average reaction time decreased exponentially with the size of the water droplet (Fig. 11.5A). Small droplets required long reaction times, and within the 4-minute time limit, only 50% elicited reactions. On the other hand, the probability that a stimulus would elicit a response within 4 minutes increased as the droplets got bigger (i.e., 4–10 μl); the reaction times were between 1 and 10 seconds. In each case in which a grooming cycle occurred, all three functional components of the cycle were observed. The mean duration of the cycles (calculated for responses to 1, 4, or 7 μl droplets) was 2.4 ($\pm$ 0.2 SEM) seconds ($n = 71$ observations for three mantids), the mean number of grooming cycles per grooming episode was 3.3 ($\pm$ 0.3 SEM) seconds ($n = 22$), and the mean duration of an episode was 8.8 ($\pm$ 1.2 SEM) seconds ($n = 22$). The times between consecutive grooming cycles were not evaluated quantitatively.

Hungry Mantids

Hungry females responded with eye grooming within 0.5 to 5.3 seconds. Compared to satiated animals, these latencies were much shorter (Fig. 11.5B). On the other hand, the mean duration of a grooming cycle and the mean number of grooming cycles per episode were the same in hungry and satiated animals. However, because the mean duration of a single grooming episode performed by a hungry mantis was shorter than that performed by a satiated mantis by the factor 1.5, the intervals between consecutive grooming cycles were shorter for hungry mantids. In other words, the individual grooming cycles followed each other more quickly.

Attempts to selectively stimulate different sectors of the eye failed because (with the exception of the dorso-anterior/lateral sector of the eye) the water droplets rolled off of the surface of the eye.

In summary, it appears that eye-grooming be-

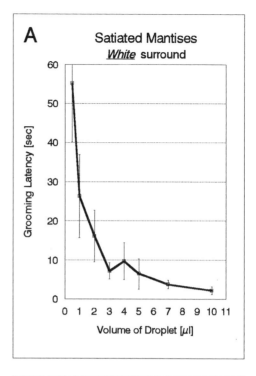

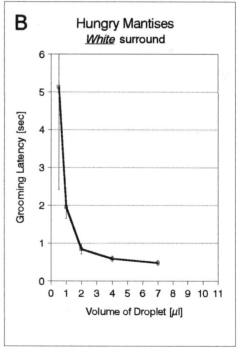

Fig. 11.5 The latency to eye groom as a function of the volume of a water droplet applied to the eye of (*A*) satiated or (*B*) hungry female mantids. Each data point represents the mean ($\pm$ SEM) of from 2–28 trials done on 3 or 4 mantids.

havior is very similar in hungry and satiated *T. a. sinensis*. This means that, once a grooming cycle is triggered, it will run according to its program and is influenced very little, if at all, by the nutritional state of the mantis. These results support the hypothesis that this behavior represents a "fixed action pattern," an idea formulated by Zack (1978) for the eye grooming by female *S. lineola*.

Saccadic Head Movements

To acquire food, the praying mantids have developed a characteristic repertoire of behaviors for capturing prey that consists of a sequence of discrete components that includes target-directed head (i.e., eye) movements followed by a rapid grasping movement of the prothoracic legs (Mittelstaedt, 1957; Cleal and Prete, 1996; Prete and Hamilton, 1996, this volume, chap. 10).

Because of the their special morphology, a group of ommatidia located within a limited frontal sector of each compound eye allows the highest possible resolution of visual stimuli (Maldonado, 1970; Rossel, 1979; Kral, this volume, chap. 7). Analogous to the vertebrate eye, this structure is called the *fovea*. By means of directed head movements the praying mantis moves a target into the binocular, foveal area of vision; this is called a *fixation*.

Foveal vision makes it possible for the mantis to fixate and track moving targets. Fixating and tracking a moving target can consist of continuous head movements; this is called *smooth tracking* (Fig. 11.6B-1), or it can appear as a series of discrete, "jerky" movements called *saccadic tracking* (Fig. 11.6B-2). In addition, the two types of tracking can alternate depending on the characteristics of the visual stimulus (e.g., its velocity; Fig. 11.6B-3).

We determined the number of head saccades that occur when a house cricket was moved horizontally several times to the left (-90°) and right (+90°). We found that for hungry mantids of both sexes, the mean number of saccadic head movements was significantly higher than in satiated animals. For females, the means were 4.5 (± 1.3 SEM) versus 1.4 (± 0.8 SEM) saccades per

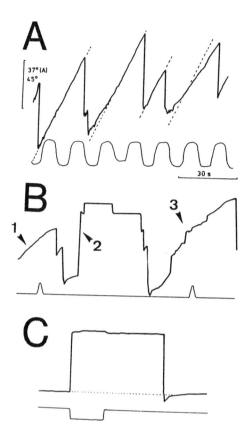

Fig. 11.6 *A (upper trace)*, Head movements associated with optokinetic nystagmus (OKN) triggered by a large moving black-and-white–striped pattern; *(lower trace)*, movement of the stimulus. The slope of the dotted lines reflects the velocity of the striped pattern. *B*, Fixated tracking movements elicited by a small moving object. The peaks in the lower trace indicate when the stimulus was located in front of the mantis. Points 1–3 in the top trace indicate different types of visual movement. *C (upper trace)*, A head movement associated with a fixation saccade elicited by the sudden appearance of prey within 70° to the right in the lateral visual field. The head returned to resting position after the saccade.

180° of stimulus movement, and for males the means were 3.4 (± 1.4 SEM) versus 1.2 (± 0.9 SEM) saccades per 180°.

Following the experiments, the prey was captured and devoured by the hungry mantids in 87% of the cases but was never attacked by the satiated mantids.

Optokinetic Behavior

The motion of a large, optically structured visual environment induces smooth tracking movements by mantids, the so-called optomotoric reaction. These movements enable mantids to compensate for the retinal shifts caused by their own movements. For instance, in mantids of both sexes, a moving pattern of vertical black-and-white stripes triggers smooth, continual tracking movements (the slow phase), which is interrupted by a fast return movement in the opposite direction (the fast phase; Fig. 11.6A). Analogous to similar behaviors in vertebrates, this alternation is called *optokinetic nystagmus* (OKN). It is of interest that the fast phases of the OKN show a striking resemblance in their velocity characteristics to fixation saccades (Liske and Mohren, 1984).

Our analysis of the optokinetic behavior failed to reveal any quantifiable differences between hungry and satiated mantids. However, there were differences between male and female mantids. In females, the velocity of continuous tracking movements turned out to be a function of the angular velocity of the striped pattern. The difference between the velocities of the head and drum was minute if the drum moved slowly (up to approximately 20°/second). In cases of higher drum velocity, however, the difference increased progressively until the head-angle velocity reached an asymptote (Fig. 11.7A, B). The maximum smooth tracking velocities for females had a mean between 43 and 60°/second. For females, different stripe widths had no effect on the optomotoric behavior.

In males, on the other hand, the smooth tracking velocities regularly exceeded the stimulus velocity (Fig. 11.8). In fact, continual smooth tracking movements by males can reach velocities of 90 to 100°/second. The mean value curves in Figure 11.8 indicate that the head tracking velocity depends not only on the velocity of the stimulus pattern, but also on the width of the black-and-white stripes. For males, the contrast frequency may be the decisive parameter for visual stabilization as it is for dipterans and hymenopterans (Kaiser and Liske, 1974). In addition, earlier studies suggest that the age of the male mantids might influence the head velocity (Liske et al., 1986).

How can one interpret these sex- (and, perhaps, age-) specific findings? It is possible that male mantids track individual moving elements (e.g., luminance edges) in the peripheral visual field with a higher tracking velocity than they use to compensate for panoramic movement. Hence, these males may have been reacting with a rapid foveal fixation behavior rather than with a slower movement compensation behavior. In addition, all previous behavioral experiments indicate that "attention" may play a role in these visually guided behaviors.

Two weeks after the adult molt, males become sexually mature (Liske and Davis, 1987). When approaching the female during courtship, any disturbances in the field of vision that might trigger only minute reactions in an older male (interrupting the courtship behavior) are responded to with directed head movements. In several species of mantid, the fovea of the male's eye is relatively larger (compared to the overall size of its eye) than it is in females (Kirmse and Kirmse, 1985). The larger fovea in males could be the cause of the stronger visual attention of these animals. Interestingly, researchers have known for some time of sex-specific differences in the optomotoric behavior (cabbage butterfly, *Pieris rapae* [Kien, 1973]) and sex-specific visual interneurons in insects (flies [Hausen and Strausfeld, 1980; Strausfeld, 1980]).

Defense Behavior

In case of danger, mantids display behavioral patterns of flight and deterrence (Edmunds and Brunner, this volume, chap. 13). Defensive behaviors are not strictly stereotyped; they are flexible depending on the particular situation (Crane, 1952, Maldonado, 1970).

Males

In the laboratory, we have observed two different defensive strategies in males ($n = 7$), fleeing and threatening behavior. Fleeing behavior always occurs when the distance to the dangerous object

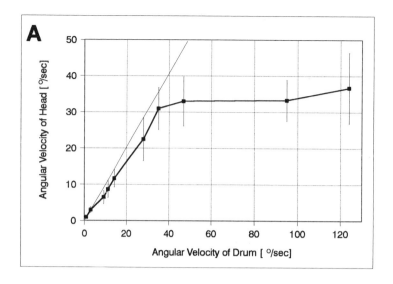

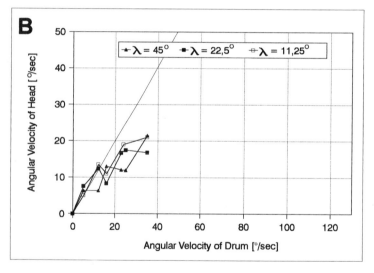

Fig. 11.7 The angular velocity of the head in the slow nystagmus phase of the optokinetic nystagmus (OKN) as function of the angular velocity of a striped pattern. A, Each data point represents the mean (± SEM) from 10–20 measurements taken on two female *Tenodera aridifolia senensis*. B, These are the mean angular head velocities of a female *T. a. senensis* elicited by striped patterns with different stripe widths. Data points are means (± SEM) of 8–57 measurements.

is reduced to 6 to 9 cm. Flight is initiated by a fast backward erection of the prothorax and extension of the mesothoracic and metathoracic legs, which increases the distance between the body and the threatening object to 11 to 12 cm, without losing the threatening object from the binocular portion of the visual field. (We found retreat velocities to be 94–116 cm/s, mean = 103.5 (± 3.5 SEM) cm/s based on six measurements in two animals.) Following its retreat, the male jumps away by means of the flying leap.

As noted above, *T. a. sinensis* assumes a characteristic defensive or threatening posture called the *deimatic reaction* (Maldonado, 1970). This complex behavior consists of erection of the prothorax, sideways turning of the prothoracic legs,

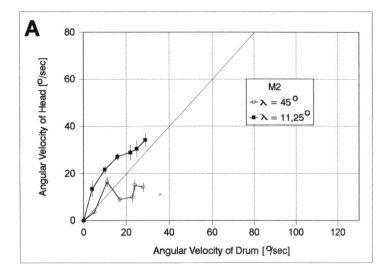

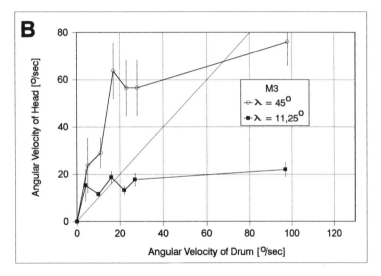

Fig. 11.8 The head tracking velocity of two male *T. a. senensis* that are watching a moving black-and-white–striped pattern with different stripe widths. Data points are means (± SEM) of 5–64 measurements.

spreading of the mandibles, and erection and spreading of the wings. The abdomen may be alternately rolled up and down. The latter creates a hissing sound caused by the cerci rubbing over the spread wings. In new adult males, escape and defensive behaviors were triggered with approximately the same frequency, but older males (>14 days) usually reacted to the presence of a larger female with courtship rather than with defensive behaviors. We could not find any influences of nutritional state on defensive behaviors in males.

Females

The threat of another female generally triggers the same defense/threatening behavior as described for the males. In addition, however, the prothoracic legs may strike in the direction of the threatening object, often causing the attacker to retreat. This kind of striking from a defensive posture was also observed in hungry females when they were approached by a courting male. Fleeing behavior was not observed in any of the seven females that we observed. It is difficult to say whether hunger level influences defensive behaviors of females. Such behavior was observed in both satiated and hungry females, but defensive striking was displayed only by hungry ones.

Interactions between Behavioral Patterns

In the previous section, the behavioral patterns that could be elicited from hungry and satiated mantids were described and quantified. Now the question is whether any of these behavioral patterns are influenced by another if two are triggered simultaneously. Theoretically, the behaviors could mutually inhibit one another, one behavior could suppress another, there could be some type of mixture of the behaviors with elements of each appearing, both behaviors could be performed simultaneously or consecutively, or a displacement behavior that matched neither of the eliciting stimuli could occur. Of course, any of these outcomes could be influenced by the intensity of a given stimulus, the time of day at which the stimulus occurred, and/or the physiological state of the mantid.

Head-Grooming Behavior before, during, and after Copulation
Males

During eye grooming and during copulation the prothoracic legs have to fulfill two different types of tasks. During grooming, the legs move forward and upward, and during copulation, they are held ventrally and grasp the female firmly.

Before copulation the males react to eye-grooming stimuli as described above, and the reaction time is inversely proportional to the intensity of the stimulus (Fig. 11.9, "before mating"). During copulation eye grooming can be triggered only to a limited extent. The reaction occurs only in the case of higher stimulus intensities and has a significantly longer reaction time. Droplets of 1 μl and 4 μl will not trigger eye grooming at all,

and only 35% of the males that we observed responded to 7 μl droplets by lifting the ipsilateral prothoracic legs (the FE1 component). Even in these latter cases, however, they never turned their head to complete the grooming cycle. The mean reaction time for 7 μl droplets was longer than it was for the same size droplet prior to copulation (10.7 [± 5.9 SEM] seconds versus 5.5 [± 1.3 SEM] seconds; n = 13 and 52, respectively). Thirty minutes after abandoning the female, males responded with eye-grooming behavior as they did prior to copulation (Fig. 11.9, "after mating"). These data demonstrate that eye-grooming behavior is not completely suppressed during copulation but the behavioral threshold is higher. No nutrition-related differences were observed.

In an additional experiment, copulation was interrupted in fifteen pairs of mantids after 48 (± 4.3 SEM) minutes, and males received a 7 μl droplet on one eye. After separation, two males groomed after a reaction time of 1 to 2 minutes, but thirteen failed to respond with eye grooming for as long as 10.8 (± 2.2 SEM) minutes. After about 11 minutes had passed, reaction times decreased rapidly, and after approximately 30 minutes precopulation reaction times were reached. This change in latency to response may represent either a lowering of the threshold for, or a reduction in inhibition, of eye grooming behavior (Liske and Davis, 1986).

Females

We found no differences in eye-grooming behavior before, during, or after copulation in females (Fig. 11.9). Both reaction times and the number of grooming cycles remained constant.

Optomotoric Behavior before, during, and after Copulation
Males

Before and 30 minutes after copulation, a moving, vertical black-and-white–striped pattern triggered compensatory head movements in 91% and 84%, respectively, of the males that we tested (n = 32). This corresponds to control levels (Fig. 11.8). During copulation, however, the same striped pattern did not trigger any optomotoric

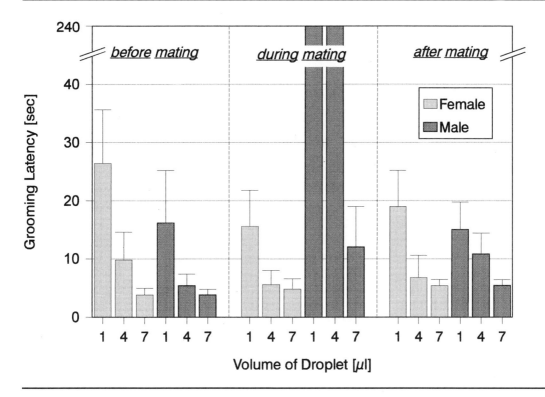

Fig. 11.9 The mean grooming latencies (± SEM) after the application of different-sized water droplets (1, 4, or 7 μl) to one compound eye of male and female *T. a. senensis*. The droplets were applied before, during, or after copulation. For males, eye grooming is inhibited during copulation (reaction time > 4 minutes). Data points are means (± SEM) of at least three measurements done on 15–20 males and 6–16 females.

reactions in 97% of the males that we tested, and in only one case did a male display slow head movements separated by long pauses.

One reason for the males' failure to react could be that they visually fixate the female prothorax during copulation, perhaps because it is a reference point, so to speak. To test this hypothesis, we covered the female's prothorax with a white container. Even under this condition, however, it was impossible to elicit head movements in response to the moving striped pattern.

Females

In contrast to males, all twenty of the females that we tested responded to the moving striped pattern before, during, and after copulation.

Saccadic Head Movements before, during, and after Copulation

Target-directed saccadic head movements often precede prey capture behavior, which consists of a predatory strike by the prothoracic legs and, depending on the distance of the prey, a simultaneous lunge effected by the mesothoracic and metathoracic legs (Cleal and Prete, 1996; Corrette, 1990; Maldonado, 1970; Prete and Cleal, 1996; Prete and Hamilton, this volume, chap. 10). During copulation, however, the male's prothoracic legs fulfill a very different role: They cling to the female and stabilize the male's position.

Males

Before and after copulation, target-directed saccadic head movements are easily elicited (see Table 11.1), and we found that significantly more saccades could be triggered in hungry than in satiated mantids ($n = 9$ and 23, respectively). Understandably, we also found that 45% of the hungry males tested captured and ate the prey used to elicit the head movements. In contrast, and with only one exception, no saccadic head movements could be triggered during copulation in hungry or satiated mantids, and none captured prey while copulating.

Females

The females that we tested responded similarly with head saccades to moving targets before, during, and after copulation (Table 11.1). Again, significantly more head movements were triggered in hungry ($n = 5$) than in satiated ($n = 9$) females, and only hungry females captured and ate prey.

Head Grooming during Optokinetic Behavior

Optokinetic behavior and eye grooming involve characteristic head movements: The former are head movements that serve to stabilize the visual field, and the latter involve movements that turn the head in the direction of the prothoracic leg that is doing the grooming. Grooming-related head movements performed in front of a black-and-white–striped pattern inevitably cause a retinal shift of the visual field, which seems counterproductive to stabilization of the visual panorama. Unfortunately, we had to exclude males from our experiments on the relationship between these two behaviors, because of their often erratic behavior.

Experiments with a Stationary Striped Pattern

SATIATED FEMALES. We found that putting a droplet of water onto one compound eye would trigger a complete eye-grooming cycle (FE1–FE3) in 60 to 80% of trials when mantids were tested in front of a stationary vertically striped black-and-white pattern. We also found that the mean reaction time decreased exponentially from 23 to 5 seconds as droplet size increased (Fig. 11.10A). The mean cycle duration coincided with the values of satiated females in a white experimental arena, but the number of grooming cycles per episode and the mean episode duration had a tendency to be shorter than they were when the mantids were in front of a white field (Fig. 11.11A).

Table 11.1 Number of saccadic head movements (mean ± SEM; one measurement per animal) made by satiated and hungry male and female mantids before, during, and after mating

	Hungry	Satiated
	Female	
Before mating	3.9 ± 1.8 ($N = 5$)	1.3 ± 0.8 ($N = 9$)
During mating	3.7 ± 1.6 ($N = 5$)	1.0 ± 0.7 ($N = 9$)
After mating	3.4 ± 1.5 ($N = 5$)	1.4 ± 0.9 ($N = 9$)
	Male	
Before mating	3.1 ± 0.7 ($N = 9$)	1.5 ± 0.2 ($N = 23$)
During mating	(1 Exception) ($N = 9$)	(1 Exception) ($N = 23$)
After mating	2.1 ± 0.3 ($N = 9$)	1.3 ± 0.3 ($N = 23$)

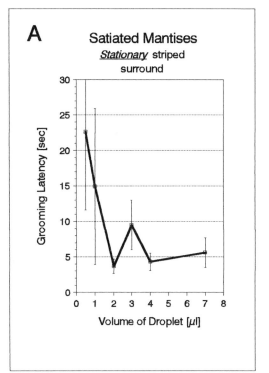
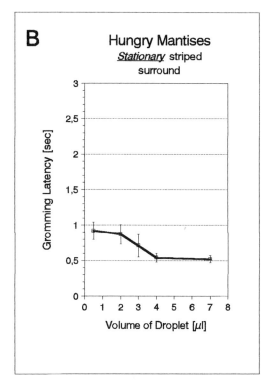
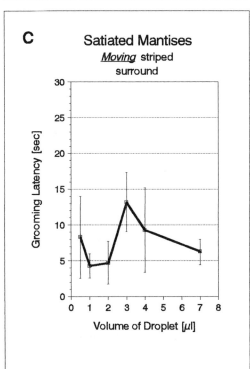
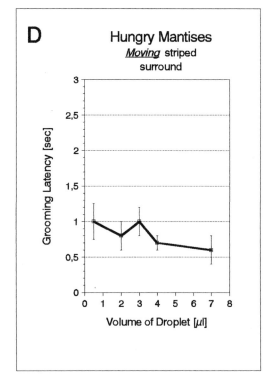

Fig. 11.10 The head-grooming latency of (A, C) satiated and (B, D) hungry female *T. a. senensis* within different visual surroundings. Eye-grooming was elicited by water droplets up to 7 μl. Each data point is the mean (± SEM) of 2–28 measurements on 3 or 4 mantids. In A and B the visual surround consisted of a stationary black-and-white striped pattern. In C and D the pattern moved at 22.5°/second.

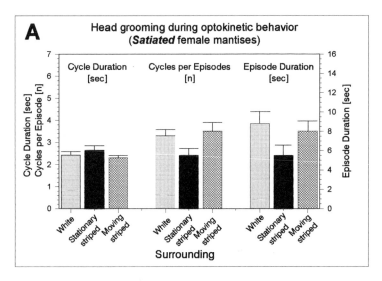

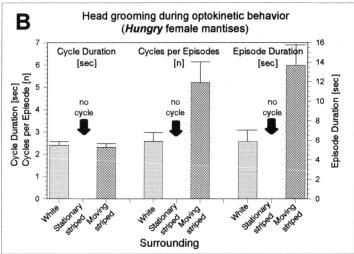

Fig. 11.11 Characteristic eye-grooming parameters (mean ± SEM) as they are affected by different physiological conditions (satiated versus hungry; 2 females) and visual surrounds. The number of measurements per column are: for duration of grooming cycle (left set of columns), 34–39; for grooming cycles per episode (center set of columns), 13–27; and for grooming episode duration (right set of columns), 17–27. These values are summed over data gathered using 1, 4, and 7 μl droplets. The velocity of the striped pattern was 22.5°/second.

HUNGRY FEMALES. In contrast, hungry females responded to water droplets in less than 1 second regardless of the droplet size (Fig. 11.10B). However, grooming did not follow a complete cycle; it consisted only of the components FE1 or FE1 and FE2. Furthermore, the characteristic head movements were missing, and in no cases was the FE3 component (cleaning of the femoral brush) observed (Fig. 11.11B).

How might one interpret the behavioral responses of these hungry females? In hungry mantids, the behavioral tendency to capture prey obviously predominates. In order to be prepared for capturing prey, the mantis must remove any "interference" from the eye within a very short time and has a low behavioral threshold to do so. Since no head movements occur when performing grooming components FE1 and FE2, the retinal image remains relatively stable—an important prerequisite for the fixation of prey. In contrast, however, cleaning the femoral brush (component FE3) is possible only with a change in the head position and, as a rule, takes the longest amount of time (Fig. 11.2E). Hence, this component would interfere the most with prey capture and is, therefore, completely suppressed in hungry animals.

EXPERIMENTS WITH A MOVING STRIPED PATTERN. When exposed to a moving striped pattern at the same time that one eye is stimulated by a droplet of water, optomotoric head movements and head turns for eye grooming are superimposed (Fig. 11.12A–C). If combined, there is a complex correlation between the reaction time of grooming, the size of the droplet, and the rotation velocity of the striped pattern. If the drum rotates slowly (e.g 10–25°/second) the reaction times for the head grooming correspond to those for the stationary striped pattern (Fig. 11.10C, D). However, mean reaction time in-

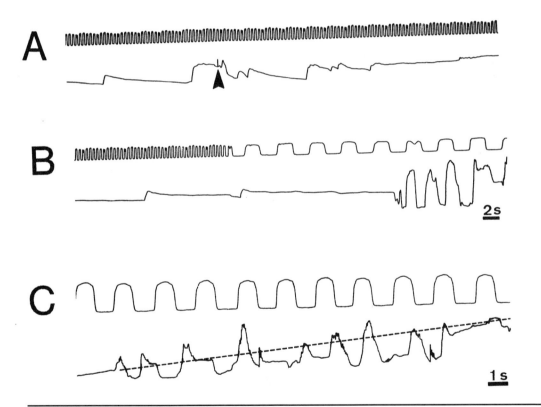

Fig. 11.12 Traces representing the eye-grooming and optokinetic head movements of female *T. a. senensis*. *Upper traces,* The velocity of the stimulus; *Lower traces,* Head movements. In *A*, a 4 µl droplet of water was applied to one compound eye (arrowhead), which triggered the eye-grooming behavior reflected in the deflections of the lower trace in *B*. Eye-grooming behavior began only after the stimulus velocity was decreased to 19°/second, indicated by the change in the upper trace in *B*. The records in *A* and *B* are separated by 3 minutes. The traces in *C* represent the simultaneous occurrence of eye grooming and optokinetic head movements. The dashed line indicates the trend in compensatory head movements induced by the moving striped pattern.

creases as rotation velocity increases; that is, the threshold for eye grooming is raised (Fig. 11.12A, B). This is especially clear when medium-sized (4 μl) droplets are applied.

SATIATED FEMALES. In satiated females the mean duration of a grooming cycle, number of grooming cycles per episode, and duration of a grooming episode coincided with the corresponding values of satiated animals in a white experimental arena (Fig. 11.11).

HUNGRY FEMALES. Regardless of the droplet size, a slowly rotating striped pattern triggered grooming reaction times of less than 1 second. Furthermore, the eye-grooming behavior intensified significantly in hungry females faced with a moving striped pattern as compared to those in a white arena (Fig. 11.11B). The mean duration of the grooming cycles remained almost the same, but the mean number of grooming cycles per episode doubled and the episode duration was prolonged by a factor of two to three. Clearly, the effect of the droplets on hungry females was intensified by the moving striped patterns. Interestingly, it is also the case that mantids react with a "wiping reflex" similar to the eye-grooming behavior if a small black point is projected onto one compound eye (i.e., the monocular cleaning reflex [Maldonado and Levin, 1967]).

So, the main results from the latter three experiments are these: First, in satiated females, the visual stimuli that we used had no decisive modulatory influence on eye-grooming behavior, and it induced orienting head movements only to a limited extent. Orienting behavior of satiated mantids is suppressed and therefore does not appear to be necessary to keep the retinal image of the environment stable. Second, in hungry females, eye-grooming behavior can be decisively influenced by visual stimulation. This finding is not surprising, given that hungry females are known to be "voracious" and respond to virtually any visual change in their field of vision with directed head movements. Finally, the sensorimotor coordination of eye-grooming behavior not only is subject to proprioceptive feedback, as postulated by Zack (1978), but is also influenced by afferent visual information and the state of nutrition of the insect (i.e., the mantid's level of motivation).

Head-Grooming and Feeding Behavior

While eating, the prothoracic legs hold the prey, and the head performs corresponding movements that enable the mandibles to mince the prey. These head movements could possibly interfere with simultaneously triggered eye-grooming behavior. We found that when a mantis is holding its prey with both prothoracic legs, an eye-grooming stimulus (4 μl and 7 μl droplets) did not trigger a response. Only after the prey had been eaten and the prothoracic legs were free would the mantis react to the droplet. If, however, the ipsilateral prothoracic leg was not occupied with prey, it showed suggestions of some movement reminiscent of the FE1 component of the behavior, but a complete grooming cycle could not be triggered during feeding. Our data show that eating suppresses eye-grooming behavior. The question of what stimulus intensity would be necessary ultimately to override this suppression could not be answered since larger water droplets applied during eating (10–20 μl) simply rolled off the eye.

Defense and Sexual Behavior
Males

Young adult males (<10–14 adult days) respond to the approach of a female with fleeing or defensive behaviors. Older adult males (10–14 adult days) do not react defensively but perform components of the courtship behavior when a female approaches, including leaping onto the female and initiating copulation.

Females

During copulation the female responds to a threat (e.g., another female or a predator) either by fleeing or with a defensive display. Copulation is not interrupted by defensive behavior, and a copulating male shows no visible changes in behavior due to the threat.

The Circadian Rhythms of the Visual System

In *Tenodera aridifolia sinensis*, the color of the eyes changes from a light green during the day to a dark brown at night. These changes in color, caused by pigment migration, indicate circadian

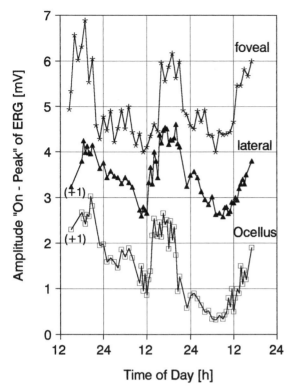

Fig. 11.13 Circadian fluctuations in the sensitivity of the compound eyes (lateral and foveal regions) and median ocellus of a dark-adapted male *T. a. senensis*. The dependent measure was the amplitude of an ERG-ON peak that was triggered by the flash (duration, 1 second) of a halogen white light. For the sake of clarity, the "lateral" and "Ocellus" curves were shifted up by one unit on the y-axis.

processes. We analyzed the physiological effects of these changes by means of ERG. These experiments were the first in which several ERG measurements were recorded simultaneously from different sectors of the eye (the lateral areas, the fovea, and the ocelli) continuously for several days (Fig. 11.13).

The summed electrical potentials derived from the mantid's eye were highlighted by a negative ON response followed by a excitatory response (Fig. 11.14). When the stimulus ended, the baseline potential returned, but often short-term positive potentials could be observed at stimulus offset. As the figure shows, ERG profiles of different eye sectors and ocelli coincided. As Figure 11.13 shows, the ERG of a dark-adapted compound eye and ocelli are subject to circadian fluctuations in amplitude. High-amplitude values in the afternoon to evening (1300 to 2100) alternate with low values in the morning (0600 to 1200).

We also determined the spectral sensitivity of the eyes using ERG measurements. We found that for both males and females the amplitude of the ON response increased with the intensity of the light at each of several wavelengths, and the slope of the curves was almost constant (Fig. 11.15). Over the course of the day, these curves shift along the intensity scale, but since the slope of the curves remains the same, no circadian wavelength-specific effects exist. Further, the shape of the spectral sensitivity curves and their position on the wavelength scale do not change with the time of day (Fig. 11.16B, C). In contrast, the relative sensitivity of the eyes does shift over the course of the day and reaches values that are approximately 2.5 to 6 times higher in the evening (Fig. 11.16A).

Conclusion: Behavioral Hierarchies in *Tenodera aridifolia sinensis*

In *T. a. sinensis*, independent behavioral patterns governing the acquisition of food, grooming, defense, and reproduction are organized hierarchically (Fig. 11.17). This behavioral hierarchy, or priority sequence, is not rigidly fixed and is sexually dimorphic.

In both hungry and satiated males, copulation suppresses target-directed saccadic head movements, prey capture, and optomotor head movements in response to a large moving striped pattern (Fig. 11.17, left diagram). If the sexual partners are separated during copulation, the eye-grooming behavior of the male is still completely suppressed for a period averaging 12 minutes. During the following 18 minutes the threshold for eye grooming is slowly reduced so that the response returns to normal in 30 minutes. On the other hand, defensive behavior has priority over all other behaviors, even copulation.

In females, the independent behavioral patterns governing the acquisition of food, reproduction, defense, and grooming can be triggered simultaneously and run their course concurrently (Fig. 11.17, right diagram). In satiated (versus hungry) females, however, target-directed head movements are suppressed. The relationship between optokinetic head movements and eye grooming is quite complex because both stationary and moving striped patterns can either suppress eye-grooming behavior or decrease the threshold necessary to elicit it. Finally, optokinetic head movements and eye-grooming behavior can be performed simultaneously.

If one compares the priority sequences of male and female mantids, the differences become obvious. In males, all behavior patterns involving head movements (saccades, OKN, eye grooming) or movements of the prothoracic legs (prey capture, eating, grooming) are suppressed during copulation. In contrast, females still respond to these stimuli.

The data presented here were gathered in the laboratory, not under conditions that mantids face in the natural environment. We have to assume that complex behavioral patterns are triggered by multisensory inputs in the natural habitat. Therefore, one should be somewhat cautious in interpreting these data. Hence, the behavioral hierarchies described here should be compared to any that can be established based on data gathered under analogous conditions in the wild.

The Influence of Nutritional Changes

Hungry mantids of both sexes readily display target-directed head movements, predatory strikes,

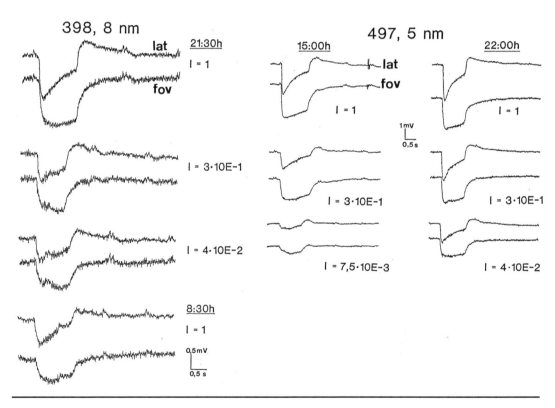

Fig. 11.14 ERGs recorded from the lateral (lat) and foveal (fov) regions of the compound eye of a dark-adapted male *T. a. senensis*. The mantid was presented with colored lights of different intensities (maximum l = 1; duration = 1 second) at different times of day (0830–2200). Maximum sensitivity was at 497.5 nm.

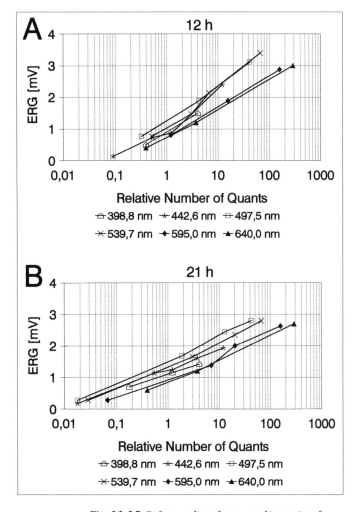

Fig. 11.15 Reference lines for spectral intensity of a dark-adapted compound eye (foveal region) of a male *T. a. senensis*. The times of day at which measurements were taken are noted above each graph. The curves for spectral sensitivities are based on calculations of the relative quanta required to induce a 1 mV change in ERG.

and eating behavior. In satiated animals, however, this sequence is completely suppressed except for some head movements.

During courtship behavior hungry mantids show significant differences from satiated mantids. First, courtship behavior is inhibited, and the potential sexual partner is struck at rather than courted. In fact, field studies of the mantis *Paratenodera* (= *Tenodera*) *angustipennis* demonstrate that nutritional deprivation can turn an inactive ambush predator into an active searching hunter (Inoue and Matsura, 1983; Matsura and Inoue, this volume, chap. 4).

The Influence of Circadian Rhythms

Zack (1978) reported that in the mantis *Sphodromantis lineola* prey capture, grooming, and moving about can be observed more frequently in the morning than in the afternoon. These qualitative observations indicate that in mantids, too, the triggering of certain behavioral patterns is subject to the time of day.

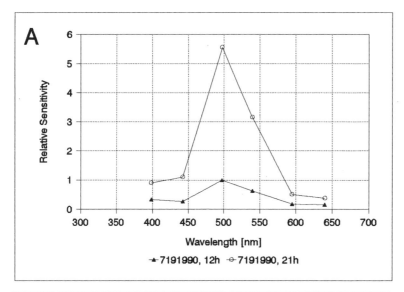
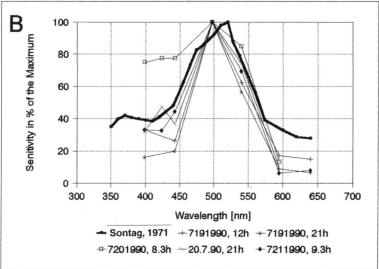
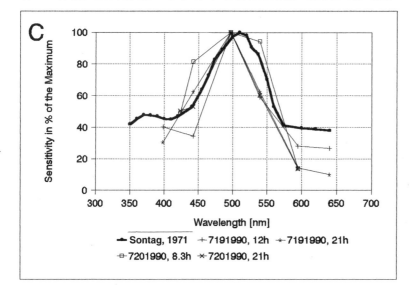

Our quantitative experiments with *T. a. sinensis*, have shown that the behaviors necessary for prey capture—visual fixation, optokinetic head movements, striking, and eating—are not subject to a circadian rhythm during a light-day of 16 hours; that is, mantids can capture and eat prey during the subjective night (also see Hurd et al., 1989). Experiments on eye grooming and sexual behavior also failed to reveal any major circadian influences. Even animals kept in a permanently illuminated environment for several days showed no significant differences in behavior. Again, when interpreting these results one must be cautious. Reliable quantitative data—for instance, on the sexual behavior of mantids in the wild—are not available despite a multitude of publications covering this subject. Observations on neotropical mantids contain clues regarding a possible circadian rhythmicity of sexual behavior: males were attracted by female pheromones only at dawn (Robinson and Robinson, 1979; Maxwell, this volume, chap. 5).

Interestingly, circadian rhythms play a role in the behaviors of other orthopteroid insects, such as crickets. Comparing laboratory and field studies of *Teleogryllus oceanicus*, Loher (1989) explained the involvement of a circadian rhythm in the organization of sexual behavior. Here, too, the dark-light change from night to day is, along with several other factors (e.g., temperature), a critical external timer for male courtship and mating and for female activity. In the female mantis, however, moving about during the day could be observed only in hungry animals; satiated mantids remained stationary (e.g., Inoue and Matsura, 1983; Matsura and Inoue, this volume, chap. 4).

Facing page

Fig. 11.16 Spectral sensitivity curves of the foveal region of (A, B) the compound eye and (C) the median ocellus of a dark-adapted male *T. a. senensis*. A, relative sensitivity. B, C, Normalized curves. Data obtained by Sontag (1971) for dark adapted female *T. a. senensis* are included for comparison.

There is no doubt that in *T. a. sinensis* the thresholds for, and the execution of, several behaviors are subject to the modulatory influences of the insect's nutritional state rather than to circadian rhythms. In honeybees, for example, the behavioral threshold for grooming the antenna can be influenced by time of day (Kaiser, 1988). At nighttime the threshold reaches its highest value and is associated with the resting state, and it is considerably lower around noon. Furthermore, physiological and anatomical experiments in a number of arthropods have demonstrated circadian changes in the sensitivity of photoreceptors and eyes (Fleissner, 1974; Page, 1985; Kaiser, 1983). The effects of such changes in sensitivity on other behaviors are still mostly unclear, but we do know that the effects exist in insects (Brady, 1975). For instance, in the cricket *G. bimaculatus* a bilateral cut through the optic tract (the optic lobes apparently have a pacemaker function) caused no change in the circadian-dependent ERG rhythm, but did desynchronize the circadian-controlled running rhythm (for a review see Loher, 1989). These findings indicate the presence of at least two separate pacemaker systems in crickets which work together to maintain circadian control over the insect's behaviors.

In *T. a. sinensis*, too, the results of the ERG experiments described here demonstrate circadian fluctuations in eye sensitivity. These findings coincide with the intracellular data derived from individual foveal and lateral retinula cells in the compound eye of *Tenodera australasiae* (Rossel, 1979; Horridge et al., 1981). At night, the acceptance angle of individual ommatidia increases along with their sensitivity (Kral, this volume, chap. 7). Electronmicrographs indicate that, aside from pigment migration within the compound eye, the cross section of the rhabdome increases due to a morphological "restructuring" of the microvilli membrane during the night hours (Horridge and Duelli, 1979, for *Ciulfina* sp.; Horridge et al., 1981, for *Orthodera* sp.).

Summary

The praying mantis is an interesting experimental animal for ethological and neurobiological research due to the ease with which a variety of be-

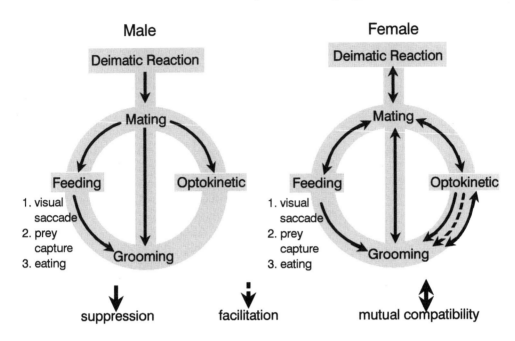

Fig. 11.17 Models of the (partial) behavioral hierarchies of male and female *T. a. senensis*.

haviors can be elicited—from simple reflexes to very complex behaviors—under controlled laboratory conditions. In spite of the mantid's rich potential, however, we still know relatively little about its behaviors. In this chapter, I have for the first time presented a summary of a large body of work done by myself and several of my students. Over the course of our research on *T. a. sinensis* we found that certain behaviors involved in prey acquisition, grooming, defense, and reproduction are structured hierarchically and are subject to a putative priority sequence. Further, these behavioral hierarchies show sex-specific differences. In males all behavior patterns involving head movements (saccades, OKN, eye grooming) or movements of the prothoracic legs (prey capture, eating, grooming) are suppressed during copulation. In contrast, females continue to respond to these stimuli while copulating. It is particularly noteworthy that these behavioral hierarchies are not rigid, but subject to modulatory influences, the most important of which is nutritional state.

DEFENSIVE BEHAVIOR

12. Ontogeny of Defensive Behaviors

Eckehard Liske, Kristian Köchy, Heinz-Gerd Wolff

Insects have been a focus of research on the ontogeny of a number of behavioral patterns and the concomitant changes in the neuronal organization underlying the behaviors (Bentley and Hoy, 1970; Kutsch, 1985; Altman and Tyrer, 1980; Altman, 1975; Kutsch and Stevenson, 1984; Stevenson and Kutsch, 1987; Sakai et al., 1990). In this chapter we are going to present data on an especially ostentatious and highly complex ontogenetic change in defensive behavior from camouflage and escape tactics to overt aggressiveness. The data were, of course, collected on praying mantids, an insect well known for all of these behaviors.

In the scientific literature mantids are mainly known as highly specialized predators (Mittelstaedt, 1957; Charnov, 1976; Copeland and Carlson, 1977; Ridpath, 1977; Corette, 1980; Prete and Hamilton, this volume, chap. 10). But very seldom do the authors mention that mantids are also prey for a variety of other predators, from reptiles to primates (Roonwal, 1937; Beier and Heikertinger, 1952; Gurney, 1950; Fenton et al., 1981). In the course of evolution, therefore, not only did mantids develop specialized behaviors for capturing prey, they also developed characteristic and complex defense mechanisms, which range from simple reflexes to intricate behavioral patterns (Goureau, 1841; Fabré, 1879; Sharp, 1899; Hingston, 1933; Varley, 1939; Cott, 1940; Crane, 1952; Clark, 1962; Robinson, 1969a, b; MacKinnon, 1969; Maldonado, 1970; Maldonado et al., 1970; Edmunds, 1972; Balderrama and Maldonado, 1971; Loxton, 1979; Edmunds and Brunner, this volume, chap. 13).

Although research on the defense behavior of the praying mantids began in the middle of the nineteenth century, it was not until a hundred years later that anyone realized that there was a specific behavioral change that occurred during postembryonic development (Balderrama and Maldonado, 1973), and until now there has been no qualitative or quantitative data about this change. So, our original objective in doing this research was to obtain a detailed description of the spectrum of defense strategies used by one mantis, *Sphodromantis viridis*, over the course of its postembryonic development. At the same time we gathered data on both times of occurrence and types of development-dependent shifts in preferences for certain behavioral strategies. In addition to the behavioral analyses, we made the first step toward finding the causes of the behavioral changes that we saw by describing the correlated changes in the mantid's physical size, coloration, external anatomy, and internal anatomy of certain body areas.

Methodology
The Mantids

Our investigations were conducted on a total of two hundred adult and larval *S. viridis*. The usual breeding conditions are described in Liske and Davis (1987). For these studies we separated hatchlings into two groups. The first (Group 1), kept in large cages (50 x 50 x 50 cm), comprised

15 to 30 mantids that had to compete for food. The second (Group 2), kept in small, individual boxes (5 x 10 cm), were fed constantly so that they did not have to compete for food. After the fifth stadium they were moved to larger individual boxes.

The lengths of the stadia differed for mantids within Group 1 because food intake affects both time of molting as well as other metabolic processes (Andrewartha and Birch, 1954; Eisenberg et al., 1981; Matsura and Nakamura, 1981; Matsura, 1981). During the early nymphal stages, this group was fed *Drosophila melanogaster*, *Musca domestica*, and *Calliphora* sp., depending upon the size of the mantids. For Group 2 the constant availability of food led to almost synchronous molting of all the mantids. This separation of the mantids into two populations yielded mantids in Group 1 that were of the same age but of different stadia (i.e., different developmental stages). On the other hand, mantids in Group 2 were always both the same age and the same development stage.

Periodically over the course of our experiments, we made several diagnostic measures of thirty mantids from Group 2. These included: measurement of the head-abdomen length; measurement of the lengths of the abdominal segments; body weight; abdominal weight; assessment of wing development; examination of the ocelli development; examination of the development of certain abdominal segments; and a description of the body coloration.

The Behavioral Experiments

Our behavioral experiments were conducted in a special study room that was separated from the observer with a glass window. The mantids were observed with a Panasonic video camera (WVP-F10 E) and recorded on tape by means of a Panasonic digital video recorder (NV-D48 EG). The behavioral sequences were measured in individual still frames (picture raster 25 ms) and were photographed with an Olympus OM-2N camera.

At the beginning of the behavioral tests artificial stimuli that could be easily standardized were tested to see if they would elicit any defensive behaviors. Such stimuli, we thought, would allow for optimum quantification of stimulus parameters and the possibility of reproducing the eliciting situation. Based on this reasoning we tried a number of moving bird dummies (stuffed birds of different sizes and shapes; cardboard dummies of different sizes, shapes, and colors; abstractly shaped objects of different sizes) all moving in different ways and at different speeds. These artificial stimuli triggered few behavioral responses. The presentation of live birds, however, was very effective: In more than 90% of the tests performed, mantids responded as expected.

Our general procedure was that a grain-eating bird (*Serinus canaria*) was caged approximately 20 cm away from a mantis that stood on a well-lit platform. This entire set-up was separated from the experimenter by a partition. The duration of each test was 3 minutes beginning when the mantis made an initial head turn toward the bird. If the mantid did not react, the test was terminated after 5 minutes. Twenty mantids at each stage of development were available for testing. In order to analyze the behaviors in detail, some nymphs were placed on a transparent platform and filmed from below during the tests. Escape velocity was measured by placing some nymphs on a surface marked with polar coordinates and then eliciting an escape run. During these tests the camera was located above the platform.

Morphological Structures Related to Specific Defensive Behaviors
Abdominal Segments

As a rule the first two abdominal segments, A1 and A2, were analyzed with regard to any development-related morphological changes. These segments were selected because of they are involved in target behavioral patterns, as we will explain below.

Cuticular Structures and Sensilla

The analysis of the postembryonic development of cuticular abdominal structures was limited to the longitudinal measurement of the first two tergites of abdominal segments A1 and A2.

The detailed analysis of the abdominal segments with regard to their cuticular sensilla was done with a Philipps scanning electron microscope (SEM 515). The air- or freeze-dried mantids of different development stages were fixed to a specimen plate generally with silver (W. Planet GmbH Marburg) and then sputter coated with gold (Balzers Union SCD 040). Photographs were made with a Contax 167 MT camera.

Neuromuscular System

Because there are already comprehensive data available on the neuromuscular system of adult *Tenodera aridifolia sinensis* (Gebauer et al., 1987; Liske et al., 1988), the abdominal neuromuscular system of our *Sphodromantis* could be analyzed and compared with these data. For this purpose, the innervation patterns of the abdominal musculature of segments A1 and A2 were examined in nymphs of different stages by means of a methylene blue staining (Plotnikova and Nevmyvaka, 1980). The methylene blue solution (0.5–5%) in mantis ringer (Berger, 1985) was injected into anesthetized mantids between the sternites of the last abdominal segments (reaction time at room temperature = 10–60 minutes). The mantids were then placed in a 12% solution of ammonium heptamolybdenite overnight at 4°C and then washed in 4% NaCl. The vital staining made it possible to follow both the general course of the peripheral nerves and their precise terminal branching patterns.

Results
Diagnosing Age-Related Changes
Body Size

Approximately 4 to 5 weeks after a cocoon has been laid, 60 to 120 larvae hatch under laboratory conditions. During hatching the mantids cast off their first larval skin; hence, they are in stadium 2 (L2) at the time of hatching.

Until the terminal molt, *S. viridis* undergo eight nymphal moltings provided they have been kept in optimum conditions. Hence, there are nine stadia (L2–L10). Intermolting intervals last 6 to 28 days, depending to a great extent on the availability of food. As noted, our Group 2 mantids had a virtually synchronous molting schedule. In fact, only in the latter stadia was there much variability. The entire developmental period was approximately 130 days.

Under optimum, consistent laboratory conditions, individual nymphs could be easily assigned to particular developmental stages based on their lengths (Fig. 12.1). The head-abdomen length measured along the longitudinal axis at the midline increases between L2 and L10 from 0.72 cm ($\pm$ 0.1 s.d.) to 4.93 cm ($\pm$ 0.25 s.d.). Between L2 and adulthood, it increases to 5.31 cm ($\pm$ 0.15 s.d.) for males and 6.17 cm ($\pm$ 0.24 s.d.) for females (L2, $n = 49$; L10, $n = 14$; adult males, $n = 8$; adult females, $n = 34$).

In addition to the parameters that we have noted, the duration of larval development depends on intensity of light. Therefore, decreased lighting and reduced food supply (only a small amount of *Drosophila melanogaster*) change both color polymorphisms and total developmental time. Under these conditions, after stage L6 some males ($n = 5$) already underwent the final adult molt, although their mean size (4.7 ± 0.13 cm) was considerably less than that of normal males (5.3 ± 0.15 cm, $n = 8$). Even more obvious was the difference in the lengths of two "dwarf females" that measured only 4 cm after a final molt in stage L7 (normal length = 6.17 ± 0.24 cm, $n = 34$). We were able to demonstrate that this deviation in developmental time and body length was caused by insufficient food rather than some genetic defect by feeding mantids from the same ootheca *ad libitum*: The latter developed normally and molted at stage L10 into normal adults.

Ocelli

The ocelli of the larvae undergo a clear development that can be detected only by conventional or electron microscopy. L2 nymphs have one slightly arched cuticular ridge between the antennal bases, which has an elliptical swelling that must be assumed to be the "prestage" frontal ocellus (fOc, Fig. 12.2). On each end of the ridge, caudal to the antennae and medial to two small hollows, there are structures that appear amber in microphotographs and that could be the lenses of the lateral ocelli. During nymphal stages L3–L5 (Fig. 12.3A) the lenses of the ocelli are clearly lighter and thicker than the surrounding cuticle.

During L7, L8 (Fig. 12.3B), and L9 adult stages, the bulges of the cuticle behind the ocellar

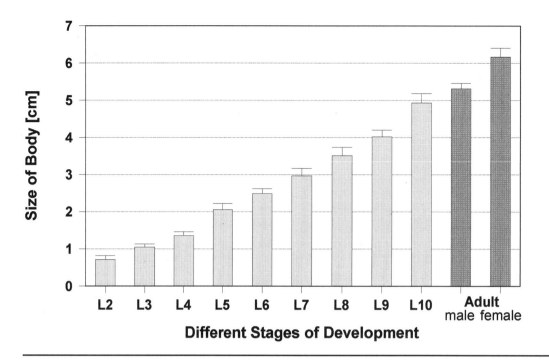

Fig. 12.1 Average lengths (± s.d.) of *Sphodromantis viridis* at different developmental stages measured along the longitudinal axis from the head to the tip of the abdomen (L2–L10, n = 14–49; males, n = 8; females, n = 34).

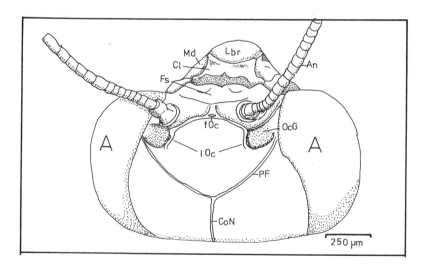

Fig. 12.2 Dorsal view of the ocellar region of the head of an L2 *Sphodromantis viridis* nymph. Abbreviations: A - compound eye, An - antenna, Cl - clypeus, CoN - coronal seam, fOc - frontal ocellus, Fs - frontal shield, Lbr - labrum, lOc - lateral ocellus, Md - mandible, OcG - ocellar hollow, PF - parietal ridge.

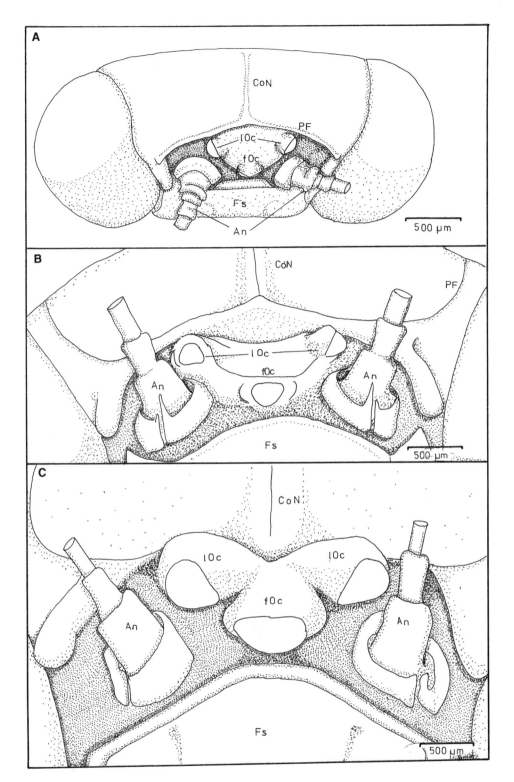

Fig. 12.3 Dorsal view of the ocellar regions of (A) an L5 nymph, (B) an L8 nymph, and (C) an adult *Sphodromantis viridis*. The abbreviations are the same as in Figure 12.2.

lenses become progressively more prominent, so much so that male adults have three very prominent ocelli that are detectable with the unaided eye (Fig. 12.3C).

Wings

Another morphological change, which is also detectable with the unaided eye, is the development of the wings. In stage L2 (Fig. 12.4A) the tergites of the mesothorax and metathorax are still very similar to the abdominal tergites in their cuticular shape. In stage L3 (Fig. 12.4B), however, the lateral and caudal tergal edges of the two segments that will bear the wings are clearly separated from the remaining tergum area. Not only is there a thin cuticular ring separating the two regions, there is also a color difference in the lateral regions. In L4 (Fig. 12.4C) and L5, the lateral area grows into pocketlike shapes, especially at the caudal end, which can reach to the posterior end of the first abdominal tergite. In addition, both thoracic segments have a median longitudinal bulge. In stage L6 (Fig. 12.4D), wing buds are found on the pterothorax for the first time. In the anterior area of these lateral protuberances the preliminary stages of the wing joint can be seen. In stage L7, the development of the wing joint continues, and the subcosta is easily detectable. In L8 and L9 (Fig. 12.4E), first the rostral (L8) and then the caudal presumptive wing (L9) become lined with wing veins. At this stage the wing joint is clearly separated from the scutellum. The development of the wings is concluded in L10. Here, the wing joint is almost completely developed, and the individual joint segments are clearly distinguishable.

Dependence of Defense Strategy on Larval Stages

The Second Stadium

With the hardening of the cuticle approximately 4 to 5 hours after hatching, mantid bodies become brownish and their compound eyes are gray or black. When at rest, the nymphs hold their abdomen so that it is tilted slightly dorsally, but when they run they stretch their abdomen out horizontally.

At this stage, any changes in the movements of the bird triggered target-directed head and antennal movements of the nymphs. The nymph's defensive strategy is to lower the prothorax and simultaneously stretch the prothoracic legs forward or pull them underneath the prothorax (Fig. 12.5). The abdomen is stretched backward slowly, appearing to the observer as if it is unrolling, and the nymph presses itself down onto the ground. In forty-seven tests conducted, 76.6% of the L2 nymphs displayed this type of "cryptic" camouflage behavior.

After such a cryptic reaction, escape behavior can be elicited in some cases (40% in our tests). The escape is characterized by a flying leap usually followed by an escape run (maximum velocity, V_{max} = 23.3 cm/s; mean velocity, V_{mean} = 13.6 cm/s). The leap and the run are in random directions. Tests on nymphs during the first 20 minutes after hatching showed that a threat will trigger a cryptic response about as often as it will trigger an escape run (68.4% [n = 40 nymphs] and 73.7% of the trials, respectively).

The Third Stadium

In contrast to the previous stage, L3 nymphs have a slightly green coloration in 75.5% of the cases (n = 120). In the remaining 24.5% of the cases the animals are brownish, with the coloration varying from a light beige to a dark brown. In some of the literature, the cause of color polymorphism in *Sphodromantis* is not considered to be due to environmental conditions (Przibram, 1907). However, we were able to change the color distribution within a cohort by varying the environmental parameters (e.g., food and lighting within a cage). Considering that the mantids came from one ootheca, we doubt that there is some genetic precondition for color polymorphism.

When threatened by the bird in the test arena, L3 nymphs performed a cryptic ducking movement similar to the L2 nymphs in 78.3% of twenty-three trials. In addition, 26% of the nymphs performed a leap or escape run with a maximum velocity of 26 cm/s (n = 39 nymphs) and an average running velocity of 12.41 (± 6) cm/s.

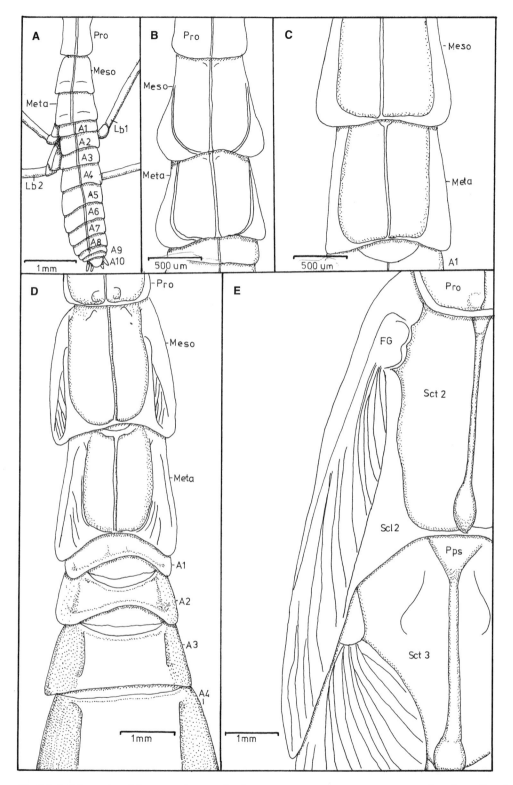

Fig. 12.4 Schematics of the postembryonic development of the wings of *Sphodromantis viridis* (dorsal view). *A*, Stage L2; *B*, Stage L3; *C*, Stage L4; *D*, Stage L6; *E*, Stage L9. Abbreviations: A1–A10 - abdominal segments 1–10; FG - wing joint; Lb1, Lb - walking legs; Meso - mesothorax; Meta - metathorax; Pps - paraside; Pro - prothorax; Scl2 - mesothoracic scutellum; Sct2, Sct3 - mesothoracic and metathoracic scutum.

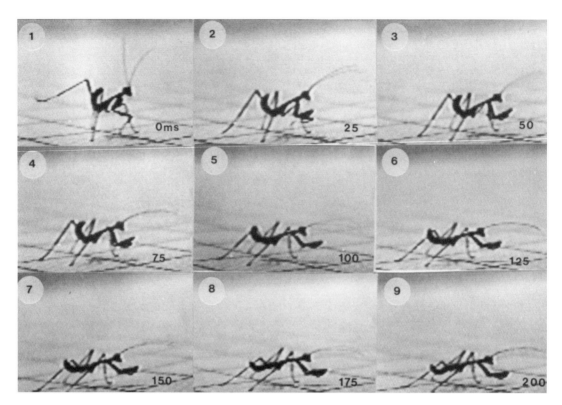

Fig. 12.5 Cryptic defense display of a *Sphodromantis viridis* L. nymph. Individual video frames beginning with the presentation of the startling stimulus (0 ms) to the complete expression of the display (200 ms). The interval between consecutive frames is 25 ms.

The Fourth Stadium

The third molting occurs 9 to 10 days after the second. L4 nymphs continue to show a color dimorphism. However, the pigmentation generally increases, causing a clear color pattern to develop, especially in the abdomen.

The body posture while resting does not obviously differ from that of younger nymphs. When confronted with a threatening situation, 67.2% ($n = 73$ nymphs) of L4 nymphs displayed the cryptic defensive response. In the majority of the cases ($n = 56$ nymphs), however, the ducking movement was not as deep (the nymph did not end up as "flat") as in the previous stages. Specifically, the final angle between the test platform and the longitudinal axis of the prothorax was approximately 15° for L2 nymphs but 35° for L4 nymphs. Also, in L4 nymphs, the prothorax was always higher than the abdomen, a position which younger larvae assume only after the threatening object was removed from their sight.

"Jerky" movements of our threatening stimuli also could at times change the course of the cryptic defensive response to a fast backward jerk of the nymph. In this behavior, the prothorax is flexed vertically, the prothoracic legs are pulled against the body, and the abdomen is held close to the thorax. This short-term body position is very similar to the posture during an offensive threatening reaction as will occur during later developmental stages. In 61.6% of the cases fast movements of the bird triggered not only such curling up, but also escape reactions that occurred in earlier stages.

The Fifth Stadium

In coloration and resting posture the L5 nymphs do not differ from L4, but they are now 2 cm long.

When L5 nymphs were exposed to the bird, 10% of them ($n = 30$) reacted for the first time with an offensive, threatening display. During this offensive reaction, also called a *deimatic response* (Maldonado, 1970), the longitudinal axis of the body and the antennae are directed toward the enemy just as they are during the cryptic behavior. The mandibles, the inner sides of which are red, are opened wide. Following the head movement the prothorax is erected, and the prothoracic legs are either pulled toward the body or stretched far up. During this display the abdomen is also stretched upward and comes quite close to the thorax.

At this stage, however, the defensive camouflage postures still prevail in approximately 87% of the cases. However, most of the time, the prothorax is not oriented horizontally and the body is not pressed toward the ground as severely as in L2 nymphs. Just as frequently (in approximately 85% of trials), the abdomen is not stretched as far backward as previously; it is tilted up and forward. In 53.3% of the cases escape runs also occurred ($V_{max} = 23.3$ cm/s, $n = 20$; $V_{mean} = 12.3$ [± 6.1] cm/s).

The Sixth Stadium

Once again, the coloration of these nymphs did not differ from the previous stage and would not until adulthood. By now, the nymphs were 2.49 (± 0.13) cm long.

While resting, the abdomens of these nymphs appeared as if they were filled full and extremely tight. Abdomens were pointed dorsally and bent at the metathorax-abdomen juncture (Fig. 12.6). Once the nymphs started running, however, the abdomen was stretched far backward during the period of acceleration.

Our analyses of the defensive behaviors at this stage show an increase in the frequency of offensive displays (in 34% of thirty nymphs), which occurred with an escape reaction. The most frequent behavioral pattern in the presence of an enemy, however, was still the cryptic camouflage position (48%). A similar ratio of defensive and offensive reactions emerged at this stage. It was interesting that there was also an increase in indifferent reactions (18%). A sign of such "indecision" was also visible in the change from a cryptic to a threatening posture and vice versa in six cases before a final threatening posture was assumed. Aside from these cases of indifference, a new defensive reaction was observed in our L6 nymphs: feigning death. This behavior was observed in two cases immediately following a very fast escape run. During their run the mantids fell on their backs and remained there motionless with their appendages directed into the air. A slight touch of the mesothoracic and metathoracic tarsi caused the nymphs to pull their legs onto the body. Mechanical stimulation of other body areas did not trigger any reaction at all. If the mantids returned to their normal body position, however, they made a fast escape run. In these cases, the maximum velocity decreased to 15.5 cm/s ($n = 6$), and the mean velocity was 12.65 (± 2.8) cm/s.

The Seventh Stadium

L7 nymphs reached a length of 2.97 (± 0.2) cm, and for the first time the offensive threatening reactions prevailed (89%, $n = 30$ nymphs). The defensive cryptic reactions were observed in only 8% of the nymphs, and escape reactions were observed in 12% of the cases. The running velocities could not be measured any longer since mantids at this stage could not be induced to perform long escape runs.

The Eighth through Tenth Stadium

In the final, preadult, nymphal stadia, the readiness to display an offensive posture remained as high as it was in the previous stadium. In L8, 88% of thirty-two nymphs displayed this behavior; in L9, practically all of the animals displayed this behavior ($n = 25$); and in L10, approximately 95% ($n = 28$) did so. Only in exceptional cases could we elicit defensive reactions. The frequency of escape runs, for instance, decreased to 5.3% in L9, and in L10 defensive reactions were elicited only in approximately 2% of the nymphs.

262 DEFENSIVE BEHAVIOR

Adults

Adult praying mantids show sex-specific differences in their body shape and physical coloration. The genders especially differ in the shape of the abdomen, which is rounder and fuller in the females than in the males. As in the nymphs, there are two adult color morphs, which are similar in coloration to the immature forms.

In behavioral tests the adults of both sexes showed an almost exclusive preparedness to display offensive threatening reactions ($n = 42$). During the reactions, both males and females first turn their head and point their antennae in the direction of the threat. Within a few milliseconds the prothorax is rapidly pulled up into the vertical position, the prothoracic legs are either stretched upward (35% of the nymphs) or pulled toward the body (60% of the nymphs), and the wing covers are lifted slightly. The coordination of wing and prothoracic leg movements can also be observed as the behavior continues. If the enemy

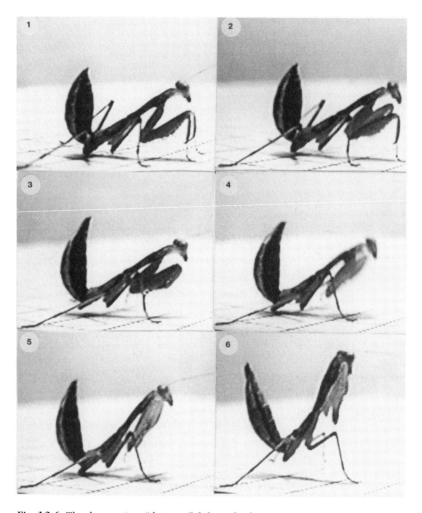

Fig. 12.6 The threatening, "deimatic" defense display of stage L6 *Sphodromantis viridis*. Individual video frames begin when the eliciting stimulus is presented (*1*) and end when the display is fully expressed (*6*). The interval between consecutive frames is 25 ms.

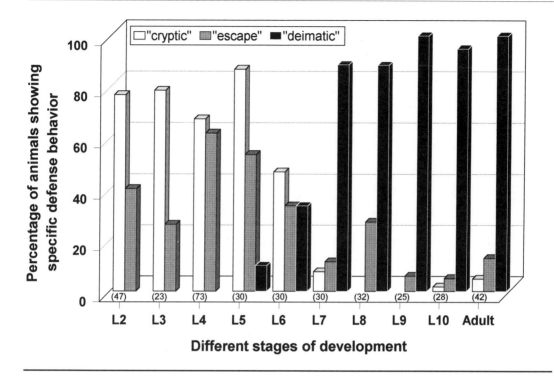

Fig. 12.7 The rates at which each of three defense behaviors are expressed at each development stage of *Sphodromantis viridis*. L2 to L10, nymphal stadia 2–10. Because individuals may express more than one behavior, totals within stadia may exceed 100%.

moves quickly, the change of the prothoracic leg position is always accompanied by lifting the wings. In addition, females hold their wings at right angles to the body.

When in a threatening posture our mantids displayed the red, inner sides of their mandibles, which are clearly visible against the contrasting face. Males and females tilt their abdomens differently, which is most likely due to their size and shape. In just 13% of the trials could we elicit escape reactions, but these were observed only in males. In 60% of these cases the escape started with a flying leap. Cryptic defensive behavior was displayed by only 5% of the mantids.

Summary of Age-Related Defensive Strategies

During the first three postembryonic stages (L2–L4) defensive strategies were elicited exclusively. At the next stage (L5), however, offensive threatening reactions occurred for the first time (Fig. 12.7). Interestingly, in L6 nymphs, defensive and threatening reactions occurred with about equal probability. Then, beginning at L7, the threatening reactions clearly prevailed. Only in rare cases did L7 nymphs display defensive camouflage postures or escape runs.

At the time that the switch from defensive to offensive reactions occurred (L5–L6), the maximum velocity of the escape run decreased dramatically from 23.3 cm/s to 15.5 cm/s. Beginning with stadium L7, it was no longer possible to trigger an escape run.

In other types of test situations mantids may display different individual defense strategies derived, ultimately, from a different set of protective mechanisms. One comes to this conclusion when observing mantids within their breeding cages. Here the insects usually hang on the ceiling of the cage. If someone moves quickly back and forth in front of the cages the mantids very often assume a cryptic ducking posture against the cage lid. Under these conditions this behavior occurs with a much higher frequency, for both older nymphs and adults, than in the experiments described above. Only in the rarest cases do the animals assume a threatening posture in this situation. These variations show how strongly the choice of behavior depends upon environmental and stimulus parameters in addition to developmental stage.

A Comparison of Same-Aged Nymphs at Different Developmental Stages

As we have explained, allowing mantids to compete for food makes it possible to split a population of same-aged insects into groups that are at different developmental stages. Comparing the behaviors of mantids in each group on the same test day makes it possible to obtain data on the influences of the age, on the one hand, and developmental stage, on the other.

When comparing the choice of strategy of mantids in the L5 to L7 transition period from defensive to offensive strategies, we could show that same-aged mantids switch their strategy once they have passed a certain developmental point (Fig. 12.8). Whereas 37-day-old animals in stadium L5 react primarily (67.1% of trials) with a defensive cryptic posture, their same-aged siblings, already in stadium L7, react primarily with deimatic displays (84.3% of the trials).

A comparison of these data with the results of previous analyses demonstrates that in both cases the critical phase for a strategy switch is always the same. Stadium L5 always is primarily defensive, and stage L7 is primarily offensive. Hence, our first group showed a behavioral pattern at 37 days old that mantids of the second group—which were already in stadium L7—had shown 17 days before.

These results clearly demonstrate that the behavioral choice between defensive reactions does not depend on the age of the animals but rather the number of moltings through which it has passed or, in other words, its developmental stage.

Position of the Prothoracic Legs

Up until now, we have described defense behaviors as a whole. Now we wish to examine individual components of the behaviors in more detail. Both in the defensive cryptic behavior and in the offensive threatening behavior, the change in the position of the prothoracic legs is an ostentatious part of the display. The cryptic behavior of young nymphs is usually marked by extending the prothoracic legs in front of the body. However, the tendency to pull the prothoracic legs under the body increases over the course of development so that by stadium L7 all nymphs will pull these legs toward the body and turn them sideways. For the offensive display, too, there are generally two positions that the prothoracic legs can take: They can be pressed tightly to the body or be extended and held forward.

Here, too, one can observe a change in the preference for the position of the prothoracic legs during development. While in L5 all animals displayed with outstretched legs, this position was taken by only 34% of adults.

In both defensive strategies (cryptic and deimatic), therefore, a switch from outstretching the prothoracic legs to pressing them to the body occurs at sometime during postembryonic development.

Position of the Abdomen

During defensive behaviors, the abdomen can either be extended posteriorly or lifted dorsally and tilted in the direction of the prothorax. During a cryptic display, both movements may occur, but in the deimatic display only the latter happens. L1 nymphs extend their abdomen posteriorly in 100% of the cases in which they display a cryptic reaction. In L3 nymphs, however, already a small number display an atypical dorsal tilting of the abdomen during the cryptic reaction; this is not observed in L4 nymphs. Interestingly, L5 nymphs display a defensive deimatic reaction with erected abdomens although their prothorax still takes a defensive cryptic position during this stage (i.e., it is pressed toward the ground). This trend continues in L6. By the time that nymphs reach the L8 to L10 stadia, the majority of the threatening displays are made with the abdomen directed dorsally.

The abdominal movements of adult mantids clearly differ between the sexes. While resting,

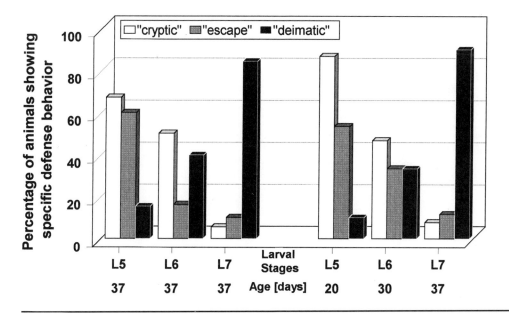

Fig. 12.8 Comparison of the behavioral choices made by same-aged (left columns, age 37 days) and different-aged (right columns, age 20, 30, 37 days) *Sphodromantis viridis* nymphs in stadia L5, L6, or L7.

the abdomen of males is stretched backward. When the animals are startled the tip of the abdomen jerks upward in several steps. Then, when the prothorax is lifted, the abdomen bends at its junction with the thorax. When startled, females always lift their abdomen slightly dorsally and, at the same time, slightly flex their wings.

Morphological Analysis of Specific Abdominal Segments during Postembryonic Development

During the developmental switch between defense behaviors, there is a clear change in the course of abdominal movement. The most obvious and severe change occurs in the movements at the thorax-abdomen junction. Whereas the cryptic behavior involves extending this region, the deimatic response requires the abdomen to be tilted dorso-cranially. One has to assume that the switch in the movement is an inherent part of the morphological changes that occur within the region of the thorax-abdomen junction. Therefore, we investigated the development of the outer cuticular structures at this site, their sense organs, and the developmental restructuring of the corresponding neuromuscular system.

Postembryonic Development of Tergal Cuticular Structures

In L2 nymphs there are no clear differences between the tergal cuticular structures of the abdominal segments (A1 and A2) and the other abdominal segments (A3– A10; Fig. 12.9A). At stadium L4, however, there is a reduction of the first two abdominal segments (Fig. 12.9B; for comparison L8 is pictured in Fig. 12.9C). This reduction is especially evident dorso-medially where the longitudinal axis of the segment is shortened. The relative tergite length (T_{rel}) of different developmental stages is listed in Figure 12.10. Here, the longitudinal tergite axes were measured at the median for each stadium, and the tergite lengths of the first abdominal segments (T1, T2) were expressed as a percentage of the entire length of the abdomen. We actually found that, in comparison to the other tergites, only the lengths of the tergites of abdominal seg-

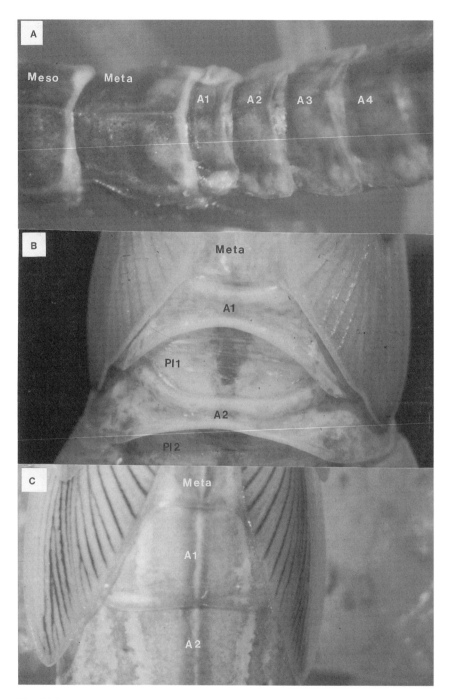

Fig. 12.9 Photographs showing the differences in the abdominal tergites of *Sphodromantis viridis* at three developmental stages. *A*, L. nymph, A1–A4: abdominal tergites 1–4 (Meso - mesothorax, Meta - metathorax). *B*, L4 nymph, PL1, PL2: intersegmental pleural region 1 and 2. *C*, L8 nymph.

ments A1 and A2 become reduced as the mantid develops. For L2 nymphs, T_{rel} for segment A1 and A2 was 8% and 8.8%, respectively. These values decrease in subsequent stadia and reach the lowest values in L7 and L8 (2–3% of the entire abdominal length). This reduction is limited to the median regions of the first two segments; that is, the lateral regions are not affected.

Due to the constrictions in the median area, the dorsal pleural regions between the segments increase. Large oval areas develop, which are covered with a thin, transparent pleura, which may be divided into two sections. Whereas the cranial section has transverse folds, the caudal section has longitudinal folds.

Changes in the Abdominal Weight: A Shift of the Center of Gravity

As the first two abdominal tergites become smaller, the abdomen as a whole becomes larger in relation to the rest of the body. In the course of development the abdominal shape changes from resembling a hose (L2) to resembling a tightly filled sack (L9). Measuring anesthetized mantids of different stadia showed that in spite of this striking change in abdominal shape there were only small variances in abdominal weight as a percentage of total body weight (41–54% of the total body weight). If, however, the physical center of gravity of nymphs of different ages is determined by laying anesthetized larvae across a bal-

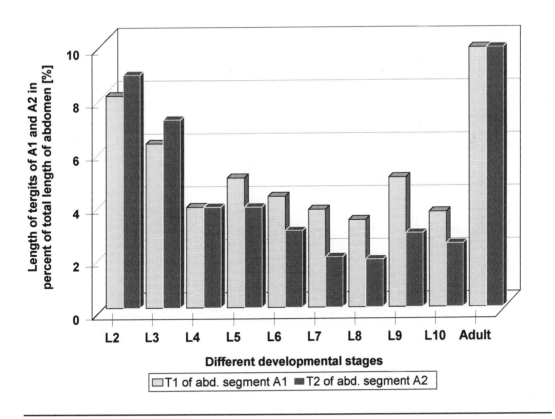

Fig. 12.10 The tergite lengths of the first (A1) and second (A2) abdominal segments expressed as a percentage of total abdominal length for L2 to L10 *Sphodromantis viridis* nymphs.

ance beam, one can see that the center of gravity shifts because of changes in the abdomen (Fig. 12.11). Whereas the center of gravity of a 0.7 cm L2 nymph is at 0.3 cm from the front end (i.e., at the point of transition from the mesothorax to the metathorax), the center of gravity of 4 cm L9 nymphs is shifted to 2.5 cm from the front end (i.e., the point of the first abdominal tergite).

When the abdomen is extended, the center of gravity shifts to between the two metathoracic legs (Fig. 12.11, point S2). When the abdomen is tilted forward, as when resting, the center of gravity moves back to its original position at the point of the mesothorax (Fig. 12.11, S1).

Sensillary Equipment of the First Two Abdominal Segments during Postembryonic Development

During stadium L2 the surfaces of the first two abdominal tergites are free of sensory hairs. Only at the anterior edge of the tergites are there microtrichae approximately 4 μm long arranged in four to five parallel rows next to 8 μm long sensilla with basal domes (i.e., trichoid sensilla).

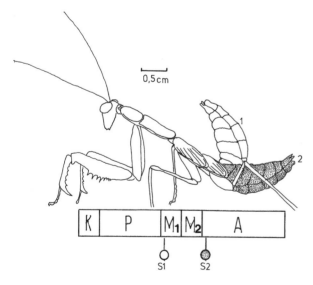

Fig. 12.11 Schematic demonstrating the compensatory shift of the nymph's center of gravity depending upon the position of its abdomen. Abbreviations: A - abdomen, K - head, M1 - mesothorax, M2 - metathorax, P - prothorax, S1 - center of gravity in abdomen position 1, S2 - physical center of gravity in abdomen position 2.

In stadium L3, almost the entire anterior tergal surface is covered with sensilla. The density decreases posteriorly, until only the region of the median fold is covered. The main part of the sensilla consists of microtrichae 4 to 6 μm long and individual trichoid sensillum up to 20 μm long. In the anterio-lateral area the number of sensilla increases to seven to eight per half tergum, mostly grouped on the side. Stadium L4 shows a distribution of the sensory hair similar to stage L3.

In the subsequent stadia the number of hairs increases despite a decrease in the available cuticular areas due to the reduction of the longitudinal tergal axes. Of course, sensilla are still located along the anterior segmental border. The number of the sensory hairs increases considerably, especially under the lateral overlapping regions of the segments, so that by stadium L8 more than fifty trichoid sensilla are located there (compared to just five or six in L3). In addition, this area is covered with approximately 360 microsensilla per square millimeter.

Adult mantids already have easily distinguishable hair fields in the latero-anterior areas of the segments A1 and A2. The hair fields are located in a region that used to be the overlapping zones of the individual segments during the larval stages. Each of the approximately ninety sensilla within each hair field is equipped with a basal socket. Furthermore, an extension of the metathoracic tergum reaches into each anterior-lateral hair field of A1. This structure, which is only found in adult animals, has no equivalent in the larval stages.

The Neuromuscular System of the First Two Abdominal Segments during Postembryonic Development

As we have noted, the abdominal segments A1 and A2 are subject to a dramatic change in the tergal structures during the postembryonic development. One may assume that this restructuring correlates with stadium-dependent changes in the neuromuscular system of these segments.

In an earlier anatomical study of the thoracic and abdominal neuromuscular system of *Tenodera* and *Hierodula*, it was shown that the first two abdominal segments of adult mantids are supplied via nerve branches of the metathoracic ganglion (Gebauer et al., 1987; Liske et al., 1988, 1989). Specifically, the posterio-lateral nerve branch III-8 (= Meta N8a in Liske et al., 1989) and the unpaired median nerve III-7.1 (= Meta N7a in Liske et al., 1989) fuse, and the resulting nerve innervates all of the muscles of the first abdominal segment. The muscles of the second abdominal segment are supplied by homologous nerves III-9 (= Meta N8b in Liske et al., 1989) and III-7.2 (= Meta N7b in Liske et al., 1989).

As a rule, each abdominal segment is equipped with eleven muscles. Of these, three muscles are designated as dorsal (tergal), two as ventral (sternal), and the other six as pleural muscles. The first abdominal segment, A1, deviates from the "normal structure" insofar as the sternum is mostly reduced in this segment (Fig. 12.10). Because of this change, the muscles also deviate from the general basic structure. Instead of the six pleural muscles present in segment A2 (i.e., M15, M16, M17, M18, M19, M20), segment A1 is equipped with only three such muscles (M4, M5, M6). The function of the frontal pleurotergite muscle (muscle M17 in A2) is performed by the metathoracic muscle M91 in the first segment. Due to the reduction of the sternite, the ventral muscles M9 and M10 of A1 (covered by M9 in Fig. 12.12) insert at the metathoracic connection points or cuticula strips, respectively, of segment A2. Least affected by the reduction are tergal muscles M1, M2, and M3.

In the course of postembryonic development practically nothing changes in the set of muscles described above, in spite of the reduction of the tergal cuticular structures in A1 and A2. Examinations of the segment A1 and A2 muscles and their innervations in nymphal stages L4 and L8 demonstrate that the number and position of the muscles are the same as they are in adults (Fig. 12.13). Only the insertion points of the muscles shift slightly. The innervation of the muscles during development is not substantively different from that of the adults either. The similarity of the nymphal innervation pattern to that of adults extends from the concurrent occurrence of intersegmental fusions (III-7.1 and III-8 in segment A1; III-7.2 and III-9 in segment A2, etc.) to an almost identical division of terminal nerve branches at the muscle. We have not as yet examined whether there are correlated developmental changes at the level of motor or sensory neurons. Unfortunately, detailed analyses of the sensory system of the tergal abdominal region are available only for adult mantids (Gebauer, 1988).

Discussion
Developmental Changes in the Defense Behavior of Sphodromantis viridis

The present investigations on the defensive behavior of *Sphodromantis viridis* demonstrate that this mantis undergoes a change in its defensive strategy during postembryonic development. Whereas young nymphs (L2–L4) use defensive strategies such as cryptic postures and escape runs only when confronted with an enemy (in this case, a bird), nymphs at stage L5 begin to display the threatening deimatic response in the same manner as adults.

The transition from the one strategy to the other does not occur suddenly, but via several intermediate steps:

1. L2 and L3 nymphs almost exclusively assume a cryptic posture.

2. Approximately two-thirds of L4 nymphs assume a cryptic posture or make an escape run.

3. In L5 threatening displays occur occasionally for the first time. Also at this stage, in the majority of cases the abdomen is erected during the cryptic posture, which decreases its effectiveness.

4. L6 constitutes the actual transition stadium from one defense strategy to another. The tendencies to display threatening or cryptic reactions are apparently balanced. The equivocal state of the nymph is also evident in the fact that there is an increased tendency to switch between strategies and to display a new strategy—pretending to be dead.

5. In L7 threatening displays clearly prevail.

6. There are differences in the individual behavioral components of the threatening displays of older nymphs and adults which, on the one

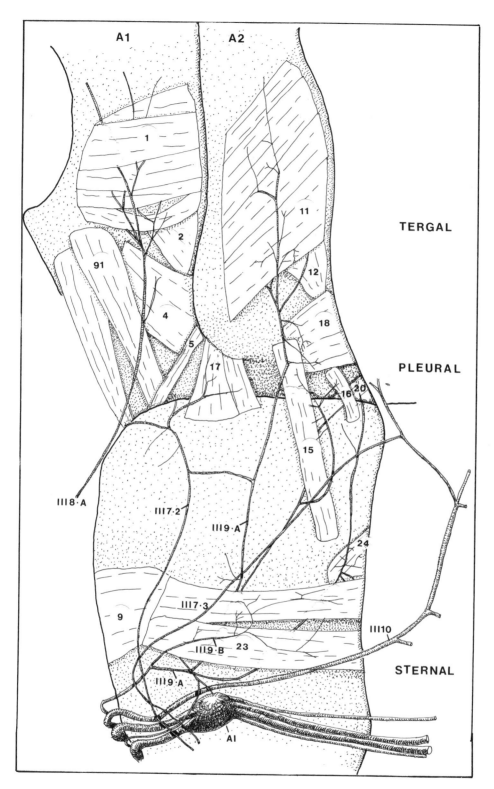

Fig. 12.12 The neuromuscular system of the first two abdominal segments, A1 and A2, of a *Sphodromantis viridis* L4 nymph. The nomenclature used for the musculature is according to Leverault (1939).

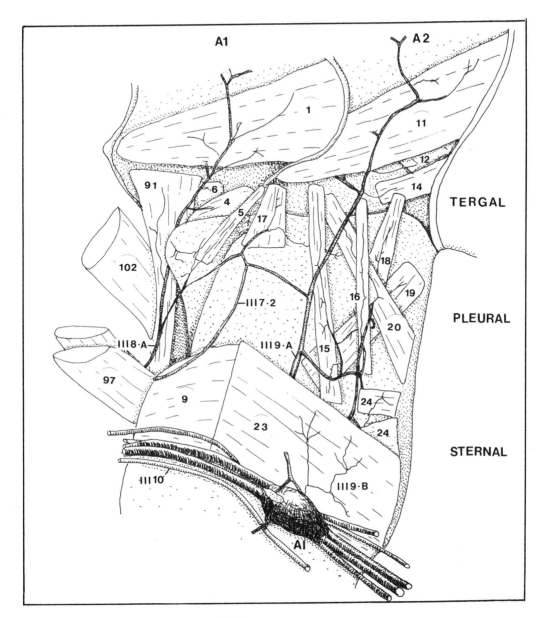

Fig. 12.13 The neuromuscular system of the first two abdominal segments, A1 and A2, of an adult *Sphodromantis viridis*. The nomenclature used for the musculature is according to Leverault (1939).

hand, are due to morphological changes (such as the use of the wings) during the display and, on the other hand, depend on the sex of the adult (the degree to which the abdomen can be bent during the display).

Developmental Changes in the Defense Behavior of Other Mantids

Of the numerous publications on the defense behavior of mantids (Sharp, 1899; Roonwal, 1937; Varley, 1939; Crane, 1952; Robinson, 1969a, b; Mackinnon, 1970; Maldonado et al., 1970; Balderrama and Maldonado, 1971, 1973; Edmunds, 1972, 1976) only that by Balderrama and Maldonado (1973) describes a complete developmental cycle. The species that they investigated, *Stagmatoptera biocellata*, also shows a strategy change that strongly depends on the distance to the threatening object. Whereas L2 to L4 *S. biocellata* nymphs perform cryptic reactions if the enemy is within 5 cm, at L5 the behavior changes to a threatening display. On the other hand, if the distance to the enemy is increased up to 45 cm—or even up to L8 (the preadult stadium)—a large percentage of cryptic reactions are observed. We did not detect any distance dependency for strategy choice in *S. viridis*.

Developmental Changes in Abdominal Position during Defense Behavior

The different abdominal positions during postembryonic development appear to be the most prominent changes in threatening displays that we observed in *S. viridis*. Comparisons of the abdominal position during cryptic and threatening displays make it clear that the change is closely related to the change in the resting position of the abdomen. In the course of postembryonic development, in general, the abdomen is increasingly erected within the context of defense, as it is for the closely related species *Sphodromantis lineola* (Kumar, 1973).

Tergites of the Abdominal Segments A1 and A2

In conjunction with the changes in abdominal position during defensive displays, *S. viridis* experiences an obvious restructuring of the first two abdominal tergites. Due to a progressive reduction of the longitudinal tergite axes of segments A1 and A2 and a concomitant dramatic enlargement of the intersegmental pleural regions, a kind of folding mechanism develops, making the new position of the abdomen possible. Such changes are not ubiquitous among mantids; *Tenodera aridifolia sinensis*, for instance, shows neither a restructuring nor the corresponding changes in abdominal position.

Based on the drawings of Kumar (1973), it appears that there is a tergite reduction in *S. lineola*, although this has never been explicitly described. The new position of the abdomen could, as Kumar assumed, allow for better mobility by increasing the mantids "compactness." This conjecture may be supported by our findings that there is a developmental shift in the nymph's center of gravity, which may be compensated for by tilting the abdomen forward (Fig. 12.11). According to our findings, if the abdomen were not so tilted, the nymphs would end up in an unfavorable position due to the shift in weight during the postembryonic development. By tilting the abdomen forward they reposition the physical center forward and therefore achieve a better ability to escape by running. In order to maintain this escape ability during the cryptic display, it would be sensible to maintain the abdomen in its forward position (although this position does diminish its efficacy). In turn, decreased efficacy of the cryptic display might make the development of a threatening display quite advantageous. So, it makes sense that, when the defensive behavior is accompanied by an erected abdomen in stadium L5, the threatening reactions begin to appear. Indirect support for this line of reasoning is the observation that *T. a. sinensis*, which does not tilt its abdomen at resting, still shows many cryptic reactions even in the later larval stages.

The forward-tilted abdomen may also create a predisposition toward making threatening displays. The idea that there is an influence of morphological parameters on the time at which threatening displays develop is supported by the fact that the behavioral preference for them does not depend on the age of the mantids but on their developmental stage and, therefore, on the degree of morphological restructuring that has occurred.

Neuromuscular System of Abdominal Segments A1 and A2

The abdominal morphological restructuring might not be limited to the cuticular structures. It might also extend to the muscles and nervous system. Such restructuring during postembryonic development is known to occur in holometabolous insects (Truman et al., 1985; Levine and Truman, 1982; Levine, 1984) and hemimetabolous insects (Bernays, 1972). Aside from muscular atrophy shortly after hatching from the egg and the disappearance of larval muscles after the imaginal molt, the possibility exists that previously rudimentary muscles begin to function.

Unfortunately, such a restructuring of the abdominal muscles of *S. viridis* could not be detected for behavior-relevant areas. Also, the general morphological analysis of the nervous system showed no differences between the innervation pattern of nymphs and adults. Both the muscles and innervation of abdominal segments A1 and A2 concur with those of *T. a. sinensis* (Gebauer et al., 1987; Liske et al., 1989), *Hierodula membranacea* (Kerry and Mill, 1987; Gebauer, 1988), *Stagmomantis carolina* (Levereault, 1939), and *Mantis religiosa* (LaGreca and Rainone, 1949).

Possible Changes of Muscle Characteristics

The morphological consistency of the abdominal neuromuscular system during development of *S. viridis* does not allow us to make any statements about possible changes in the contractile characteristics of the muscles or their responsiveness. Investigations providing information about the changes in these characteristics so far exists only for the orthopteroids *Achaeta domestica* (Chudakova and Bocharova-Messner, 1965), *Schistocerca gregaria* (Tyrer, 1969; Kutsch, 1971), *Gryllus campestris* (Bentley and Hoy, 1970), *Chortiocetes termifera* (Altman, 1975), and *Locusta migratoria* (Kutsch and Stevenson, 1984). And most of these publications refer to the development of the flight system.

Today it is known that there are complex modulatory effects on neuromuscular activity in invertebrates working at a number of levels. Substances that circulate in the body fluids are released by modulatory neurons or are already located in the motor neurons of the system to be modulated and are responsible for the effects. Due to the influence of these modulators the amplitude, velocity, and duration of the normal muscle contraction can be changed. A well-characterized example of such modulation is the role activity of the posterior extensor tibia muscle of *Schistocerca* (Burrows, 1996; Hoyle and Barker, 1975; Evans and O'Shea, 1978).

Neuronal Circuits

As is the case for muscle characteristics, our findings do not permit statements about possible postembryonic changes in neural circuitry. To this end, staining the participating neurons during the defensive behavior would be necessary. In the orthoptera, the ganglia are complete when the insect hatches (Panov, 1966; Gymer and Edwards, 1967; Sbrenna, 1971; Bentley, 1973). However, because the sensory neurons are already developed during the postembryonic phase (Wigglesworth, 1953; Edwards and Palka, 1971; Sviderský, 1969; Bernays, 1972), they may play a role in changing behavioral patterns (Wilson, 1968; Bentley and Hoy, 1970; Kutsch, 1971). On the other hand, development-dependent changes could occur at the level of sensory–motor neuron synapses (Burrows, 1973; Levine, 1984).

A further possibility would be that existing neuronal circuits, for instance, are switched on by hormones during postembryonic development (Bentley and Hoy, 1970; Altman, 1975; Haskell and Moorhouse, 1963; Kutsch and Stevenson, 1984). Such a modulating effect tightly linked to the patterns of aggressive behavior is known to occur in the crustaceans via the neuro-hormones octopamine and serotonin. Application of octopamine in *Homarus* triggers a stretching of the limbs and the abdomen, and serotonin triggers a tilting (Harris-Warrick and Kravitz, 1984; Kravitz, 1986). Since both neuromodulators play an important roll in insects and are ubiquitous in those insects most closely related to the mantids (Agricola et al., 1988), they might modulate the defense behaviors of mantids.

Mechanoreceptors

The first clues regarding a change in the abdominal sensory equipment of *S. viridis* were found when we examined the sensilla of the frontal abdominal tergites at different developmental stages. During postembryonic development of the tergites not only do the number of cuticular hairs increase, the sensory fields shift. For instance, in nymphs the anterior and posterior segmental borders are equipped with sensilla, but in adults there are, apparently, no sensilla in the anterior region of segment A2. This change may be involved in the shift in the position of the abdomen. If, in the nymphal stadia, the sensilla register the position of the abdomen when it is tilted forward, they would not be necessary in the adult due to the change in abdominal position.

The location and the type of the larval sensilla in *S. viridis* suggest that their main function is registering the movement of the abdomen based on the shifting of the overlapping posterior-lateral cuticular bulges. On the other hand, the long filiform hairs of the adults appear to be suitable for registering the position of the newly formed wings, since these sensilla are located mostly in the lateral area of the first abdominal segment, directly below the wings.

Only very few data are available on the postembryonic development of the hair sensilla of *S. viridis*. However, what we do know suggests that these associated nerves remain unchanged over development. Unfortunately, there is no information available on the number of sensory neurons, their central connections, or whether there are any changes in their arborizations during development.

Which Stimuli Trigger Defensive Behavior?

Crane (1952) investigated the influences of visual input on the defensive behavior of mantids and noticed that neither the compound eyes nor the ocelli are necessary for triggering the behavior. However, our first experiments with *S. viridis* whose compound eyes were covered with black lacquer showed that no threatening displays could be triggered when both eyes were covered and that monocular mantids displayed less readily. We also found that the ocelli have no effect on defense behaviors in *S. viridis*.

Aside from visual input, airborne vibration caused, for instance, by the movements of a potential predator could influence abdominal muscle activity via the cercal system. Projections of cercal interneurons project as far as the subesophageal ganglion (Boyan, personal communication). The paths and anatomy of these so-called giant interneurons have been well investigated in orthoptera (Roeder, 1948; Hess, 1958; Callec and Boistel, 1966; Farley and Milburn, 1969; Milburn and Bentley, 1970; Camhi and Tom, 1978; Camhi et al., 1978; Ritzmann and Pollack, 1986), but their role in mantid behavior remains unclear.

Data on the wind-sensitive cercal system in *Archimantis* (Boyan and Ball, 1986) and *T. a. sinensis* (Liske, 1991) suggest differences in the role of this system in fast escape behaviors in mantids versus crickets and cockroaches. Some observations on *S. viridis* in glass cages have shown that, when the direct influence of aerial vibrations as well as other vibration stimuli have been interrupted but the visual system remains intact, the defense pattern takes its course. In other words, the visual system is necessary in order to elicit defense behavior.

A third sensory system that has been discovered only during the last couple of years and that could influence muscle activity during defense behavior is the auditory system (Yager and Hoy, 1986, 1987, 1989). Whereas Crane (1952) assumed that no sense equivalent to hearing is important to defensive behavior, Yager and Hoy (1989) have demonstrated that the median ear, which is sensitive to ultrasound, plays a role in escape systems (Yager, this volume, chap. 6).

Plasticity of the Defense Behavior

The general importance of integrative influences on mantid defense behavior—mediated, perhaps, by both ascending and descending neurons—is still not clear. Investigations in *Stagmatoptera biocellata* (Maldonado, 1970), for instance, showed that after cutting the connection between the prothoracic and subesophageal ganglion only the head reaction (antennae and mandibles motion) of the defense behavior can be triggered by visual stimuli, while tactile stimulation of the abdomen induced all of the other behavioral components (with the exception of the head reaction). When the connection between the prothoracic and the mesothoracic ganglion was cut, a visual stimulus sometimes also triggered the prothoracic leg component of the defensive reaction in addition to the head component. These findings caused Maldonado (1970) to assume that the movement patterns of defense behaviors are an uncoordinated "closed-circuit" system lacking the integration by "higher" centers such as that seen during the antennal grooming reflex of crickets (Huber, 1955).

However, some observations appear to argue against such a stereotyped organization of defense behavior, one based simply on closed neuronal circuits. If, for instance, mantids are fixed so that it is impossible for them to carry out threatening displays with both prothoracic legs, they perform the display with only one leg while supporting themselves with the other. The plasticity of the behavior also is demonstrated by the reaction of feeding mantids, which will hold on to their prey with one prothoracic leg while using the other for the threatening display. These types of compensatory abilities, as well as the variability in the defensive strategies themselves, demonstrate the high degree of flexibility in defensive behaviors. The opinion of Maldonado (1970) that the defensive behaviors of the praying mantids are stereotyped was based on the belief that the behaviors of insects are based on a rigid reflex system. Today we know that the output of the neuronal networks is not rigidly fixed (Marder and Hooper, 1985; Cohen et al., 1988; Marder, 1989). Thanks to the progress in biophysics and biochemistry the effects of the factors that modulate neuronal networks can be followed all the way down to the level of subcellular components (Kaczmarek and Levitan, 1987).

13. Ethology of Defenses against Predators

Malcolm Edmunds, Dani Brunner

Nineteenth-century British naturalists were fascinated by the colors and forms of the insects, birds, and mammals found in the exotic tropics of America, Asia, and Africa. Butterflies, in particular, caught the traveler's attention, because of their magnificent colors and because of the finding that similar color patterns occur in unrelated species. Naturalists were delighted with the many cryptic and mimetic insects, and they wrote detailed chronicles of their experiences, as recounted by Beddard:

> Every naturalist traveller appears to have some instance to relate of how he was taken in by a protectively-colored insect. The stories are told with a curiously exaggerated delight at the deception, and often with a framework of details tending to throw the deception into still greater prominence. Professor Drummond . . . tells us that he went to (Africa) resolved to be proof against the frauds of insects, and suspicious that "the descriptions of Wallace and the others were somewhat highly colored." The insect which succeeded in deceiving Professor Drummond was one of the Mantids. (1892, 110)

Critics, however, saw "adaptive" coloration as the product of the florid imaginations of naturalists. In contrast, others saw resemblances everywhere: for instance, some claimed that giraffes are well protected by imitating tree trunks (Baker, quoted in Beddard, 1892), and pink flamingos are perfectly colored to match the sunset (Thayer, 1909). The voices of both the critics and the fanatics were eventually quieted by a steadily increasing body of experimental evidence that demonstrated the real defensive value of both mimicry and crypsis. Early experiments by Lloyd-Morgan, Guy Marshall, and Poulton (described in Sykes, 1904) showed that predators do not avoid distasteful animals innately, but they quickly learn to do so through experience. Experiments also showed that not every color has a defensive or communicative function and that some coloring plays unexpected roles. In the latter case, for example, the black peritoneum of many reptiles and fish, once thought to be nonadaptive (Beddard, 1892), apparently shields the animal from incident ultraviolet light (Burtt, 1979). So, today only a handful of scientists remain unconvinced and maintain that mimicry is a "figment of the imagination" (Urquhart, 1960, 1987, cited in Brower, 1988).

After almost a hundred and fifty years of research the study of animal defense is still mostly descriptive (Endler, 1986), and little is known about the origins of specific defense mechanisms. It has been suggested, for example, that distastefulness may have originated through the appropriation of chemicals from the diet (Rosenthal and Berenbaum, 1991), through the accumulation of excreta in the skin (which just happened to be brightly colored [Eisig, cited in Beddard, 1892]), or as a secondary product of skin secretions involved in water retention (Endler, 1986). Although the origin of such adaptations is an important question, there are other issues that deserve our attention here. These include the interactions between defensive mechanisms and other functions (e.g., reproduction), the perceptual basis of crypsis, and the population dynamics of mimic complexes.

Primary and Secondary Defenses

Successful predation involves several stages—detection, identification, approach, subjugation, and consumption (Endler, 1986)—and defensive mechanisms have evolved to counteract every one of these stages. The most effective antipredator defenses are those which stop predation early on, at the detection and identification stages. We call these *primary defensive adaptations* (Robinson, 1969a). Secondary defensive adaptations (Robinson, 1969a) are designed to prevent the later stages of an attack.

Defenses are needed at every stage of a mantid's life and must be used against conspecifics as well as other species. Defenses of nymphs are for the most part similar to those of adults but with fewer strategies that rely on large size for their effect (e.g., Liske, this volume, chap. 11). Table 13.1 shows all of the defensive mechanisms known in mantids. Most of the mechanisms listed relate to predators that hunt by sight, primarily because very little is known about mantid defenses against predators that hunt using olfactory or auditory cues. The notable exception is Yager's excellent work on bat avoidance by flying mantids (Yager, this volume, chap. 6). This bias toward vision-based defenses in mantids may reflect our greater reliance on vision than on chemical senses, and such a bias could be a serious shortcoming for the study of mantid defense. For instance, we know that insect prey is detected by echolocation by bats and some omnivorous primates (Erickson, 1991) and by chemical cues by lizards (Huey and Bennet, 1986) and parasitic wasps (Lewis and Tumlinson, 1988). This fault would prove especially damaging if it were true that prey have their best defenses attuned to the sensory mode of their major predator (Endler, 1986). However, as phenotypic plasticity is constrained, specialization against one predator may leave a prey defenseless against another, which may then become the major predator. Thus, for animals like mantids, whose predators belong to different taxa, there must be general defense mechanisms as well as specialized, predator-specific defenses.

Historical Perspectives on Defensive Coloration

For more than a century the study of protective adaptations has contributed to our understanding of natural selection. Together with the signals used in sexual advertisement and interspecific or intraspecific communication, aposematic signals (which indicate unpalatability) have been the focus of a considerable amount of research (e.g., Zahavi, 1987; Guilford, 1990). It has even been suggested that there are two distinct types of evolutionary mechanisms: natural selection and signal selection (Zahavi, 1991). However, many colors may actually act as masks or attenuators because they function to reduce the informational content of the signals that disclose the palatability of the prey. Thus, color signals and color masks may have been subjected to very different selective mechanisms because they affect information transfer in opposite ways.

Some years after Darwin (1859) outlined his ideas on natural selection, Wallace (1867) pointed out the importance of coloration in defense, and Bates (1862) and Müller (1878) distinguished two types of mimicry. Batesian mimicry occurs when a palatable species resembles an unpalatable one. So, it involves two species in conflict: a model that pays the cost of producing the real defense (e.g., spines, sting, or a noxious substance) and a mimic that benefits from the model's investment. Müllerian mimicry, which involves two or more unpalatable species, is unknown in mantids. The complex interactions between mimics, models, and predators have made the study of mimicry central to both classical and neo-Darwinism. In Wallace's words: "Among the numerous applications of the Darwinian theory in the interpretation of the complex phenomena presented by the organic world, none have been more successful, or are more interesting, than those which deal with the colors of animals and plants" (quoted in Beddard, 1892).

Even now, mimicry remains a central topic of debate among Darwinists. For example, Turner

Table 13.1 Defensive strategies in Mantodea

Primary Defenses

Against detection: crypsis
  A. Stable evolutionary equilibrium. Not stabilizing selection
  B. Rarity: dispersion, reduces random encounter
  C. Apparent rarity: differences between prey and predator activity periods, safe resting place, increased variability of characters, and polymorphism
  D. Cryptic behavior: immobility, random movements in a moving environment, concealment of body appendages, active search for safe place (chantlitaxia)
  E. Cryptic coloration: resemblance to substrate
  F. Cryptic morphology: concealment of insect outline
  G. Confusion: disruptive coloration and movement between contrasting sensory background

Against identification
Special resemblance: resemblance to an inedible object
  A. Convergence on model image, some variability of characters preserved
  B. Stable evolutionary equilibrium. Not stabilizing selection
  C. Special behavior: position and mobility according to model, concealment of body appendages, active search of appropriate substratum
  D. Special coloration: resemblance to model
  E. Special morphology: resemblance to model, concealment of insect outline
  F. Disruptive coloration breaks typical insect form

Batesian mimicry: resemblance to an unpalatable animal
  A. Convergence on prototypical model image, reduced variability of characters
  B. Model "escapes" mimicry: unstable evolutionary equilibrium, stabilizing selection
  C. Special behavior: position and mobility according to model
  D. Special coloration: resemblance to model
  E. Special morphology: resemblance to model

Secondary Defenses

Against approach
  A. Escape: running, flying
  B. Evasive maneuvers: aerial power turns, dives, and spirals
  C. Rush for cover
  D. Confusion: dash-then-freeze, startle display, flash coloration
  E. Deception: thanatosis

Against subjugation
  Attack: jaws and spines

Source: After Robinson, 1969a, and Endler, 1986.

(1988) used data on Müllerian mimicry to argue that the punctuated equilibrium theory of Eldredge and Gould (1972) does not preclude evolution by small steps. Mimicry has also played a role in our understanding of individual, kin, and green-beard selection (Guilford, 1988). In individual selection, the individual benefits from its own genetic makeup. In both kin and green-beard selection, however, there is an altruistic element. In the former, the individual's benefits depend on its relatedness to the altruistic donor; in green-beard selection individuals benefit from their genotypical or phenotypical similarity to the donor. In Batesian mimicry the altruistic donor is the aposematic model which is attacked and damaged by the inexperienced predator. Thus in cases of Batesian mimicry, a palatable mimic with the best phenotypic copy of the aposematic model is better protected than less exact mimics (green-beard selection).

Crypsis, Special Resemblance, and Mimicry

Crypsis is the resemblance of an animal to its background, and Batesian mimicry is the resemblance of a palatable animal to an aposematic animal. These two strategies appear to be distinct, but leaflike insects, for example, are described by some workers as being cryptic and by others as being mimetic, so that Cott (1940) was led to recognize a third category, which he called *special resemblance* or *disguise*, which includes those animals that have a specific morphological resemblance to an inedible object. There has been some debate over the precise definition and classification of defensive adaptations (Vane-Wright, 1976, 1980; Cloudesley-Thompson, 1981; Edmunds, 1981; Endler, 1981; Robinson, 1981; Rothschild, 1981; Zabka and Tembrock, 1986; Starret, 1993). *Choeradodis rhombicollis* illustrates the dilemma: it resembles an inedible leaf (either mimicry or special resemblance), and its predators often fail to recognize it (crypsis). So, the key points in debate are that (1) leaflike insects have evolved from morphologically "normal," cryptic ancestors; (2) leaflike insects produce signals that are of no interest to a predator, whereas mimics produce signals that are of interest to potential predators (albeit they are counterfeit, signaling "do not eat me"); (3) leaflike insects when placed on a neutral background still look like a leaf, whereas simply cryptic insects look like an insect. The different processes are represented in Figure 13.1 where the central mantid represents a "normal" or "primitive" condition. On the left, grass and bark mantids are cryptic and are not perceived as different from their background. At the bottom, stick, leaf (and flower) mantids are not perceived as being edible (though they may be recognized as sticks or leaves). Finally, on the right, ant mimicry conveys the false message of unpalatability.

One distinction between crypsis, special resemblance, and mimicry is the kind of population dynamics produced by the species interactions. Usually the model must be more common than its Batesian mimic. This is because if the model is rare then predators will be more likely to learn that the mimic is edible rather than that the model is nasty, and so predation on both model and mimic will increase. In Müllerian mimicry a high frequency of mimics (which are also aposematic) helps the model by "teaching" the predator about their similar chemical protection (Cott, 1940). In special resemblance and crypsis the similarity of the animal to part of its environment is less likely to have such dramatic consequences. Resembling a leaf, for example, must have negligible effects on the environment. However, there are exceptions: flower mantids are examples of special resemblance with both a defensive and an aggressive function. If these mantids were very common, and their predation on pollinating insects were strong, then the flowers that they mimic would suffer greatly.

Defenses that have a negligible impact on the part of the environment that they resemble generate stable evolutionary equilibria: if the environment does not change, the defense does not have to change. This is true of crypsis and special resemblance, although (as we noted above) flower mantids may be an exception. Batesian mimicry is normally assumed to result in an unstable evolutionary equilibrium, because the model escapes from the mimic (Huheey, 1988). This is because predation on the typical palatable mimic will increase predation on typical unpalatable models. Hence, deviant models will be preyed on less and will be at a selective advantage compared to the typical model. In turn, this will cause the model pattern to evolve away from that of the typical mimics. On the other hand, predators will weed out mimics that depart too much from the optimal model pattern. This process of elimination of deviant mimics canalizes the mimic population toward the optimal (but constantly changing) model pattern (stabilizing selection, Mayr, 1970). Ant-mimicking mantids, however, might result in little evolutionary instability because the mantids are rare compared to ants. In this case, then, we might expect stabilizing evolution in the mantids without much change in the ant population. This contrasts with crypsis and special resemblance wherein intraspecific form and color variation makes search image formation and learning by predators more difficult and, consequently, selection favors diversity.

Our argument, then, is that defensive strategies may not fall into natural categories and that their classification may therefore depend ultimately on one's particular scientific point of

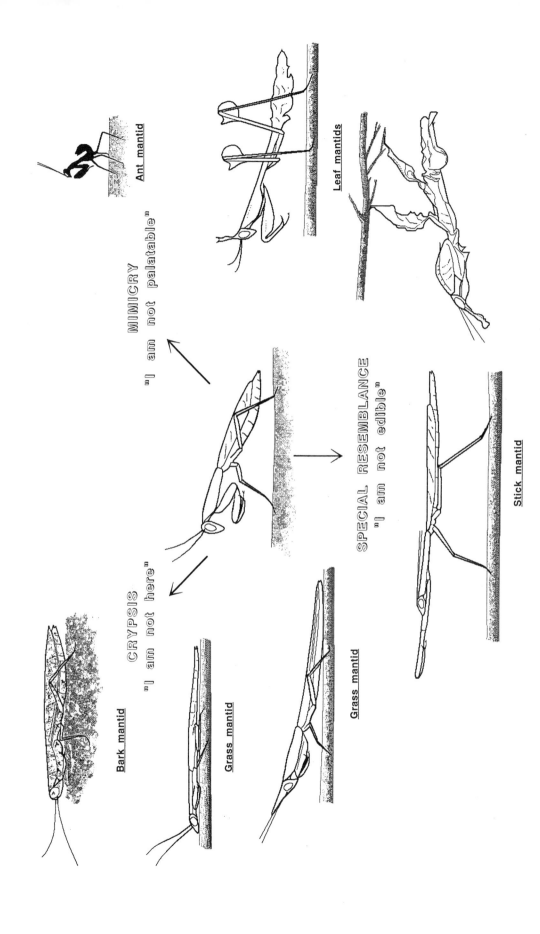

view. In our review of mantid defenses, we include leaf, flower, stick, and grass mimics under the heading of special resemblance.

Primary Defense
Crypsis

The simplest form of crypsis is matching one's color to the background, and many species of praying mantids are camouflaged in this way. The African *Negromantis modesta* is commonly green and well camouflaged on green vegetation; *Statilia apicalis* and *Oxypilus hamatus* are brown and camouflaged on twigs and branches; *Amorphoscelis* spp. are mottled green and gray, which gives excellent camouflage on tree bark in tropical rain forests; and *Eremiaphila* spp. are sandy colored and well camouflaged on desert soils (Edmunds, 1972, 1976; Cott, 1940; Crawford, 1981).

Cryptic animals cannot become too abundant (Table 13.1); otherwise predators that find one may acquire a "search image" and prey more intensively on them (Croze, 1970; Edmunds, 1991). Croze showed that pressure from predators that hunt in this way selectively leads to a scattered distribution of cryptic prey. Some predators also prey apostatically, that is, take more of common than of rare prey (frequency- and density-dependent predation; Curio, 1976; Endler, 1986). What is important, however, is not absolute rarity so much as apparent rarity, and prey that is nocturnally active and/or hides under cover can be much more abundant, though less visible, than a diurnal species that lives in the open. Cryptic animals may therefore actively search for a safe place to rest (chantlitaxia, Franco and Cervantes-Perez, 1990). Cryptic animals must also rest motionless to avoid detection, but if they have to move they often do so slowly with a rocking of the body such that their movement resembles disturbances caused by the wind.

Because light falls from above, the ventral part of an animal's body is typically in shadow, and this provides predators with a cue to the location of prey that otherwise match the background. Animals as different as caterpillars and fish have evolved countershading such that their lower surface is lighter than their upper, which erases the ventral shadow (Cott, 1940; Edmunds, 1974a; Edmunds and Dewhirst, 1994). Countershading has not been reported in cryptic mantids, probably because they either have flattened bodies (e.g., in bark-living species), or because their wings are held at an angle laterally so that the ventral curvature of the abdomen is hidden, or because they adopt a variety of postures—horizontal, vertical, or upside down—so that effective countershading when in one position would be quite conspicuous in another.

Crypsis also can be improved by disruptive colors which break up the outline of the body. For instance, many large mantids of the genera *Polyspilota*, *Prohierodula*, and *Tenodera* are green and brown; the African *Pseudoharpax virescens* is green with disruptive black marks on head and thorax; the Central American *Phyllovates cornutus* has two red-brown streaks on green forewings; the East African *Omomantis zebrata* has a cream and brown stigma and five brown disruptive bands on green forewings; the Malayan flower mantis *Hymenopus coronatus* is pink or cream with a bar of green on the thorax that divides the insect into a front and back half, each of which resembles part of a flower head (e.g., *Melastoma polyanthum*), and the ocelluslike black and yellow markings on the wings of the flower mantids *Pseudocreobotra* and *Chlidonoptera* spp. are also apparently disruptive when the insect is resting on flowers (Fig. 13.10E; Edmunds, 1972, 1976; Cott, 1940).

Facing page

Fig. 13.1 The possible evolution of the various primary defensive strategies of mantids. *Center*, A typical, relatively unspecialized mantid. *Left*, Three examples of mantids that conceal the body contour and limbs, a cryptic bark mantis, and two short-bodied grass mantids. The grass mantids are certainly cryptic but are treated here as examples of special resemblance. *Below* and *right*, Three more examples of special resemblance to an inedible part of the environment: a long-bodied stick mantis and two leaf mantids. *Top right*, A mantid that mimics another animal, an ant, which is well protected because it bites and stings.

Predators also may detect cryptic prey by their shadow, so an animal that loses its shadow will be better off (i.e., the Peter Pan effect, Portmann, 1956). Shadows are less important for an animal resting among vegetation well away from a continuous substrate, but for mantids living on tree trunks, bare earth, or sand, shadows could be critical. Such mantids minimize their shadow by having a flattened body. Bark mantids, for example, are dorso-ventrally flattened and spend much time resting motionless on tree trunks (Edmunds, 1972, 1976). In the African genus *Tarachodes* the head is flexed in the opisthognathous position (with mouthparts directed backward) so that in side view the contour line of tree to head is imperceptible. The wings of adult males are gray, like bark, but are still fairly conspicuous. Adult females are brachypterous, and the abdomens of the females and nymphs are corrugated, resembling the texture of bark (Fig. 13.2A). *Theopompa* and *Theopompella* are similarly flattened African bark mantids, but they hold their head in the prognathous position (with the mouthparts directed forward), and the wings are broad and angled laterally so that the contour with the substrate is smoothed (Fig. 13.2E). The wings of these mantids are also mottled cream and brown so that they resemble lichens and other markings on the surface of bark. *Liturgusa* is a South American genus of bark mantids that are brown with patches of green and yellow that resemble green and yellow lichens on the bark (Hingston, 1932). The morphological and behavioral changes that must have occurred in the evolution of bark mantids from less specialized cryptic mantids have been outlined by Edmunds (1972).

Some species of mantid are polymorphic with two or more different color forms. Polymorphism is an evolutionary prey response to the requirement for rarity in cryptic animals, requiring predators to learn to recognize two or more forms instead of just one (Croze, 1970; Edmunds, 1991). Polymorphisms may be determined genetically as in the classic case of the peppered moth *Biston betularia*, in which a single gene controls the melanic form (Kettlewell, 1973) or, environmentally, in which some physical or chemical factor determines the morph. For mantids no polymorphism has been unequivocally shown to be genetically determined, but one possible example is the bark mantis *Tarachodes afzelii*. This species is normally gray-brown, but there is a form with black bars on the pronotum and abdomen (Fig. 13.2A) and another with a black line mid-dorsally on the pronotum and abdomen (Edmunds, 1976). These abdominal marks are conspicuous in nymphs and females, but they are hidden by the wings in males. In males these two morphs had frequencies of 0.5 and 1.5% at Legon, Ghana. These two morphs may be better camouflaged on trees with dark grooves in the bark, where the black marks disrupt the form of the insect. The bark mantis *Astape denticollis* from Java is a similar species with a brachypterous female and is trimorphic. The three morphs occur with different frequencies on different species of trees which they resemble in color (Lieftinck, 1953). However, it is not known if this is a genetic or an environmentally determined polymorphism, nor if the different frequencies are caused by the mantid's choice of tree trunk or by selective predation of the more conspicuous insects.

Environmental Polymorphism

Environmentally determined polymorphisms occur in many species of Mantidae belonging to the genera *Mantis, Sphodromantis, Miomantis, Polyspilota, Prohierodula, Tenodera, Galepsus,* and *Pyrgomantis*. Nymphs are able to change color at each molt so that an individual may change from green to brown or vice versa several times during its development. The factor determining if a nymph will change color in *Sphodromantis lineola* is light intensity (Barnor, 1972 [summarized in Edmunds, 1974a]). High intensity causes nymphs to change to brown; low intensity causes them to turn green at the next molt. Edmunds argued that at Legon, Ghana, this results in good camouflage: trees shed leaves and flush with new ones during the dry season, so a well-camouflaged green nymph may find its background changing from green to brown within a few days, rendering it very conspicuous. Similarly, a brown nymph on a leafless tree will quickly become conspicuous when the new leaves flush. An adult insect can fly to another tree on which it

Fig. 5.1 Female *Acanthops falcata* in the pheromone-release posture. She is hanging upside down from a leaf, with her abdomen curled ventrally and her wings uplifted. She is facing to the right. Photograph courtesy of Michael Robinson.

Fig. 5.2 Female *Stagmomantis limbata* cannibalizing a male in the field. She is holding him by the thorax.
Photograph by the author.

Fig. 5.3 Approach and mount sequence in *Stagmomantis limbata*. *A*, Male approaching female from behind. *B*, Male has mounted but has yet to make genital contact with the female. Photographs by the author.

Fig. 5.4 A headless male *Stagmomantis limbata* copulating with a female in the field. Photograph by the author.

Fig. 5.5 Copulation in *Stagmomantis limbata*. The black markings on the adults' pronota are identification codes. Photograph by the author.

A

B

Fig. 13.2 *A,* Female bark mantis *Tarachodes afzelii* with black bars on thorax and abdomen from Ghana resting on and guarding ootheca. *B,* Male grass mantis *Pyrgomantis nasuta* from South Africa. *C,* Female brown leaf mantis *Phyllocrania paradoxa* from Ghana.

All photographs in this chapter by Malcolm Edmunds.

C

Fig. 13.2 (*continued*) *D*, Male long-bodied grass mantis *Hemiempusa capensis* from Ghana giving deimatic display. *E*, Male bark mantis, *Theopompella westwoodi* from Ghana. *F*, Third-instar red ant mimicking nymph of *Miomantis aurea* from Ghana. *G*, First- or second-instar black ant mimicking nymph of *Tarachodes afzelii* from Ghana.

Fig. 13.5 *A*, Female brown leaf mantis *Deroplatys truncata* from Malaysia. *B*, Female brown leaf mantis *Panurgica compressicollis* from Ghana. *C*, Male stick mantis *Popa spurca* from South Africa (this genus has many of the adaptations of long-bodied stick mantids but without the elongated body).

Fig. 13.5 (*continued*) *D*, Male short-bodied stick mantis *Catasigerpes occidentalis* from Ghana. *E*, Male long-bodied stick mantis *Danuria thunbergi* from South Africa in cryptic posture with forelegs extended in line with head and body. *F*, Female long-bodied stick mantis *Heterochaeta strachani* from Ghana in characteristic resting posture with forelegs projecting sideways.

A

B

C

D

Fig. 13.8 Deimatic displays of six species of mantids to show variation in position and markings of wings and forelegs (all except F are from Ghana). A, Female *Polyspilota aeruginosa*. B, Female *Plistospilota guineensis*. C, Female *Tarachodes afzelii* with reduced but brightly colored wings. D, Male *Pseudocreobotra ocellata* with ocelli on wings.

E

F

Fig. 13.8 (*continued*) Deimatic displays. *E*, Female *Miomantis aurea* with slightly reduced wings (she cannot fly). *F*, Female *Deroplatys truncata* from Malaysia.

A

B

Fig. 13.10 *A*, Deimatic display of female *Heterochaeta strachani* from Ghana. *B*, Deimatic display of male *Sphodromantis lineola* from Ghana with red spots on coxae (most individuals have yellow spots). *Facing page: C,* Male long-bodied grass mantis *Oxyothespis longipennis* from Ghana. *D*, Female *Tenodera superstitiosa* from Ghana, a mantis with normal body shape that extends its front legs anteriorly like a long-bodied stick mantis.

C

D

E F

G

Fig. 13.10 (*continued*) *E*, Male flower mantis *Chlidonoptera chopardi* from Ghana. *F*, Female green leaf mantis *Neomantis australis* from Queensland. *G*, Final-instar nymph of *Pseudocreobotra ocellata* from Ghana with pink body after being reared on red *Bougainvillea* flowers.

is cryptic, but a nymph cannot. However, since nymphs use light intensity as the trigger for color change, they can regain good color matching to background after a few days when they molt next. A collection of 684 male *S. lineola* collected over nearly six years showed that the twelve brown insects were all found in the dry season (October to March), with just one at the start of the rains (April). Furthermore, green *S. lineola* preferentially rest on green rather than brown leaves (chantlitaxia), so on a tree with mixed live and dead leaves they are likely to be well camouflaged. In contrast, brown mantids do not prefer to rest on brown (Fig. 13.3). Their choice of background differs neither from random nor from that of green mantids, perhaps because a tree that loses its leaves is entirely brown, with no chance of the mantid nymph finding a green spot, so there could be no selective pressure for background-matching behavior.

The European *Mantis religiosa* also changes color in response to light intensity (Jovančić, 1960), but more rigorously designed experiments are required to confirm the nature of this response. *M. religiosa* also occurs in Ghana, where Edmunds (1976) found 58% of mantids were green in the wet season compared to a significantly smaller 28% in the dry season. We know of no data on the frequencies of green and brown morphs of *M. religiosa* during different seasons or habitats in southern Europe, where it is often abundant.

Miomantis paykullii is another species that occurs in the savanna at Legon, Ghana, where it lives on grass. For this species relative humidity is the key factor determining whether it becomes green or brown at the next molt. Nymphs reared in dry conditions become brown, and those reared in humid conditions become green (Barnor, 1972; Edmunds, 1976). Grass can change from brown to green in just a few days following rain, so if the humidity caused by this rain also induces nymphs to switch to green at the next molt, there is a good chance that they will remain well camouflaged. There is a statistically significant correlation between the number of days in the month on which rain fell and the percentage of adult males that were green, and if the number of rainy days in the preceding month is compared with the percentage of males that were green (Fig. 13.4), this correlation is even stronger. This is presumably because it takes time for rain to stimulate fresh green grass and for humidity to induce a response in nymphs at their next molt. In addition, *M. paykullii* nymphs preferentially choose to rest on either green or brown crepe paper, depending on which they match (Fig. 13.3). The mechanism for this choice may be that the mantids are comparing their color with that of the background because if the rear part of the eyes are painted black (so they cannot see the rest of their body), the substrate preference of green insects switches from green to brown. However, if the front of the eyes is blackened, green nymphs also prefer brown, so the mechanism of background choice remains unclear.

Bush fires in African grasslands induce another type of color change in which normally green or brown insects (e.g., grasshoppers) change to black (Hocking, 1964). Two mantids, *Galepsus toganus* and *Pyrgomantis pallida*, also occasionally exhibit fire melanism (Edmunds, 1976), with the melanic forms occurring mainly toward the end of the dry season (January to March) when most fires occur. It is assumed that fire melanism is an environmentally induced polymorphism, although the hypothesis has never been tested.

There are three more forms of color change that require further investigation. First, an adult of the Chinese mantis *Sinomantis denticulata* was seen to change from green to straw-colored in 24 hours following captivity (Edmunds and Dudgeon, 1991). Second, *Eremiaphila* spp. commonly match the color of their desert substrate (Cott, 1940; Preston-Mafham, 1990), and when juveniles molt in the laboratory they change color to match their background (Lev Fishelson, personal communication). Third, when nymphs of the African flower mantids *Pseudocreobotra wahlbergi* and *P. ocellata* are reared on *Bougainvillea* flowers they change color over a period of several days to become almost entirely pink like the flowers (MacKinnon, 1970; Edmunds, 1972) (Fig. 13.10G). The Southeast Asian flower mantis *Hymenopus coronatus* can also become pink or

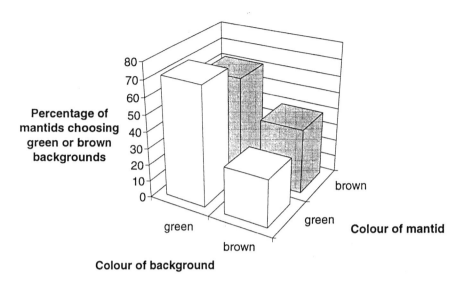

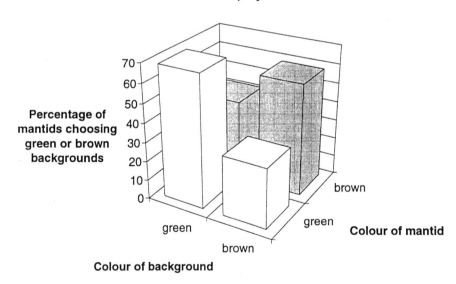

Fig. 13.3 Choice of resting on green or brown backgrounds by green and brown mantid nymphs. *Above, Sphodromantis lineola;* below, *Miomantis paykullii*. For both species, green insects prefer green to brown ($p < .01$) while brown *Miomantis* prefer brown to green ($p < .05$), but the choice of brown *Sphodromantis* differs neither from random nor from the choice of green insects. (Data from Barnor, 1972, after Edmunds, 1974a.)

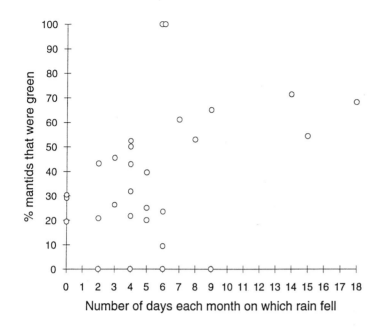

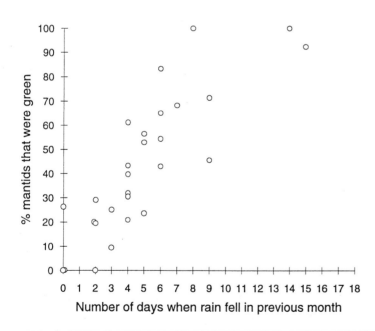

Fig. 13.4 Relationship between rainfall and frequency of green or brown male *Miomantis paykullii* at Legon, Ghana. *Above*, Percentage of green mantids compared with the number of rainy days in the same month ($r_s = .65$, $p < .01$); *below*, percentage of green mantids compared with the number of rainy days in the previous month ($r_s = .74$).

white depending on the color of the flower on which it rests (Fogden and Fogden, 1974). This color matching is presumably brought about by optical input somehow inducing a change in pigment deposition in the cuticle or epidermis. (Many insects, especially Lepidoptera, are known to sequester carotenoids from their food [Feltwell and Rothschild, 1974; Rothschild, 1975], and these may then form the principal color of the insect [Grayson et al., 1991].) Another possible mechanism is the redox reaction of ommochromes, pigments found in some Diptera, Lepidoptera, Orthoptera, and Odonata. A reduced form of this pigment is deposited, for example, in the integument of locusts and appears pink (Fox and Vevers, 1960). In dragonflies the reduced form of the pigment gives males a scarlet color, and an oxidized form gives females a yellowish tone.

The Function of Crypsis in Mantids

Crypsis can be either defensive or aggressive in function. Evidence that crypsis reduces the probability of being detected by a predator was collected almost a century ago by Cesnola (1904) and later by Beljajeff (1927, summarized by Cott, 1940). Cesnola (1904) tethered green and brown *M. religiosa* to green and brown vegetation and observed the reduction in their numbers on successive days. After eighteen days he found that all of the cryptic insects were alive and all of the conspicuous ones were eaten. Beljajeff (1927) carried out a similar experiment with green, yellow, and brown mantids on a brown background and obtained similar results (but see Croze, 1970).

It is also possible that crypsis in mantids reduces the probability that a mantid will be detected by its prey, but no experiments have been done to test this idea.

Special Resemblance

The main types of special resemblance found in mantids are resemblance to leaves, flowers, sticks, and grass. Such mantids are often called leaf mimics, flower mimics, and so on. However, since we are treating them as examples of special resemblance rather than of Batesian mimicry, we will refer to them as leaf mantids, flower mantids, etc. The morphological and behavioral adaptations of long-bodied and short-bodied stick and grass mantids are very different, so we treat these in separate categories.

Leaf Mantids

Leaf mantids bear a close resemblance either to living green leaves or to dead brown leaves. A superb green leaf mantis is the neotropical *Choeradodis rhombicollis* in which the body is flattened, the head is tucked under the enlarged pronotum in the opisthognathous position, the pronotum is exceptionally broad, and the wings together with the pronotum resemble the leaves of some of the many epiphytic green plants that commonly grow on tree trunks and branches (Robinson, 1969b). A mantis that could represent an early stage in the evolution of a green leaf mantis is *Sinomantis denticulata* from China and Hong Kong (Edmunds and Dudgeon, 1991). In this species the nymphs are green and flattened, with the head held in the prognathous position. It normally rests on the underside of leaves where it is highly cryptic to the human eye, and when disturbed it darts quickly round the leaf to rest flattened on the other side. The adult male has normal wings and has no specific resemblance to leaves. An intermediate stage between *Sinomantis* and *Choeradodis* is represented by the tropical Australian *Neomantis australis* (Fig. 13.10F). This species is also green and flattened, with a prognathous head and broad, flat wings. It also rests on the underside of leaves and switches quickly to the other side when disturbed (M. E., personal observation).

Brown leaf mantids resembling dead leaves occur in most tropical regions: *Acanthops* lives in the neotropics, *Deroplatys* in Southeast Asia, and *Panurgica* and *Phyllocrania* in Africa. *Panurgica compressicollis* has a short, stout body with enormous forelegs, the femora of which extend when at rest from the head to the middle legs (Fig. 13.5B). It resembles a rolled leaf suspended by a thread from an overhead twig, and if disturbed it sways sideways pivoting from its middle legs and hindlegs (Edmunds, 1976). *Acanthops falcata* also hangs from a twig like a dead leaf, but its body is held flexed with the abdomen hanging down vertically and the head and forelegs held so they appear as a single oval structure (Crane, 1952; Robinson, 1969a). *Deroplatys truncata* is a large, heavy-bodied mantis with a broad pronotum and with both pronotum and wings marked with shades of brown like dead leaves (Fig. 13.5A; M. E., personal observation). The African *Phyllocrania paradoxa* is perhaps the most modified of all the leaf mantids (Edmunds, 1972, 1976). The female is either dark brown or straw in color, with prominent wing veins that resemble the veins on a dead leaf. This mantid also has a twisted projection on the vertex, frills on the sides of the abdomen and on the walking legs, and a broad, flattened pronotum. These all break up the outline of a typical mantis head, abdomen, legs, and thorax, respectively. The forelegs are held close beneath the head so that the profile of head and forelegs in typical "praying" position is lost. The male is not quite so leaflike and has a narrower pronotum and wings that are transparent instead of translucent (Fig. 13.2C). *Empusa* spp. have similar adaptations, including a foliate vertex, but we do not know of any observations on their resting posture and behavior (Fig. 13.9A). *Sybilla limbata* has a brown body with elongated vertex and small foliations on the legs, but its relatively "normal" wings are green. Presumably its mix of green and brown disrupts its outline when resting on vegetation, but it represents a possible intermediate stage in the evolution of such perfect leaf mantids as *Phyllocrania* (Edmunds, 1976).

Flower Mantids

The nymphs of several Old World hymenopodid mantids habitually rest on flowers to which they bear a close resemblance, and they capture insects that come to visit the flowers (Fig. 13.10G). In Africa they belong to the genera *Pseudocreobotra* and *Chlidonoptera* and in Asia to the genus *Creobroter*. Background color matching and color change in these insects have been discussed above. The nymphs are mottled yellow and green (or pink and green) and normally rest with the abdomen reflexed over the thorax. The forelegs are held close beneath the head so that the head-thorax profile of a typical praying mantis is obscured. The adults have yellow, black, and green ocelli on the forewings, which are used in a dramatic deimatic display (see below), but when resting on flowers only one ocellus is visible. These markings are disruptive, and the insects seem to the human eye to be well camouflaged (Fig. 13.10E; Edmunds, 1976). The Asian *Hymenopus coronatus* nymphs are even more perfectly camouflaged on pink or white flowers with enlarged mid-femora and hind femora resembling flower petals (Annandale, 1900; see also illustrations in Fogden and Fogden, 1974). The adults are less well camouflaged than the nymphs with the white forewings forming a conspicuous recognition feature. Annandale reports that one mantis explored several nonflowering branches before finally resting on an inflorescence of *Melastoma polyanthum*, on which it is highly cryptic (and on which it frequently occurs in Malaya). When the flowers began to wither the mantid lowered its normally reflexed abdomen, and eventually jumped to the ground, still with the dorsal surface of the abdomen exposed. It has brown lines dorsally on the otherwise creamy white abdomen so that it resembles a withered flower that has fallen to the forest floor. These observations suggest highly adaptive behavior but require experimental confirmation.

A final, but more controversial, possible example of a flower mantis is *Idolium diabolicum*. Sharp (1899) quotes observations on this species in Mozambique by Mr. Muir. The mantid hangs from branches with its forelegs held wide apart, displaying bright colors on the inner surfaces of the enlarged coxae. Flies are reported to be attracted to these colors, and as they approach they are grabbed and eaten. Wickler (1968) illustrates the remarkable posture of this mantid in color. Sharp believed that the colors on the forelegs of certain other mantids (*Gongylus* and *Empusa*) are also flower lures for insect prey (i.e., "floral simulators"), but we now know that many of these colors are actually used in defensive displays. Carpenter (1921) also made observations on *Idolium diabolicum*: he never observed the flower lure posture, but he did observe a dramatic deimatic display. Edmunds (1976) also pointed out that in many species it is impossible to elicit a predatory strike from a mantis that is giving a deimatic display, so it is difficult to see how such a display could be transformed into a flower lure for prey capture. Clearly the behavior of this large and remarkable species requires re-examination.

Short-Bodied Stick and Grass Mantids

Grass, shrubs, and trees all have many narrow cylindrical structures. Short-bodied mantids that live in grass and in trees have therefore evolved very similar morphological and behavioral adaptations. *Orthoderella ornata* is a South American grass mantis that can be either green or brown, and it rests closely apposed to grass stems. The head is prognathous with two lateral processes behind the eyes which conceal the typically broad mantid head profile (Fig. 13.6), the forelegs are held close beside the pronotum, and the middle legs and hindlegs are short and also held very close to the body and substrate. It adopts a flattened resting position and shows a dash-then-freeze strategy when disturbed. All of these features help to smooth the contour of the mantis so that it is a barely noticeable swelling on the grass (Brunner and Gandolfo, 1990). Females are brachypterous, so there are no conspicuous wings to provide a recognition cue for a predator, but the males have functional wings.

An African grass mantis with very similar adaptations is *Pyrgomantis* spp. Here too the female is brachypterous, and middle legs and hindlegs are short, but the head is opisthognathous with a prominent pointed vertex, and the forelegs fit precisely beneath the head and the substrate so that here too there is no obvious head contour (Fig. 13.2B). The pointed vertex means that if the head is lifted from the grass substrate then it resembles a pointed blade of grass (Edmunds, 1972). Other mantids with similar though less extreme adaptations toward resembling sticks or grass, and which show how the above insects may have evolved, include *Paramorphoscelis gondokorensis* and *Galepsus toganus* (Edmunds, 1972, 1976).

Catasigerpes occidentalis is an African short-bodied stick mantis (Edmunds, 1972, 1976). In this and in related genera the body is held at an angle to the substrate so that the insect resembles

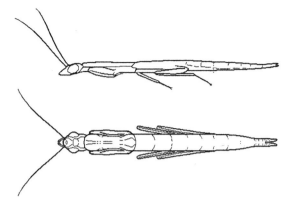

Fig. 13.6 Resting posture of the short-bodied South American grass mantis *Orthoderella ornata* showing prognathous head with postero-lateral processes that conceal the neck, narrow body held close to a grass stem, forelegs held close beneath the head and thorax, and short, inconspicuous middle legs and hindlegs held close to body.

a broken twig arising from a branch. The body is gray, there is a prominent vertex, the forelegs fit neatly between head and body, and the wings are truncated posteriorly so that they appear to arise directly from the branch on which it is resting (Fig. 13.5D).

Long-Bodied Stick and Grass Mantids

Because grass, trees, and shrubs contain many long, thin elements, any mantid that lives here that is long, thin, and of the same color as the grass or twigs is likely to be difficult for predators to see. *Angela guianensis* and *Danuria buchholzi* are long-bodied stick mantids from tropical America and Africa, respectively. Both respond to slight disturbances by extending the forelegs in front of the head so that they appear to be a continuation of the body-head unit and thereby increase the resemblance to a long, thin stick (Robinson, 1969b; Edmunds, 1972). In both species the female is brachypterous, with absence of long wings further increasing the similarity to a stick. In *Danuria* the forecoxa is notched so that the head fits neatly into it, and small projections at the rear of the head conceal the narrow "neck" between head and thorax so that the contour from forelegs to head and thorax is smoothed (Fig. 13.5E). The closely related *Popa undata* is of more typical mantis shape rather than being very long and thin (Fig. 13.5C), but it has all of the other morphological and behavioral adaptations described here for *Danuria* (Edmunds, 1976). It therefore suggests that the morphological and behavioral adaptations of stick mantids found in *Danuria* evolved first, and the body became elongated subsequently.

Another elongated stick mantis is the giant African *Heterochaeta* (= *Stenovates*), but here the forelegs are normally held extended laterally where they resemble side branches of a thick twig, and the projecting eyes resemble buds at the apex of the twig (Fig. 13.5F; Edmunds, 1972).

Many grass mantids are also elongated, but most of them lack the full range of morphological and behavioral adaptations to perfect the resemblance to grass that occurs in long-bodied stick mantids. *Tenodera superstitiosa* has only a moderately elongated body (it is much thicker than a stem or leaf of grass), but it does sometimes extend its forelegs directly in front of its head when disturbed, just like *Danuria* (Fig. 13.10D). *Oxyothespis longipennis* (Fig 13.10C) and *Leptocola phthisica* both have exceptionally long, narrow bodies but lack any other morphological or behavioral adaptations to perfect the resemblance to grass. *Idolomorpha lateralis* and *Hemiempusa capensis* are two other grass mantids with elongated bodies. In these species the head has a pointed vertex (Fig. 13.9B) and the forelegs are held tucked beneath the head so that they are concealed in lateral profile and the front half of the insect resembles a blade of grass. But the wings are broad and quite conspicuous (Edmunds, 1972, 1976). A possible reason for the very precise resemblance of stick mantids to sticks and the less precise resemblance of grass mantids to grass may be because there are so many stems and leaves of grass that an elongated grass-colored insect is likely to be well concealed, but there is much more space between twigs in shrubs and trees so an insect with only a general similarity to twigs may still be detected by a predator.

Aposematism

Tarachodula pantherina from East Africa is thought to be a possible example of aposematic (warning) coloration (Preston-Mafham, 1990). The female is colored purplish gray with orange bands on the abdomen, and by contrast the legs are cream with small black spots. However, the beautiful illustration of this species (Preston-Mafham, 1990, Plate 33) shows that it adopts precisely the same resting posture on tree trunks as the west African *Tarachodes afzelii*, which is highly cryptic. It is likely that when it rests on appropriately colored tree trunks it is similarly cryptic with the contrasting colored legs and orange marks disrupting the typical body outline of a resting mantid. Furthermore, aposematic insects do not normally adopt cryptic behaviors (as *T. pantherina* does) but advertise their presence (Cott, 1940). We therefore consider that there are no unequivocal examples of warning coloration in the Mantodea.

Batesian Mimicry

Since we have treated the various plant-part mimics as examples of special resemblance, the only examples of Batesian mimicry among mantids are of ant mimicry in the early instars of various species. Edmunds (1976) summarized the results of his own studies and those of Kumar (1973), which showed that ten out of fifteen species of African mantid studied had ant-mimicking first instars. Ant mimicry is widespread in the Mantidae and also occurs in many species of Hymenopodidae (Fig. 13.1). The young mantids resemble ants in size and color, but they have only four instead of six legs extending out from the body. Young *Tarachodes afzelii* are black and have a large head, very similar to the common black ant *Camponotus acvapimensis* (Fig. 13.2G). As they grow they lose the black color and become gray bark mimics. The early instars of *Pseudocreobotra ocellata* and *Panurgica compressicollis* are glossy black and resemble glossy black ants such as *Crematogaster* spp. Young *Sphodromantis lineola* and *Miomantis aurea* are brownish red with black eyes, very like the common and vicious weaver ant *Oecophylla longinoda* (Fig. 13.2F). After the third instar they are too large to mimic this ant and become green or pale brown. Each shrub at Legon, Ghana, is overrun with one species of dominant ant including *Oecophylla lon-*

ginoda and *Camponotus acvapimensis*. In a survey of 331 shrubs at Legon, Edmunds (1976) found 102 plants overrun with the red *Oecophylla* and 229 plants without it. Five early-instar ant-mimicking *S. lineola* were on shrubs with the red ant, and just one was on a different plant. Although the numbers are small, this difference is statistically significant, indicating a positive association between the mantids and *Oecophylla* ants. However, the way in which this association arises is not known. It could arise through female mantids ovipositing close to weaver ants, or mantids resting preferentially close to weaver ants, or it could be due to selective predation of mantids that are not associated with weaver ants. There is clearly scope for further work exploring the resemblance and association of young mantids with ants.

Secondary Defense

Active Escape

Adults of small mantid species usually respond to a simulated or actual attack by running or flying whereas adults of large species usually respond with a startle (deimatic) display (Figs. 13.7, 13.8). This difference is probably due to the relationship between speed of response and body size: small insects can run or fly so quickly that a predator may be unable to catch them, while large insects may need to "warm up" their wing muscles before they can fly, and their initial movement is much slower because of their bulk (Edmunds, 1972, 1976; Liske et al., this volume, chap. 12). Species that attempt to escape by running rather than flying include the primitive cockroach-eating *Metallyticus semiaeneus* from Malaysia (Shelford, 1903) and the brachypterous *Eremiaphila* from Old World deserts. Many species of flying mantids can carry out a variety of aerial maneuvers comparable to those displayed by moths in response to bat sonar (Yager, this volume, chap. 6). Nymphs typically respond actively by running or jumping; jumping off a leaf may enable a small insect to escape because a predator loses track of it among the vegetation.

Flash coloration may occur in some mantids, but the evidence is inconclusive. Flash colors are exposed when an animal runs or flies but are hidden when it is at rest. It is thought that predators follow the bright colors and when they disappear, the predator loses sight of its prey (Cott, 1940; Edmunds, 1974a). There is no experimental demonstration that flash colors really do protect animals in this way, but it seems likely. The African mantis *Pseudoharpax virescens* is green and cryptic, but when disturbed both sexes sometimes run with the wings held partly open, exposing orange on the dorsal abdomen and purple (male) or purple and black (female) on the hindwings. When they stop running these colors disappear. Many other species of mantid also have bright colors on the hindwings, but in most of these the colors are exposed during deimatic displays, not when they attempt to run. *P. virescens* has never been observed to give a deimatic display, so it may be that in this species the orange and purple are flash colors.

Deflection of Attack

Satyrid and other butterflies have small eye spots distally on the wings which deflect attacks away from the body and toward the less vulnerable wings (Edmunds, 1974a, b). Similar deflection marks may occur in mantids. The center of gravity of an adult mantid is probably on the thorax or front part of the abdomen, which is normally covered by the wings. This is the point of the body that has the most predictable path when the animal moves because it is the point that moves the least, and so it may be the best target for a predator (Webb, 1986). Color spots elsewhere on the wings may be important for a flying or running insect to hide the center of gravity. Such color spots may act as "supernormal" stimuli which predators cannot help but follow. As such they need to be of a contrasting color, to help detectability and discriminability. The conspicuousness of a color spot may be augmented by sharp edges and contrast with the surrounding colors. Many of the Mantinae (e.g., *Mantis, Sphodromantis*) have small stigma spots on the wings, while *Parhierodula coarctata* and *Chloroharpax modesta* have much larger black and yellow spots which might be deflection marks. However, butterfly deflection marks are typically near wing edges where an attack can do little damage, but in resting mantids

Ethology of Defenses against Predators 291

these spots are on each side of the body not far behind the center of gravity. It is difficult, therefore, to see how they could deflect an attack away from the mantid's body. They might, however, be more effective in the flying insect, but there is no evidence of wing damage by predators to mantids that have survived such attacks, so we conclude that the presence of deflective markings in mantids remains unproved.

Deimatic Display

A startle or deimatic display (Edmunds, 1974a) is the response to a nearby predator of exhibiting marks and colors which function to intimidate the predator and thereby reduce the probability of an attack (Fig. 13.7). In a survey of the defenses of forty-two species of mantids from Ghana, Edmunds (1972, 1976) found that deimatic

Fig. 13.7 Deimatic display of the large mantid *Stagmatoptera biocellata* and the response of a shiny cowbird (*Molothrus bonariensis*). A, B, The cowbird approaches and attacks the mantid. C, D, The mantid responds by repeatedly lunging at the bird, such that (D, E, F) the cowbird eventually flies off. (Redrawn from a series of motion picture frames from Maldonado, 1970b.)

displays occurred most frequently in the larger species, and active escape by running or flying was more frequent in smaller species. There is considerable variation in the details of the display, but typically the mantid turns toward the aggressor with the forelegs held close to the thorax at an angle of 180° to each other and the wings partially raised (Fig. 13.10B). The insect thus appears to be larger than it does when at rest. The posture is held for several seconds or even half a minute, sometimes with occasional lunges toward the aggressor. Figure 13.7 shows a sequence of frames taken from a film (Maldonado, 1970b) where a mantid successfully wards off the attack of a shiny cowbird (*Molothrus bonariensis*). Maldonado (1970b) showed that the display of the neotropical *Stagmatoptera biocellata* is effective in preventing attacks from several species of birds. Bright colors, which are hidden when the insect is at rest, are exposed during the display. These are either on the inner surfaces of the forelegs, the ventral surface of the thorax or the abdomen, the hindwings, or the gaping jaws. The dramatic display of *Stagmatoptera biocellata* is unusual in that there are conspicuous large eye spots on the forewings, but these are on the hindmost part that is hidden in the resting insect (Maldonado, 1970b). Some species stridulate by flexing the abdomen up and then down repeatedly between the wings, making a hissing noise. Crane (1952) argued that this display has evolved because it gives protection against predators, but it rarely succeeds against mammals (Edmunds, 1976). Maldonado's experiments therefore suggest that the display has evolved as a defense against birds.

Praying mantids, however, have no effective retaliatory defense against a bird, and none is known to be aposematic, so the deimatic display is a bluff. Hence, a bird that encounters several praying mantids should quickly learn of the bluff, develop a "search image" for the display, and then proceed to attack any mantid that displays. However, some species never use the display, some display occasionally, and others display frequently when attacked, but the display is unpredictable and does not occur every time a particular stimulus is encountered. Furthermore, each species' display is unique in terms of the colors and patterns exposed during the display and whether or not it stridulates (Figs. 13.2D, 13.8, 13.10A, B). There is also considerable intraspecific variation in the color patterns exposed in the display in different populations of *Paramantis prasina* and in a single population of *Sphodromantis lineola* (Edmunds, 1976; Roy, 1965). This variation may reflect the evolutionary pressure against a standard deimatic display, which could be used by an experienced predator as a foraging cue.

Some of the most dramatic deimatic displays occur in species of *Pseudocreobotra* and *Chlidonoptera* (Cott, 1940; MacKinnon, 1970; Edmunds, 1976) where the most conspicuous color markings are, unusually, on the upper surface of the forewings. One would normally expect this exposed part of the insect to be camouflaged and the bright colors to be hidden in the resting insect. However, in these genera the green, yellow, black, and white on the forewings may be disruptive when the insect rests on a flower head. During a display the two forewings are opened above the abdomen, and these colored markings resemble two large ocelli (Fig. 13.8D). Blest (1957) established experimentally that the sudden appearance of two large "eyes" intimidates small birds, so we assume that this display is directed at insectivorous birds. Possible stages in the evolution of these elaborate and precise ocelli have been suggested by Edmunds (1976), to which additional material can now be added. The south African flower mantis *Harpagomantis discolor* has yellow blotches on green wings, like all of these genera, but with no trace of black or white markings; the Asian *Theopropus elegans* has two parallel curved black lines separated by yellow, clearly disruptive but not similar to an eye spot; *Chlidonoptera lestoni* from Africa has two black semicircles partially enclosing an oval white space; and *Creobroter urbana* from Asia and *Pseudocreobotra* spp. from Africa have black ringed white spots with a small black "pupil," giving an even better resemblance to a vertebrate eye.

Another quite remarkable display occurs in the African *Idolium diabolicum*. This large species has the forelegs modified into green leaflike plates that provide superb camouflage, but when attacked the brightly colored lower surfaces of these plates are displayed in so dramatic a way that they frightened a monkey (Carpenter, 1921).

Thanatosis

An alternative defense for some animals is to feign death (thanatosis). Predators such as cats and praying mantids typically attack only prey that moves, so if the potential prey remains motionless it may be ignored. Thanatosis has been recorded in the neotropical *Angela guianensis* and in five species from Ghana (Robinson, 1969b; Edmunds, 1976). One might predict that thanatosis would be more effective in larger species that are too bulky to escape an alert predator by running or flying and in species with a special resemblance to a plant rather than in those that look like a typical mantid, because these might more easily be mistaken for something unpalatable. The available evidence supports these predictions: five of the six species that exhibit thanatosis are large (only *Catasigerpes occidentalis* is small), and four of them have a special resemblance to leaves or sticks (*Phyllocrania paradoxa* is a leaf mantis, *Catasigerpes occidentalis* is a short-bodied stick mantis, and *Danuria buchholzi* and *Angela guianensis* are long-bodied stick mantids). Of the remaining two species, *Polyspilota aeruginosa* has no particular resemblance to either sticks or leaves, but *Tenodera superstitiosa*, while not as extreme in its adaptations as *D. buchholzi* or *Angela guianensis*, is a relatively elongated species with some of the adaptations of a long-bodied grass mantis.

Larry Hurd (personal communication) has also observed thanatosis in North American *Tenodera sinensis* and *T. angustipennis* when these are knocked off vegetation onto the ground. They remain sticklike for a minute or two. However, Dr. Hurd was unable to elicit this response predictably to any particular stimulus. It seems to be a much less frequent response than just running away.

Retaliation

The final defense of many animals when seized by a predator is to retaliate with whatever weapons they possess (Edmunds, 1974a). In the larger mantids this means scratching or gripping the predator with the raptorial foreleg spines and biting with the jaws. This can be painful for a human if the spines are fine and sharp, but there is no evidence that they induce a natural predator to release the mantid. In an encounter between a *Mantis religiosa* and a gecko reported by Gerald Durrell (1956) the mantid seized the gecko with its forelegs and actually drew blood, but this did not prevent the gecko from consolidating its grip on the body of the insect and then eating it.

Defenses of Eggs

The ootheca is believed to protect the eggs and developing embryos from environmental forces and predators. During the process of egg laying, a foamy, viscous white liquid is secreted over and around the eggs. This hardens to form the typically brown or cream-colored ootheca which in the larger species of Mantidae can be two or more centimeters in diameter. In cockroaches the hardened wall of the ootheca contains sclerotin, which confers resistance to digestive enzymes and acids (Fox and Vevers, 1960), so this may also be true for praying mantids.

In India, mantid oothecae are sometimes tunneled by minute ants of the genus *Crematogaster*. In one of these oothecae one hundred mantids emerged successfully, but many were subsequently killed by the ants (Daniels et al., 1989). Chalcid Hymenoptera frequently parasitize oothecae although a few of the mantid eggs may still hatch successfully (Daniels et al., 1989; Ene, 1962; Brunner, personal observation). In a few species (*Tarachodes afzelii, Tarachodula pantherina, Galepsus toganus,* and *Pyrgomantis pallida* from Africa and *Astape denticollis, Theopropus elegans,* and *Hymenopus coronatus* from Southeast Asia), the female rests on and guards the ootheca (Shelford, 1903; Lieftinck, 1953; Ene, 1962; Edmunds, 1972; Preston-Mafham, 1990). The female African bark mantis *Tarachodes afzelii* lays a flattened ootheca on bark where it is well camouflaged, and she rests on it until the nymphs emerge (Fig. 13.2A). Ants and parasitic Hymenoptera that approach are struck at with the forelegs and either captured, hit, or driven away. Guarded oothecae suffered lower frequencies of parasitization than unguarded ones (Ene, 1962).

Constraints on the Evolution of Specific Defenses

Each species of mantid has several of the defenses that we have discussed in its defensive repertoire; for example *Sphodromantis lineola* is cryptic, can change color when it molts from green to brown (or vice versa), has a dramatic startle display, and will bite and scratch when seized. Adult males often try to escape by running or flying, though females do so less readily. Early-instar nymphs are red ant mimics and often attempt to escape by jumping. Predation pressure should lead to progressive improvements in the effectiveness of these defenses, so why is it that all species do not have similar defenses? There are basically two reasons: because the requirements for greater efficiency in one defense may conflict with the effectiveness of other defenses or with other essential activities (such as feeding), and because each species will encounter a different spectrum of predators with different hunting strategies. The conflict between the requirements for greater efficiency in one defense with other essential activities of the animal has been discussed by Robinson (1969a, b) and Edmunds (1972, 1974a).

Devices used in crypsis have a cost to the animal, just as honest signals do (Zahavi, 1987). This is so because they act to hide information that the insect would give away if it had no disguise. In other words, efficiency for functions other than camouflage must be sacrificed (and vice versa) for the achievement of cryptic adaptation. In the desert, for example, insects must balance cryptic coloration with their needs to regulate temperature and conserve water. For instance, a light color that matches the sand may result in a very short activity period because it takes longer to warm up in the morning. Further, light colors do not protect against ambient temperature, which can rise to lethal levels at midday. So, contrary to intuition and despite an increased probability of detection, black insects are better able to survive in the hottest, driest deserts because they become active earlier in the morning and remain active for longer at night (Hamilton, 1971).

All mantids have a primary defense in which they rest motionless for much of the time, but if they are to capture prey they must either sit and wait in a posture suitable for striking, or must stalk slowly until close enough to lunge or pounce on prey. However, predators of mantids will be watching for the key features by which an otherwise well-concealed mantid can be detected. These probably include movement, the shape of the mantid's head and forelegs when in the "praying" posture, the walking legs, and the wings (Robinson, 1969a, b, 1973; Edmunds, 1972, 1974a, 1976). There are ways by which all of these features can be concealed, but the cost of so doing may make the mantid less effective at some other essential activity.

Concealment of Movement

Predators, including praying mantids, use movement as a cue to detect prey (Curio, 1976; Prete, this volume, chap. 8). Robinson (1973) found that rufous-naped tamarins ignore motionless sticks but fixate on sticks that move and attack if the sticks have legs attached to them. For widely foraging or hunter species that actively pursue their prey, cryptic behavior, in particular immobility, is at conflict with successful foraging. On the other hand, for sit-and-wait or ambush predators like most mantids, cryptic behavior is more suited to their foraging strategy. *Sphodromantis lineola*, for example, spends most of the time in an inactive state (Zack, 1978), consistent with both its cryptic strategy and its way of foraging. Mantids, however, face a dilemma because they need to move in order to capture prey and find a mate. Like other highly cryptic animals, including phasmids and chameleons, they often move in a curious, jerky way. This rocking movement may help to conceal the animal because it resembles the movement of leaves disturbed by a gentle breeze (Robinson, 1969a).

Concealment of the Head and the "Praying" Profile

Robinson (1973) showed that tamarins specifically attack the head of sticks and other models to which heads were attached, indicating that the head may be one of the visual cues by which they recognize prey. As we suggested earlier, predators may recognize mantids by the unique shape of their head and forelegs. Stick and grass mantids would be more sticklike if they had eyes that did not project from the slender contour of the body, but widely spaced eyes are more effective for binocular vision used in judging prey distance, so there is an apparent conflict between the requirements for optimal defense and optimal prey capture. *Orthoderella, Danuria, Popa, Hoplocorypha*, and other stick mantids have come to a compromise with eyes that project slightly laterally but are concealed by projections on the back of the head that hide the "neck" between head and thorax (Figs. 13.5C, E, 13.6; Edmunds, 1972; Brunner and Gandolfo, 1990). *Heterochaeta* retains widely spaced eyes but adopts a unique resting posture with forelegs held extended laterally like branches arising from its sticklike body. The eyes resemble buds at the tip of the branch (Fig. 13.5F; Edmunds, 1972).

The "praying" profile of the mantid can be avoided by adjusting the head to the prognathous or the opisthognathous position. Figure 13.1 shows how the opisthognathous head of the bark mantis *Tarachodes* is lowered and the forelegs are held close so as to minimize any shadow and obliterate the "praying" profile. It also shows how *Orthoderella* and *Danuria* extend the forelegs anteriorly in line with the prognathous head and thorax so that together they resemble a thin stick (in *Danuria*) or a thin part of a stem of grass (in *Orthoderella*) (Edmunds, 1972; Brunner and Gandolfo, 1990). In *Danuria* and *Popa* the forecoxa has a notch into which the head fits so that the contour of head and extended forelegs is smooth (Fig. 13.5C, E). However, the cost of these adaptations is probably impaired hunting success because the mantid is not ready to strike at prey. The compromise adopted by all of these mantids is to rest in the praying posture most of the time, ready to strike at any prey that comes near, but to adopt the more cryptic posture if they detect movements that might be a predator. The "praying" profile can also be concealed by developing a vertex so that the contour of head, forelegs, and vertex resembles the tip of a blade of grass (as in *Pyrgomantis* and *Idolomorpha*, Fig. 13.9). This does not appear to affect the optimal posture for prey capture (Edmunds, 1972).

Concealment of the Middle Legs and Hindlegs

Predators often attack models with legs sticking out much more readily than models without legs or with legs held close to a substrate (Robinson, 1973). Limbs can be concealed more easily if they are short rather than long, as in *Orthoderella ornata* (Brunner and Gandolfo, 1990), but short legs decrease the effective prey-capture distance, prevent the mantid's typical rocking movements (Edmunds, 1972), and decrease maximum running speed. Fast, cursorial mantids such as *Amorphoscelis* and *Eremiaphila* have long legs, but mantids with short legs that are easily hidden are usually short bodied and cryptic on sticks and grass. The latter do not try to outrun predators; they try to dodge them by darting around to the other side of the stick or leaf on which they are perched (e.g., *Orthoderella, Pyrgomantis, Galepsus, Paramorphoscelis,* Edmunds, 1972, 1976; Brunner and Gandolfo, 1990). For long-bodied stick and grass mantids, long legs may be less of a giveaway to a predator because they live in places where there is a mass of confusing long, narrow sticks or blades of grass. Many such mantids (e.g., *Leptocola, Danuria, Angela, Heterochaeta*) have extremely long legs that are far too long and cumbersome for rapid running. They have probably evolved because they increase the distance from which the mantids can strike successfully at prey and because they make rocking movement more convincing. Legs are also concealed in some species by leaflike appendages on the leg segments as in *Empusa gongyloides, Hemiempusa capensis, Phyllocrania paradoxa,* and *Pseudocreobotra wahlbergi*.

Concealment of Wings

Wings must retain a fairly well-defined shape if they are to function effectively in flight. They can be concealed to some extent by means of color or pattern, but they may still be an easy recognition feature for a predator. The naked abdomen of a

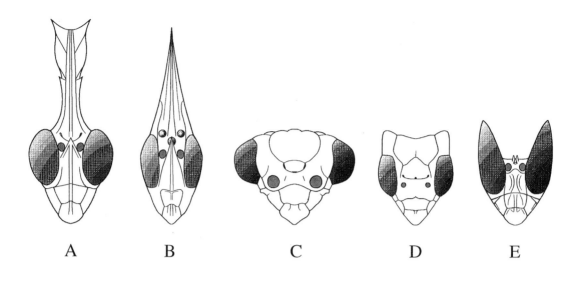

Fig. 13.9 Variation in head morphology in mantids. A, *Empusa* sp. is opisthognathous with a foliate vertex typical of leaf mantids. B, *Idolomorpha lateralis* is opisthognathous with a pointed vertex typical of grass mantids. C, *Coptopterix argentina* has a typical, orthognathous mantid head. D, *Orthoderella ornata* is prognathous, typical of many short-bodied stick or grass mantids which lack a pointed vertex. E, *Pseudoharpax virescens* is opisthognathous with protuberant eyes whose significance is not known.

mantid is easier to conceal than are wings because in addition to altering its color and pattern it is also possible to modify its shape and texture. But if adult mantids were to lose their wings there would be considerable problems in finding a mate (especially for males) and, perhaps, in entering new habitats. Loss of wings in both sexes is common in island insects where the danger of being blown out to sea is more critical than the ability to fly. For continental mantids, however, wings are usually lost only by females (e.g., *Hoplocorypha nigerica*, a stick mantis, and *Oxypilus hamatus*, a small brown species lacking any special resemblance).

Why is it only the female that loses or has reduced wings? Obviously, it is important that one sex can fly so that the sexes can find each other, and in mantids it is the male that is attracted to and searches for the female (Maxwell, this volume, chap. 5). Females are also heavier than males, because of the weight of their developing eggs, so they are likely to be weaker fliers than males. So apterous females have sacrificed the ability to fly for better camouflage and, presumably, a longer survival time. Males, on the other hand, have retained wings because this enables them to find females quickly even though the cost of flying may be increased vulnerability to predators. Data from Côte d'Ivoire by Gillon and Roy (1968) have been used by Edmunds (1976) to support this contention. First, species where both sexes are fully winged have a higher proportion of males than species in which the females have reduced wings. This suggests that females of African mantids live longer than males, but this difference is more marked where females have reduced wings. Second, bird predation by two species of raptor was greater on females that are winged than on females with reduced wings, suggesting that flightless females are less easy to find (and presumably more cryptic) than fully winged females (Edmunds, 1976).

However, comparatively few mantids have lost their wings completely. In a study of forty-two species from Ghana, Edmunds (1976) found just two in which the females were apterous and eight in which they were brachypterous. The short wings of brachypterous mantids are useless for flight but might be retained because they play a role in thermoregulation, in dispersal of volatile pheromones (Hurd, personal communication), or in deimatic displays. The defensive display of female *Tarachodes afzelii* exposes the underside of the forewings, which are black with cream-colored veins, and the hindwings, which are red (Fig. 13.8C). Both sexes of the highly cryptic desert mantis *Eremiaphila braueri* have similarly reduced wings, whose function appears to be to intimidate predators by a dramatic defensive display (Roonwal, 1938).

Differences in Defenses between the Two Sexes

The female *Acanthops falcata* gives a deimatic display more readily than the male, and it exposes brighter, more startling colors (Preston-Mafham, 1990). The experience of one of us (M. E.) confirms that, in general, females are more likely to display than are males and they show brighter, more conspicuous colors when they do display. Certainly, the eye spots and wing colors of female *Chlidonoptera lestoni* are more conspicuous than those of the male (Edmunds, 1976). This suggests that in many species females have more effective deimatic defenses, perhaps because they are heavier than males and so are less able to respond by means of active escape (i.e., flight). Analogously, females of species with special resemblance are more cryptic than are their males (e.g., *Phyllocrania paradoxa*).

Geographical Distribution of Defensive Behaviors

Praying mantids are widely distributed from temperate to tropical regions on every continent but are absent from boreal and tundra regions, and they show the greatest diversity in the tropics. The more bizarre species (to the human eye), those which are the most extreme examples of special resemblance, occur primarily (if not exclusively) in tropical forests and savannas. Examples are the leaf mantids *Choeradodis* and *Phyllocrania*; the long stick mantids *Angela*, *Danuria*, and *Heterochaeta*; the long grass mantid *Leptocola*; the short grass mantid *Pyrgomorpha*; and the flower mantids *Hymenopus* and *Pseudocreobotra*. Similarly, the most dramatic deimatic displays are made by tropical species (e.g., *Pseudocreobotra* and *Idolium*). This is partly because there are more species of mantid in the tropics, but it is also because the environment is more complex, with more species of plants and potential prey, so there are more niches available for mantids to occupy. Nevertheless, there are some anomalies. The New World appears to lack empusid and hymenopodid mantids, two taxa that include a great many highly specialized and "unusual" species.

Absence of Certain Defense Strategies

Secretion of noxious substances or irritants is unknown in Mantodea, so it is not surprising that aposematism and Müllerian mimicry appear to be absent in this group. Defensive glands occur in some cockroaches, and both defensive glands and aposematic colors have evolved several times within the Orthoptera (Whitman, 1990), so why not in the Mantodea? The absence of aposematism may be related to Lindeman's ratio (Krebs, 1985), implying that predators can sustain only a tenth of the biomass of their prey, and such a relatively small population may be unable to withstand the losses associated with predators learning to recognize aposematic colors. The absence of chemical defenses is harder to explain. One might expect that a chemically defended mantid would be less likely to fall prey than a palatable one, but unless the defense was so potent that the mantid could deflect and ultimately survive an attack, there would be little advantage gained by producing, storing, and deploying the chemical. Some of these costs could be avoided by acquisition of the chemical through food. However, we know of only three cases in which insects acquire defensive chemicals from a carnivorous diet. These are a moth, a beetle, and a fly larva which sequester carminic acid from the cochineal bug (Eisner et al., 1994). Although possible, it seems that acquisition of chemical defenses from a carnivorous diet is rare in insects.

Batesian mimicry in mantids seems to be restricted to mimicry of ants. This restriction is equally difficult to explain, especially as some mantispids, which have a very similar life style to mantids, mimic aposematic wasps. For mantids such as *Tarachodes afzelii*, ant mimicry results in a curious association between ants (the model) and mantids (the mimic), because the juvenile mantids gain protection from the chemical defense of the ants, while the larger nymphs and adults specialize in feeding on ants.

It can be argued that the existence of mimicry may be more likely in those species with great plasticity for the production of colors, contrasts, and patterns. Wings with scales, such as those of butterflies, may be the perfect color palette for such variation. Wings without scales, such as those of Orthoptera and Mantodea, may be less plastic in that respect, and evolution may have favored strategies that rely more on subdued hues and morphological variation. However, bright colors occur in aposematic Orthoptera and in the deimatic displays of Mantodea, so the potential for evolving aposematic and mimetic colors must surely be present in mantids.

Protective Coloration from the Predator's Point of View

Critics of adaptationist explanations of animal coloration were justified in pointing out that forms and patterns apparent to the human observer may be unseen by the nonhuman eye, and vice versa. Indeed, comparative studies have shown that animal vision varies enormously across groups. Arthropods and vertebrates have eyes that are constructed very differently so they must perceive objects differently. There are also big differences within vertebrates: some are dichromats (e.g., squirrels, rabbits), some are trichromats (e.g., humans, some fishes), some seem to be tetrachromats (e.g., goldfish, turtle) or pentachromats (e.g., pigeons, ducks) (Thompson, Palacios, and Varela, 1992), and some are even sensitive to ultraviolet (UV) light (Gruber, 1979; Tovee, 1995), which could be especially useful for detection of UV reflectance patterns of otherwise cryptic arthropods (Parrish, Ptacek, and Will, 1984; Bennett and Cuthill, 1994; Viitala et al., 1995). Color vision also varies within species and even within individuals over time. This pattern of variation may be particularly important for the evolution of defense mechanisms given that differences in many visual abilities seem to correlate better with ecological niche than with phylogenetic relationships (Partridge, 1989, cited in Thompson et al., 1992).

Variation is so widespread, and plasticity so common, that the critics' argument against anthropocentric interpretations of animal coloration is turned upside down. Let's take cryptic coloration as an example. Green and brown are the most common colors in vegetation, so cryptic patterns that include green and brown will be most effective. Hence, we should expect strong selective pressures on all insectivores for the ability to recognize these colors in potential prey. Understandably, then, despite the differences between predator visual systems we do find common denominators. For instance, the range of visible light overlaps considerably between quite disparate taxa (400–700 nm for primates, 310–590 nm for insects, and 350–720 nm for birds); both color constancy and color induction have been found in bees, pigeons, and humans (Srinivasan, Lehrer, and Wehner, 1987); and pigeons group wavelengths of light, natural objects, and geometric forms into categories just as we do (Thompson et al., 1992). So it makes sense that predators hunting for the same prey will all evolve visual systems that are good at detecting the organisms that they are after. In turn, prey should evolve such that they are generally cryptic to all of the predators that hunt them. Although we are omnivorous rather than insectivorous, our color vision seems to be as good as that of any other animal, at least within the portion of the spectrum that is visible to us (Snodderly, 1979). Hence, if an insect seems cryptic or conspicuous to us, we have good reason to believe that it will appear so to predators as well. This is not to say that we should fail to rigorously test our hypotheses regarding mantid defenses. Only detailed studies of an animal's behavior and its interactions with both conspecifics and predators can help us decide whether a particular suite of characteristics acts as an effective defense (Portmann, 1956).

Directions for Future Research

There is an enormous literature on all aspects of mimicry, both theoretical and experimental: in contrast, the theoretical implications of crypsis for the evolution of coloration and for evolution in general have received little attention.

There are several specific questions that can be asked about mantids, especially polymorphic species. First, for both genetically and environmentally determined polymorphisms, do the different morphs suffer different levels of predation? Do they choose different backgrounds, and if so, by what mechanisms do they do so? In the cases of environmentally induced polymorphisms further work is required to determine the precise factors that induce each color morph, particularly for species that seem to display fire melanism and those which change color to match specific flowers. Further, nothing seems to be known about the physiological mechanisms by which environmental changes give rise to color changes in mantids.

We have suggested that some color markings may be disruptive or deflect an attack, but there is no experimental evidence to support these suggestions.

As we have explained, mantids with special resemblances to leaves, flowers, and the like have characteristic postures and behaviors. One such species that would be most interesting to study is *Idolium diabolicum*, because there seems to be no consensus in the literature regarding its natural history. Unfortunately, it appears to be exceedingly rare, and with increasing destruction of forests in central Africa this remarkable species could well become extinct.

It would be interesting to explore whether insectivorous predators are really deceived by ant mimicry. While experiments of this type are sometimes difficult to perform because the species concerned are rare, first-instar ant mimics can be obtained in large numbers so long as the oothecae can be acquired. Further work is also required on the nature of the association between mantid and ant: is it fortuitous, or are there specific behaviors that bring about the association?

The effectiveness of deimatic displays would repay careful investigation, but designing suitable experiments is likely to prove a considerable challenge. We have argued that interspecific variation in markings and in the details of the display have evolved because that makes it harder for predators to learn that the display is bluff, but this has never been tested experimentally with any predator, and certainly not for the spectrum of predators that are presumably relevant to any one species of mantid.

Finally, we have suggested that many predators use specific cues to detect mantids, but there are very few experiments that support this contention.

ACKNOWLEDGMENTS

We are grateful to Lev Fishelson and Larry Hurd for permission to report unpublished observations, to Jim Carpenter and Judith Marshall, who allowed us to examine material in the collections of the American Museum of Natural History and the Natural History Museum, London, respectively, and to Larry Hurd and Janet Edmunds for comments on earlier drafts of the manuscript.

TECHNIQUES

14. Rearing Techniques, Developmental Time, and Life Span Data for Lab-Reared *Sphodromantis lineola*

Frederick R. Prete

One of the contributors to this book quipped in an e-mail message that "rearing mantises has to be one of the most frustrating jobs—at least sometimes." He was right.

When I was a graduate student, I began my venture into mantid husbandry by rearing *Tenodera aridifolia sinensis*, but after about a year I switched to *Sphodromantis lineola*. Since then I have raised a number of species, though not as many as some of my colleagues (such as Dr. Yager). My favorite remains *S. lineola*. The most difficult part of my early work was inventing all of the stuff that I needed to house and care for my mantids as they grew. Unfortunately, back then I knew of only a few people who raised mantids, and they did not raise them in quantities as large as I needed to supply an ongoing research project. I had to take a guess at everything and see what worked. The second most difficult part of my early research was figuring out how long my adult mantids would be viable, experimentally speaking. There were virtually no data on the life span of lab-reared mantids, and none that I could find on *S. lineola*. I had no idea how long I could expect my mantids to live, how long their behavior was "normal" (for a mantid), and if they went through a period of senescence or just dropped dead all of a sudden. So, what I want to do here is explain the basics of mantid rearing (as I understand them, anyway) and share some data on the life span of *S. lineola*.

As you read this chapter, you must remember that raising mantids is, in part, an art. There are no strict formulas for success. Things vary from lab to lab and from person to person. Some people have a "green thumb" for certain species and have disastrous results with others. Dr. Yager has done a wonderful job in his chapter explaining these points. What I want to do is give you a starting point from which to begin your mantid husbandry program so that you do not have to reinvent the wheel as did I. In addition, I want to give you some sense of what to expect if you raise *S. lineola* or, alternatively, how you might go about making sense of the life span data that you will accumulate raising your favorite species.

Hatching and Rearing

I think that there are three key issues to keep in mind if you want to raise any sizable quantity of mantids: (1) The chambers must be abundantly (but not excessively) supplied with food, (2) *Drosophila* (the standard food for hatchlings) must not be able to escape from the rearing chambers, and (3) humidity must be kept at an appropriate level.

In my lab, we find that the limiting factor in raising mantids is the amount of *Drosophila virilis* that we can raise. We use this species because it is large (i.e., each fly supplies a new hatchling with a good-sized meal), and all of the mantids that we have raised (even the small banded mantis, *Theopropus elegans*) can handle *D. virilis* as soon as they hatch. We have to maintain about 115 active fly cultures (in standard *Drosophila* tubes) to supply enough flies for just *one* of the chambers described below. I have found that if a chamber is not thick with flies, we lose many more nymphs than we would otherwise. In fact, despite the fact that we always have 650 to 750 fly cultures going at any one time, I have never been able to overfeed my hatchlings. Our cultures yield the most flies when we use plain (not blue) *Drosophila* medium from Wards Scientific (Rochester, NY, USA).

Although I have tried a lot of alternatives, if you are interested in raising mantids in quantity, I cannot recommend anything better than the chambers that my student and I described in Prete and Mahaffey (1993a). These chambers are virtually indestructible and are large enough to accommodate the hatchlings from as many as eight oothecae at one time. Basically, they are constructed by bolting a clear Lucite top onto a standard 51 × 41 × 22 cm polycarbonate rodent cage (e.g., from Fisher Scientific). I cut the top in half widthwise and join the halves with an aluminum piano hinge. This way, one half can be unbolted and used as a door while the other remains permanently attached to the box.

Both the top and the sides of the chamber should have a number of 3 cm holes drilled in them. Four in the top and one or two in each side are sufficient. These can be filled with foam rubber plugs or covered with a piece of screen held in place with tape, depending upon how humid the chamber is to be kept. Fresh, humid air is supplied to the chamber by a standard aquarium pump connected to plastic wash bottle that is partially filled with distilled water. The bottle's nozzle is inserted into the chamber through a hole in one of the foam plugs in the side of the chamber. Generally, any high-quality mid-sized aquarium pump delivering about 2500 cm^3/minute at 4 psi is sufficient to aerate two to four chambers.

We use a heavy plastic mesh (approximately 5 × 5 mm grid) to create perch sites for our mantids. The mesh, sold in lumberyards and hardware stores as a covering for rain gutters, is generically referred to as *gutter guard* (e.g., Thermwell Products Co., Patterson, NJ 07524, USA).

Drosophila can be added to the chamber by dumping anesthetized or chilled flies into a funnel inserted into one of the holes in the chamber lid. We used to put jars of fly food in our chambers for the *Drosophila,* but I think that doing so increases the amount of bacteria in the chambers and, hence, the number of sick mantids. Now we just put a 100 ml plastic bottle of water with a filter paper wick in the chamber so that the flies can drink.

At some point, the young mantids will have to be switched from *Drosophila* to larger prey. If you switch them to larger flies nothing has to change. If you switch to small crickets as we do, then you will have to add a shallow container of water and cricket food to the chamber. We suggest powdered rodent food for the crickets. This can be obtained from any feed store or pet supply retailer or from a manufacturer such as PMI Feeds (St. Louis, MO 63144, USA). Remember, if your crickets are thirsty or hungry (especially for a protein-rich meal) they will eat the mantids.

Mantids are kept in these chambers until they reach the fourth stadium or so. Then we switch them to clear, 22 gallon plastic buckets (e.g., Rubbermaid® brand from Consolidated Plastics, Twinsburg, OH 44087, USA). The buckets can be outfitted just like the chambers with foam rubber plugs, screens, humidifiers, and plastic mesh perch sites. We use the buckets because larger mantids need the extra height in which to molt. Each day we harvest adults from the buckets and place then in individual clear plastic containers the size of which depends on the species. In general, we raise *S. lineola* in an ambient temperature of about 22 to 24°C.

Life Span of Lab-Reared *S. lineola*

Although there is a wonderful body of literature on the basic biology and life histories of mantids in the wild (e.g., Bartley, 1983; Barrows, 1984; Eisenberg et al., 1981, 1992; Hurd and Eisenberg, 1984; Lawrence, 1992; Moran and Hurd, 1994; Fagan and Hurd, 1994), the data cover only a handful of the approximately two thousand known species, and there is even less information available on the lab-reared species with which many researchers work (but see Copeland, 1975; Corette, 1980; Matsura and Morooka, 1983; Krombholz, 1977; Suckling, 1984).

The genus *Sphodromantis* is widely distributed in Africa south of the Sahara and one species, *S.*

viridis, is also present in northern Africa, southernmost Europe, and western Asia (Roy, 1987). The species with which I work, *Sphodromantis lineola* (Burmeister), is commonly found in western Africa (Edmunds, 1972; Miller, 1972; Kumar, 1973) and is a hardy, relatively large insect (the average female body length is 67 mm) that requires no additional water if sufficiently well fed (Prete et al., 1992). Little has been published about this species' life history: Kumar (1973) reported that *S. lineola* reaches adulthood in nine to ten stadia (based on three observations) and that developmental time (hatching to final ecdysis) takes 228 to 295 days (mean = 255.7 ± 27.3 s.d.). On the other hand, Gillon and Roy (1968) reported that developmental time was only 75 to 86 days. To the best of my knowledge, there are no published life span data for *S. lineola* except that Phipps (1960) noted one female that lived at least 204 days.

In addition to increasing our general knowledge about mantids, life span studies on these creatures may offer researchers unique opportunities to explore several critical behavioral correlates of aging (and their underlying physiological mechanisms) that are either absent or not easily studied in other insect models of aging. These correlates include changes in visual perception (e.g., Prete, this volume, chap. 8), visuomotor coordination (e.g., Kral, this volume, chap. 7; Rossel, 1991; Rossel et al., 1992; Mathis et al., 1992), and the execution of complex motor patterns (e.g., Copeland and Carlson, 1979; Lockshin and Wadewitz, 1986; Prete and Hamilton, this volume, chap. 10; also see Collatz and Sohol, 1986).

Over the last six years, my students and I have collected longevity data on 667 adult female and 130 adult male *S. lineola* that were arbitrarily selected from our laboratory colony and that died apparently natural deaths (i.e., did not succumb to external events such as cannibalism or trauma or to disease). In addition, two cohorts were raised separately, and developmental time was recorded for 113 mantids (67 females) that reached adulthood. A arbitrarily selected 39 of the females were followed until their death.

As a rule, life span studies exclude from their analyses embryonic time for vertebrates and larval or nymphal time for holometabolous and hemimetabolous insects, respectively (e.g., Arking, 1991; Schneider and Rowe, 1990; Collatz and Sohol, 1986). In the case of insects, this exclusion is particularly critical in that developmental time can be dramatically affected by ambient temperature, food availability, and non–age-related losses—particularly at ecdysis (e.g., Fagan and Hurd, 1994; Moran and Hurd, 1994; Paradise and Stamp, 1991; Phipps, 1960; Sauer et al., 1986; Yager, personal communication; also note the range of developmental times reported by Copeland [1975] and Corrette [1980] for *Tenodera aridifolia sinensis*, by Rau and Rau [1923] for *Stagmomantis carolina*, by Matsura et al. [1984] for *Paratenodera angustipennis*, and by Roberts [1937] for *Stagmomantis limbata*). However, because of the dearth of information about *S. lineola*, I am going to include data on developmental time here, but I will keep these data separate from life span data.

Data on developmental time are plotted in Figure 14.1. Under my laboratory conditions, the mean time from hatching to final ecdysis was 93 (± 36 s.d.) days for females (min = 62, max = 184) and 88 (± 40 s.d.) days for males (min = 57, max = 183). Discounting the outliers that you see at the far right of these graphs (>120 days for 9 females and 7 males), the mean time from hatching to final ecdysis was 80 (± 13 s.d.) days for females (max = 110 days) and 72 (± 12 s.d.) days for males (max = 99 days). Mann-Witney U tests revealed a significant difference between the median times to adulthood for females and males irrespective of whether all cases were considered ($p = .018$; median = 82 and 72 days, respectively) or outliers were removed ($p = .002$; median = 80 and 69 days, respectively). These data are consistent with those reported by Gillon and Roy (1968) for *S. lineola*, and they support Kumar's (1973) conjecture that low ambient temperature slowed the development of his mantids.

Longevity data for mantids that reached the final stadium are presented in Figure 14.2. Females lived an average of 93 (± 60 s.d.) days (median = 91 days), and males lived an average of 57 (± 50 s.d.) days (median = 39 days) after the final

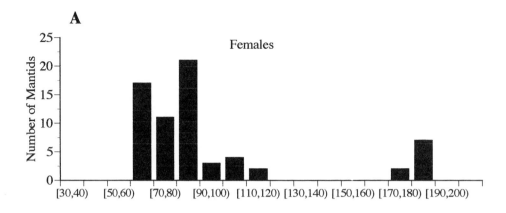

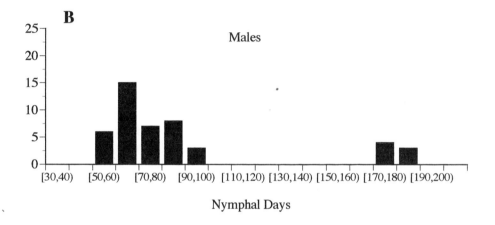

Fig. 14.1 Developmental time for (A) female and (B) male *S. lineola*. Overall, males reach adulthood sooner than do females, and in both sexes there is some small percentage of individuals for whom the final molt is delayed.

ecdysis (i.e., adult days). The columns in graphs A and B indicate the distributions of adult ages at death (d_x, left axes), and the curves indicate the proportions of insects surviving (l_x, right axes) at each time interval for females and males, respectively, all normalized to a population of one thousand individuals. (This is a standard format for this type of data [Comfort, 1956; Schneider and Rowe, 1990; Arking, 1991].) Note that both the survival curves (l_x) and the underlying distributions of ages at death (d_x) are considerably different for the two sexes. Specifically, up to day 180 (approximately twice the median age at death),

the female survival curve shows what Comfort (1956) called an *arith-linear* decay, which is characteristic of a senescing population with a constant number of individuals dying in each interval. Thereafter, there is a decrease in the raw number of females that died in each time interval, which reflects the fact that there were many fewer individuals remaining that could die within an interval (e.g., Arking, 1991; Comfort, 1956). In contrast, the survival curve for males approximates an exponentially decaying curve, which is characteristic of a nonsenescing population with an constant age specific death rate, that is, one in which a constant percentage of survivors dies in each time interval (Comfort, 1956; Arking, 1991; Schneider and Rowe, 1990). This theoretical interpretation supports my anecdotal observations: Females senesce prior to death; males just drop dead.

The survival rates of a wide range of organisms from invertebrates, such as *Caenorhabditis elegans* and *Drosophila*, to a variety of vertebrates (including *Homo sapiens*) have been accurately described by a mathematical model originally formulated by Gompertz (1825; Arking, 1991;

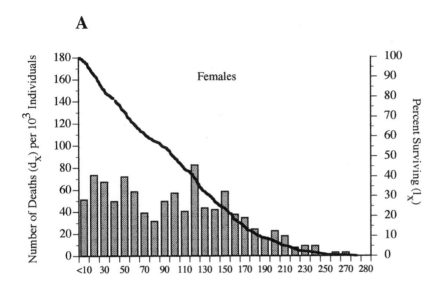

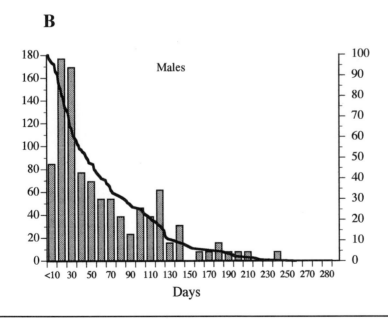

Fig. 14.2 The distributions of adult ages at death (d_x, columns associated with the left axes) and the proportions of mantids surviving (l_x, heavy lines associated with the right axes) at each time interval (days) for (*A*) female and (*B*) male *S. lineola*, normalized to a population of 1000 individuals.

Schneider and Rowe, 1990). The standard (linear) form of that model is $\ln(q_x) = \ln(q_o) + x$, where ln is the natural logarithm, q_x is the age-specific death rate (i.e., the proportion of organisms alive at the beginning of an interval that die during that interval), q_o is the initial vulnerability of the organisms (i.e., the y-intercept), and x (i.e., the slope of the line) is the rate of aging of the population under consideration. I have plotted the mantid data according to this convention in Figure 14.3. Interestingly, the (linear) Gompertz model accurately describes the rate of aging for males ($r^2 = .76, p \leq .001$), but not for females. I think that the survival rate for females is better characterized by two regression lines, the first describing the data from days 1 to 180 (see above; $r^2 = .39, p \leq .001$) and the second describing the data for days 181 to 266 ($r^2 = .79, p \leq .001$). (I know that a polynomial would be even better, but I want to emphasize that there are two separate trends in these data.)

As you can see in this figure, the log of the age-specific death rate for males increased linearly over time as it does for a wide variety of other organisms. However, the rate of aging for females was different: Besides the obvious fact that the data for females are not linear when plotted according to the Gompertz convention, an analysis of covariance confirmed a highly significant effect of sex ($F = 155.35, p \leq .0001$) and a time by sex interaction ($F = 14.83, p \leq .0001$): Although the overall rate of aging for females was lower than that for males, there was a dramatic (3.68 times) increase in the female aging rate that began at about adult day 180 and continued thereafter.

Figure 14.4 indicates the relationship between developmental time and days lived as an adult for thirty-nine females randomly selected from two

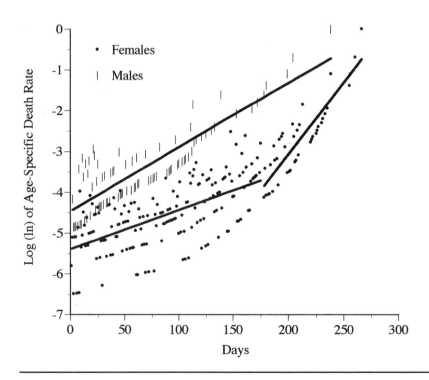

Fig. 14.3 The rate of aging for adult male and female *S. lineola* plotted in terms of the convention originated by Gompertz (1825). Adult males have a higher initial vulnerability (indicated by the y intercept) and, initially, age more quickly than females. At about 180 days of adult life, however, female death rate increases dramatically.

cohorts reared separately. Overall, these thirty-nine mantids did not differ from the experimental population as a whole in developmental time (100 ± 39 s.d. days; median = 85) or time lived as adults (127 ± 61 s.d. days; median = 120). Once again, however, two distinct subpopulations are apparent, one reaching adulthood within 110 days and the other taking as long as 173 to 184 days to do so. Although there is no *overall* relationship between developmental time and adult life span, there is such a relationship for that group reaching the final stadium within 120 days ($r^2 = 31, p \leq .001$). Within that group, females that reached the final stadium more quickly lived longer as adults. The pattern seen in these data is probably not due to pathology in those insects taking the longest time to reach adulthood. If it were, one would expect the group with the longest developmental time to be particularly short lived, but they were not. Their adult life spans ranged from 45 to 208 days (120 ± 52 s.d. days; median = 108), which is indistinguishable from the rest of the females. It seems more likely that delayed developmental time (which is also seen in other hemimetabolous insects) is either a product of random variability or an evolved tactic to distribute the population of adults over time. (But I probably should leave matters ecological to my more knowledgeable colleagues such as Larry Hurd.)

The Biology of Aging in Mantids and Others

The study of the biology and evolution of the aging process—presumably with an eye to postponing its effects in humans—is receiving renewed interest (Rose and Nusbaum, 1994). However, the mechanisms underlying aging and senescence still remain elusive or only partially understood. Recent progress in our understanding of aging has come from work with invertebrates such as the nematode (*Caenorhabditis elegans*) and the fly (*Drosophila melanogaster*) (Rose, 1991) with minor roles played by a number of other arthropods and nonmammalian vertebrates. Casting such a wide net is certainly appropriate in that the comparative biology of senescence suggests that the aging process is complex, appears to have evolved independently several times, and varies considerably between species (Reznick, 1993, 80; Finch, 1991). Hence, elucidating the underlying mechanisms will require studying a variety of organisms at many levels of analysis, from the molecular to the organismal (Finch, 1990, 1991).

It makes sense, evolutionarily speaking, that male mantids reach adulthood sooner and have shorter adult life spans than females given the facts that high mortality rates create selective pressure for both early maturity and increased reproductive effort early in life and that these two characteristics are correlated with decreased performance later in life and/or a shorter life span (e.g., Finch, 1991; Reznick, 1993). Male mantids in this study had both a higher initial mortality and a higher overall rate of mortality than females. In addition, in the wild, adult male *S. lineola* are probably much more vulnerable to predation than females in that only males can fly, and they may have to do so in order to locate suitable mates. During flight, of course, males are vulnerable to predators to which they would not be if they remained relatively stationary and cryptic like females (e.g., Yager and May, 1990; Yager et al., 1990). In addition, adult male *S. lineola* are vulnerable to being preyed upon by females as they approach them to mate (although this phenomenon has been somewhat oversold [Prete, 1995; Hurd et al., 1994]). Further, as the season progresses, the probability that a female has mated increases, and, because mated females are supposedly less attractive to males (Maxwell, this volume, chap. 5), the probability of finding a mate decreases as time goes on while the probability of falling prey to cannibalism increases as other prey diminishes and gravid females become progressively more hungry (e.g., Lawrence, 1992). All of these factors create selection pressures for early maturity and early reproductive effort in males: The sooner a male is responsive to female pheromones, the more likely it is that he will find a mate and the less likely it is that he will fall prey to the female before he can mate with her.

On the other hand, selective pressures on females are quite different (e.g., Moran and Hurd, 1994), and one would expect some of them to select for longer life span. Whereas the probability

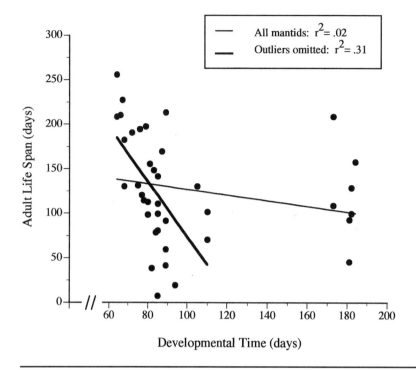

Fig. 14.4 The relationship between adult life span and developmental time for female *S. lineola*. Overall, there is no relationship between the two variables. However, you will remember from Figure 14.1 that some mantids take an inordinately long time to reach the adult stadium, and the life span of those that take the longest to mature has as large a variance as does that of all other females. On the other hand, there is a relationship between life span and developmental time for those females maturing within 120 days: Those that reach adulthood soonest tend to live the longest.

that a male will successfully mate decreases as the season progresses, the probability that a female will lay additional oothecae increases the longer that she can stay alive (Krombholz, 1977; Lawrence, 1992; Matsura and Morooka, 1983).

Unfortunately, mantid senescence has received no attention in any of the literature of which I am aware. It would be of great interest to understand the physiological bases of the sex differences in aging in this species and to know something about the changes that occur in females as they age. It saddens me that I will probably never have the time to pursue these questions. Perhaps a graduate student reading this chapter will.

ACKNOWLEDGMENTS

I would like to thank Dr. David Yager for his comments and suggestions and for bringing several critical sources of information to my attention. The data presented here were collected, with the help of a number of hard-working students, when I was at the University of Chicago, Youngstown State University, and Denison University.

15. Comparative Aspects of Rearing and Breeding Mantids

David D. Yager

Although rearing mantises can be exceptionally rewarding, researchers contemplating the endeavor should go in with their eyes open to the challenges. Even modest rearing projects will require substantial commitments of space, manpower, and imagination. Over the last fifteen years, we have kept approximately seventy species in our colony—some for many generations, others relatively briefly—and have tried to develop efficient rearing practices general enough to suit most species while still allowing us to cater to the special needs of some. We have specifically avoided diapausing species. This chapter briefly outlines the procedures and considerations that we have found important for maintaining a diverse colony of several hundred adult mantids and their offspring.

Mantis Life Histories

Mantids follow the typical hemimetabolous pattern of an egg stage followed by several instars of nymphal development followed by a molt to the adult form. However, the timetable for these events varies considerably. Smaller mantis species spend less time at each life stage and have shorter lives as adults than larger species. For small species, eggs hatch in 3 to 4 weeks instead of 4 to 6 weeks for the largest species; there may be five to six nymphal instars instead of eight to ten; adults live 4 to 8 weeks compared to 4 to 6 months. Life history data such as this must always be viewed skeptically, however, because of the dominant role that environmental conditions can play. Light, temperature, and humidity can all have significant effects on juvenile hormone levels in addition to their direct effects on metabolism. This is especially—but not exclusively—important during nymphal development. For example, under varying conditions, we have reared *Miomantis ehrenbergi* with as few as four nymphal instars and as many as seven. Food supply is also very important, and insufficient food slows development substantially. In short, it is necessary, through careful record keeping, to establish a life history timetable for each species under the particular conditions of each colony facility.

Approximately 30% of mantis species show substantial sexual dimorphism. In many respects, this is like dealing with two separate species. Males may have fewer nymphal instars than females and become adults sooner; they also have substantially shorter adult lives than females. In dimorphic species, males are smaller and more fragile and often have different housing and food requirements.

The Environment

The goal in designing the physical environment of the colony is to mimic as closely as possible conditions natural for the animals while meeting your objectives (number of species, number of adults, etc.) and staying within the realm of practicality.

Light

Accurately reproducing sunlight indoors—especially the UVA (320–400 nm) and UVB (290–320 nm) components—is difficult, but it can be approximated with fluorescent full-spectrum bulbs (not plant lights). Ultraviolet light does not pass through glass and most plastics. Incandescent bulbs are not suitable substitutes. We provide the brightest light possible by having separate fixtures above every shelf, realizing that this is still orders of magnitude dimmer than the sun outdoors.

A consistent daily light cycle is critical for the well being of all animals. Since the vast majority of mantids come from equatorial regions, we have standardized on a 12-hour day length (the range of day lengths at the Tropics of Cancer and Capricorn is 11 to 13 hours). However, optimal breeding by demanding species from higher latitudes may require cycling through days up to 15 hours long (maximum day length at 45° latitude). There is anecdotal evidence that inappropriate day length may increase the number of eggbound females in temperate species. We also simulate dawn and dusk in our colony with two banks of lights on offset timers.

Temperature

Common sense and habitat data should guide decisions on temperature ranges for each mantis species. Some species from hot, arid localities like *Yersiniops solitarium*, *Empusa pennata*, and *Eremiaphila brunneri* do best with daytime temperatures of 35°C or even higher. The majority of species seem to do well with daytime temperatures of 28 to 30°C, and we have successfully kept more temperate species like *Mantis religiosa* and *Tenodera aridifolia sinensis* at 24 to 28°C. Temperature stratification can cause ceiling-to-floor differences in the colony room of greater than 5°C, depending on the heater type and position, so a circulating fan may be advantageous.

A convenient way to house high-temperature species within a room at lower temperature is to create an enclosed microhabitat on a shelf or other area with insulation (flexible, aluminized foam insulation is especially convenient and effective; styrofoam sheeting also works) on five of the six sides. Very fine ("no-see-um"–proof) mosquito netting fastened with Velcro tape can be used over the open side—you can see through the netting, and it substantially reduces heat loss from the enclosure. Heat can be provided by ceramic infrared sources or high-wattage incandescent bulbs connected to a thermostat (a proportional control thermostat is better than the normal type—it offers more precise temperature control and longer life of the heating elements, but it is more expensive). Digital thermometers with remote probes are inexpensive and extremely useful for calibrating the thermostat.

We have the heating systems for the colony on a timing system linked to the day cycle so that there is a nighttime temperature drop of 5 to 7°C. This is especially easy and inexpensive to arrange using the newer digital thermostats.

Humidity

Although important for all life stages, appropriate humidity is one of the most critical factors in rearing nymphs successfully. The optimal range appears to be 50 to 70% relative humidity (RH) for nymphs and somewhat lower for adults. This is complicated, however, by the equally important requirement for adequate ventilation. Nymphs kept in an aquarium with a glass top will have sufficient humidity, but are not likely to prosper because of poor air turnover. Excessive humidity will also cause problems, including increased risk of disease, and should be avoided.

One approach to managing humidity is to keep the entire colony room at 30 to 50% RH using a console humidifier. All animals live in containers with the tops covered only by the coarsest mesh screen consistent with preventing escape of mantis and prey. They are misted once or twice a day to give a period of high microhabitat humidity. Adults are misted less than nymphs, and we don't mist species from arid locales.

Housing

Mantises need plenty of "personal space" to do well—which can create a severe conflict with the colony space limitations most laboratories face. Species with adults over 70 mm long demand too much room to make rearing significant numbers practical in our facility.

Our container armamentarium comprises round, clear plastic cups of three sizes (diameter × height in centimeters: 8 × 8, 12 × 13, and 15 × 18) and glass aquaria (20, 40, and 80 L). Tops for the plastic cups have the largest screen area possible (coarse or fine mosquito netting or window screen); aquarium tops are netting secured with Velcro tape. We usually include a strip of screen or wood spanning the diagonal height of the cup to serve as a perch. Color-coded cafeteria trays greatly simplify managing large numbers of containers.

Fortunately, mantises can be reared communally in aquaria for part of their lives. Immediately after hatching, nymphs are transferred to an aquarium where they stay until about half grown. A complex three-dimensional environment in the aquarium is essential to space out the animals; we use a maze made of window screen strips. An excess of food is also critical—if the nymphs can't find smaller prey, they will start eating one another. Either of two criteria indicate when it is time move the nymphs to individual containers: The number of nymphs declines substantially or some nymphs (usually females) get much larger than the others. If the nymphs remain together too long, there will be an increasing sex ratio bias favoring females. In some species, nymphs cohabit with minimal cannibalism (*Heterochaeta* and *Creobroter* are examples), but other species (like *Sphodromantis* and *Miomantis*) have to be monitored closely. With a large enough space, plenty of environmental complexity, and ample food, some laboratories have even reared *T. a. sinensis* to adulthood communally.

We choose the proper size container for individual nymphs following the rule that container height must be *at minimum* two to three times the body length of the nymph. The extra space is necessary for successful molting. This means that we sometimes keep last-instar nymphs in bigger containers than we use for adults of the same species.

Nutrition
Food

Most mantids are not picky predators and will take any prey that moves (beetles and ants are rarely suitable, however). The staples of mantis diet in the laboratory are flies, crickets, and cockroaches. It is generally practical to raise flies and cockroaches in the lab. Crickets (*Acheta domesticus*) up to 6 to 8 mm can also be reared easily in a minimum of space, but it is much more practical to buy larger crickets from a commercial supplier. A reputable supplier is important, however, since crickets in poor health or nutritional state pose a risk to the mantids. It is best to provide the crickets with high-quality food (commercial dog/cat chow or chicken mash; high-protein fish food is excellent, but expensive) for several days before giving them to the mantises.

A key to successful feeding is choosing the proper size food for the size and strength of the mantis. First-instar nymphs do well on *Drosophila melanogaster* (wild type or vestigial-winged). However, newly hatched nymphs of very small species (some Amelinae, Oligonychinae, Thespinae, and others) cannot handle even these small flies. It may be possible to use springtails or aphids in these cases. We have never been able to keep such small nymphs alive. *D. virilis* is extremely useful for larger nymphs since it is intermediate in size between *D. melanogaster* and houseflies and survives very well in the mantids' containers. *D. hydei sturdevant* has the same advantages and does not fly. Houseflies (*Musca* sp.) are excellent for larger nymphs, small adults, and males of sexually dimorphic species. Blowflies also make very good mantis food and are larger than houseflies, but they are prohibitively inconvenient for many laboratories. Crickets and cockroaches (especially *Blatella germanica* and *Periplaneta americana*) of appropriate sizes can be used at any stage and are necessary for large adults. It is also mandatory to choose prey whose behavior makes them vulnerable to predation. For instance, crickets are unsuitable in aquaria, since they stay at the bottom whereas the mantids perch nearer the top.

Although largely unexplored, there are several anecdotal reports of mantis feeding preferences. *Metallyticus splendens* is said to starve to death in captivity if not offered a diet of cockroaches. *Deroplatys* sp. and *Hymenopus coronatus* will take "normal" lab food, but feed most enthusiastically on Lepidoptera. Honeybees are known to make up a large proportion of the diet of several species that are sit-and-wait predators perching on flowers. It is tempting to speculate that some of the species that have done very poorly in the colony (various Oligonychinae, for instance) succumbed because they were not provided an appropriate diet.

In our routine feeding regime for all species, twice a week each mantis housed individually receives the amount of prey it can consume in one to two days. We leave one to three large, active cultures of fruitflies in aquaria with nymphs at all times. Three common exceptions to this routine are these: nymphs do not eat for 24 to 36 hours before molting and for a variable period afterward; females need extra food after laying an eggcase; and females should be fed heavily for several days before a mating attempt.

Prey can be dangerous to mantis nymphs. Too many flies in a container can fatally disrupt molting attempts by their constant activity. We try to minimize the use of crickets with nymphs—or at least feed only what the nymph will consume in a few hours—since the vulnerable molting mantis all too often becomes cricket prey.

Water

The misting discussed above also provides an opportunity for the mantis to drink water. Males and small nymphs are especially susceptible to desiccation and need this opportunity each day. Dishes of water and soaked cotton are ineffective. We always use distilled or deionized water instead of tap water because of the chlorine and unknown substances the latter can contain.

Breeding

Successful mantis breeding requires as much art as science. In some trying instances, candlelight and soft music make as much sense as any "rational" approach I know of.

It is especially interesting that species show such wide variability in their willingness to mate in the laboratory and in their ability to produce viable, fertilized eggs even if they do. *Sphodromantis, Hierodula, Stagmomantis, Parasphendale,* and *Taumantis,* for instance, mate readily, and the resulting eggcases almost invariably produce one hundred to two hundred hatchlings. *Creobroter,* on the other hand, can be induced to mate, but very often produces infertile eggs. It can be very difficult to entice some *Deroplatys* to mate at all.

This variability can be reduced—but not eliminated—by carefully choosing the animals to pair. Age matters: males are not sexually active for 1 to 2 weeks after their final molt, and females need to be even older. Females should have the enlarged abdomen that signals active egg production. In fact, we tend to select females that have already laid an unfertilized eggcase, a common occurrence in most species. If at all possible, choose animals to minimize inbreeding. Repeatedly pairing offspring from a single original eggcase (unfortunately a very common situation) is inviting reproductive failure between the second and fourth generations as deleterious recessive genes are "uncovered"—though this is certainly not the only cause of reproductive failure.

Much of the strategy for successfully staged matings revolves around protecting the male, particularly in strongly dimorphic species. Females selected for breeding should be fed to satiety every day for several days before being placed with a male. The mating should take place in the largest enclosure available. We use screen cages 30 x 30 x 30 cm for most species. We also use "group mating" with success: three to four females and five to seven males are placed in a 1.3 x 0.7 x 0.7 m screen cage. Food in the mating cage also helps distract the female from the male. While it is not practical to watch the animals continuously, they can be checked frequently and separated as soon as copulation is finished. It is common for the male to ride around on the female's back for extended periods before and/or after copulation. We separate the animals after 24 hours even if no copulation has been observed.

Matings are staged in the colony room so the animals do not have to contend with any environmental disruption. However, for species very reluctant to mate, some workers have had better success placing the cages outdoors in mixed shade and bright sun.

After the mating, the females should be given a suitable place to attach their oothecae. For the majority of species, the walls or top of their container or their perch will do fine. Some species (*Litaneutria* and *Creobroter* are examples) seem more comfortable using the underside of a branch or wood slat. Arid country, cursorial forms like *Yersiniops* may attach the oothecae to the underside of rocks, and a few species (*Ameles,* for example) form the eggcase in the sand or soil (or possibly in leaf litter for some Oligonychinae). If in doubt, provide the females with several options.

Since females often lay unfertilized eggcases, it is impossible to know until long after the fact if a mating attempt has actually been successful. However, eggs laid within two days after a mating attempt usually indicate a completed copulation. Females can store sperm and will continue to produce fertile eggcases for long periods after a single copulation. The percentage of fertilized eggs decreases with successive oothecae, so it is sometimes advantageous to remate females after the second or third laying.

Many of the familiar large species (*Hierodula, Sphodromantis, Tenodera,* etc.) produce large oothecea containing more than one hundred eggs. Smaller species like the Amelinae and some hymenopodids adopt the contrasting strategy of producing a greater number of oothecae, each containing fewer eggs.

If an eggcase cannot be left where laid, it can be removed and glued with a very small drop of silicone sealant to a strip of screen or a stick. Eggs should be kept in the colony room under the same environmental conditions as the adults and misted once a day. They will hatch after 3 to 8 weeks depending on the species (4 to 6 weeks is most common). In many species, the hatchlings all hatch within 1 or 2 hours, but we have had hymenopodid eggcases continue to produce young over a 2-day period.

Miscellaneous Topics
Choosing a Species to Rear

Choice of a species to adopt for research depends on three factors: (1) the requirements of the particular project—for instance, does the mantis need to be large? winged? dimorphic?; (2) its suitability as a lab animal—Is it easy to maintain in the available colony facility? Does it breed prolifically?; and (3) its availability—Is it currently obtainable? Will it be convenient (or possible) to renew or replace the culture at some later date?

Table 15.1 addresses the second factor. Since this table is based on the experience in my laboratory with the approximately seventy species that have come our way—out of some two thousand mantis species extant—and since our successes and failures derive from our own particular facility and techniques, the information in the table can be taken only as a starting point in choosing a species of mantis for laboratory studies.

Causes of Mortality

Three causes account for most of the mortality in our colony. The most prevalent is molting disasters. Common wisdom has it that inadequate humidity promotes molting difficulties. While this is likely true, *anything* that disrupts the normal hormonal patterns of nymphal development can also cause problems. This underscores the need for stable and appropriate lighting, temperature, humidity, and food supply. Cannibalism also claims a number lives, both of nymphs and of males attempting to mate. The third, and fortunately least common, is "unexplained death."

Disease can pose a threat, and some facilities have lost entire cultures to infections (in some cases, possibly introduced by crickets from commercial suppliers). Disease is unusual, however, and reasonable cleanliness, moderate humidity, and quarantine of new arrivals will minimize the risk.

Identifications

Accurate species identification can pose a significant dilemma since there are few published keys and even fewer taxonomists specializing in mantids. The best identifications rely on access to original descriptions in the literature, comparisons with reliably determined museum specimens, and the talents of an experienced taxonomist (even if not a mantis specialist). The museums housing the major mantis collections (in Philadelphia, London, Paris, and Vienna) can provide this service, but it may require several months and some expense. Identification to the

Table 15.1 Suggestions for species to use in research

Recommended[a]	Possible[b]	Not recommended[c]
Hierodula bipapilla	*Blepharopsis medica*	Any desert species
Miomantis spp.[d]	*Creobroter* spp.	Any diapausing species
Parasphendale spp.	*Empusa* spp.	Any Oligonychinae
Sphodromantis viridis	*Hestiasula brunneriana*	*Ameles heldreichi*
Stagmomantis spp.	*Heterochaeta strachani*	*Coptopteryx gayii*
Taumantis ehrmannii (sigiana)	*Hierodula membranacea*	*Deroplatys* spp.
Sphodromantis aurea	*Humbertiella ceylonica*	*Hymenopus coronatus*
Iris oratoria	*Phasmomantis sumichrasti*	*Litaneutria minor*
	Polyspilota aeruginosa	*Mantis religiosa*
	Pseudocreobotra wahlbergi	*Stagmatoptera hyaloptera*
	Rivetina fasciata	*Tenodera aridifolia sinensis*
	Sphodromantis lineola	*Theopropus elegans*

[a] *Recommended* species are hardy, medium-sized, breed well, and don't require special care.

[b] *Possible* indicates species than may be suitable but have significant drawbacks such as their large size, unreliable reproduction, or unusual environmental requirements.

[c] *Not recommended* means that we have had a great deal of difficulty keeping the species in culture.

[d] *Spp.* means that we have reared multiple species in the genus and find them uniform in suitability.

genus rather than species level may be satisfactory for some purposes and is less demanding.

All laboratories publishing on mantids should maintain (or deposit in a museum) properly preserved voucher specimens of their animals.

Collaborations

Proactively distributing eggcases and information to others in the mantis community brings substantial benefits to all involved. The most important benefit is assuring that several labs or hobbyists have the same species so that a flagging or crashed culture in any one laboratory can be renewed or replaced. Collaboration also maintains a constant influx of new species for the whole group. Finally, the flow of information is crucial to overall improvement of mantis rearing practices.

Permits and Responsibilities

Most national and regional governments have and enforce regulations on the importation of insects. Even though mantises are not considered pests, they do require permits. Some agencies may allow "blanket permits" so that mantises can be imported without repeating the paperwork for every species.

Maintaining a colony of foreign insects carries with it the responsibility to assure that they never have the opportunity to become established in the wild. In a practical sense, this means taking adequate precautions to prevent escape and never providing insects to any individual or organization unable or unwilling to take the same precautions. Transfer of exotic insects within a country may also be regulated by the government and require permits.

Resources
Rearing Equipment, Supplies, and Information

Supplies and information useful for the rearing of mantids are available from a variety of sources.

- The herpetology pet trade is a rich resource for ideas and technology applicable to mantis rearing. Several periodicals (for instance, *Reptiles* and

The Vivarium) regularly list sources for lighting, temperature-control devices, and feeder-insect cultures.

- The food service industry is an excellent source for rearing containers and trays.

- Camping supply stores often carry both regular and very fine ("no-see-um"–proof) mosquito netting.

- Several on-line services provide useful discussions. Especially recommended are: entomo-L (listserve), sci.bio.entomology.misc (newsgroup), and the reptile and amphibian bulletin boards of America Online (under Pets) where there are detailed discussions on rearing prey species. There are also a growing number of very useful and entertaining web sites featuring mantids.

Museums Housing Major Mantis Collections

The following museums house major mantis collections.

- British Museum of Natural History, Department of Entomology, Cromwell Road, London SW7 5BD, England. Contact person: Mrs. Judith Marshall (jam@nhm.ac.uk)

- Museum Natl. D'Histoire Naturelle, Entomologie, 45 rue Buffon, Paris 75005, France. Contact person: Dr. Roger Roy

- The Academy of Natural Sciences, Department of Entomology, 19th St. at The Parkway, Philadelphia, PA, USA 19103. Contact person: Dr. Donald Azuma (azuma@say.acnatsci.org)

- Naturhistorisches Museum Wein, Zweite Zoologische Abteilung (Insekten), Postfach 417, Burgring 7, A-1014 Vienna, Austria. Contact person: Dr. Maximilian Fischer. Though retired, Dr Alfred Kaltenbach is still affiliated with the museum and is an expert on mantids.

ACKNOWLEDGMENTS

The rearing techniques discussed here have evolved through many discussions with colleagues, including Syril Blondheim, Jon Copeland, Frank Elia, Tom Mason, Randy Morgan, and Leslie Saul. Stephan Kallas, Volker Schwenk, and the members of my lab made many valuable suggestions that improved the manuscript. Special thanks are due Stephan Kallas and his colleagues in the German entomological hobbyist community for sharing their valuable insights on mantis rearing and for bailing my colony out of trouble on many occasions.

16. Mantids in Ecological Research

L. E. Hurd

For those scientists primarily interested in praying mantids, the scope of research is determined by the specific biology of their favorite beasts. Here, the process of investigation follows a very focused path of discovery into the autecology, physiology, or behavior of one or more species. Mantid specialists forge their own techniques, peculiar to their model species, and it would be presumptuous of me to offer a cookbook of recipes for handling mantids in the lab or observing them in the field. Rather, I want to discuss features of some mantids that make them suitable tools with which to probe experimentally questions of broad ecological interest: population regulation, predator-prey relationships, and the community niche of generalist predators.

Experimental ecology requires that subject species be amenable to manipulation and that their populations can be monitored easily in the field. Since field ecosystems are necessarily complex, it is also highly desirable that experimental animals can be kept in the laboratory in order to test variables one at a time. I can think of no terrestrial predator more ideally suited to experimental ecology than a praying mantis, which is why I chose to work with them twenty years ago. My discussion pertains to those temperate zone, old-field species which I have been using to date: *Tenodera sinensis*, *T. angustipennis*, and *Mantis religiosa*. I have begun to work with *Stagmomantis carolina* and *Bruneria borealis*, which show similar promise as experimental tools, and I expect that many other species can be used to gain insight into ecological processes (e.g., *Iris oratoria* and *Stagmomantis limbata*, Maxwell, 1995). Since most species in this taxon are tropical, I hope to see population- and community-level studies of such species in the near future.

Population density, the currency of demography, can be manipulated in the field. Emergence of nymphs can be predicted from the mass of oothecae (Eisenberg and Hurd, 1977), which makes it easy to introduce known densities into replicated field plots. For relatively large-scale studies such as Hurd and Eisenberg (1984a), oothecae can be tied to wooden dowels and labeled as to mass, and the dowels stuck into the soil according to any desired spacing pattern. Placement of ootheca in the field will normally take place early in the spring, before normal hatch time, which in early successional old fields is a time when vegetation cover is minimal. This turns out to be of signal importance with respect to experimental design. The first time I tried this method, I got a lesson in the highly developed pattern-search strategy of crows. I witnessed a flock of these birds systematically destroying oothecae in plot after plot as I stood waving my arms in the chill March air. This failure prompted me to cover the doweled egg masses with paper cups. Bird predation was thus avoided, but spiders tended to colonize the cups. Thus, I unwittingly instigated a spider-feeding experiment without appropriate controls. The best solution turned out to be small cages of half-inch hardware mesh placed over groups of oothecae, which discouraged birds while not inviting spider webs.

For field studies on a smaller scale (e.g., Fagan and Hurd, 1994), a more exact method is to introduce counted first instars as soon as eggs hatch. The logistical problem here is that oothecae do not all hatch at the same time. Therefore, since an experiment usually entails simultaneous release of hundreds of nymphs, it is necessary to collect a large number of oothecae to ensure that enough will hatch on a given day to produce a suitable cohort size.

Not only is the packaging of mantid eggs into oothecae handy from the standpoint of manipulating numbers of hatchlings, it also makes possible accurate estimates of egg production at the end of the season. The oothecae of *T. sinensis* are large and relatively conspicuous in the field and can be collected, weighed, and even returned to the field if the investigator wishes to follow the population for another season. Thus, both spring recruitment and autumn fecundity can be estimated with reasonable accuracy in a field population, even if one never sees a mantid. For some other species, establishing fecundity is more problematic. The oothecae of *T. angustipennis* are much smaller and more cryptic than those of its congener. Eggs of *Mantis religiosa* often are deposited under low grasses (even under rocks), which makes them harder to find.

Another useful feature of oothecae is that timing of egg hatch can be manipulated by refrigeration and incubation. This allows one to assess the role of egg hatch phenology on guild interactions and community dynamics. It is possible to change the normal order of appearance of two species in the field (Snyder and Hurd, 1995) or to play with the impact of egg phenology on interspecific predation and cannibalism in the laboratory (Hurd, 1988). After hatching, rate of development of nymphs in the laboratory can be manipulated by varying food level (Hurd and Rathet, 1986; Liske, this volume, chap. 11) or ambient temperature (Hurd et al., 1989; Yager, this volume, chap. 15). It is therefore possible, by manipulating egg phenology and/or growth rate, to place into the field virtually any stage of mantid at any time or, for that matter, all stages at once.

For many animals, it can be difficult to keep track of population density following egg hatch. Either the terrain is too difficult to sample in a representative fashion (e.g., stratified forest canopies) or juveniles disperse to new habitats, remote from the site of hatch (e.g., ballooning of young spiders). Old-field mantid species, in contrast, have poor interhabitat dispersal abilities (Eisenberg et al., 1992). Further, emigration from experimental field plots can be quantitatively monitored by using sticky trap plot barriers (e.g., Hurd and Eisenberg, 1984a; Fagan and Hurd, 1994). These same barriers can be used to assess cursorial spider movement (Moran and Hurd, 1994b). Although the barriers catch many kinds of arthropods that attempt to cross them, their effects on population levels of these arthropods within the plots have been negligible in our experiments judging by comparisons between plots with and without barriers. Nonetheless, it is wise to incorporate a set of barrier-free control plots into such experiments to avoid criticism (e.g., by reviewers of manuscripts) on this account.

Sampling of early stadia mantid nymphs within experimental plots can be accomplished by D-Vac suction sampling along transects, which method also serves to sample the above-ground arthropod assemblage (Hurd and Wolf, 1974; Hurd and Eisenberg, 1984a, b). For later stages (fifth to seventh instars), individuals can be picked up by hand in the wake of the D-Vac, disturbance of the vegetation causing them to move conspicuously.

Though open-field studies are preferred for most applications, sometimes it is important to limit movement of mantids and/or other arthropods during an experiment while retaining ambient climate, vegetation structure, etc. Nylon mesh cages placed over vegetation in the field can be used to manipulate densities of mantids and spiders for short-term experiments (Hurd and Eisenberg, 1990a). The top side of each cage is zippered on three edges, so that sampling can be done by D-Vac, by hand, or using a combination of techniques. This permits exhaustive sampling of arthropods within the enclosures over a period of days without the confounding influence of emigration or immigration.

Manipulation of predator density must be done with an eye toward the question being asked. As Price (1987) pointed out, elevating predators above background (control) level does not by itself tell us how important they are to a community at normal densities. Predator removal is a better technique for this purpose—the difference between experimental plots in which predators have been eliminated and those in which predator density is unmanipulated is a measure of their natural effect on a community. However, there are two instances where predator elevation is both valid and necessary. First, if one wishes to find out what will happen to a naive community when a predator invades and colonizes it, the appropriate treatment is to add predators to portions of that community, at a density which naturally occurs in similar areas where the predator already exists (e.g., Hurd and Eisenberg, 1984a, b; Fagan and Hurd, 1994; Moran and Hurd, 1994; Moran et al., 1996). This is not *predator augmentation* in the sense of Price (1987). The second case is when one wishes to find the carrying capacity (K in the Verhulst logistic model of population growth) of a predator, that is, whether the established density in a community is a saturation level. If elevated predator density does not damp to control level (e.g., Hurd and Eisenberg, 1984a; Fagan and Hurd, 1994), then control does not represent a saturation level for this species.

In addition to ready manipulation of all life history stages in the field, many aspects of mantid ecology can be investigated in the laboratory. This permits testing variables one at a time to compare with field results, for example, the separate effects of food limitation and density on survivorship, growth, and cannibalism (Hurd and Eisenberg, 1984a; Hurd, 1989), or predatory relationships of different-sized mantids and cursorial spiders in pairwise tests (Moran and Hurd, 1994b). Laboratory studies also can be used to assess the relationship of behavior to components of fitness (Liske and Davis, 1987; Hurd and Rathet, 1986; Hurd, 1991).

As with any organism, laboratory habitats are not identical to field habitats, and some things observed indoors may not accurately reflect in quality or quantity what routinely happens in the field. For instance, cannibalism among adult female *T. sinensis* is not uncommon when they are housed in the same cage, but very rarely occurs in nature. The laboratory is best viewed as a theater of the possible, not necessarily of the probable.

Another important consideration is the ease with which experimental animals acclimate to artificial conditions. Further, if interactions among different species are the focus of study, are their environmental needs both met in the same arena? *Tenodera sinensis* acclimates readily to many kinds of cages and climbs even glass surfaces well, so that the small vials used by Hurd and Rathet (1986) are quite adequate for early instars. However, *M. religiosa* is a poorer climber and prefers to perch on the underside of low-lying vegetation. Therefore, when studying interactions between these two species, two kinds of perches must be provided in the same cage. In fact, both species can be kept in numbers, in the same aquarium, if vertical wooden dowels are inserted into a low platform made from half-inch hardware cloth fitted to the bottom of the tank. *Mantis religiosa* nymphs tend to hang upside down under the mesh at the bottom, whereas *T. sinensis* (or *T. angustipennis*) climb up and down the dowels or hang upside down on the screen cover of the tank. Even though both species can negotiate the dowels to climb to the top or to the bottom of the tank, if ample food is provided they seldom interact. The point to be made here is, perhaps, an obvious one: In order to provide suitable housing in the laboratory, one must pay attention to the habits of different mantid species in the field.

It may turn out to be an axiom that those species that are most widespread and abundant are, by virtue of their tolerance to the variety of environments occurring over a wide geographic range, the best experimental animals for both lab and field studies. However, I hope ecologists will not confine their studies to such species. After all, the question of why some species are common and robust can also be approached by asking why others are rare and delicate. Most mantids are tropical, and of limited distribution, so scientists who ask questions about distribution and abundance cannot afford to confine their studies to temperate species that are more readily accessible.

17. Histological Techniques for Mantid Research

David D. Yager

The scientific literature contains little information dealing with histological procedures specific to mantids. A rational general strategy for laboratories doing histology on mantids is to take as a starting point formulations that work with cockroaches or even more distantly related orthopteroids (see, for instance, Guthrie and Tyndall, 1968; Barbosa, 1974; Blagburn and Sattelle, 1990). Many less demanding procedures (general fixation, simple monochrome stains, etc.) will carry over to mantids well, while more critical applications (fixation for electron microscopy and trichrome staining are examples) may have to be tailored not only for mantids, but for individual species. Presented here are general guidelines and a few specific techniques—with an emphasis on neuroanatomy—that have proved their value in my laboratory with a range of commonly encountered genera such as *Tenodera*, *Hierodula*, *Sphodromantis*, and *Mantis*.

Note: Some of the chemicals required in the procedures given below pose safety/health risks and should not be used without appropriate precautions. Refer to the Material Safety Data Sheet provided with each chemical and to a standard histology text like Presnell and Schreibman (1997).

Physiological Salines

All surgeries require a bathing solution for the tissues that maintains their anatomical and physiological integrity. The ideal saline would duplicate the role of hemolymph in providing proper ionic environment, pH, nutrients, osmolarity, lipids, and proteins and at the same time be easy to prepare, inexpensive, and slow to deteriorate. Fortunately, mantis tissue tolerates divergence from this unattainable ideal reasonably well. Mantis blood has received little study (Meyer, 1936), so both recipes below derive from cockroach research (for a review of cockroach blood composition, see Bell and Adiyodi, 1981).

The simplest salines attempt only to mimic the osmolarity, pH, and ionic content of extracellular fluid and, thus, are be best suited to relatively brief (1- to 2-hour) procedures. Guthrie and Tyndall (1968) summarize a range of traditional cockroach salines; the recipe from Brecht et al. (1960) is representative.

	mM	g/l
NaCl	159	9.32
KCl	10	0.77
$CaCl_2 \cdot 2H_2O$	4.5	0.66
$NaH_2PO_4 \cdot H_2O$	0.1	0.01
pH = 7.2		

More complex saline solutions additionally provide nutrients, generally in the form of trehalose and/or glucose. They are suitable both for general use and for longer tissue immersion. Because they contain sugar, these salines readily support bacterial and fungal growth—they should be made in small quantities and must be refrigerated. The recipe below is based on a general orthopteroid saline (Strausfeld et al., 1983), but we have modified it to match more closely the composition of cockroach—and presumably mantis—hemolymph (Pichon, 1970; Bell and Adiyodi, 1981). TES (N- tris[hydroxymethyl]-methyl-2-aminoethane-sulfonic acid) is a buffer. The sucrose is not an energy source, but is included to raise the osmolarity to 409 mOs.

	mM	g/l
NaCl	155	9.06
KCl	12	0.89
$CaCl_2 \cdot 2H_2O$	4	0.59
$MgCl_2 \cdot 6H_2O$	5	1.02
$NaHCO_3$	4	0.34
TES	5	1.15
sucrose	15	6.49
trehalose	20	7.57
pH = 7.0–7.2		

Fixatives

Proper fixation is the most critical requirement for adequate tissue preservation in any type of histology. Choice of fixative is determined by the quality of tissue preservation required (for light versus electron microscopy, for instance) and by the requirements of the staining or histochemical procedure to follow. Presnell and Schreibman (1997), Kiernan (1990), and other manuals of histological techniques outline strategies for selecting an appropriate fixative and give many examples.

General-Purpose Fixatives

Listed below are recipes (Presnell and Schreibman, 1997; Kiernan, 1990) for three fixatives broadly useful for light microscopy. Except in a few special cases, we have avoided fixatives containing mercuric chloride since they require extra processing steps.

Carnoy's fixative penetrates tissue very rapidly—30 to 45 minutes generally proves sufficient for a 1 mm³ block of tissue. Its drawback is significant shrinkage and tissue hardening; we use it primarily for tissue in whole mount. For best results, make the fixative up fresh each week.

Carnoy's fixative

glacial acetic acid	10.0 ml
absolute ethyl alcohol	60.0 ml
chloroform	30.0 ml

Alcoholic Bouin's fixative is slower acting than Carnoy's—8- to 24-hour fixation times—but produces less shrinkage and hardening. This is an excellent general-purpose histological fixative for light microscopy. It keeps in the refrigerator for at least two weeks. If necessary, any residual picric acid can be removed from the tissue prior to staining by treating the sections for 5 minutes with 70% ethanol to which a few drops of a saturated solution of lithium carbonate have been added.

Alcoholic Bouin's fixative

glacial acetic acid	15.0 ml
formalin (37% solution)	60.0 ml
80% ethyl alcohol	150.0 ml
picric acid	1.0 g

Aldehyde fixation yields excellent tissue preservation with fixation times of 12 to 48 hours depending on the size of the tissue block (the tissue can be left in the fixative indefinitely). A simple 10% buffered formalin solution works well.

Lillie's buffered formalin

10% formalin	1000 ml
$NaH_2PO_4 \cdot H_2O$	4.0 g
Na_2HPO_4	6.5 g

Fixative for Transmission Electron Microscopy

Proper fixation is even more critical for transmission electron microscopy (TEM) than for light microscopy. The most common strategy is to use a carefully buffered mixed aldehyde fixative that is slightly hypertonic to the tissue. We have used the fixative below with good results in examining *Sphodromantis lineola* central nervous system. The glutaraldehyde must be EM-grade and absolutely fresh. The paraformaldehyde is dissolved in 60°C buffer; the other ingredients are added after the solution cools. The osmolarity is approximately 520 mOs.

TEM fixative

paraformaldehyde	1 g
25% glutaraldehyde	10 ml
0.15 M Sörenson's buffer (pH 7.17)	90 ml
sucrose	12.0 g

Embedding and Sectioning

The overriding challenge in histology on insect tissue is dealing with the tough, sclerotized cuticle. The matrix (plastic or paraffin) used to support the tissue for cutting does not penetrate and hold the cuticle well. During sectioning, the mixed hard and soft tissue can cause extensive tears and distortion so that serial reconstruction of internal structures becomes virtually impossible. If cuticle must be included in the block to be sectioned, the strategy that best preserves the tissue architecture is to embed the tissue in one of the hard epoxy plastics such as Epon. Relatively thin sections (0.5–2 μm) are then cut with a glass or diamond knife. Somewhat thicker sections are possible using softer formulations of plastics such as Spurr's embedding medium. Many, but not all, staining procedures work with these plastics.

Our studies of the mantis ear require a trichrome stain not compatible with plastics and relatively thick sections (5–15 μm) of tissue blocks containing cuticle, so we have developed a different strategy that should be generally useful in mantis laboratories (see also DeGiusti and Ezman, 1955; Barrós-Pita, 1971). First, we use only very freshly molted animals (<1 hour after the molt) so that cuticle is relatively soft. Second, we use a dehydration/embedding protocol (given below) modified from Gatenby and Beams (1950) that substitutes butanol for the usual ethanol. Tissue processed in this way can be embedded in paraffin and cut on a steel knife. Tearing and distortion are greatly reduced, and we can do serial reconstructions in more than half of these preparations. The times given are for a 10 mm^3 block of mixed tissue and should be adjusted for other size blocks. Tissue can be held in pure n-butanol for at least 24 hours without hardening. The optional step colors the tissue for better visibility during embedding. Infiltration under vacuum is essential, and a period of vacuum for all steps is recommended (Table 17.1).

Stains

We have successfully used a broad range of standard histological stains on mantis tissue, and in most cases only minor modifications of the protocols were required. Trichrome stains are among the most useful for seeing tissue relationships. However, they also tend to be the least consistent in performance across species, and a period of experimentation with staining times is generally necessary. Given below are protocols for two stains we use frequently with a broad range of mantis species. Luxol fast blue/cresyl violet is a central nervous system stain that nicely differentiates fiber tracts and cell bodies. The variant of Masson's trichrome (modified from Pantin, 1946) emphasizes color contrast between nerve and other tissues and stains scolopales and nucleoli bright, cherry red.

Luxol Fast Blue/Cresyl Violet Stain
Solutions

LUXOL FAST BLUE

luxol fast blue (= solvent blue #38)	0.1 g
10% acetic acid	0.5 ml
95% ethanol	100 ml

CRESYL VIOLET

cresyl violet	0.5 g
acetate buffer	100 ml

ACETATE BUFFER

sodium acetate	2.7 g
glacial acetic acid	4.8 ml
distilled water	495 ml

Procedure for 10 μm Sections

1. Remove paraffin and rehydrate to 95% ethanol.
2. Stain in luxol fast blue for 8 minutes.

3. Dip briefly in 70% ethanol three times.
4. Stain in cresyl violet for 2 minutes.
5. Leave in acetate buffer for 3 minutes.
6. Destain in 70% ethanol for a maximum of 90 seconds—adjust time for optimal contrast.
7. Dehydrate, clear in xylene, and mount.

Masson Trichrome Stain (Baker modification)
Solutions

1. Hansen's iron trioxyhaematin

Part A

ammonium ferric sulfate	10 g
ammonium sulfate	1.4 g
distilled water	150 ml

Part B

hematoxylin	1.6 g
distilled water	75 ml

Mixing procedure:

a. Make Part A and Part B separately by dissolving the ingredients with gentle heating.
b. Add Part A to Part B (not vice versa) with constant stirring. Heat slowly without stirring just to the boiling point. Cool solution very rapidly in ice. Filter cooled solution into a well-stoppered glass bottle. Solution can be reused. Keeps for about six months.

2. Xylidine de ponceau

Xylidine de ponceau (= ponceau 2R)	0.25 g
glacial acetic acid	1.0 ml
distilled water	100 ml

3. Phosphomolybdic acid

phosphomolybdic acid	1.0 g
distilled water	100 ml

4. Light green

light green SF	2.0 g
glacial acetic acid	2.0 ml
distilled water	100 ml

Procedure for 10 μm Sections

1. Remove paraffin and rehydrate to distilled water.
2. Stain in Hansen's iron trioxyhaematin for 3 minutes.

Table 17.1 Dehydration/embedding protocol

	Solution	Ratio	Time
Fixation	Alcoholic Bouin's	—	12–24 h
Dehydration	50% Ethanol	—	1 h
	70% Ethanol: *n*-butanol	4:1	2 h
	80% Ethanol: *n*-butanol	3:1	2 h
(Optional step)	(0.5% Eosin-Y in 90% ethanol: *n*-butanol)	1:1.2	1 h
	90% Ethanol: *n*-butanol	1:1.2	>5 h
	n-Butanol	—	>4 h
Infiltration	*n*-Butanol: paraffin	1:1	12 h
	n-Butanol: paraffin	1:2	12 h
	Paraffin	—	2 h
	Paraffin	—	2 h
Embed			
Stain			

Source: Modified from Godenby and Beams (1950).

3. Wash for 15 minutes in gently running tap water.
4. Stain in xylidine de ponceau until very dark (1 minute to >5 minutes).
5. Rinse in distilled water.
6. Differentiate in phosphomolybdic acid for 4 minutes (note that this solution must be made up fresh each day).
7. Rinse in distilled water.
8. Stain with light green for 2 to 6 minutes (monitor at regular intervals in distilled water)
9. Dehydrate, clear in xylene, and mount.

ACKNOWLEDGMENTS

Marie Read and Alex Pettyjohn brought skill and creativity to the development of many of the techniques described here.

References

Abrams, P. 1987. On classifying interactions between populations. *Oecologia* 73: 272–81.

Adam, J. P., and J. Lepointe. 1948. Recherches sur la morphologie des sternites et des pleurites des mantes. *Bull. Mus. Hist. Nat. Paris* (2d series). 20: 169–73.

Addicott, J. F. 1974. Predation and prey community structure: An experimental study of the effect of mosquito larvae on the protozoan communities of pitcher plants. *Ecology* 55: 475–92.

Agricola, H., W. Hertel, and H. Penzlin. 1988. Octopamine—Neurotransmitter, neuromodulator, neurohormone. *Zool. Jb. Physiol.* 92: 1–45.

Ahmad, M., S. N. Vijayachandran, and J. C. Basu Choudhuri. 1985. Biology of *Hestiasula brunneriana* Saussure (Dictyoptera: Mantidae). *Ind. Forest.* 111: 333–38.

Alcock, J. 1994. Postinsemination associations between males and females in insects: The mate–guarding hypothesis. *Ann. Rev. Ent.* 39: 1–21.

Aldrovandi, Ulisse. 1602. *De animalibus insectis libri VII.* Bologne.

Allee, W. C., A. E. Emerson, O. Park, T. Park, and K. P. Schmidt. 1949. *Principles of animal ecology.* Philadelphia: Saunders.

Altegrim, O. 1989. Exclusion of birds from bilberry stands: Impact on insect larval density and damage to bilberry. *Oecologia* 79: 136–39.

Altman, J. S. 1975. Changes in the flight motor pattern during the development of the Australian plague locust, *Chortiocetes termifera. J. Comp. Physiol.* 97: 127–42.

Altman, J. S., and N. M. Tyrer. 1980. Filling selected neurons with cobalt through cut axons. In *Neuroanatomical techniques, insect nervous systems,* ed. N. J. Strausfeld and T. A. Miller, 373–403. Berlin: Springer-Verlag.

Anderson, F. J. 1974. Responses to starvation in the spiders *Lycosa lenta* (Hentz) and *Flistata hibernalis* (Hentz). *Ecology* 55: 576–86.

Andersson, M. 1994. *Sexual selection.* Princeton, N. J.: Princeton University Press.

Andrade, M. C. B. 1996. Sexual selection for male sacrifice in the Australian redback spider. *Science* 271: 70–72.

Andrewartha, H. G., and L. C. Birch. 1954. *The distribution and abundance of animals.* Chicago: University of Chicago Press.

Anholt, B. R. 1990. An experimental separation of interference and exploitative competition in a larval damselfly. *Ecology* 71: 1483–93.

Anholt, B. R., and E. E. Werner. 1995. Interaction between food availability and predation mortality mediated by adaptive behavior. *Ecology* 76: 2230–34.

Annandale, N. 1900. Observations on the habits and natural surroundings of insects made during the "Skeat Expedition" to the Malay Peninsula, 1899–1900. *Proc. Zool. Soc. London,* 837–69.

Anonymous. 1841. The mantis. In *The penny magazine of the society for the diffusion of useful knowledge,* 435–36. London: Charles Knight and Co.

Anonymous. 1890. Notes from the zoo—the praying mantis. *Sat. Rev.* 69: 735–36.

Aoki B., and S. Takeishi. 1927. On the copulatory behavior of the Japanese mantis (*Tenodera aridifolia,* Stoll.), and the seat of its nerve center. *Dobutsugaku Zasshi* 39: 114–29.

Arkin, R. 1991. *Biology of aging: Observations and principles.* Englewood Cliffs, New Jersey: Prentice Hall.

Arnqvist, G., and S. Henriksson. 1997. Sexual cannibalism in the fishing spider and a model for the evolution of sexual cannibalism based on genetic constraints. *Evol. Ecol.* 11:225–73.

Audinet–Serville, J.-G. 1831. Revue méthodique des insectes des l'ordre des Orthoptères. *Ann. Sci. Nat.* 22: 18–29.

Audinet-Scrville, J.-G. 1839. *Histoire naturelle des insectes Orthoptères.* Paris: Librairie de Roret.

Baars, M. A., and Th. S. Van Dijk. 1984. Population dynamics of two carabid beetles at a Dutch heathland. II. Egg production and survival in relation to density. *J. Anim. Ecol.* 53: 389–400.

Bacon, J. 1980. An homologous interneuron in a locust, a cricket and a mantid. In *Verhandlungen der Deutschen Zoölogischen Gesellschaft*, 300. Stuttgart: Gustav Fischer Verlag.

Bailey, W. J. 1991. *Acoustic behavior of insects.* London: Chapman and Hall.

Baker, P. S., and R. J. Cooter. 1979. The natural flight of the migratory locust, *Locusta migratoria* L. 1. Wing movements. *J. Comp. Physiol.* 131: 79–87.

Baker, R. L. 1982. Effects of food abundance on growth, survival, and use of space by nymphs of *Coenagrion resolutum* (Zygoptera). *Oikos* 38: 47–51.

Balderrama, N., and H. Maldonado. 1971. Habituation of the deimatic response in the mantid (*Stagmatoptera biocellata*). *J. Comp. Physiol. Psychol.* 75: 98–106.

Balderrama, N., and H. Maldonado. 1973. Ontogeny of the behaviour in the praying mantis. *J. Insect Physiol.* 19: 319–36.

Baldi, E. 1922. Studi sulla fisiologia del sistema nervoso negli insetti. *J. exp. Zool.* 36: 211–288.

Baldus, K. 1926. Experimentelle Untersuchungen über die Entfernungslokalisation der Libellen (*Aeshna cyanea*). *Z. vergl. Physiol.* 3: 475–505.

Barlow, H. B. 1953. Summation and inhibition in the frog's retina. *J. Physiol. (London)* 119: 69–88.

Barnor, J. L. 1972. Studies on colour dimorphism in praying mantids. Master's thesis, University of Ghana.

Barrow, J. 1805. *Travels in China.* Philadelphia: W. F. McLaughlin.

Barrows, E. M. 1984. Perch sites and food of adult Chinese mantids (Dictyoptera: Mantidae). *Proc. Entomol. Soc. Wash.* 86: 898–901.

Bartley, J. A. 1982. Movement patterns in adult male and female mantids *Tenodera aridifolia sinensis* Saussure (Orthoptera: Mantodea). *Environ. Ent.* 11: 1108–11.

Bartley, J. A. 1983. Prey selection and capture by the Chinese mantid (*Tenodera sinensis* Saussure). Ph.D. diss., University of Delaware.

Bateman, A. J. 1948. Intra–sexual selection in *Drosophila*. *Heredity* 2: 349–68.

Bates, H. W. 1862. Contributions to an insect fauna of the Amazon Valley. Lepidoptera: Heliconidae. *Trans. Linn. Soc. London* 23: 495–566.

Bateson, H. 1913. *Insects: their life-histories and habits.* London: T. C. & E. C. Jack.

Beddard, F. E. 1892. *Animal Coloration.* London: Swan Sonnenschein.

Beebe, W., J. Crane, and S. Hughes–Schrader. 1952. An annotated list of the mantids (Orthoptera, Mantoidea) of Trinidad. *Zoologica, N. Y.* 37: 245–58.

Beier, M. 1934a. Mantodea, fam. Mantidae, subfam. Hymenopodinae. In *Genera insectorum, fasc. 196e*, ed. P. Wytsman, 1–40. Brussels: Louis Desmet–Verteneuil.

Beier, M. 1934b. Mantodea, fam. Mantidae, subfam. Sibyllinae und Empusinae. In *Genera insectorum, fasc. 197e*, ed. P. Wytsman, 1–10. Brussels: Louis Desmet–Verteneuil.

Beier, M. 1934c. Mantodea, fam. Mantidae, subfam. Toxoderinae. In *Genera insectorum, fasc. 198e*, ed. P. Wytsman, 1–10. Brussels: Louis Desmet–Verteneuil.

Beier, M. 1935a. Mantodea, fam. Mantidae, subfam. Thespinae. In *Genera insectorum, fasc. 200e*, ed. P. Wytsman, 1–34. Brussels: Louis Desmet–Verteneuil.

Beier, M. 1935b. Mantodea, fam. Mantidae, subfam. Orthoderinae–Choeradodinae–Deroplatynae. In *Genera insectorum, fasc. 201e*, ed. P. Wytsman, 1–10. Brussels: Louis Desmet–Verteneuil.

Beier, M. 1935c. Mantodea, fam. Mantidae, subfam. Mantinae. In *Genera insectorum, fasc. 203e*, ed. P. Wytsman, 1–150. Brussels: Louis Desmet-Verteneuil.

Beier, M. 1964. Blattopteroidea, Mantodea. In *Klassen und Ordnungen des Tierreichs.* Fünfter Band: Arthropoda. III. Abteilung: Insecta, 6 Buch, 849–970. Leipzig: Geest & Portig.

Beier, M. 1968. Mantodea (Fangheuschrecken). In *Handbuch der Zoologie, IV.* Band: Arthropoda, 2 Hälfte: Insecta, Zweite Auflage, 1–47. Berlin: Walter de Gruyter.

Beier, M. 1973. The early naturalists and anatomists during the Renaissance and seventeenth century. *Ann. Rev. Entomol.* 18: 81–93.

Beier, M., and F. Heikertinger. 1952. *Fangheuschrecken. Neue Brehm Bücherei 64.* Leipzig: Academische Verlagsgesellschaft.

Beljajeff, M. M. 1927. Ein Experiment über die Bedeutung der Schutzfärbung. *Biol. Zbl.* 47: 107–13.

Bellamy, E. W. 1890. A queer pet. *Pop. Sci. Month.* 37: 528–33.

Bender, E. A., T. J. Case, and M. E. Gilpin. 1984. Perturbation experiments in community ecology, theory and practice. *Ecology* 65: 1–13.

Benke, A. C. 1976. Dragonfly production and prey turnover. *Ecology* 57: 915–27.

Bennett, A. T. D., and I. C. Cuthill. 1994. Ultraviolet vision in birds: What is its function? *Vision Res.* 34: 1471–78.

Bentley, D. R. 1973. Postembryonic development of insect motor systems. In *Developmental neurobiology of arthropods*, ed. D. Young, 147–77. London: Cambridge University Press.

Bentley, D. R., and R. R. Hoy. 1970. Postembryonic development of adult motor pattern in crickets: A neuronal analysis. *Science* 170: 1409–11.

Berger, F. 1985. Morphologie und Physiologie einiger visueller Interneurone in den optischen Ganglien der Gottesanbeterin, Mantis religiosa. Ph.D. diss., University of Düsseldorf, Germany.

Bernays, E. A. 1972. The muscles of newly hatched *Schistocerca gregaria* larvae and their possible functions in hatching, digging and ecdysial movements (Insecta: Acrididae). *J. Zool.* 166, 141–58.

Bernays, E. A. 1985. Regulation of feeding behavior. In *Comprehensive insect physiology, biochemistry and pharmacology*, vol 4., ed. G. A. Kerkut, G. A., and C. I. Gilbert, 1–32. Oxford: Pergamon Press.

Bernays, E. A., and R. F. Chapman. 1974. The regulation of food intake by acridids. In *Experimental analysis of insect behaviour*, ed. L. Barton-Browne, 48–59. Berlin: Springer Verlag.

Bernays, E. A., and S. J. Simpson. 1982. Control of food intake. In *Advances in insect physiology, 16*, ed. M. J. Berridge, J. E. Treherne, and V. B. Wigglesworth, 59–117. New York: Academic Press.

Bernays, E., and M. Graham. 1988. On the evolution of host specificity in phytophagous arthropods. *Ecology* 69: 886–92.

Bertness, M. D., and R. Callaway. 1994. Positive interactions in communities. *Trends Ecol. Evol.* 9: 191–93.

Bethe, A. 1898. Nervous system of arthropods. *J. Comp. Neur.* 8: 232–38.

Betts, C. R. 1986. Functioning of the wings and axillary sclerites of Heteroptera during flight. *J. Zool. London* 1: 283–301.

Betts, C. R., and R. J. Wootton. 1988. Wing shape and flight behaviour in butterflies (Lepidoptera: Papilionoidea and Hesperioidea): A preliminary analysis. *J. Exp. Biol.* 138: 271–88.

Birkhead, T. R., K. E. Lee, and P. Young. 1988. Sexual cannibalism in the praying mantis *Hierodula membranacea*. *Behav.* 106: 112–18.

Blaney, W. M., and A. M. Duckett. 1975. The significance of palpation by the maxillary palps of *Locusta Migratoria* (L.): An electrophysiological and behavioral study. *J. Exp. Biol.* 63: 701–12.

Birkhead, T., and S. Lawrence. 1988. Life and loves of a sexual cannibal. *New Sci.* 118: 63–66.

Blatchley, W. S. 1896. Miscellaneous notes. *Can. Ent.* 28: 265–66.

Blest, A. D. 1957. The function of eyespot patterns in the Lepidoptera. *Behaviour* 11: 209–56.

Bodenheimer, F. S. 1928. *Materialien zur Geschichte der Entomologie bis Linne,* Band I. Berlin: Junk.

Bodenheimer, F. S. 1960. *Animal and man in Bible lands.* Leiden: Brill.

Boggs, C. L. 1995. Male nuptial gifts: Phenotypic consequences and evolutionary implications. In *Insect reproduction*, ed. S. R. Leather and J. Hardie, 215–42. New York: CRC Press.

Bonfils, J. 1967. Une espèce nouvelle du genre Oligonyx Saussure: Description et notes biologiques (Dict.). *Bull. Soc. ent. France* 72: 244–47.

Boudreaux, H. B. 1979. *Arthropod phylogeny with special reference to insects.* New York: J. Wiley.

Bowdish, T. I., and T. L. Bultman. 1993. Visual cues used by mantids in learning aversion to aposematically colored prey. *Am. Midl. Nat.* 129: 215–22.

Bowers, M. A., and C. F. Sacchi. 1991. Fungal mediation of a plant-herbivore interaction in an early successional plant community. *Ecology* 72: 1032–37.

Boyan, G. S. 1983. Postembryonic development in the auditory system of the locust. *J. Comp. Physiol.* 151: 499–513.

Boyan, G. S. 1993. Another look at insect audition: The tympanic receptors as an evolutionary specialization of the chordotonal system. *J. Insect Physiol.* 39: 187–200.

Boyan, G. S., and E. E. Ball. 1986. Wind-sensitive interneurones in the terminal ganglion of praying mantids. *J. Comp. Physiol.* 159: 773–89.

Brackenbury, J. H. 1990. Wing movements in the bush-cricket *Tettigonia viridissima* and the mantis *Ameles spallanziana* during natural leaping. *J. Zool. London* 220: 593–602.

Brackenbury, J. H. 1991a. Kinematics of take-off and climbing flight in butterflies. *J. Zool. London* 224: 251–70.

Brackenbury, J. H. 1991b. Wing kinematics during natural leaping in the mantids *Mantis religiosa* and *Iris oratoria*. *J. Zool. London* 223: 341–56.

Brackenbury, J. H. 1992. *Insects in flight*. London: Blandford.

Brackenbury, J. H. 1994a. Wing-folding and free-flight kinematics in Coleoptera: A comparative study. *J. Zool. London* 232: 253–83.

Brackenbury, J. H. 1994b. Hymenopteran wing kinematics: A qualitative study. *J. Zool. London* 233: 523–40.

Brackenbury, J. H., and A. Dack. 1992. Semi-automated system for photographing wing motion in free flying insects. *Biol. Eng. Comput.* 30: 230–34.

Brackenbury, J. H., and R. Z. Wang. 1995. Ballistics and visual targeting in flea-beetles (Alticinae). *J. Exp. Biol.* 198: 1931–42.

Brady, J. 1975. Circadian changes in central excitability—the origin of behavioral rhythms in tsetse flies and other animals. *J. Ent.* 50: 79–95.

Brauer, F. 1869. Beschreibung der Verwandlungsgeschichte der *Mantispa styriaca* Poda und Betrachtungen über die sogenannte Hypermetamorphose Fabre's. *Verh. k. k. Zool. Bot. Ges. Wien* 19: 831–40.

Brighton, Mrs. 1895. *Inmates of my house and garden*. New York: Macmillan.

Brodsky, A. K. 1979a. Evolution of the flight apparatus of stoneflies (Plecoptera). I. Functional morphology of the wings. *Ent. Rev.* 58: 31–36.

Brodsky, A. K. 1979b. Evolution of the flight apparatus of stoneflies (Plecoptera). II. Functional morphology of the axillary region, skeleton and musculature. *Ent. Rev.* 58: 16–26.

Brodsky, A. K. 1981. Evolution of the flight apparatus of stoneflies (Plecoptera). III. Wing deformation of *Isogenus nubecula*. Newman during flight. *Ent. Rev.* 60: 25–36.

Brodsky, A. K. 1982. Evolution of the flight apparatus of stoneflies (Plecoptera). IV. The kinematics of the wings and the general conclusion. *Ent. Rev.* 61: 34–43.

Brodsky, A. K. 1985. Kinematics of insect wings in horizontally maintained flight: A comparative study. *Ent. Rev.* 64: 56–73.

Brodsky, A. K., and D. L. Grodnitsky. 1986. Aerodynamics of fixed flight of *Thymelicus lineola* (Lepidoptera: Hesperiidae). *Ent. Rev.* 65: 60–69.

Brodsky, A. K., and V. D. Ivanov. 1983. Functional assessment of wing structure in insects. *Ent. Rev.* 62: 32–51.

Brower, L. P. 1988. Avian predation on the monarch butterfly and its implications for mimicry theory. In *Mimicry and the evolutionary process*, ed. L. P. Brower, 4–6. Chicago: University of Chicago Press.

Brown, W. L., Jr. 1982. Mantodea. In *Synopsis and classification of living organisms*, vol. 2, 347–349. New York: McGraw–Hill.

Brown, S. M. 1986. Of mantises and myths. *Bioscience* 36: 421–23.

Brown, V. K. 1985. Insect herbivores and plant succession. *Oikos* 44: 17–22.

Bruce, W. G. 1958. Bible references to insects and other arthropods. *Bull. Ent. Soc. Am.* 4: 75–78.

Brunner de Wattenwyl, C. 1893. Révision du système des Orthoptères et description des espèces rapportées par M. Leonardo Fea de Birmanie. *Ann. Mus. civic. Stor. nat. Genova* 17: 54–76.

Brunner, D., and D. Gandolfo. 1990. Morphological adaptations to an unusual defensive strategy in the mantid *Orthoderella ornata* (Insecta: Mantodea). *J. Zool. London* 222: 129–36.

Buddenbrock, W. von. 1921. Rhythmus der Schreitbewegungen der Stabheuschrecke, *Dixippus morosus*. *Biol. Zbl. Bd.* 41: 47–48.

Burdohan, J. A., and C. M. Comer. 1990. An antennal-derived mechanosensory pathway in the cockroach descending interneurons as a substrate for evasive behavior. *Brain Res.* 535: 347–52.

Burghagen, H., and J.-P. Ewert. 1982. Question of "head preference" in response to worm–like dummies during prey–capture of toads, *Bufo bufo*. *Behav. Processes* 7: 295–306.

Burmeister, H. 1838. *Handbuch der Zoologie*, 2. 2. Berlin: Theod. Ehr. Fried. Enslin.

Burmeister, H. C. C. 1838. Kaukerfe. Gymnognatha. Erste Hälfte; vulgo Orthoptera. In *Handbuch der Entomologie*, 397–756. Berlin.

Burrows, M. 1973. The role of delayed excitation in the coordination of some metathoracic flight motoneurons of a locust. *J. Comp. Physiol.* 83: 135–64.

Burrows, M. 1996. *The neurobiology of an insect brain*. London: Oxford University Press.

Burtt, E. H. J. 1979. Tips on wings and other things. In *The behavioral significance of color*, ed. E. H. J. Burtt, 75–110. New York: Garland STPM Press.

Buskirk, R. E., C. Frolich, and K. G. Ross. 1984. The natural selection of sexual cannibalism. *Am. Nat.* 123: 612–25.

Butler, L. 1966. Oviposition in the Chinese mantid (*Tenodera aridifolia sinensis* Saussure) (Orthoptera: Mantidae). *J. Georgia Ent. Soc.* 1: 5–7.

Cade, W. H. 1985. Insect mating and courtship behavior. In *Comprehensive insect physiology, biochemistry and pharmacology*, vol. 9: Behaviour, ed. G. A. Kerkut and L. I. Gilbert, 591–619. Oxford: Pergamon Press.

Callec, J. J., and J. Boistel. 1966. Etude de divers types d' activitès électriques enregistrées par microélectrodes capillaires au niveau du dernier ganglion abdominal de la blatte *Periplaneta americana* L. *R. Soc. Biol. Paris* 160: 1943–47.

Camhi, J. M., and A. Levy. 1988. Organization of a complex movement: Fixed and variable components of the cockroach escape behavior. *J. Comp. Physiol. A* 163: 317–28.

Camhi, J. M., W. Tom, and S. Volman. 1978. The escape behaviour of the cockroach *Periplaneta americana*. II. Detection of natural predators by air displacement. *J. Comp. Physiol. A* 128: 203–12.

Cariaso, B. J. 1967. Biology of the black widow spider, *Latrodectus hasselti* Thorell (Araneida, Theriidae). *Philippine Agri.* 51: 171.

Carpenter, G. D. H. 1921. Experiments on the relative edibility of insects, with special reference to their coloration. *Trans. R. Ent. Soc. London* 54: 1–105.

Carpenter, W. B. 1867. *Zoology*. London: Bell & Daldy.

Cesnola, A. P. di. 1904. Preliminary note on the protective value of colour in *Mantis religiosa*. *Biometrika* 3: 58–59.

Chapman, R. F. 1982. *The insects*. Cambridge: Harvard University Press.

Charnov, E. L. 1976. Optimal foraging: Attack strategy of a mantid. *Am. Nat.* 110: 141–51.

Charnov, E. L. 1976. Optimal foraging: The marginal value theorem. *Theor. Pop. Biol.* 9: 129–36.

Chopard, L. 1914. Sur la vitalité de *Mantis religiosa*, L. (Orth. Mantidae); ponte après décapitation. *Bull. Soc. ent. France* 19: 481–82.

Chopard, L. 1949. Ordre des Dictyoptères. In *Traité de Zoologie IX*, ed. P. P. Grassé, 355–407. Paris: Masson.

Chudokova, I. V., and O. M. Bocharova–Messner. 1965. Changes in the functional structure features of the wing muscles of the domestic cricket *Achaeta domesticus* L. *Doklady Akad. Nauk SSSR* 164: 656–59.

Clark, B. 1962. Balanced polymorphism and the diversity of sympatric species. *Nichols Taxon. Geog. London* 4: 47–70.

Clarke, R. D., and P. R. Grant. 1968. An experimental study of the role of spiders as predators in a forest litter community. Part 1. *Ecology* 49: 1152–54.

Cleal, K., and F. R. Prete. 1996. The predatory strike of free ranging praying mantises, *Sphodromantis lineola* (Burr.). II. Strikes in the horizontal plane. *Brain Behav. Evol.* 48: 191–204.

Clements, F. E. 1936. Nature and structure of the climax. *J. Ecol.* 24: 252–84.

Cloudsley–Thompson, J. L. 1981. Comments on the nature of deception. *Biol. J. Linn. Soc.* 16: 11–14.

Collatz, K.-G., and R. S. Sohol, eds. 1986. *Insect aging strategies and mechanisms*. Berlin: Springer-Verlag.

Cohen, A. H., S. Rossignol, and S. Grillner, eds. 1988. *Neural control of rhythmic movements in vertebrates*. New York: Wiley.

Cohen, J. E., S. L. Pimm, P. Yodzis, and J. Saladana. 1993. Body sizes of animal predators and animal prey in food webs. *J. Anim. Ecol.* 62: 67–78.

Collett, T. S. 1965. The control of the rocking response in hemileucine moths by sensory stimulation. *J. Insect Physiol.* 11: 1407–25.

Collett, T. S. 1978. Peering—a locust behavior pattern for obtaining motion parallax information. *J. Exp. Biol.* 76: 237–41.

Collett, T. S. 1987. Binocular depth vision in arthropods. *TINS* 10: 1–2.

Collett, T. S., and M. F. Land. 1975. Visual control of flight behaviour in the hoverfly *Syritta pipiens*. *J. Comp. Physiol.* 99: 1–66.

Collett, T. S., and C. J. Paterson. 1991. Relative motion parallax and target localization in the locust, *Schistocerca gregaria*. *J. Comp. Physiol.* 169: 615–21.

Comer, C. M., E. Mara, K. A. Murphy, M. Getman, and M. C. Mungy. 1994. Multisensory control of escape in the cockroach *Periplaneta americana*. II. Patterns of touch-evoked behavior. *J. Comp. Physiol. A,* 174: 13–26.

Comfort, A. 1956. *The biology of senescence*. New York: Rinehart & Co.

Compton, S. G., J. H. Lawton, and V. K. Rashbrook. 1989. Regional diversity, local community structure, and vacant niches: The herbivorous arthropods of bracken in South Africa. *Ecol. Ent.* 14: 356–73.

Connell, J. H. 1961. The influence of interspecific competition and other factors on the distribution of the barnacle *Chthamalus stellatus*. *Ecology* 42: 710–23.

Connell, J. H., and E. Orias. 1964. The ecological regulation of species diversity. *Am. Nat.* 98: 399–414.

Copeland, J., and A. D. Carlson. 1977. Prey capture in mantis: Prothoracic tibial flexion reflex. *J. Insect Physiol.* 23: 1151–56.

Copeland, J., and A. D. Carlson (1979) Prey capture in mantids: A non–stereotyped component of lunge. *J. Insect Physiol.* 25: 263–69.

Cornell, H. V., and J. H. Lawton. 1992. Species interactions, local and regional processes, and limits to the richness of ecological communities. *J. Anim. Ecol.* 61: 1–12.

Cornilleau-Pérès, V., and C. C. A. M. Gielen. 1996. Interactions between self–motion and depth perception in the processing of optic flow. *TINS* 19: 196–202.

Corrette, B. J. 1980. Motor control of prey capture in the praying mantis *Tenodera aridifolia sinensis*. Ph.D. diss., University of Oregon.

Corrette, B. J. 1990. Prey capture in the praying mantis *Tenodera aridifolia sinensis*: Coordination of the capture sequence and strike movements. *J. Exp. Biol.* 148: 147–80.

Cott, H. B. 1940. *Adaptive coloration in animals.* London: Methuen.

Covie, R. J. 1977. Optimal foraging in great tits (*Parus major*). *Nature* 268: 137–39.

Cox, C. B., and P. D. Moore. 1993. *Biogeography—an ecological and evolutionary approach.* Oxford: Blackwell Scientific Publications.

Crane, J. 1952. A comparative study of innate defensive behaviour in Trinidad mantids (Orthoptera, Mantoidea). *Zoologica* 37: 259–93.

Crawford, C. S. 1981. *Biology of desert invertebrates.* Berlin: Springer-Verlag.

Croze, H. 1970. Searching image in carrion crows. *Z. Tierpsychol. Suppl.* 5: 1–85.

Cukier, M., G. A. Guerrero, and M. C. Maggese. 1979. Parthenogenesis in *Coptopteryx viridis*, Giglio Tos (1915) (Dyctioptera, Mantidae). *Biol. Bull.* 157: 445–52.

Cukier, M., G. A. Guerrero, and M. C. Maggese. 1986. Oogenesis in *Coptopteryx viridis* (Dyctioptera, Mantidae). I. Characteristics of the mature oocyte envelopes. *Rev. Bras. Biol.* 46: 3–10.

Cumming, G. S. 1996. Mantis movements by night and the interactions of sympatric bats and mantises. *Can. J. Zool.* 74: 1771–74.

Curio, E. 1976. The ethology of predation. *Zoophysiology and ecology*, vol. 7, ed. D. S. Farner. Berlin: Springer-Verlag.

Curley, M. J. 1979. *Physiologus.* Austin: University of Texas Press.

Cusson, M., and J. N. McNeil. 1989. Ovarian development in female armyworm moths, *Pseudaletia unipuncta*: Its relationship with pheromone release activities. *Can. J. Zool.* 67: 1380–85.

Cutting, J. E., C. Moore, and R. Morrison. 1988. Masking the motions of human gait. *Percept. Psychophys.* 44: 339–47.

Cuvier, B. 1832. *The Class Insecta*, vol. 2. London: Whittaker, Treacher, and Co.

Daniels, R. J. R., M. Hegde, and C. Vinutha. 1989. Observations on the biology of the praying mantis *Creobater urbana* Fabr. (Orthoptera: Mantidae). *J. Bombay Nat. Hist. Soc.* 86: 329–32.

Darwin, C. 1859. *On the origin of species by means of natural selection or the preservation of favoured races in the struggle for life.* London: John Murray.

Daufer, B. 1927. Insect–musicians and cricket champions of China. *Field Mus. Nat. Hist. Chicago, Anthrpol. Leaf.* 22: 1–28.

Davis, W. J. 1976. Behavioral and neuronal plasticity in mollusks. In *Sompler networks and behavior*, ed. I. C. Fentress, 224–38. Sunderland: Sinauer.

Davis, W. J., and E. Liske. 1985. Der Balztanz der Gottesanbeterin—Anatomie eines wissenschaftlichen Mythos. *Naturwiss. Rundschau* 38: 223–30.

Davis, W. J., and E. Liske. 1988. Cerci mediate mating movements in the male praying mantis. *Zool. Jb. Physiol.* 92: 47–55.

Dayton, P. K., and R. R. Hessler. 1972. Role of biological disturbance in maintaining diversity in the deep sea. *Deep-sea Res.* 19: 199–208.

DeAngelis, D. S., R. A. Goldstein, and R. V. O'Neill. 1975. A model for trophic interaction. *Ecology* 56: 881–92.

Delamare–Deboutteville, C. 1952. L'homogénéité de la morphologie sternale des Blattoptéroïdes. *Trans. Ninth Int. Cong. Ent.* 1: 147–50.

Denny, H. 1867. Hatching of the mantis in England. *Mag. Nat. Hist.* 19: 144.

Deruntz, P., C. Palevody, and M. Lambin. 1994. Effect of dark rearing on the eye of *Gryllus bimaculatus* crickets. *J. Exp. Zool.* 268: 421–27.

Dethier, V. G. 1969. Feeding behavior of the blowfly. In *Advances in the study of behavior*, vol. 2, ed. D. S. Lehrman, R. A. Hinde, and E. Shaw, 111–266. New York: Academic Press.

Dethier, V. G. 1987. Discriminative taste inhibitors affecting insects. *Chem. Senses* 12: 251–63.

Dial, R., and J. Roughgarden. 1995. Experimental removal of insectivores from rainforest canopy: Direct and indirect effects. *Ecology* 76: 1821–34.

Didlake, M. 1926. Observations on life–histories of two species of praying mantis (Orthopt.: Mantidae). *Ent. News* 37: 169–75.

Diehl, S. 1993. Relative consumer sizes and the strengths of direct and indirect interactions in omnivorous feeding relationships. *Oikos* 68: 151–57.

Dittrich, W., F. Gilbert, P. Green, P. McGregor, and D.

Grewcock. 1993. Imperfect mimicry: A pigeon's perspective. *Proc. R. Soc. London* 251: 195–200.

Dow, R. P. 1917. Studies in the Old Testament. *Bull. Brooklyn Ent. Soc.* 12: 1–13, 64–69.

Dow, R. P. 1918. Studies in the Old Testament. *Bull. Brooklyn Ent. Soc.* 13: 90–93.

Dubois, R. 1893. Sur l'innervation réflexe chez la mante religieuse. *Ann. Soc. Linn. Lyon* 39: 205–7.

Duncan, C. J., and P. M. Sheppard. 1965. Sensory discrimination and its role in the evolution of Batesian mimicry. *Behaviour* 24: 269–82.

Duncan, J. 1843. *The naturalist's library,* vol. XXXI. Entomology, ed. W. Jardine. Edinburgh: W. H. Lizars.

Duncan, P. M. 1870. *The transformations (or metamorphoses) of insects.* Philadelphia: Claxton, Remsen, and Haffelfinger.

Durrell, G. 1956. *My family and other animals.* London: Rupert Hart–Davis.

Dussé, K. M., and L. E. Hurd. 1997. Food limitation and body length in the mantid *Tenodera sinensis* (Mantodea: Mantidae): How nymphal feeding affects fitness. *Proc. Ent. Soc. Wash.* 99: 490–93.

Eberhard, W. G., and C. Cordero. 1995. Sexual selection by cryptic female choice on male seminal products: A new bridge between sexual selection and reproductive physiology. *Trends Ecol. Evol.* 10: 493–96.

Edmunds, M. 1972. Defensive behaviour in Ghanaian praying mantids. *Zool. J. Linn. Soc.* 51: 1–32.

Edmunds, M. 1974a. *Defence in animals: A survey of anti–predator defences.* Harlow: Longman.

Edmunds, M. 1974b. Significance of beak marks on butterfly wings. *Oikos* 25: 117–18.

Edmunds, M. 1975. Courtship, mating, and possible sex pheromones in three species of Mantodea. *Ent. Month. Mag.* 111: 53–57.

Edmunds, M. 1976. The defensive behaviour of Ghanaian praying mantids with a discussion of territoriality. *Zool. J. Linn. Soc.* 58: 1–37.

Edmunds, M. 1981. On defining "mimicry." *Biol. J. Linn. Soc.* 16: 9–10.

Edmunds, M. 1986. The phenology and diversity of praying mantids in Ghana. *J. Trop. Ecol.* 2: 39–50.

Edmunds, M. 1988. Sexual cannibalism in mantids. *Trends Ecol. Evol.* 3: 77.

Edmunds, M. 1991. The evolution of cryptic coloration. In *Insect defenses,* ed. D. L. Schmidt, and J. O. Evans, 3–21. New York: SUNY Press.

Edmunds, M., and R. A. Dewhirst. 1994. The survival value of countershading with wild birds as predators. *Biol. J. Linn. Soc.* 51: 447–52.

Edmunds, M., and D. Dudgeon. 1991. Cryptic behaviour in the oriental leaf mantis *Sinomantis denticulata* Beier (Dictyoptera, Mantodea). *Ent. Month. Mag.* 127: 45–48.

Edwards, D. H. 1982a. The cockroach DCMD neuron. I. Lateral inhibition and the effects of light– and dark–adaptation. *J. Exp. Biol.* 99: 61–90.

Edwards, D. H. 1982b. The cockroach DCMD neuron. II. Dynamics of response habituation and convergence of spectral inputs. *J. Exp. Biol.* 99: 91–107.

Edwards, J. S., and J. Palka. 1971. Neural regeneration delayed formation of central contacts by insect sensory cells. *Science* 172: 511–14.

Egelhaaf, M. 1985a. On the neuronal basis of figure–ground discrimination by relative motion in the visual system of the fly. I. *Biol. Cybern.* 52: 123–40.

Egelhaaf, M. 1985b. On the neuronal basis of figure–ground discrimination by relative motion in the visual system of the fly. II. *Biol. Cybern.* 52: 195–209.

Egelhaaf, M. 1985c. On the neuronal basis of figure–ground discrimination by relative motion in the visual system of the fly. III. *Biol. Cybern.* 52: 267–80.

Egelhaaf, M., K. Hausen, W. Reichardt, and C. Wehrhahn. 1988. Visual control in flies relies on neuronal computation of object and background motion. *TINS* 11: 351–58.

Eggenreich, U., and K. Kral. 1990. External design and field of view of the compound eyes in a raptorial neuropteran insect, *Mantispa styriaca.* *J. Exp. Biol.* 148: 353–65.

Ehrmann, R. 1986. Standorttreue von Mantis religiosa (L.) (Mantodea: Mantidae). *Ent. Z. Insek.* 96: 63–64.

Eibl–Eibesfeldt, I. 1972. *Grundriss der vergleichenden Verhaltensforschung.* München: Piper und Co. Verlag.

Eisenberg, R. M., and L. E. Hurd. 1977. An ecological study of the emergence characteristics for egg cases of the Chinese mantis (*Tenodera aridifolia sinensis* Saussure). *Am. Midl. Nat.* 97: 478–82.

Eisenberg, R. M., and L. E. Hurd. 1990. Egg dispersion in two species of praying mantids (Mantodea: Mantidae). *Proc. Ent. Soc. Wash.* 92: 808–10.

Eisenberg, R. M., and L. E. Hurd. 1993. Relative egg success and implications for distribution of three sympatric mantids (Mantodea: Mantidae). *Proc. Ent. Soc. Wash.* 95: 271–77.

Eisenberg, R. M., L. E. Hurd, and J. A. Bartley. 1981. Ecological consequences of food limitation for adult mantids (*Tenodera aridifolia sinensis* Saussure). *Am. Midl. Nat.* 106: 209–18.

Eisenberg, R. M., L. E. Hurd, and R. B. Ketcham. 1989. The cellular slime mold guild and its bacterial prey: Growth rate variation at the inter- and intraspecific levels. *Oecologia* 79: 458–62.

Eisenberg, R. M., L. E. Hurd, W. F. Fagan, K. J. Tilmon, W. E. Snyder, K. S. Vandersall, S. G. Datz, and J. D. Welch. 1992a. Adult dispersal of *Tenodera aridifolia sinensis* (Mantodea: Mantidae). *Environ. Ent.* 21: 350–53.

Eisenberg, R. M., E. Pilchik, L. L. Bedwell, S. Winram. M. Rodgers, S. McFalls, C. T. Kessler, M. Gross, A. Walter, E. Wadman, K. M. Dougherty, W. Smith, C. L. Berman, and L. E. Hurd. 1992b. Comparative egg ecology of two sympatric mantids (Mantodea: Mantidae). *Proc. Ent. Soc. Wash.* 94: 366–70.

Eisner, T., R. Ziegler, J. L. McCormick, M. Eisner, E. R. Hoebeke, and J. Meinwald. 1994. Defensive use of an acquired substance (carminic acid) by predaceous insect larvae. *Experientia* 50: 610–15.

Eldredge, N., and S. J. Gould. 1972. Punctuated equilibria: An alternative to phyletic gradualism. In *Models in paleobiology*, ed. J. M. Schopf, 82–115. San Francisco: Freeman Cooper.

Elgar, M. A. 1992. Sexual cannibalism in spiders and other invertebrates. In *Cannibalism: Ecology and evolution among diverse taxa*, ed. M. A. Elgar and B. J. Crespi, 128–55. Oxford: Oxford University Press.

Ellington, C. P. 1980. Vortices and hovering flight. In *Instationare Effekte an Schwingenden Tierflugeln*, ed. W. Nachtigall, 64–101. Wiesbaden: Franz Steiner Verlag.

Ellington, C. P. 1984a. The aerodynamics of hovering insect flight. III. Kinematics *Phil. Trans. Roy. Soc. London B* 305: 41–78.

Ellington, C. P. 1984b. The aerodynamics of hovering insect flight. IV. Aerodynamic mechanisms. *Phil. Trans. Roy. Soc. London B* 305: 79–113.

Elton, C. 1946. Competition and the structure of communities. *J. Anim. Ecol.* 15: 54–68.

Endler, J. A. 1981. An overview of the relationships between mimicry and crypsis. *Biol. J. Linn. Soc.* 16: 25–31.

Endler, J. A. 1986. Defense against predators. In *Predator–prey relationships. Perspectives and approaches from the study of lower vertebrates*, ed. M. E. Feder and G. V. Lauder, 109–34. Chicago: University of Chicago Press.

Ene, J. C. 1962. Parasitisation of mantid oothecae in West Africa. *Int. Congr. Ent., XI, Wien*, 2: 725–27.

Ene, J. C. 1964. The distribution and post-embryonic development of *Tarachodes afzelii* (Stål), (Mantodea: Eremiaphilidae). *Ann. Mag. Nat. Hist.* 7: 493–511.

Engelmann, F. 1970. *The physiology of insect reproduction.* Oxford: Pergamon Press.

Ennos, A. R. 1988a. The importance of torsion in the design of insect wings. *J. Exp. Biol.* 140: 137–60.

Ennos, A. R. 1988b. The inertial cause of wing rotation in Diptera. *J. Exp. Biol.* 140: 161–69.

Ennos, A. R. 1989. The kinematics and aerodynamics of the free flight of some Diptera. *J. Exp. Biol.* 142: 49–85.

Ennos, A. R., and R. J. Wootton. 1989. Functional wing morphology and aerodynamics of *Panorpa germanica* (Insecta: Mecoptera). *J. Exp. Biol.* 143: 267–84.

Erichson, W. F., ed. 1843. Zur Naturgeschichte der Mantis Carolina: Aus einem Schreiben des Dr. Zimmermann. Cited in Kevan, D. K. M. 1985. The mantis and the serpent. *Ent. Month. Mag.* 121: 1–8.

Erickson, C. J. 1991. Percussive foraging in the aye–aye, *Daubentonia madagascariensis*. *Anim. Behav.* 41: 793–801.

Evans, P. D., and M. O`Shea. 1978. The identification of an octopaminergic neurone and the modulation of a myogenic rhythm in the locust. *J. Exp. Biol.* 73: 235–60.

Evans, S., and B. Tallmark. 1985. Niche separation within the mobile predator guild on marine shallow soft bottoms. *Mar. Ecol. Prog. Ser.* 23: 279–86.

Everson, P. R., and J. F. Addicott. 1982. Mate selection strategies by male mites in the absence of intersexual selection by females: A test of six hypotheses. *Can. J. Zool.* 60: 2729–36.

Ewert, J.-P. 1987. Neuroethology of releasing mechanisms: Prey catching in toads. *Behav. Brain Sci.*, 10: 337–405.

Ewert, J.-P. 1982. Effects of background structure on the discrimination of configural moving prey dummies by toads *Bufo bufo* (L.). *J. Comp. Physiol.* 147: 179–87.

Ewert, J.-P. 1989. The release of visual of visual behavior in toads: Stages of parallel/hierarchical information processing. In *Visuomotor coordination*, ed. J.-P. Ewert and M. A. Arbib, 9–120. New York: Plenum.

Ewert, J.-P., H. Burghagen, and E. Schürg-Pfeiffer. 1983. Neuroethological analysis of the innate releasing mechanism for prey-catching behavior in toads. In *Advances in vertebrate neuroethology*, ed. J.-P. Ewert, R. R. Capranica, and D. J. Ingle, 413–75. New York: Plenum.

Ewing, H. Z. 1904. The functions of the nervous system with special regard to respiration in Acrididae. *Univ. Kans. Sci. Bull.* 2: 305–19.

Exner, S. 1891. *Die Physiologie der facettierten Augen von Krebsen und Insekten.* Berlin: Springer-Verlag.

Fabré, J. H. 1897a. Das offenbare Geheimnis, In *Dem Lebenswerk des Insektenforschers,* ed. K. Guggenheim and A. Portman. Zürich: Artemis Verlag.

Fabré, J. H. 1897b. *Souvenirs entomologiques,* vol. 5. Paris: Delagrave.

Fabré, J. H. 1901. *Insect life.* London: Macmillan.

Fabré, J. H. 1912. *Social life in the insect world,* trans. B. Miall. London: T. Fisher Unwin.

Fabré, J. H. 1914. *Social life in the insect world,* trans. B. Miall. New York: Century.

Fagan, W. F., and G. M. Odell. 1996. Size-dependent cannibalism in praying mantids: Using biomass flux to model size-structured populations. *Am. Nat.* 147: 230–68.

Fagan, W. F., and L. E. Hurd. 1991a. Direct and indirect effects of generalist predators on a terrestrial arthropod community. *Am. Midl. Nat.* 126: 380–84.

Fagan, W. F., and L. E. Hurd. 1991b. Late season food level, cannibalism, and oviposition in adult mantids (Orthoptera: Mantidae): Sources of variability in a field experiment. *Proc. Ent. Soc. Wash.* 93: 956–61.

Fagan, W. F., and L. E. Hurd. 1994. Hatch density variation of a generalist arthropod predator: Population consequences and community impact. *Ecology* 75: 2022–32.

Farley, R. D., and N. S. Milburn. 1969. Structure and function of giant fibre system in the cockroach Periplaneta americana. *J. Insect Physiol.* 15: 457–76.

Feltwell, J., and M. Rothschild. 1974. Carotenoids in thirty-eight species of Lepidoptera. *J. Zool. London* 174: 441–65.

Fenton, M. B. 1985. The feeding behavior of insectivorous bats: Echolocation, foraging strategies, and resource partitioning. *Transvaal Mus. Bull.* 21: 5–16.

Fenton, M. B., and J. H. Fullard. 1981. *Nysteris grandis* (Nycteridae): An African carnivorous bat. *J. Zool. London* 194: 461–65.

Figuier, L. 1869. *The insect world.* New York: D. Appleton.

Figuier, L. 1874. The mantis or praying insect. *Pop. Sci. Month.* 4: 710–13.

Finch, C. E. 1990. *Longevity, senescence and the genome.* Chicago: University of Chicago Press.

Finch, C. E. 1991. New models for new perspectives in the biology of senescence. *Neurobiol. Aging* 12: 625–34.

Fleissner, G. 1974. Circadiane Adaptation und Schirmpigmentverlagerung in den Sehzellen der Medianaugen von *Androctonus australis* L. (Buthidae, Scorpiones). *J. Comp. Physiol.* 91: 399–416.

Fogden, M., and P. Fogden. 1974. *Animals and their colours: Camouflage, warning colouration, courtship and territorial display, mimicry.* London: Peter Lowe.

Folsom, T. C., and N. C. Collins. 1984. The diet and foraging behavior of the larval dragonfly *Anax junius* (Aeshnidae), with an assessment of the role of refuges and prey activity. *Oikos* 42: 105–13.

Formanowicz, D. R., Jr. 1982. Foraging tactics of larvae of *Dytiscus verticalis* (Coleoptera: Dytiscidae): The assessment of prey density. *J. Anim. Ecol.* 51: 757–67.

Formanowicz, D. R., Jr. 1987. Fluctuations in prey density: Effects on the foraging tactics of scolopendrid centipedes. *Anim. Behav.* 35: 453–61.

Forster, L. 1992. The stereotyped behaviour of sexual cannibalism in *Latrodectus hasselti,* Thorell (Araneae: Theridiidae), the Australian redback spider. *Aust. J. Zool.* 40: 1–11.

Fox, H. M., and G. Vevers. 1960. *The nature of animal colours.* London: Sidgwick & Jackson.

Fox, L. R. 1975. Cannibalism in natural populations. *Ann. Rev. Ecol. Syst.* 6: 25–33.

Fox, P. 1979. *Mantis.* New York: St. Martin's Press.

Franco, A., and F. Cervantes–Perez. 1990. Experimental and theoretical studies on visuomotor coordination in the praying mantis. *Ciencia* 41: 237–64.

Frank, M. E. 1930. An observation on the diet of the praying mantis. *Zool. J. Linn. Soc.* 9: 321–22.

Fraser, H. M. 1931. *Beekeeping in antiquity.* London: University of London Press.

Frazier, J. L., T. E. Nebeke, R. F. Mizell, and W. H. Calvert. 1981. Predatory behavior of the clerid beetle *Thanasimus dubius* (Coleoptera: Cleridae) on the southern pine beetle (Coleoptera: Scolytidae). *Can. Ent.* 113: 35–43.

Fullard, J. H. 1985. Sensory ecology and neuroethology of moths and bats: Interactions in a global perspective. In *Recent advances in the study of bats,* ed. M. B. Fenton, P. A. Racey, and J. M. V. Rayner, 244–72. Cambridge: Cambridge University Press.

Fullard, J. H., and J. E. Yack. 1993. The evolutionary biology of insect hearing. *Trends Ecol. Evol.* 8: 248–52.

Gause, G. F. 1934. *The struggle for existence.* Baltimore: Williams and Wilkins.

Gebauer, D. 1988. Vergleichende neuroanatomische Untersuchungen an dem Abdomen der männlichen Gottesanbeterinnen *Tenodera aridifolia sinensis* und *Hierodula membranacea.* Master's thesis, Technische University Braunschweig, Germany.

Gebauer, D., E. Liske, K. Köchy, and H. G. Wolff. 1987. The praying mantis nervous system: Thoracic and abdominal ganglia. In *New frontiers in brain research.*

Proc. 15th Göttingen Neurobiol. Conf., ed. N. Elsner and O. Creutzfeldt, 66. Stüttgart: Georg Thieme Verlag.

Gelperin, A. 1968. Feeding behavior of the praying mantis: A learned modification. *Nature* 219: 399–400.

Gerhardt, U. 1914. Copulation und Spermatophoren von Grylliden und Locustiden. *Zool. Syst. Geog. Biol.* 37: 1–64.

Giglio-Tos, E. 1917. Note al catalogo dei Mantidi di Kirby. *Bull. Soc. ent. Ital.* 48: 139–63.

Giglio-Tos, E. 1919. Saggio di una nova classificazione dei Mantidi. *Bull. Soc. ent. Ital.* 49: 50–87.

Giglio-Tos, E. 1927. Orthoptera, Mantidae. In *Das Tierreich,* 50, I–XL; 1–708. Berlin: Walter de Gruyter.

Gillispie, C. C., ed. 1970. *Dictionary of scientific biography,* vols. V, IX, XIV. New York: Charles Scribner's Sons.

Gillon, Y., and R. Roy. 1968. Les Mantes de Lamto et des savanes de Côte d'Ivoire. *Bull. Inst. fond. Afr. noire* 30: 1038–1151.

Gillott, C. 1980. *Entomology.* New York: Plenum.

Gleason, H. A. 1926. The individualistic concept of the plant association. *Bull. Torrey Bot. Club* 53: 7–26.

Gomez, J. M., and R. Zamora. 1994. Top-down effects in a tritrophic system: Parasitoids enhance plant fitness. *Ecology* 75: 1023–30.

Gompertz, B. 1825. On the nature of the function expressive of the law of human mortality, and on a new mode of determining the value of life contingincies. *Phil. Trans. R. Soc. London* 513–85.

Gould, J. L. 1982. *Ethology.* New York: W. W. Norton.

Gould, S. J. 1977. *Ontogeny and phylogeny.* Cambridge: Harvard University Press.

Gould, S. J. 1979. The spandrels of San Marco, *Proc. R. Soc. London B* 205: 581–98.

Gould, S. J. 1984. Only his wings remained. *Nat. Hist.* 93: 10–18.

Goulet, M., R. Campan, and M. Lambin. 1981. The visual perception of relative distances in the woodcricket, *Nemobius sylvestris. Physiol. Ent.* 6: 357–67.

Grandcolas, P. 1994. Phylogenetic systematics of the subfamily Polyphaginae, with the assignment of Cryptocercus Scudder, 1862 to this taxon (Blattaria, Blaberoidea, Polyphagidae). *System. Ent.* 19: 145–58.

Grasshoff, M. 1964. Die Kreuzspinne Araneus pallidus: Ihr Netzbau und ihre Paarungsbiologie. *Nat. Mus.* 94: 305–14.

Gratshev, V. G., and V. V. Zherikhin. 1993. New fossil mantids (Insecta, Mantida). *Paleontol.* 27: 148–65.

Grayson, J., M. Edmunds, E. H. Evans, and G. Britton. 1991. Carotenoids and colouration of poplar hawkmoth caterpillars (*Laothoe populi*). *Biol. J. Linn. Soc.* 42: 457–65.

Gressitt, J. L. 1946. Entomology in China. *Ent. Soc. Am.* 39: 153–64.

Gressitt, J. L. 1969. Oriental caged insects. *Ent. News* 80: 138.

Grey, P. T. A., and P. J. Mill. 1985. The musculature of the prothoracic legs and its innervation in *Hierodula membranacea* (Mantidea). *Phil. Trans. R. Soc. London B* 309: 479–503.

Gruber, S. H. 1979. Mechanisms of color vision: An ethologist primer. In *The behavioral significance of color,* ed. E. H. J. Burtt, 183–236. New York: Garland STPM Press.

Grüsser, O.-J, and U. Grüsser-Cornehls. 1970. Die Neurophysiologie visueller gesteuerter Verhaltensweisen bei Anuren. *Verh. Dtsch. Zool. Ges. Köln* 64: 201–18.

Guilford, T. 1988. The evolution of conspicuous coloration. In *Mimicry and the evolutionary process,* ed. L. P. Brower, 7–21. Chicago: University of Chicago Press.

Guilford, T. C. 1990. The evolution of aposematism. In *Insect defenses: Adaptive mechanisms and strategies of prey and predators,* ed. D. L. Evans and J. O. Schmidt, 23–61. New York: SUNY Press.

Gupta, M. L. 1975. A list of chromosome numbers and sex chromosome mechanisms in mantids. *Ind. J. Zool.* 3: 27–34.

Gurney, A. B. 1950. Praying mantids of the United States, native and introduced. *Ann. Rep. Smithson. Inst.* 1950: 339–62.

Gustafson, M. P. 1993. Intraguild predation among larval plethodontid salamanders: A field experiment in artificial stream pools. *Oecologia* 96: 271–75.

Gwynne, D. T. 1981. Sexual difference theory: Mormon crickets show role reversal in mate choice. *Science* 213: 779–80.

Gwynne D. T. 1983. Male nutritional investment and the evolution of sexual differences in Tettigoniidae and other Orthoptera. In *Orthopteran mating systems,* ed. D. T. Gwynne and G. K. Morris, 337–66. Boulder: Westview Press.

Gwynne, D. T. 1984. Sexual selection and sexual differences in Mormon crickets (Orthoptera: Tettigoniidae, *Anabrus simplex*). *Evolution* 38: 1011–22.

Gwynne, D. T. 1985. Role–reversal in katydids: Habitat influences reproductive behavior (Orthoptera: Tettigoniidae: *Metaballus* sp.). *Behav. Ecol. Sociobiol.* 16: 355–61.

Gwynne, D. T. 1990. Testing parental investment and the control of sexual selection in katydids: The operational sex ratio. *Am. Nat.* 136: 474–84.

Gwynne, D. T. 1993. Food quality controls sexual selection in Mormon crickets by altering male mating investment. *Ecology* 74: 1406–13.

Gwynne, D. T., and L. W. Simmons. 1990. Experimental reversal of courtship roles in an insect. *Nature* 346: 172–74.

Gymer, A., and J. Edwards. 1967. The development of the insect nervous system. I. An analysis of postembryonic growth in the terminal ganglion of *Achaeta domesticus*. *J. Morphol.* 123: 191–97.

Hadden, F. C. 1927. A list of insects eaten by the mantis *Paratenodera sinensis* (Sauss.). *Proc. Hawaii Entomol. Soc.* 6: 385.

Hairston, N. G., Sr. 1989. *Ecological experiments*. New York: Cambridge University Press.

Hairston, N. G., Sr., F. E. Smith, and L. B. Slobodkin. 1960. Community structure, population control, and competition. *Am. Nat.* 94: 421–25.

Hamilton, W. J. 1971. Competition and thermoregulation behavior of the Namib Desert tenebrionid beetle genus Cardiosis. *Ecology* 52: 810–22.

Hardin, G. 1960. The competitive exclusion principle. *Science* 131: 1292–97.

Harmer, S. F., and A. E. Shipley, ed. 1922. *The Cambridge natural history*, vol. 5. London: Macmillan.

Harpaz, I. 1973. Early entomology in the Middle East. *Ann. Rev. Entomol.* 18: 21–36.

Harper, J. L. 1969. The role of predation in vegetational diversity. *Brookhaven Symp. Biol.* 22: 48–62.

Harris, M. O., S. Rose, and P. Malsch. 1993. The role of vision in the host plant finding behavior of the Hessian fly. *Physiol. Ent.* 18: 31–42.

Harron, A. L., and D. D. Yager. 1996. Juvenile hormone reduces auditory sensitivity in the praying mantis *Taumantis ehomanni*. *Soc. Neurosci. Abstr.* 22: 1144.

Harrus–Warrick, R. M., and E. A. Kravitz. 1984. Cellular mechanisms for modulation of posture by octopamine and serotonin in the lobster. *J. Neurosci.* 4: 1976–93.

Haskell, P. T., and J. E. Moorhouse. 1963. A blood–borne factor influencing the activity of the central nervous system of the desert locust. *Nature* 197: 56–58.

Hassell, M. P. 1978. *The dynamics of arthropod predator-prey systems*. Princeton, N. J.: Princeton University Press.

Hatziolos, M. E., and R. L. Caldwell. 1983. Role reversal in courtship in the stomatopod *Pseudosquilla ciliata* (Crustacea). *Anim. Behav.* 31: 1077–87.

Hausen, K., and N. J. Strausfeld. 1980. Sexually dimorphic interneuron arrangements in fly visual systems. *Proc. R. Soc. London* 208: 57–71.

Haynes, D. L., and P. Sisojevic. 1966. Predatory behavior of *Philodromus rufus* Walkneaer (Aranae: Thomisidae). *Can. Ent.* 98: 113–33.

Heath, G. L. 1980. Rearing and studying the praying mantids. *Amateur Ent. Soc. Leaflet* 36: 1–15.

Heessen, H. J. L., and A. M. H. Brunsting. 1981. Mortality of larvae of *Pterostichus oblongopunctatus* (Fabricius) (Col., Carabidae) and *Philonthus decorus* (Gravenhorst) (Col., Staphylinidae). *Nether. J. Zool.* 31: 729–45.

Helfer, J. R. 1963. *How to know the grasshoppers, cockroaches and their allies*. Dubuque: W. C. Brown.

Helmholtz, H. von. 1868. *Handbuch der physiologischen Optik*. Hamburg: Voss.

Henneguy, L. F. 1904. *Les insectes*. Paris: Masson.

Hess, W. R. 1954. *Das Zwischenhirn*, 2. Aufl. Basel: Schwabe.

Hess, W. R. 1957. Die Formatio reticularis des Hirnstammes im verhaltensphysiologischen Aspekt. *Arch. Psychiatr. Nervenkr.* 196: 329–36.

Hess, A. 1958. Experimental anatomical studies of pathways in the severed central nerve cord of the cockroach. *J. Morph.* 103: 479–85.

Hingston, R. 1932. *A naturalist in the Guiana Forest*. London: Edward Arnold.

Hirsch, H. V. B., D. Potter, D. Zawierucha, T. Choudhri, A. Glasser, R. K. Murphey, and D. Byers. 1990. Rearing in darkness changes visually–guided choice behavior in Drosophila. *Visual Neurosci.* 5: 281–89.

Hocking, B. 1964. Fire melanism in some African grasshoppers. *Evolution* 18: 332–35.

Holling, C. S. 1959. The components of predation as revealed by a study of small mammal predation of the European pine sawfly. *Can. Ent.* 91: 293–320.

Holling, C. S. 1964. The analysis of complex population processes. *Can. Ent.* 96: 335–47.

Holling, C. S. 1966. The functional response of invertebrate predators to prey density. *Mem. Ent. Soc. Can.* 48: 5–86.

Holmes, S. J. 1916. *Studies in animal behavior*. Boston: Richard G. Badger.

Holst, E. von. 1969. *Zur Verhaltensphysiologie bei Tieren und Menschen*. München: R. Piper & Co. Verlag.

Holst, E. von, and U. von Saint Paul. 1960. Vom Wirkungsgefüge der Triebe. *Naturwissenschaften* 47: 409–22.

Holt, R. D. 1977. Predation, apparent competition, and the structure of prey communities. *Theor. Pop. Biol.* 12: 197–229.

Holt, R. D., and J. H. Lawton. 1994. The ecological consequences of shared natural enemies. *Ann. Rev. Ecol. Syst.* 25: 495–520.

Hopper, K. R., P. H. Crowley, and D. Kielman. 1996. Density dependence, hatching synchrony, and within–cohort cannibalism in young dragonfly larvae. *Ecology* 77: 191–200.

Horridge, G. A. 1977. The compound eye of insects. *Sci. Am.* 237: 108–20.

Horridge, G. A. 1980. Apposition eyes of large diurnal insects as organs adapted to seeing. *Proc. R. Soc. London B* 207: 287–309.

Horridge, G. A. 1986. A theory of insect vision: Velocity parallax. *Proc. R. Soc. London B* 229: 13–27.

Horridge, G. A., and P. Duelli. 1979. Anatomy of the regional differences in the eye of the mantis Ciulfina. *J. Exp. Biol.* 80: 165–90.

Horridge, G. A., J. Duniec, and L. Marcelja. 1981. A 24-hour cycle in single locust and mantis photoreceptors. *J. Exp. Biol.* 91: 307–22.

Howard, L. O. 1886. The excessive voracity of the female mantis. *Science* 8: 326.

Hoy, R. R. 1992. The evolution of hearing in insects as an adaptation to predation from bats. In *The evolutionary biology of hearing*, ed. D. B. Webster, R. R. Fay, and A. N. Popper, 115–29. Heidelberg: Springer-Verlag.

Hoy, R. R., and D. Robert. 1996. Tympanal hearing in insects. *Ann. Rev. Ent.* 41: 433–50.

Hoyle, G. 1984. The scope of neuroethology. *Behav. Brain Sci.* 7: 367–412.

Hoyle, G., and D. L. Barker. 1975. Synthesis of octopamine by insect dorsal median unpaired neurones. *J. Exp. Zool.* 193: 433–39.

Huber, F. 1952. Verhaltensstudien am Männchen der Feldgrille (Gryllus campestris L.) nach Eingriffen am Zentralnervensystem. *Zool. Anz. Suppl.* 46: 138–49.

Huber, F. 1955. Sitz und Bedeutung nervöser Zentren für Instinkthandlungen bei Männchen der Gryllus campestris. *Z. Tierpsychol.* 12: 12–48.

Huber, F. 1960a. Untersuchungen zur nervösen Atmungsregelung der Orthopteren (Saltatoria: Gryllidae). *Z. Vergl. Physiol.* 43: 341–59.

Huber, F. 1960b. Untersuchungen über die Funktion des Zentralnervensystems und insbesondere des Gehirns bei der Fortbewegung und Lauterzeugung der Grillen. *Z. Vergl. Physiol.* 44: 60–132.

Huber, F. 1965a. Brain controlled behavior in orthopterans. In *Physiology of the central nervous system*, ed. J. E. Treherne and J. W. L. Beament, 233–46. New York: Academic Press.

Huber, F. 1965b. Aktuelle Probleme in der Physiologie des Nervensystems der Insekten. *Naturwiss. Rundschau* 18: 143–56.

Huber, F. 1970. Nervöse Grundlagen der akustischen Kommunikation bei Insekten. *Rheinisch-Westfäl. Akad. Wiss. Nat. Ing. Wirtschaftswiss., Vorträge N* 205: 41–84.

Huey, R. B., and A. F. Bennet. 1986. A comparative approach to field and laboratory studies in evolutionary biology. In *Predator-prey relationships: Perspectives and approaches from the study of lower vertebrates*, ed. M. E. Feder and G. V. Lauder, 82–98. Chicago: University of Chicago Press.

Hughes-Schrader, S. 1950. The chromosomes of mantids (Orthoptera: Manteidae) in relation to taxonomy. *Chromosoma* 4: 1–55.

Hughes-Schrader, S. 1953. Supplementary notes on the cyto-taxonomy of mantids (Orthoptera: Mantoidea). *Chromosoma* 6: 79–90.

Huheey, J. E. 1988. Mathematical models of mimicry. In *Mimicry and the evolutionary process*, ed. L. P. Brower, 22–41. Chicago: University of Chicago Press.

Hunter, M. D., and P. W. Price. 1992. Playing chutes and ladders: Heterogeneity and the relative roles of bottom-up and top-down forces in natural communities. *Ecology* 73: 724–32.

Hurd, L. E. 1985a. Ecological considerations of mantids as biocontrol agents. *Antenna* 9: 19–22.

Hurd, L. E. 1985b. On the importance of carrion to reproduction in an omnivorous estuarine neogastropod, *Ilyanassa obsoleta* (Say). *Oecologia* 65: 513–15.

Hurd, L. E. 1988. Consequences of divergent egg phenology to predation and coexistence in two sympatric, congeneric mantids (Orthoptera: Mantidae). *Oecologia* 76: 547–50.

Hurd, L. E. 1989. The importance of late season flowers to the fitness of an insect predator, *Tenodera sinensis* Saussure (Orthoptera: Mantidae) in an old field community. *Entomologist* 108: 223–28.

Hurd, L. E. 1991. Growth efficiency in juvenile mantids: Absence of selection for optimization in a food–limited environment. *Proc. Ent. Soc. Wash.* 93: 748–50.

Hurd, L. E., and R. M. Eisenberg. 1984a. Experimental density manipulations of the predator *Tenodera sinensis* (Orthoptera: Mantidae) in an old–field community. I. Mortality, development, and dispersal of juvenile mantids. *J. Anim. Ecol.* 53: 269–81.

Hurd, L. E., and R. M. Eisenberg. 1984b. Experimental density manipulations of the predator *Tenodera sinensis* (Orthoptera: Mantidae) in an old-field community. II. The influence of mantids on arthropod community structure. *J. Anim. Ecol.* 53: 955–67.

Hurd, L. E., and R. M. Eisenberg. 1989a. A midsummer comparison of sizes and growth rates of three sympatric mantids (Mantodea: Mantidae) in two old-field habitats. *Proc. Ent. Soc. Wash.* 91: 51–54.

Hurd, L. E., and R. M. Eisenberg. 1989b. The temporal distribution of hatching times in three sympatric mantids (Mantodea: Mantidae) with implications for niche separation and coexistence. *Proc. Ent. Soc. Wash.* 91: 55–58.

Hurd, L. E., and R. M. Eisenberg. 1990a. Arthropod community responses to manipulation of a bitrophic predator guild. *Ecology* 76: 2107–14.

Hurd, L. E., and R. M. Eisenberg. 1990b. Experimentally synchronized phenology and interspecific competition in mantids. *Am. Midl. Nat.* 124: 390–94.

Hurd, L. E., R. M. Eisenberg, W. F. Fagan, K. J. Tilmon, W. E. Snyder, K. S. Vandersall, S. G. Datz, and J. D. Welch. 1994. Cannibalism reverses male-biased sex ratio in adult mantids: Female strategy against food limitation? *Oikos* 69: 193–98.

Hurd, L. E., R. M. Eisenberg, M. D. Moran, T. P. Rooney, W. J. Gangloff, and V. M. Case. 1995. Time, temperature, and food as determinants of population persistence in the temperate mantid *Tenodera sinensis* (Mantodea: Mantidae). *Environ. Ent.* 24: 348–53.

Hurd, L. E., R. M. Eisenberg, and J. O. Washburn. 1978. Effects of experimentally manipulated density on field populations of the Chinese mantis (*Tenodera aridifolia sinensis* Saussure). *Am. Midl. Nat.* 99: 58–64.

Hurd, L. E., and W. F. Fagan. 1992. Cursorial spiders and succession: Age or habitat structure? *Oecologia* 92: 215–21.

Hurd, L. E., P. E. Marinari, and R. M. Eisenberg. 1989. The influence of temperature and photoperiod on early developmental rate of *Tenodera sinensis* Saussure (Mantodea: Mantidae). *Proc. Ent. Soc. Wash.* 91: 529–33.

Hurd, L. E., M. V. Mellinger, L. L. Wolf, and S. J. McNaughton. 1971. Stability and diversity at three trophic levels in terrestrial successional ecosystems. *Science* 173: 1134–36.

Hurd, L. E., and I. H. Rathet. 1986. Functional response and success in juvenile mantids. *Ecology* 67: 163–67.

Hurd, L. E., and L. L. Wolf. 1974. Stability in relation to nutrient enrichment in arthropod consumers of old-field successional ecosystems. *Ecol. Monogr.* 44: 465–82.

Hutchinson, G. E. 1957. Concluding remarks. *Cold Spring Harbor Symp. Quant. Biol.* 22: 415–27.

Hutchinson, G. E. 1959. Homage to Santa Rosalia, or why are there so many kinds of animals? *Am. Nat.* 93: 145–59.

Hutchinson, G. E. 1978. An introduction to population ecology. New Haven, Conn.: Yale University Press.

Ibbotson, M. R., and L. J. Goodman. 1990. Response characteristics of four wide-field motion-sensitive descending interneurons in Apis mellifera. *J. Exp. Biol.* 148: 255–79.

Ingle, D. 1968. Visual releasers of prey–catching behavior in frogs and toads. *Brain Behav. Evol.* 1: 500–518.

Ingle, D. 1976. Spacial vision in anurans. In *The amphibian visual system: A multidisciplinary approach*, ed. K. V. Fite, 119–202. New York: Academic Press.

Ingle, D. 1983. Brain mechanisms of visual localization by frogs and toads. In *Advances in vertebrate neuroethology*, ed. J.-P. Ewert, R. R. Capranica, and D. J. Ingle, 177–226. Plenum: New York.

Inoue, T. 1983a. Foraging strategy of a non-omniscient predator in a changing environment. I. Model with a data window and absolute criterion. *Res. Popul. Ecol.* 25: 81–104.

Inoue, T. 1983b. Foraging strategy of a non-omniscient predator in a changing environment. II. Model with two data windows and a relative comparison criterion. *Res. Popul. Ecol.* 25: 264–79.

Inoue, T., and T. Matsura. 1983. Foraging strategy of a mantid, *Paratenodera angustipennis* S.: Mechanisms of switching tactics between ambush and active search. *Oecologia* 56: 264–71.

Iwasaki, T. 1990. Predatory behavior of the praying mantis *Tenodera aridifolia sinensis*. I. Effect of prey size on prey density. *J. Ethol.* 8: 75–79.

Iwasaki, T. 1991. Predatory behavior of the praying mantis *Tenodera aridifolia sinensis*. II. Combined effect of prey size and predator size on prey recognition. *J. Ethol.* 9: 77–81.

Iwasaki, T. 1995. Habitat segregation in two praying mantises, *Tenodera aridifolia* (Stoll) and *Tenodera angustipennis* Saussure. Ph.D. diss., University of Osaka Prefecture.

Iwasaki, T. 1996. Comparative studies on the life histories of two praying mantises, *Tenodera aridifolia* (Stoll) and *Tenodera angustipennis* Saussure (Mantodea: Mantidae). I. Temporal pattern of egg hatch and nymphal development. *Appl. Ent. Zool.* 31: 345–56.

Jakobs, A.-K. 1993. Postembryonale Entwicklung des Beutefangverhaltens der Gottesanbeterin *Tenodera sinensis*. Master's thesis, University of Graz.

Jensen, A. 1956. Biology and physics of locust flight. III. The aerodynamics of locust flight. *Phil. Trans. Roy. Soc. (B)* 239: 511–52.

Johns, P. M., and M. R. Maxwell. 1997. Sexual cannibalism: Who benefits? *Trends Ecol. Evol.* 12: 127–28.

Johnson, E. 1937. *The praying mantis.* New York: Stackpole Sons.

Johnson, M. D. 1976. Concerning the feeding habits of the praying mantis *Tenodera aridifolia sinensis*, Saussure. *J. Kansas Ent. Soc.* 49: 164.

Jolly, G. M. 1965. Explicit estimates from capture–recapture data with both death and immigration—stochastic model. *Biometrika* 52: 225–47.

Jones, T. H., M. D. Moran, and L. E. Hurd. 1997. Cuticular extracts of five common mantids (Mantoidea: Mantidae) of the Eastern United States. *Comp. Biochem. Physiol.* 116: 419–22.

Joulin, P. 1983. L'élevage de Mantis religiosa à partir de spécimens capturés dans la nature. *Bull. Soc. Sci. Nat.* 37: 12–14.

Jovančić, L. 1960. Genèse des pigments tégumentaires et leur rôle physiologique chez la mante religieuse et chez d'autres espèces animales. *Mus. Hist. Nat. Beograd.* 29: 1–114.

Juelez, B. 1971. *The foundations of cyclopean perception.* Chicago: University of Chicago Press.

Kaczmarek, L. K., and L. Levitan, ed. 1987. *Neuromodulation.* New York: Oxford University Press.

Kaiser, W. 1983. Effects of non-visual and circadian inputs to visual interneurones in the honey bee. In *Fortschritte der Zoologie, 28: Horn (Hrsg.), Multimodal Convergences in Sensory Systems,* ed. E. Horn,149–66. Stuttgart: Gustav Fischer Verlag.

Kaiser, W. 1988. Busy bees need rest, too: Behavioural and electromyographical sleep signs in honey bees. *J. Comp. Physiol.* 163: 565–84.

Kaiser, W., and E. Liske. 1974. Die optomotorischen Reaktionen von fixiert fliegenden Bienen bei Reizung mit Spektrallichtern. *J. Comp. Physiol.* 89: 391–408.

Kaltenbach, A. P. 1996. Unterlagen für eine Monographie der Mantodea des südlichen Afrika: 1. Artenbestand, geographische Verbreitung und Ausbreitungsgrenzen (Insecta: Mantodea). *Ann. Naturhist. Mus. Wien* 98B: 193–346.

Kambhampati, S. 1995. A phylogeny of cockroaches and related insects based on DNA sequence of mitochondrial ribosomal RNA genes. *Proc. Nat. Acad. Sci. USA* 92: 2017– 20.

Karlson, R. H., and L. E. Hurd. 1993. Disturbance, coral reef communities, and changing ecological paradigms. *Coral Reefs* 12: 117–25.

Kaushik, R. K. 1985. Dual pattern in the mating behaviour of the mantid *Humbertiella similis* G. Tos (Mantodea, Mantidae). *Entomon. (Trivandrum, India)* 10: 335–37.

Kellog, V. J. 1905. *American insects.* New York: Henry Holt.

Kelner-Pillault, S. 1957. Attirance sexuelle chez *Mantis religiosa* (Orth.). *Bull. Soc. ent. France* 62: 9–11.

Kerfoot, W. C., and A. Sih. 1987. *Predation: Direct and indirect effects on aquatic communities.* Boston: Univ. Press of New England.

Kerry, C. J., and P. J. Mill. 1987. An anatomical study of the abdominal musculature, nervous and respiratory systems of the praying mantid, *Hierodula membranacea* (Burmeister). *R. Soc. London B* 229: 415–38.

Kettlewell, B. 1973. *The evolution of melanism.* Oxford: Clarendon Press.

Kevan, D. K. M. 1978. The land of the locusts: Being some further verses on grigs and cicadas, pt. I. *Mem. Lyman Ent. Mus. Res. Lab.* 6: 1–530.

Kevan, D. K. M. 1983. The land of the locusts: Being some further verses on grigs and cicadas, pt. II. *Mem. Lyman Ent. Mus. Res. Lab.* 10: 1–554.

Kevan, D. K. M. 1985. The land of the locusts: Being some further verses on grigs and cicadas, pt. III. *Mem. Lyman Ent. Mus. Res. Lab.* 16: 1–446.

Kevan, D. K. M. 1985. The mantis and the serpent. *Ent. Month. Mag.* 121: 1–8.

Kick, S. A. 1982. Target-detection by the echolocating bat, *Eptesicus fuscus. J. Comp. Physiol.* 145: 431–35.

Kien, J. 1973. A difference between the sexes in an optomotor response in the cabbage white butterfly, *Pieris rapae* L. *Experientia* 29: 492.

Kingsley, J. S., ed. 1884. *The standard natural history.* Boston: S. E. Cassino.

Kirby, W. F. 1904. *A synonymic catalogue of Orthoptera I.* London: British Museum of Natural History.

Kirby, W., and W. Spence. 1815–26. *An introduction to entomology,* vols. 1–4. London: Longman, Hurst, Rees, Orme and Brown.

Kirmse, R., and W. Kirmse. 1985. Struktur und Funktion der fovealen Differenzierung bei Fangschrecken (Mantodea). *Zool. Jb. Physiol.* 89: 169–80.

Kneib, R. T. 1988. Testing for indirect effects of predation in an intertidal soft-bottom community. *Ecology* 69: 1795–1805.

Köck, A. 1992. Morphologische und optische Indizien für eine postembryonale Entwicklung des binokularen Mechanismus der Entfernungsmessung bei der Gottesanbeterin *Tenodera sinensis.* Master's thesis, University of Graz.

Köck, A., A.-K. Jakobs, and K. Kral. 1993. Visual prey discrimination in monocular and binocular praying mantis *Tenodera sinensis* during postembryonic development. *J. Insect Physiol.* 39: 485–91.

Kopec, S. 1912. Über die Funktionen des Nervensystems der Schmetterlinge während sukzessiven Stadien ihrer Metamorphose. *Zool. Anz.* 40: 353–60.

Kral, K. 1989. Fine structure of the larval eyes of *Mantispa* sp. (Neuroptera: Planipennia, Mantispidae). *Int. J. Insect Morphol. Embryol.* 18: 135–43.

Kral, K. 1990. The Planipennia eye using *Mantispa styriaca* (Poda, 1761) (Mantispidae) as an example. *Neuroptera Int.* 6: 51–56.

Kral, K., and M. Poteser. 1997. Motion parallax as a source of distance information in locusts and mantids. *Insect Behav.* 10: 145–63.

Kral, K., K. Herbst, and M.-A. Pabst. 1990. The compound eye of *Mantispa styriaca* (Neuroptera: Planipennia). *Zool. Jb. Physiol.* 94: 333–43.

Kravitz, E. A. 1986. Serotonin, octopamine, and proctolin: Two amines and a peptide, and aspects of lobster behaviour. In *Fast and slow chemical signalling in the nervous system*, ed. L. L. Iversen and E. Goodman, 244–59. Oxford: Oxford Science Publications.

Krebs, C. J. 1985. *Ecology*, 3d ed. New York: Harper & Row.

Krebs J. R., and A. Kacelnik. 1991. Decision-making. In *Behavioural ecology*, 3d ed., ed. J. R. Krebs and N. B. Davies, 105–36. Oxford: Blackwell Scientific Publications.

Kristensen, N. P. 1995. Forty years' insect phylogenetic systematics. *Zool. Beitr. N. F.* 36: 83–124.

Krombholz, P. H. 1977. Food consumption in female mantids. Ph.D. diss., Tufts University.

Kruess, A., and T. Tscharntke. 1994. Habitat fragmentation, species loss, and biological control. *Science* 264: 1581–84.

Kumar, R. 1973. The biology of some Ghanaian mantids (Dictyoptera: Mantodea). *Bull. Inst. Fond. Afr. Noire* 35: 551–78.

Kutsch, W. 1985. Pre-imaginal flight motor pattern in locusta. *J. Insect Physiol.* 31: 581–86.

Kutsch, W., and R. Kittmann. 1991. Flight motor pattern in flying and non–flying Phasmida. *J. Comp. Physiol.* 168: 483–90.

Kutsch, W., and P. Stevenson. 1984. Manipulation of the endocrine system of locusta and the development of the flight motor pattern. *J. Comp. Physiol.* 155: 129–38.

Kynaston, S. E., P. McErlain–Ward, and P. J. Mill. 1994. Courtship, mating behaviour and sexual cannibalism in the praying mantis, *Sphodromantis lineola*. *Anim. Behav.* 47: 739–41.

Kyriacou, C. P. 1987. Death and reproduction in mantids. *Trends Ecol. Evol.* 2: 349–50.

La Greca, M. 1949. L'evoluzione della pleure pterotoraciche degli Insetti Blattotteroidi. *Boll. Zool., Torino.* 16: 119–29.

La Greca, M., and A. Rainone. 1949. Il dermascheletro e la muscolatura dell addome di *Mantis religiosa*. *Ann. Inst. Mus. Zool. Università Napoli* 1: 1–43.

Lampa, S. 1894. Engendomliga vanir hos Mantidernas honor. *Ent. Tidskrift.* 15: 118.

Land, M. F., and T. S. Collett. 1974. Chasing behaviour of houseflies (*Fannia canicularis*): A description and analysis. *J. Comp. Physiol.* 89: 331–57.

Laurent, P. 1898. A species of Orthoptera. *Ent. News* 9: 144–45.

Lawler, S. P., and P. J. Morin. 1993. Temporal overlap, competition and priority effects on larval anurans. *Ecology* 74: 174–82.

Lawrence, S. E. 1991. Sexual cannibalism in praying mantids. Ph.D. diss., University of Sheffield, England.

Lawrence, S. E. 1992. Sexual cannibalism in the praying mantid, *Mantis religiosa*: A field study. *Anim. Behav.* 43: 569–83.

Lawrence, W. S. 1986. Male choice and competition in *Tetraopes tetraophthalmus:* The effect of local sex ratio variation. *Behav. Ecol. Sociobiol.* 18: 289–96.

Lawton, J. H. 1971. Maximum and actual feeding rates in larvae of the damselfly *Pyrrhosoma nymphula* (Sulzer) (Odonata: Zygoptera). *Fresh. Biol.* 1: 99–111.

Lawton, J. H., and M. P. Hassell. 1981. Asymmetrical competition in insects. *Nature* 289: 793–95.

Lawton, J. J., and D. R. Strong, Jr. 1981. Community patterns and competition in folivorous insects. *Am. Nat.* 118: 317–38.

Lea, J. Y., and C. G. Mueller. 1977. Saccadic head movements in mantids. *J. Comp. Physiol.* 114: 115–28.

Lefroy, H. M. 1923. *Manual of entomology*. New York: Longmans, Green & Co.

Leitinger, G. 1994. Frühe postembryonale Entwicklung des Komplexauges und der Lamina ganglionaris der Gottesanbeterin nach Photodegeneration der akuten Zone mit Sulforhodamin. Master's thesis, University of Graz.

Leitinger, G., M.-A. Pabst, and K. Kral. 1994. Foveale Applikation von Sulforhodamin hat strukturelle Auswirkungen auf die postembryonale Entwicklung des Komplexauges der Gottesanbeterin. *Verh. Dtsch. Zool. Ges.* 87: 252.

Lenski, R. E. 1984. Food limitation and competition: A field experiment with two Carabus species. *J. Anim. Ecol.* 53: 203–16.

Lettvin, J. Y., H. R. Maturana, W. S. McCulloch, and W. H. Pitts. 1959. What the frog's eye tells the frog's brain. *Proc. Inst. Radio Eng. N. Y.* 47: 1940–51.

Lettvin, J. Y., H. R. Maturana, W. S. McCulloch, and W. H. Pitts. 1961. Two remarks on the visual system of the frog. In *Sensory communication,* ed. W. Rosenblith, 757–76. Cambridge: MIT Press.

Levereault, P. 1936. The morphology of the Carolina mantis. *Univ. Kansas Sci. Bull.* 24: 205–59.

Levin, L., and H. Maldonado. 1970. A fovea in the praying mantis eye. III. The centering of prey. *Z. Vergl. Physiol.* 67: 93–101.

Levine, R. B. 1984. Changes in neuronal circuits during insect metamorphosis. *J. Exp. Biol.* 112: 27–44.

Levine, R. B., and J. W. Truman. 1982. Metamorphosis of the insect nervous system: Changes in morphology and synaptic interactions of identified cells. *Nature* 299: 250–52.

Lewis, B. 1983. Directional cues for auditory localization. In *Bioacoustics—a comparative approach,* ed. B. Lewis, 233–60. New York: Academic Press.

Lewis, W. J., and J. H. Tumlinson. 1988. Host detection by chemically mediated associative learning in a parasitic wasp. *Nature* 331: 257–59.

Leyhausen, P. 1965. Über die Funktion der relativen Stimmungshierarchie (dargestellt am Beispiel der phylogenetischen und ontogenetischen Entwicklung des Beutefangs von Raubtieren). *Z. Tierpsychol.* 22: 412–94.

Lichtenstein, A. A. H. 1802. A dissertation on two natural genera hitherto confounded under the name of Mantis. *Trans. Linn. Soc.* 6: 1–39.

Lieftinck, M. A. 1953. Biological and ecological observations on a bark haunting mantid in Java (Orthopt., Mantodea). *Int. Congr. Ent.,IX, Amsterdam* 2: 125–34.

Linnaeus, C. 1758. *Systema Naturae,* 10th ed. Holmiae: L. Salvius.

Linsley, E. G., and R. L. Usinger. 1966. Insects of the Galapagos Islands. *Proc. Calif. Acad. Sci. (fourth series)* XXXIII 7: 113–96.

Liske, E. 1977. The influence of head position on the flight behaviour of the fly, *Calliphora erythrocephala. J. Insect Physiol.* 23: 377–79.

Liske, E. 1989. Behavioral hierarchies of the praying mantis, with particular reference to neuroethological studies of the mating behavior. In *Neural mechanisms of behaviour. Proc. of 2nd Int. Congr. of Neuroethology,* ed. J. Erber, R. Menzel, H. J. Pflüger, and D. Todt. Stuttgart: Georg Thieme Verlag.

Liske, E. 1991a. Abdominal mating movements of the male praying mantis: Motor programs and their modulation by sensory organs. In *Synapse, transmission, modulation. Proc. 19th Gittingen Neurobiol. Conf.,* ed. N. Elsner, and H. Penzlin. Stuttgart: Georg Thieme Verlag.

Liske, E. 1991b. Sensorimotor control of abdominal mating movements in the male praying mantis. *Zool. Jb. Physiol.* 95: 465–73.

Liske, E. 1991c. Verhaltenshierarchie der Chinesischen Gottesanbeterin *Tenodera aridifolia sinensis* (Insecta, Mantodea) unter besonderer Berücksichtigung des Sexualverhaltens. Braunschweig, Germany: Habilitationsschrift der TU Braunschweig.

Liske E., and W. J. Davis. 1984. Sexual behaviour of the Chinese praying mantis. *Anim. Behav.* 32: 916–17.

Liske, E., and W. J. Davis. 1986. Behavioral suppression of head grooming in the male praying mantis during mating. *Naturwissenschaften* 73: 333–34.

Liske, E., and W. J. Davis. 1987. Courtship and mating behaviour of the Chinese praying mantis, *Tenodera aridifolia sinensis. Anim. Behav.* 35: 1524–37.

Liske, E., U. Heckele, and H. G. Wolff. 1986. Verhaltensanalyse visuell induzierter Kopfbewegungen bei der Gottesanbeterin, *Tenodera aridifolia sinensis. Verh. Dtsch. Zool. Ges.* 79: 225.

Liske, E., K. Köchy, and H. G. Wolff. 1988. The thoracic nerves and muscles in the praying mantis *Ternodera aridifolia sinensis.* In *Sense organs, Proc. 16th Göttingen Neurobiol. Conf.,* ed. N. Elsner and F. G. Barth. Stuttgart: Georg Thieme Verlag.

Liske, E., K. Köchy, and H. G. Wolff. 1989. The thoracic nervous system of the Chinese praying mantis, *Tenodera aridifolia sinensis:* Peripheral distribution of the nerves and the musculature they supply. *Zool. Jb. Anat.* 118: 191–99.

Liske, E., and W. Mohren. 1984. Saccadic head movements of the praying mantis, with particular reference to visual and proprioceptive information. *Physiol. Ent.* 9: 29–38.

Liu, G. 1939. Some extracts from the history of entomology in China. *Psyche* 46: 23–28.

Loher, W. 1989. Temporal organization of reproductive behavior. In *Cricket behavior and neurobiology,* ed. F. Huber, T. E. Moore, and W. Loher, 83–113. Ithaca: Cornell University.

Loher, W., and M. Dambach. 1989. Reproductive behavior. In *Cricket behavior and neurobiology,* ed. F. Huber, T. E. Moore, and W. Loher, 43–82. Ithaca: Cornell University Press.

Long, T. F., and L. L. Murdock. 1983. Stimulation of blowfly feeding behavior by octopaminergic drugs. *Proc. Nat. Acad. Sci.* 80: 4159–63.

Lorenz, K. 1950. The comparative method in studying innate behavior patterns. *Symp. Soc. Exp. Biol.* 4: 421–68.

Lorenz, K. 1981. *The foundations of ethology.* New York: Springer-Verlag.

Louda, S. M. 1982. Inflorescent spiders: A cost/benefit analysis for the host plant, *Haplopappus venetus* Blake (Asteraceae). *Oecologia* 55: 185–91.

Lovelock, J. E. 1988. *The ages of Gaia: A biography of our living earth.* Norton: New York.

Loxton, R. G. 1979. On display behaviour and courtship in the praying mantis *Ephestiasula amoena* (Bolivar). *Zool. J. Linn. Soc.* 65: 103–10.

Loxton, R. G., and I. Nicholls. 1979. The functional morphology of the praying mantis forelimb (Dictyoptera: Mantodea). *J. Zool. Linn. Soc.* 66: 185–203.

Lucas, J. R. 1985. Metabolic rates and pit–construction costs of two antlion species. *J. Anim. Ecol.* 54: 295–309.

Macfadyen, A. 1963. *Animal ecology*, 2d ed. New York: Pitman.

MacKinnon, J. 1970. Indications of territoriality in mantids. *Z. Tierpsychol.* 27: 150–55.

Maguire, B., Jr., D. Belk, and G. Wells. 1968. Control of community structure by mosquito larvae. *Ecology* 49: 207–10.

Maldonado, H. 1970a. A fovea in the praying mantis eye. II. Some morphological characteristics. *Z. Vergleich. Physiol.* 67: 79–92.

Maldonado, H. 1970b. The deimatic reaction in the praying mantis *Stagmatoptera biocellata. Z. vergl. Physiol.* 68: 60–71.

Maldonado, H., and J. C. Barros-Pita. 1970. A fovea in the praying mantis eye. I. Estimation of the catching distance. *Z. vergl. Physiol.* 67: 58–78.

Maldonado. H., M. Benko, and M. Isern. 1970. A study of the role of the binocular vision in mantids to estimate long distances, using the deimatic reaction as experimental situation. *Z. Vergl. Physiol.* 68: 72–83.

Maldonado, H., and L. Levin. 1967. Distance estimation and the monocular cleaning reflex in praying mantis. *Z. Vergl. Physiol.* 56: 258–67.

Maldonado, H., L. Levin, and J. C. Barros-Pita. 1967. Hit distance and the predatory strike of the praying mantis. *Z. Vergl. Physiol.* 56: 237–57.

Maldonado, H., E. Rodriguez, and N. Balderrama. 1974. How mantids gain insight into the new maximum catching distance after each ecdysis. *J. Insect Physiol.* 20: 591–603.

Mann, T. 1984. *Spermatophores: Development, structure, biochemical attributes and role in the transfer of spermatozoa.* Berlin: Springer-Verlag.

Manning, J. T. 1980. Sex ratio and optimal male time investment strategies in *Asellus aquaticus* (L.) and *A. meridianus* Racovitsza (Crustacea: Isopoda). *Behaviour* 74: 265– 73.

Marder, E. 1989. Modulation of neural networks underlying behaviour. *Sem. Neurosci.* 1: 3–4.

Marder, E., and S. L. Hooper. 1985. Neurotransmitter modulation of the stomatogastric ganglion of decapod crustaceans. In *Model neural networks and behavior,* ed. A. I. Selverston, 319–37. New York: Plenum.

Margalef, R. 1963. On certain unifying principles in Ecology. *Am. Nat.* 97: 357–74.

Margulis, L., and K. V. Schwartz. 1982. *Five kingdoms.* San Francisco: W. H. Freeman.

Marler P., and W. Hamilton. 1967. *Mechanisms of animal behavior.* New York: John Wiley & Sons.

Marquis, R. J., and C. J. Whelan. 1994. Insectivorous birds increase growth of white oak through consumption of leaf-chewing insects. *Ecology* 75: 2007–14.

Mathis, U., and S. Rossel. 1993. Binocular projection areas in the brain of the praying mantis. A possible anatomical correlate for stereoscopic vision? *Proc. 21th Göttingen Neurobiol. Conf.* 523.

Mathis,U., S. Eschbach, and S. Rossel. 1992. Functional binocular vision is not dependent on visual experience in the praying mantis. *Visual Neurosci.* 9: 199–203.

Mathur, R. N. 1946. Notes on the biology of some Mantidae. *Indian J. Ent.* 8: 89–106.

Matsuda, R. 1970. Morphology and evolution of the insect thorax. *Mem. Ent. Soc. Canada.* 76: 1–431.

Matsura, T. 1981. Responses to starvation in a mantis, *Paratenodera angustipennis* (S.). *Oecologia* 50: 291–95.

Matsura, T. 1982. Ecological studies on the feeding of the mantid, *Paratenodera angustipennis.* Ph.D. diss., University of Kyoto.

Matsura, T. 1987. An experimental study on the foraging behavior of a pit–building antlion larva, *Myrmeleon bore. Res. Popul. Ecol.* 29: 17–26.

Matsura, T., T. Inoue, and Y. Hosomi. 1975. Ecological studies of a mantid, *Paratenodera angustipennis* de Saussure. I. Evaluation of the feeding condition in natural habitat. *Res. Popul. Ecol.* 17: 64–76.

Matsura, T., and K. Morooka. 1983. Influences of prey density on fecundity in a mantis, *Paratenodera angustipennis* (S.). *Oecologia* 56: 306–12.

Matsura, T., and T. Murao. 1994. Comparative study on the behavioral response to starvation in three species of antlion larvae (Neuroptera: Myrmeleontidae). *J. Insect Behav.* 7: 873–84.

Matsura, T., and S. Nagai. 1983. Estimation of prey

consumption of a mantid, *Paratenodera angustipennis* (S.) in a natural habitat. *Res. Pop. Ecol.* 25: 298–308.

Matsura, T., and K. Nakamura. 1981. Effects of prey density on mutual interferences among nymphs of a mantis, *Paratenodera angustipennis* (S.). *Jpn. J. Ecol.* 31: 221–23.

Matsura, T., H. Yoshimaya, and T. Nagai. 1984. Growth, prey consumption and food assimilation efficiency in a mantid, *Paratenodera angustipennis*. *Kontyu (Tokyo)* 52: 37–49.

Matsura, T., T. Satomi, and K. Fujiharu. 1991. Control of the life cycle in a univoltine antlion, *Myrmeleon bore* (Neuroptera). *Jpn. J. Ent.* 59: 275–87.

Maturana, W. S., J. Y. Lettvin, W. S. McCulloch, and W. H. Pitts. 1960. Anatomy and physiology of vision in the frog (*Rana Pipens*). *J. Gen. Physiol.* (Suppl. 2) 43: 129–75.

Maxwell, M. R. Submitted–a. Does a single meal affect female reproductive output in the sexually cannibalistic mantid, *Iris oratoria*? *Behaviour.*

Maxwell, M. R. Submitted–b. Male choice and sexual cannibalism.

Maxwell, M. R. In press. The risk of cannibalism and male mating behaviour in the Mediterranean praying mantid, *Iris oratoria*. *Behaviour.*

Maxwell, M. R. 1998a. Lifetime mating opportunities and male mating behaviour in sexually cannibalistic praying mantids. *Anim. Behav.* 55: 1011–28.

Maxwell, M. R. 1998b. Seasonal adult sex ratio shift in the praying mantid *Iris oratoria* (Mantodea: Mantidae). *Environ. Ent.* 27: 318–23.

Maxwell, M. R. 1995. Sexual cannibalism and male mating behavior in praying mantids. Ph.D. diss., University California, Davis.

Maxwell, M. R., and O. Eitan. 1998. Range expansion of an introduced mantid *Iris oratoria* and niche overlap with a native mantid *Stagmomantis limbata* (Mantodea: Mantidae). *Ann. Ent. Soc. Am.* 91: 422–29.

Mayer, H., and K. Kral. 1993. Electrophysiological and optical studies of the spectral sensitivity of *Mantispa styriaca* (Neuroptera: Planipennia). *Mitt. Dtsch. Ges. Allg. Angew. Ent.* 8: 709–13.

Maynard Smith, J. 1977. Parental investment: A prospective analysis. *Anim. Behav.* 25: 1–9.

Mayr, E. 1970. *Populations, species and evolution.* Cambridge: Harvard University Press.

Mayr, E. 1982. *The growth of biological thought.* Cambridge, Mass.: Belknap Press.

McBrien, H., R. Harmsen, and A. Crowder. 1983. A case of insect grazing affecting plant succession. *Ecology* 64: 1035–39.

McCulloch, F. 1960. *Mediaeval Latin and French bestiaries.* Chapel Hill: University of North Carolina Press.

McKittrick, F. A. 1964. Evolutionary studies of cockroaches. *Memoir 389. Cornell Univ. Agricultural Expt. Sta.*, Ithaca, NY.

McLaren, B. E., and R. O. Peterson. 1994. Wolves, moose, and tree rings on Isle Royale. *Science* 266: 1555–58.

McNaughton, S. J., and L. L. Wolf. 1979. *General Ecology,* 2d ed. New York: Holt, Rinehart and Winston.

McNeil, J. N. 1991. Behavioral ecology of pheromone-mediated communication in moths and its importance in the use of pheromone traps. *Ann. Rev. Ent.* 36: 407–30.

McNeill, J. 1901. Entomological results (4): Orthoptera. Papers from the Hopkins Stanford Galapagos Expedition, 1898–99. *Proc. Wash. Acad. Sci.* 111: 487–506.

Meissner, O. 1909. Biologische Beobachtungen an der indischen Stabheuschrecke *Dixippus morosus* Br. (Phas. Orth.). *Z. Wiss. Insekten Biol.* 5: 14–21, 55–61, 87–95.

Mellinger, M. V., and S. J. McNaughton. 1975. Structure and function of successional vascular plant communities in central New York. *Ecol. Monogr.* 34: 161–82.

Menzel, R. 1983. Neurobiology of learning and memory: The honeybee as a model system. *Naturwissenschaften* 70: 504–11.

Michelsen, A., and H. Nocke. 1974. Biophysical aspects of sound communication in insects. *Adv. Insect Physiol.* 9: 247–96.

Michelsen, A., and O. N. Larsen. 1985. Hearing and Sound. In *Comprehensive insect physiology, biochemistry, and pharmacology,* ed. G. A. Kerkut and L. I. Gilbert, 495–556. Oxford: Pergamon.

Milburn, N. S., and D. R. Bentley. 1971. On the dendritic topology and activation of cockroach giant interneurons. *J. Insect Physiol.* 17: 607–23.

Miles, R. N., Robert, D., and Hoy, R. R. 1995. Mechanically coupled ears for directional hearing in the parasitoid fly *Ormia ochracea*. *J. Acoust. Soc. Am.* 98: 3059–70.

Miller, R. G. (1981) *Simultaneous statistical inference.* New York: Springer.

Miller, P. L. 1972. Swimming in mantids. *J. Entolmol.* 46: 91–97.

Mimura, K. 1986. Development of visual pattern discrimination in the fly depends on light experience. *Science* 232: 83–85.

Mimura, K. 1987. The effect of partial covering of the

eye on the results of selective deprivation of visual pattern in the fly. *Brain Res.* 437: 97–102.

Minnis, P., E. F. Harrison, L. L. Stowe, G. G. Gibson, F. M. Denn, D. R. Doelling, and W. L. Smith. 1993. Radiative climate forcing by the Mount Pinatubo eruption. *Science* 259: 1411–15.

Mittelstaedt, H. 1957. Prey capture in mantids. In *Recent advances in invertebrate physiology*, ed. B. T. Scheer, 57–71. Eugene: University Oregon Press.

Mittelstaedt, H. 1962. Control systems of orientation in insects. *Ann. Rev. Ent.* 7: 177–98.

Mittelstaedt, H. 1971. Reafferenzprinzip-Apologie und Kritik. In *Erlanger Physiologentagung*, ed. W. D. Keidel and K.-H. Plattig. Heidelberg: Springer-Verlag.

Möbius, K. 1877. *Die Auster und die Austernwirtschaft.* Berlin: Wiegundt, Hempel, and Parey, Rep. U.S. Fish. Comm. 1880 (transl.).

Moffett, T. 1634. *Insectorvm sive minimorum animalium theatrvm.* London: Thomas Cotes.

Mook, L. J., and D. M. Davies. 1966. The European praying mantis (*Mantis religiosa* L.) as a predator of the red–legged grasshopper (*Melanoplus femurrubrum* De Geer). *Can. Ent.* 98: 913–18.

Moore, N. H. 1901. The *mantis religiosa* in Rochester, N. Y. *Sci. Am.* 84: 105–6.

Moran, M. D. 1995. Intraguild predation between sympatric species of mantids (Mantodea: Mantidae). *Proc. Ent. Soc. Wash.* 97: 634–38.

Moran, M. D., and L. E. Hurd. 1994a. Environmentally determined male–biased sex ratio in a praying mantid. *Am. Midl. Nat.* 132: 205–8.

Moran, M. D., and L. E. Hurd. 1994b. Short–term responses to elevated predator densities: Noncompetitive intraguild interactions and behavior. *Oecologia* 98: 269–73.

Moran, M. D., and L. E. Hurd. 1997. Relieving food limitation reduces survivorship of a generalist predator. *Ecology* 78: 1266–70.

Moran, M. D., and L. E. Hurd. 1998. A trophic cascade in a diverse arthropod community caused by a generalist arthropod predator. *Oecologia* 113: 126–32.

Moran, M. D., T. P. Rooney, and L. E. Hurd. 1996. Top down cascade from a bitrophic predator in an old field community. *Ecology* 77: 2219–27.

Morge, G. 1973. Entomology in the western world in antiquity and in Medieval times. *Ann. Rev. Entomol.* 18: 37–80.

Morin, P. J. 1986. Interactions between intraspecific competition and predation in an amphibian predator–prey system. *Ecology* 67: 713–20.

Morin, P. J., S. P. Lawler, and E. A. Johnson. 1988. Competition between aquatic insects and vertebrates: Interaction strength and higher order interactions. *Ecology* 69: 1401–9.

Morse, D. H., and R. S. Fritz. 1982. Experimental and observational studies of patch choice at different scales by the crab spider *Misumena vatia*. *Ecology* 63: 172–82.

Mourgue, M. 1909. Un reptile chassé et tué par un insecte. La Feuille des jeunes. *Naturalistes* 39: 87.

Müller, F. 1878. Notes on Brazilian Entomology. *Trans. Ent. Soc. London* 211–23.

Murdoch, W. W., J. Chesson, and P. L. Chesson. 1985. Biological control in theory and practice. *Am. Nat.* 125: 344–66.

Nachtigall, W. 1981a. The influence of changes in geometrical wing shape on the aerodynamic function of the fore-wing in the Desert locust: Further analysis of Jensen's investigations. *Zool. Jb. Anat.* 106: 1–11.

Nachtigall, W. 1981b. The forewings of large locusts as generators of aerodynamic forces. I. Model measurements of aerodynamic effects of various wing profiles. *J. Comp. Physiol.* 142: 127–34.

Natural history of insects, The. 1835. Vol. 2. London: John Murray.

Nel, A., and R. Roy. 1996. Revision of the fossil "mantid" and "ephemerid" species described by Piton from the Palaeocene of Menat (France) (Mantodea: Chaeteessidae, Mantidae; Ensifera: Tettigonioidea. *Eur. J. Ent.* 93: 223–34.

Nesbitt, H. H. J. 1941. A comparative morphological study of the nervous system of the orthoptera and related orders. *Annal. Entomol. Soc. Am.*, 34: 51–81.

Newman, E. 1874. A living mantid exhibited. *Entomologist* 7: 188.

Newman, J. A., and M. A. Elgar. 1991. Sexual cannibalism in orb–weaving spiders: An economic model. *Am. Nat.* 138: 1372–95.

Nickle, D. A. 1981. Predation on a mouse by the Chinese Mantid *Tenodera aridifolia sinensis*, Saussure (Dictyoptera: Mantoidea). *Proc. Ent. Soc. Wash.* 83: 802–3.

Norsgaard, E. J. 1975. Connubial cannibalism. *Nat. Hist.* 84: 58–63.

Nye, S. W., and R. E. Ritzmann 1992. Motion analysis of leg joints associated with escape turns of the cockroach, *Periplaneta americana*, *J. Comp. Physiol. A*, 171: 183–94.

O'Shea, M., and C. H. F. Rowell. 1976. The neuronal basis of a sensory analyzer, the acridid movement detector system. II. Response decrement, convergence, and the nature of the excitatory afferents to the fan-like dendrites of the LGMD. *J. Exp. Biol.* 65: 289–308.

Olberg, R. M. 1981. Object- and self-movement detectors in the ventral nerve cord of the dragonfly. *J. Comp. Physiol.* 141: 327–34.

Otto, D. 1971. Untersuchungen zur zentralnervösen Kontrolle der Lauterzeugung von Grillen. *Z. Vergl. Physiol.* 74: 227–71.

Pacala, D., and J. Roughgarden. 1984. Control of arthropod abundance by Anolis lizards on St. Eustatius (Neth. Antilles). *Oecologia* 64: 160–62.

Packard, A. S. 1869. *Guide to the study of insects.* Salem: Naturalist's Book Agency.

Page, T. L. 1985. Clocks and circadian rhythms. In *Comprehensive insect physiology, biochemistry and pharmacology,* vol. 6, ed. G. A. Kerkut and L. I. Gilbert, 578–652. London: Pergamon Press.

Paine, R. T. 1966. Food web complexity and species diversity. *Am. Nat.* 100: 65–75.

Paine, R. T. 1969. A note on trophic complexity and community stability. *Am. Nat.* 103: 91–93.

Paine, R. T. 1992. Food–web analysis through field measurement of per capita interaction strength. *Nature* 355: 73–75.

Panov, A. A. 1966. Correlations in the ontogenetic development of the central nervous system in the house cricket *Gryllus domesticus* L. and the mole cricket *Gryllotalpa gryllotalpa* L. (Orthoptera, Grylloidea). *Ent. Rev.* 45: 179–85.

Paradise, C. J., and N. E. Stamp. 1993. Episodes of unpalatable prey reduce consumption and growth of juvenile praying mantids. *J. Insect Behav.* 6: 155–66.

Parker, G. A. 1979. Sexual selection and sexual conflict. In *Sexual selection and reproductive competition in insects,* ed. M. S. Blum and N. A. Blum, 123–66. New York: Academic Press.

Parker, G. A., and L. W. Simmons. 1989. Nuptial feeding in insects: Theoretical models of male and female interests. *Ethology* 82: 3–26.

Parker, G. A., L. W. Simmons, and H. Kirk. 1990. Analysing sperm competition data: Simple models for predicting mechanisms. *Behav. Ecol. Sociobiol.* 27: 55–65.

Parrish, J. W., J. A. Ptacek, and K. L. Will. 1984. The detection of near–ultraviolet light by nonmigratory and migratory birds. *Auk* 101: 53–58.

Partridge, J. C. 1989. The visual ecology of avian cone oil droplets. *J. Comp. Physiol.* 165: 415–26.

Partridge, L., and P. H. Harvey. 1988. The ecological context of life history evolution. *Science* 241: 1449–54.

Pearson, D. L., and C. B. Knisley. 1985. Evidence for food as a limiting resource in the life cycle of tiger beetles (Coleoptera: Cicindelidae). *Oikos* 45: 161–68.

Pemberton, R. W. 1990. The selling of Gampsocleis gratiosa, Brunner (Orthoptera: Tettigoniidae) as singing pets in China. *Pan–Pacif. Ent.* 66: 93–95.

Pfau, H. K. 1977. Zur Morphologie und Funktion des Vorderflügels und Vorderflügelgelenks von *Locusta migratoria* L. *Fortschr. Zool.* 24: 341–45.

Pfau, H. K. 1978. Funktions-anatomische Aspekte des Insektenflugs. *Zool. J. (Anat).* 99: 99–108.

Pflüger, H.-J. 1977. The control of the rocking movements of the phasmid *Carausius morosus* Br. *J. Comp. Physiol.* A 120: 181–202.

Pflumm, W. 1968. Zum Verhalten nektarsammelnder Honigbienen an der Futterquelle. *Verh. Dtsch. Zool. Ges.* 61: 381–87.

Pflumm, W. 1969. Die Beziehungen zwischen Putzverhalten und Sammelbereitschaft bei der Honigbiene. *Z. Vergl. Physiol.* 64: 1–36.

Phipps, J. 1960. The breeding biology of *Sphodromantis lineola* Burm. (Dictyoptera, Mantidae) in Sierra Leone. *Ent. Month. Mag.* 96: 192–93.

Pianka, E. R. 1966. Latitudinal gradients in species diversity. *Am. Nat.* 100: 33–46.

Pimm, S. L., J. H. Lawton, and J. E. Cohen. 1991. Food web patterns and their consequences. *Nature* 350: 669–74.

Plotnikova, S. I., and G. A. Nevmyvaka. 1980. The methylene blue technique: Classic and recent applications to the insect nervous system. In *Neuroanatomical techniques, insect nervous system,* ed. N. J. Strausfeld and T. A. Miller, 1–19. Berlin: Springer-Verlag.

Polis, G. A. 1981. The evolution and dynamics of intraspecific predation. *Ann. Rev. Ecol. Syst.* 12: 225–51.

Polis, G. A., C. A. Myers, and R. D. Holt. 1989. The ecology and evolution of intraguild predation. *Ann. Rev. Ecol. Syst.* 20: 297–330.

Polis, G. A., ed. 1990. *The biology of scorpions.* Stanford: Stanford University Press.

Pollack, A. J., R. E. Ritzmann, and J. T. Watson. 1994. Dual pathways for tactile sensory information to thoracic interneurons in the cockroach. *J. Neurobiol.* 26: 33–46.

Pollard, S. D. 1993. Little murders. *Nat. Hist.* 102: 58–65.

Portmann, A. 1956. *Animal camouflage.* Ann Arbor: University of Michigan Press.

Post, W. M., and C. C. Travis. 1979. Quantitative stability in models of ecological communities. *J. Theor. Biol.* 79: 547–53.

Poteser, M., and K. Kral. 1995. Visual distance discrimination between stationary targets in praying mantis: An index of the use of motion parallax. *J. Exp. Biol.* 198: 2127–37.

Power, M. E. 1992. Top–down and bottom–up forces in food webs: Do plants have primacy? *Ecology* 73: 733–46.

Prazdny, K. 1985. Detection of binocular disparities. *Biol. Cybern.* 52: 93–99.

Presnell, J. K., and M. P. Schreibman. 1997. *Humason's Animal Tissue Techniques.* 5th ed. Baltimore: The Johns Hopkins University Press.

Preston–Mafham, K. 1990. *Grasshoppers and mantids of the world.* London: Blandford.

Preston–Mafham, R. A., and K. G. Preston–Mafham. 1993. *The encyclopedia of land invertebrate behaviour.* Cambridge: MIT Press.

Prete, F. R. 1990a. Prey capture in mantids: The role of the prothoracic tibial flexion reflex. *J. Insect Physiol.* 36: 335–38.

Prete, F. R. 1990b. Configural prey selection in the praying mantis, *Sphodromantis lineola* (Burm.). *Brain Behav. Evol.* 36: 300–6.

Prete, F. R. 1992a. The discrimination of visual stimuli representing prey versus non-prey by the praying mantis, *Sphodromantis lineola* (Burm.). *Brain Behav. Evol.* 39: 285–88.

Prete, F. R. 1992b. The effects of background pattern and contrast on prey discrimination by the praying mantis *Sphodromantis lineola* (Burm.). *Brain Behav. Evol.* 40: 311–20.

Prete, F. R. 1993. Stimulus configuration and location in the visual field affect appetitive responses to computer generated stimuli by the praying mantis *Sphodromantis lineola* (Burr.). *Vis. Neurosci.* 10: 997–1005.

Prete, F. R. 1995. Designing behavior: A case study. In *Perspectives in ethology, vol. 11: Behavioral design,* ed. N. S. Thompson, 255–77. New York: Plenum Press.

Prete, F. R., and K. Cleal. 1996. The predatory strike of free ranging praying mantises, *Sphodromantis lineola* (Burm.). I. Strikes in the midsagittal plane. *Brain Behav. Evol.* 48: 173–90.

Prete, F. R., C. A. Klimek, and S. P. Grossman. 1990. The predatory strike of the praying mantis, *Tenodera aridifolia sinensis* (Sauss.). *J. Insect. Physiol.* 36: 561–65.

Prete, F. R., H. Lum and S. P. Grossman. 1992. Non–predatory ingestive behaviors of the praying mantids *Tenodera aridifolia sinensis* (Sauss.) and *Sphodromantis lineola* (Burm.). *Brain Behav. Evol.* 39: 124–32.

Prete, F. R., and R. J. Mahaffey. 1993a. A chamber for mass hatching and early rearing of praying mantids (Orthoptera: Mantidae). *Entomol. News* 104: 47–52.

Prete, F. R., and R. J. Mahaffey. 1993b. Appetitive responses to computer generated visual stimuli by the praying mantis *Sphodromantis lineola* (Burm.). *Visual Neurosci.* 10: 669–79.

Prete, F. R., and T. McLean. 1996. Responses to moving small–field stimuli by the praying mantis, *Sphodromantis lineola* (Burmeister). *Brain Behav. Evol.* 47: 42–54.

Prete, F. R., P. J. Placek, M. A. Wilson, R. J. Mahaffey, and R. R. Nemcek. 1993. Stimulus speed and order of presentation effect the visually released predatory behaviors of the praying mantis *Sphodromantis lineola* (Burm.). *Brain Behav. Evol.* 42: 281–94.

Prete, F. R., and M. M. Wolfe. 1992. Religious suppliant, seductive cannibal, or reflex machine? In search of the praying mantis. *J. Hist. Biol.* 25: 91–136.

Price, P. W. 1987. The role of natural enemies in insect populations. In *Insect outbreaks,* ed. P. Barbosa and J. Shultz, 287–312. New York: Academic Press.

Price, P. W., C. E. Bouton, P. Gross, B. A. McPheron, J. W. Thompson, and A. E. Weis. 1980. Interactions among three trophic levels: Influence of plants on interactions between insect herbivores and natural enemies. *Ann. Rev. Ecol. Syst.* 11: 41–65.

Przibram, H. 1907. Die Lebensgeschichte der Gottesanbeterinnen (Fangheuschrecken). *Zeit. Wiss. Insektenbiol.* 3: 117–22 and 147–53.

Przibram, H. 1930. Wachstumsmessungen an *Sphodromantis bioculata* Burm. Zunahme der Facettengrösse und –anzahl. *Wilhelm Roux's Arch. Entw. mechan. Organ.* 122: 280–99.

Pulliam, R. H. 1988. Sources, sinks, and population regulation. *Am. Nat.* 132: 652–61.

Quesnel, V. C. 1967. Observations on the reproductive behaviour of the mantis, *Acontiothespis multicolor. J. Trinidad Field Nat. Club.* 53–56.

Rabaud, E. 1916. Accouplement d'un mâle décapité de *Mantis religiosa* L. (Orth.). *Bull. Soc. ent. France* 21: 57–59.

Ragge, D. R. 1955. *The wing-venation of the Orthoptera Saltatoria with notes on the Dictyopteran venation.* London: British Museum of Natural History.

Ragge, D. R., and R. Roy. 1967. A review of the praying mantises of Ghana (Dictyoptera Mantodea). *Bull. Inst. fond. Afr. Noire* 29A: 586–644.

Ramsay, G. W. 1984. *Miomantis caffra*, a new mantid record (Mantodea: Mantidae) for New Zealand. *New Zealand Ent.* 8: 102–4.

Ramsay, G. W. 1990. *Mantodea (Insecta), with a review of aspects of functional morphology and biology.* Wellington: DSIR Publishing.

Rathet, I. H., and L. E. Hurd. 1983. Ecological relationships among three co–occurring mantids, *Tenodera sinensis* (Saussure), *T. angustipennis* (S.), and *Mantis religiosa* (L.). *Am. Midl. Nat.* 110: 240–48.

Rau, P., and N. Rau. 1913. The biology of Stagmomantis carolina. *Trans. Acad. Sci. St. Louis* 22: 1–58.

Richards, L. J. 1983. Feeding and activity patterns of intertidal beetle. *J. Exp. Mar. Biol. Ecol.* 73: 213–24.

Ricklefs, R. E. 1987. Community diversity: Relative roles of local and regional processes. *Science* 235: 167–71.

Ridley, M. 1989. The incidence of sperm displacement in insects: Four conjectures, one corroboration. *Biol. J. Linn. Soc.* 38: 349–67.

Ridpath, M. G. 1977. Predation on frogs and small birds by *Hierodula werneri* (G. T.) (Mantidae) in tropical Australia. *J. Aust. Ent. Soc.* 16: 153–54.

Riechert, S. E., and L. Bishop. 1990. Prey control by an assemblage of generalist predators: Spiders in garden test systems. *Ecology* 71: 1441–50.

Riley C., and L. O. Howard. 1892. The female rear–horse versus the male. *Insect Life* 5: 145.

Rilling, S., H. Mittelstaedt, and K. D. Roeder. 1959. Prey recognition in the praying mantis. *Behaviour* 14: 164–84.

Rind, F. C., and P. J. Simmons. 1992. Orthopteran DCMD neuron: A reevaluation of responses to moving objects. I. Selective responses to approaching objects. *J. Neurophysiol.* 68: 1654–66.

Ritzmann, R. E., and A. J. Pollack. 1986. Identification of thoracic interneurons that mediate giant interneuron–to–motor pathways in the cockroach. *J. Comp. Physiol.* 159: 639–54.

Ritzmann, R., A. J. Pollack, S. E. Hudson, and A. Hyvonen. 1991. Convergence of multi–modal sensory signals at thoracic interneurons of the cockroach, *Periplaneta americana*. *Brain Res.* 563: 175–83.

Robert, D., M. P. Read, and R. R. Hoy. 1994. The tympanal hearing organ of the parasitoid fly *Ormia ochracea* (Diptera, Tachinidae, Ormiini). *Cell Tissue Res.* 275: 63–78.

Roberts, R. A. 1937a. Biology of the bordered mantid, *Stagmomantis limbata* Hahn (Orthoptera, Mantidae). *Ann. Ent. Soc. Am.* 30: 96–109.

Roberts, R. A. 1937b. Biology of the minor mantid, *Litaneutria minor* Scudder (Orthoptera, Mantidae). *Ann. Ent. Soc. Am.* 30: 111–19.

Robinson, M. H. 1969a. Defenses against visually hunting predators. In *Evolutionary biology 3*, ed. T. Dobzhansky, M K. Hecht, and W. C. Steere, 225–59. New York: Meredith Corporation.

Robinson, M. H. 1969b. The defensive behaviour of some orthopteroid insects from Panama. *Trans. R. Ent. Soc. London* 121: 281–303.

Robinson, M. H. 1973. Insect anti–predator adaptations and the behavior of predatory primates. *Acta. IV Congr. Latinamer. Zool.* 2: 811–36.

Robinson, M. H. 1981. A stick is a stick and not worth eating: On the definition of mimicry. *Biol. J. Linn. Soc.* 16: 15–20.

Robinson, M. H., and B. Robinson. 1978. Culture techniques for *Acanthops falcata*, a neotropical mantid suitable for biological studies (with notes on raising web building spiders). *Psyche* 85: 239–47.

Robinson, M. H., and B. Robinson. 1979. By dawn's early light: Matutinal mating and sex attractants in a neotropical mantid. *Science* 205: 825–27.

Roeder, K. D. 1935. An experimental analysis of the sexual behavior of the praying mantis (*Mantis religiosa*, L.). *Biol. Bull.* 69: 203–20.

Roeder, K. D. 1937. The control of tonus and locomotor activity in the praying mantis (*Mantis religiosa*, L.). *J. Exp. Zool.* 76: 353–74.

Roeder, K. D. 1948. Organization of the ascending giant fiber system in the cockroach (*Periplaneta americana*). *J. Exp. Zool.* 108: 243–61.

Roeder, K. D. 1963. *Nerve cells and insect behavior.* Cambridge: Harvard University Press.

Roeder, K. D. 1959. A physiological approach to the relation between prey and predator. *Smithsonian Misc. Collns.* 137: 286–311.

Roeder, K. D. 1967. *Nerve cells and insect behavior,* rev. ed. Cambridge: Harvard University Press.

Roeder, K. D. 1975. Neural factors and evitability in insect behavior. *J. Exp. Zool.* 194: 75–88.

Roeder, K. D., L. Tozian, and E. A. Weiant. 1960. Endogenous nerve activity and behaviour in the mantis and cockroach. *J. Insect Physiol.* 4: 45–62.

Römer, H., V. Marquart, and M. Hardt. 1988. Organization of a sensory neuropile in the auditory pathway of two groups of Orthoptera. *J. Comp. Neurol.* 275: 201–15.

Rooney, T. R., A. T. Smith, and L. E. Hurd. 1996. Global warming and the regional persistence of a temperate zone insect. *Am. Midl. Nat.* 136: 84–93.

Roonwal, M. L. 1937. Studies on the embryology of the

African migratory locust *Locusta migratoria* m. II. Organogeny. *Phil. Trans. R. Soc. London* 227 (543): 175–244.

Roonwal, M. L. 1938. The frightening attitude of a desert mantid, *Eremiaphila braueri* Kr. (Orthoptera, Mantodea). *Proc. R. Ent. Soc. London* 13: 71–72.

Root, R. B. 1967. The niche exploitation pattern of the blue–gray gnatcatcher. *Ecol. Monogr.* 37: 317–50.

Root, R. B. 1973. Organization of a plant–arthropod association in simple and diverse habitats: The fauna of collards (*Brassica oleracea*). *Ecol. Monogr.* 43: 95–124.

Rösel von Rosenhof, J. A. 1749. *Insecten–Belustigung.* Nürnberg: Zweiter Teil.

Rösel von Rosenhof, J. A. 1769. *Insecten–Belustigung.* Nürnberg: Vierter Teil.

Rosenheim, J. A., L. R. Wilhoit, and C. A. Armer. 1993. Influence of intraguild predation among generalist insect predators on the suppression of an herbivore population. *Oecologia* 96: 439–49.

Rosenthal, G. A., and M. R. Berenbaum. 1991. *Herbivores: Their interaction with secondary metabolites.* New York: Academic Press.

Rossel, S. 1979. Regional differences in photoreceptor performance in the eye of the praying mantis. *J. Comp. Physiol.* 131: 95–112.

Rossel, S. 1980. Foveal fixation and tracking in the praying mantis. *J. Comp. Physiol.* 139: 307–31.

Rossel, S. 1983. Binocular stereopsis in an insect. *Nature* 302: 821–22.

Rossel, S. 1986. Binocular spatial localization in the praying mantis. *J. Exp. Biol.* 120: 265–81.

Rossel, S. 1991. Spatial vision in the praying mantis. Is distance implicated in size detection? *J. Comp. Physiol. A* 169: 101–8.

Rossel, S. 1996. Binocular vision in insects: How mantids solve the correspondence problem. *Proc. Nat. Acad. Sci., USA* 93: 3229–32.

Rossel, S., U. Mathis, and T. Collett. 1992. Vertical disparity and binocular vision in the praying mantis. *Vis. Neurosci.* 8: 165–70.

Rothschild, M. 1975. Remarks on carotenoids in the evolution of signals. In *Coevolution of animals and plants*, ed. L. E. Gilbert and P. H. Raven, 20–50. Austin: University of Texas Press.

Rothschild, M. 1981. The mimicrats must move with the times. *Biol. J. Linn. Soc.* 16: 21–23.

Rowell, C. H. F. 1971. The Orthopteran descending movement detector (DMD) Neurons: A characterization and a review. *Z. vergleich. Physiol.* 73: 167–94.

Rowell, C. H. F., and M. O'Shea. 1976a. The Neuronal basis of a sensory analyser, the acridid movement detector system. I. Effects of simple incremental and decremental stimuli in light and dark adapted animals. *J. Exp. Biol.* 65: 273–88.

Rowell, C. H. F., and M. O'Shea. 1976b. The Neuronal basis of a sensory analyser, the acridid movement detector system. III. Control of response amplitude by tonic lateral inhibition. *J. Exp. Biol.* 65: 617–25.

Rowell, C. H. F., M. O'Shea, and J. L. D. Williams. 1977. The Neuronal basis of a sensory analyser, the acridid movement detector system. IV. The preference for small field stimuli. *J. Exp. Biol.* 68: 157–85.

Roy, R. 1965. Les Mantes de la Guinée forestière. *Bull. Inst. f. Afr. noire* 27: 577–612.

Roy, R. 1967. Contribution à la connaissance des genres Mantis Linné et Paramantis, nov. [Mantidae]. *Bull. Inst. f. Afr. noire* 29A: 126–49.

Roy, R. 1971. Contribution à L'étude biologique du Sénégal septentrional, IV. Dictyoptères Mantodea. *Bull. Inst. f. Afr. noire* 32A: 1019–33.

Roy, R. 1973. Premier inventaire des Mantes du Gabon. *Biol. Gabon.* 8: 235–90.

Roy, R. 1987. General observations on the systematics of Mantodea. In *Evolutionary biology of the Orthopteroid insects*, ed. B. Baccetti, 483–88. New York: Halsted Press (John Wiley & Sons).

Roy, R. 1987. Overview of the biogeography of African mantids. In *Evolutionary biology of Orthopteroid insects*, ed. B. M. Baccetti, 489–95. New York: Halsted Press (John Wiley & Sons).

Roy, R., and D. Leston. 1975. Mantodea of Ghana: New species, further records and habitats. *Bull. Inst. f. Afrique noire, Series A* 29: 586–644.

Rupprecht, R. 1971. Bewegungsmimikry bei *Carausius morosus* Br. (Phasmida). *Experientia* 27: 1437–38.

Ryan, M. J. 1985. *The túngara frog: A study in sexual selection and communication.* Chicago: University of Chicago Press.

Rypstra, A. L. 1983. The importance of food and space in limiting web-spider densities: A test using field enclosures. *Oecologia* 59: 317–20.

Sachs, L. 1974. *Angewandte Statistik.* Berlin: Springer-Verlag.

Sakai, M., T. Katayama, and Y. Taoda. 1990. Postembryonic development of mating behavior in the male cricket *Gryllus bimaculatus* DeGeer. *J. Comp. Physiol.* 166: 775–84.

Salt, R. W., and H. G. James. 1947. Low temperature as a factor in the mortality of eggs of *Mantis religiosa*. *Can. Ent.* 79: 33–36.

Sandeman, D. C. 1968. A sensitive position measuring device for biological systems. *Comp. Biochem. Physiol.* 24: 635–38.

Sandness, J. N., and J. A. McMurtry. 1972. Prey consumption behavior of *Amblyseius largoensis* in relation to hunger. *Can. Ent.* 102: 692–704.

Sathe, A. A., and P. V. Joshi. 1986. Spermathecal histology of virgin and mated females of the mantid, *Hierodula coarctata*, West (Dictyoptera: Mantidae). *Current Sci.* 55: 1042–44.

Sauer, K. P., C. Grüner, and K.-G. Collatz. 1986. Critical points in Time and their influence on life cycle. In *Insect aging strategies and mechanisms*, ed. K.-G. Collatz and R. S. Sohol, 9–22. Berlin: Springer-Verlag.

Saussure, H. de. 1869. Essai d'un système des Mantides. *Mittheil. Schw. ent. Gesellsch.* 3: 49–73.

Saussure, H. de. 1870. Additions au système des Mantides. *Mittheil. Schw. ent. Gesellsch.* 3: 221–44.

Sbrenna, G. 1971. Postembryonic growth of the ventral nerve cord in *Schistocerca gregaria* F. (Orthoptera: Acrididae). *Bull. Zool.* 38: 49–74.

Scanlan, J. J., trans. 1987. *Albert the Great, man and the beasts, de animalibus*. Binghamton: Medieval & Renaissance Texts & Studies.

Schauff, M. E., and J. C. Jones. 1978. The sexual behavior of *Tenodera sinensis* (Saus.) (Orthoptera: Mantidae). *J. New York Ent. Soc.* 86: 319.

Schneider, E. L., and J. W. Rowe. 1990. *Handbook of the biology of aging*. New York: Academic Press.

Schoener, T. W. 1982. The controversy over interspecific competition. *Am. Sci.* 70: 586–95.

Schürg-Pfeiffer, E., and J.-P. Ewert. 1981. Investigation of neurons involved in the analysis of Gestalt prey features in the frog *Rana temporaria*. *J. Comp. Physiol.* 141: 139–52.

Schwind, R. 1989. Size and distance perception in compound eyes. In *Facets of vision*, ed. P. G. Stavenga and R. C. Hardie, 425–444. Berlin: Springer-Verlag.

Scudder, S. H. 1893. The Orthoptera of the Galapagos Islands. *Bull. Mus. Comp. Zoology Harvard* XXV 1: 1–24.

Seber, G. A. F. 1965. A note on the multiple recapture census. *Biometrika* 52: 249–59.

Sharov, A. G. 1962. Redescription of *Lithophotina floccosa* Cock. (Manteodea) with some notes on Manteod wing venation. *Psyche.* 69: 102–6.

Sharp, D. 1899. The modification and attitude of *Idolium diabolicum*, a mantis of the kind called "floral simulators." *Proc. Camb. Phil. Soc.* 10: 175–80.

Shaw, G. 1806. *General zoology*. London: G. Kearley.

Shaw, S. R. 1994. Re–evaluation of the absolute threshold and response mode of the most sensitive known "vibration" detector, the cockroach's subgenual organ: A cochlea-like displacement threshold and a direct response to sound. *J. Neurobiol.* 25: 1167–85.

Shelford, R. 1903. Bionomical notes on some Bornean Mantidae. *Zoologist* 4: 293–304.

Shelford, V. E. 1911. Physiological animal geography. *J. Morph.* 22: 551–618.

Shelly, T. E., and W. J. Bailey. 1992. Experimental manipulation of mate choice by male katydids: The effect of female encounter rate. *Behav. Ecol. Sociobiol.* 30: 277–82.

Shmida, A., and M. Wilson. 1985. Biological determinants of species diversity. *J. Biogeog.* 12: 1–20.

Sholes, O. D. 1984. Responses of arthropods to the development of goldenrod inflorescences (Solidago: Asteraceae). *Am. Midl. Nat.* 112: 1–14.

Sih, A., L. B. Kats, and R. D. Moore. 1992. Effects of predatory sunfish on the density, dirft, and refuge use of stream salamander larvae. *Ecology* 73: 1418–30.

Sih, A., P. Crowley, M. McPeek, J. Petranka, and L. Strohmeier. 1985. Predation, competition, and prey communities: A review of field experiments. *Ann. Rev. Ecol. Syst.* 16: 269–311.

Simberloff, D., and W. Boecklen. 1981. Santa Rosalia reconsidered: Size ratios and competition. *Evolution* 35: 1206–28.

Simmons, L. W., and W. J. Bailey. 1990. Resource influenced sex roles of zaprochiline tettigoniids (Orthoptera: Tettigoniidae). *Evolution* 44: 1853–68.

Simmons, L. W., and G. A. Parker. 1989. Nuptial feeding in insects: Mating effort versus paternal investment. *Ethology* 81: 332–43.

Simmons, P. J., and F. C. Rind. 1992. Orthopteran DCMD neuron: A reevaluation of responses to moving objects. II. Critical cues for detecting approaching objects. *J. Neurophysiol.* 68: 1667–82.

Simon, H.-R. 1966. Vorstellungen über die Konkurrenz von Verhaltensweisen. *Ent. Blätter* 62: 121–24.

Simpson, S. J., and P. R. White. 1990. Associative learning and locust feeding: Evidence for a "learned hunger" for protein. *Anim. Behav.* 40: 506–13.

Slifer, E. H. 1968. Sense organs on the antennal flagellum of a praying mantis, *Tenodera angustipennis*, and of two related species (Mantodea). *J. Morph.* 124: 105–16.

Smedes, G. W., and L. E. Hurd. 1981. An empirical test of community stability: Resistance of a fouling community to a biological patch-forming disturbance. *Ecology* 62: 1561–72.

Smith, J. E. 1807. *A sketch of a tour on the continent*, vol. 1. London: Longman, Hurst, Rees, and Orme.

Smith, R. B., and T. P. Mommsen. 1984. Pollen feeding in an orb-weaving spider. *Science* 226: 1330–32.

Snodderly, M. D. 1979. Visual discriminations encountered in food foraging by a neotropical primate: Implications for the evolution of color vision. In *The behavioral significance of color*, ed. E. H. J. Burtt, 237–79. New York: Garland STPM Press.

Snodgrass, R. E. 1935. *Principles of insect morphology*. New York: McGraw–Hill.

Snyder, A. W. 1977. Acuity of compound eyes: Physical limitations and design. *J. Comp. Physiol. A* 116: 161–82.

Snyder, A. W., D. G. Stavenga, and S. B. Laughlin. 1977. Spatial information capacity of compound eyes. *J. Comp. Physiol. A* 116: 183–207.

Snyder, W. E., and L. E. Hurd. 1995. Egg–hatch phenology and intraguild predation between two mantid species. *Oecologia* 104: 496–500.

Sobel, E. C. 1990. The locust's use of motion parallax to measure distance. *J. Comp. Physiol.* 167: 579–88.

Sontag, C. 1971. Spectral sensitivity of the visual system of the praying mantis, *Tenodera sinensis*. *J. Gen. Physiol.* 57: 93–112.

Spangler, H. G. 1988. Moth hearing, defense, and communication. *Ann. Rev. Ent.* 33: 59–81.

Spight, T. M. 1967. Species diversity: A comment on the role of the predator. *Am. Nat.* 101: 467–74.

Spiller, D. A. 1984. Competition between two spider species: Experimental field study. *Ecology* 65: 909–19.

Spiller, D. A. 1986. Interspecific competition between spiders and its relevance to biological control by generalist predators. *Environ. Ent.* 15: 177–81.

Spiller, D. A., and T. W. Schoener. 1990. A terrestrial field experiment showing the impact of eliminating top predators on foliage damage. *Nature* 347: 469–72.

Spiller, D. A., and T. W. Schoener. 1994. Effects of top and intermediate predators in a terrestrial food web. *Ecology* 75: 182–96.

Srinivasan, M. V. 1992. Distance perception in insects. *Curr. Direct. Psychol. Sci.* 1: 22–25.

Srinivasan, M., M. Lehrer, and R. Wehner. 1987. Bees perceive illusory colours induced by movement. *Vision Res.* 27: 1285–89.

Srinivasan, M. V., M. Lehrer, S. W. Zhang, and G. A. Horridge. 1989. How honey bees measure their distance from objects of unknown size. *J. Comp. Physiol.* 165: 605–13.

Stäger, R. 1928. *Mantis religiosa* L. als Musikerin. *Z. Insekt Biol.* 23: 162–64.

Stål C. 1873. Recherches sur le système des Mantides. *Bihang till K. Svenska vet. Akad. Handlingar* 1: 1–26.

Stål, C. 1877. Systema Mantodeorum. Essai d'une systèmatisation nouvelle des Mantodées. *Bihang till K. Svenska vet. Akad. Handlingar* 4: 1–91.

Starret, A. 1993. Adaptive resemblance: A unifying concept for mimicry and crypsis. *Biol. J. Linn. Soc.* 48: 299–317.

Stephens, D. W., and J. R. Krebs. 1986. *Foraging theory*. Princeton, N. J.: Princeton University Press.

Stephenson, E. M., and C. Stewart. 1955. *Animal camouflage*, 2d ed. London: Adam & Charles Black.

Stevenson, P. A., and W. Kutsch. 1987. A reconsideration of central pattern generator concept for locust flight. *J. Comp. Physiol.* 161: 115–29.

Stoll, C. 1787. *Natuurlyke, en naar 't leeven nauwkeurig gekleurde Afbeeldingen, en Beschryvingen der Spooken, Wandelnde Bladen, Zabelspringhanen, Krekels, Trekspringhaanen en Kakkerlakken*. Amsterdam: J. C. Sepp.

Strausfeld, N. J. 1980. Male and female visual neurons in dipteran insects. *Nature* 283: 381–83.

Stierle, I. E., M. Getman, and C. M. Comer. 1994. Multisensory control of escape in the cockroach *Periplaneta americana*. I. Initial evidence from patterns of wind-evoked behavior. *J. Comp. Physiol. A* 174: 1–11.

Strong, D. R. 1992. Are trophic cascades all wet? Differentiation and donor-control in speciose ecosystems. *Ecology* 73: 747–54.

Strong, D. R., and D. Simberloff. 1981. Straining at gnats and swallowing ratios: Character displacement. *Evolution* 35: 810–12.

Suckling, D. M. 1985. Laboratory studies on the praying mantis Orthodera ministralis (Mantodea: Mantidae). *New Zealand Ent.* 8: 96–101.

Svensson, B. G., and E. Petersson. 1988. Non-random mating in the dance fly *Empis borealis:* The importance of male choice. *Ethology* 79: 307–16.

Svidersky, V. L. 1969. Receptors of the forehead of the locust Locusta migratoria in ontogenesis. *Zh. evol. Biokhim. Fiziol.* 5: 482–90.

Sykes, M. L. 1904. *Protective resemblance in the Insecta*. Manchester: Hinchliffe.

Takafuji, A., and D. J. Chant. 1976. Comparative studies of two species of predaceous mites (Acarina: Phytoseiidae), with special reference to their response to the density of their prey. *Res. Pop. Ecol.* 17: 255–310.

Teale, E. W. 1935. Dinosaur of the insect world. *Travel* 64: 22–25, 47.

Terra, P. S. 1992. Zelo materno em Cardioptera brachyptera (Mantodea, Vatidae, Photininae). *Revta bras. Ent.* 36: 493–503.

Terra P. S. 1995. Revisão sistematica dos gêneros de louva-a-Deus da região neotropical (Mantodea). *Revta bras. Ent.* 39: 13–94.

Thayer, A. H. 1896. The law which underlies protective coloration. *Auk* 13: 124–29.

Thayer, G. H. 1909. *Concealing-coloration in the Animal Kingdom. An exposition of the laws of disguise through color and pattern.* New York: Macmillan.

Thompson, E., A. Palacios, and F. J. Varela. 1992. Ways of coloring: Comparative color vision as a case study for cognitive science. *Behav. Brain Sci.* 15: 1–26.

Thorne, B. L., and J. M. Carpenter. 1992. Phylogeny of the Dictyoptera. *Syst. Ent.* 17: 253–68.

Thornhill, R. 1976. Sexual selection and paternal investment in insects. *Am. Nat.* 110: 153–63.

Thornhill, R. 1984. Alternative female choice tactics in the scorpionfly *Hylobittacus apicalis* (Mecoptera) and their implications. *Am. Zool.* 24: 367–83.

Thornhill, R., and J. Alcock. 1983. *The evolution of insect mating systems.* Cambridge: Harvard University Press.

Thorpe, W. H. 1979. *The origins and rise of ethology.* London: Heinemann.

Tilman, D. 1996. Biodiversity: Population versus ecosystem stability. *Ecology* 77: 350–63.

Tinbergen, N. 1951. *The study of instinct.* London: Oxford University Press.

Tinkham, E. R. 1938. Western Orthoptera attracted to lights. *J. New York Ent. Soc.* 46: 339–53.

Topsell, E. 1608. *The historie of four-footed beasts and serpents.* London: William Jaggard.

Topsell, E. 1658. *The historie of four-footed beasts, serpents and insects.* London: E. Cotes.

Trimarchi, J. R., and A. M. Schneiderman. 1995. Initiation of flight in the unrestrained fly *Drosophila melanogaster. J. Zool. London* 235: 211–22.

Trivers, R. L. 1972. Parental investment and sexual selection. In *Sexual selection and the descent of man,* ed. B. Campbell, 139–79. Chicago: Aldine.

Truman, J. W., J. C. Weeks, and R. B. Levine. 1985. Developmental plasticity during the metamorphosis of an insect nervous system. In *Comparative neurobiology,* ed. M. Cohen and F. Strumwasser, 25–44. New York: Wiley.

Tsai, J. H. 1982. Entomology in the People's Republic of China. *N. Y. Ent. Soc.* 90: 186–212.

Tulk, A. 1844. Habits of the mantis. *Ann. Mag. Nat. Hist.* 14: 78.

Turner, J. R. G. 1988. The evolution of mimicry: A solution to the problem of punctuated equilibrium. In *Mimicry and the evolutionary process,* ed. L. P. Brower, 42–66. Chicago: University of Chicago Press.

Tyrer, N. M. 1969. Time course of contraction and relaxation in embryonic locust muscle. *Nature* 224: 815–17.

Uexküll, J. von. 1909. *Umvelt und Innenwelt.* Berlin.

Vance, R. R. 1979. Effects of grazing by the sea urchin *Centrostephanus coronatus,* on prey community composition. *Ecology* 60: 537–46.

Vandermeer, J. H. 1980. Indirect mutualism: Variations on a theme by Stephen Levine. *Am. Nat.* 116: 441–48.

Vane–Wright, R. I. 1976. A unified classification of mimetic resemblances. *Biol. J. Linn. Soc.* 8: 25–56.

Vane–Wright, R. I. 1980. On the definition of mimicry. *Biol. J. Linn. Soc.* 12: 1–6.

Varley, G. C. 1939. Frightening attitudes and floral simulation in praying mantids. *Proc. R. Ent. Soc. London* 14: 91–96.

Via, S. E. 1977. Visually mediated snapping in the bulldog-ant: A perceptual ambiguity between size and distance. *J. Comp. Physiol.* 121: 33–51.

Vickery, V. R., and D. K. M. Kevan. 1983. A monograph of the Orthopteroid insects of Canada and adjacent regions. *Lyman Ent. Mus. Res. Lab.* 13: 216–37.

Vickery V., and D. K. M. Kevan. 1985. *The insects and arachnids of Canada.* Agriculture Canada.

Viitala, J., E. Korpimäki, P. Palokangas, and M. Kolvula. 1995. Attraction of kestrels to vole scent marks visible in ultraviolet light. *Nature* 373: 425–27.

Vonnegut, K., Jr. 1963. *Cat's cradle.* New York: Delacorte.

Walcher, F., and K. Kral. 1994. Visual deprivation and distance estimation in the praying mantis larva. *Physiol. Entomol.* 19: 230–40.

Waliczky, Z. 1991. Guild structure of beetle communities in three stages of vegetational succession. *Acta Zool. Hung.* 37: 313–24.

Walker, W. F. 1980. Sperm utilization strategies in nonsocial insects. *Am. Nat.* 115: 780–99.

Wallace, A. R. (1867). On the Pieridae of the Indian and Australian regions. *Trans. Entomol. Soc. London* 4: 301–415.

Waloff, N. 1968. Studies on the insect fauna on scotch broom *Sarothamnus scoparius* (L.) Wimmer. *Adv. Ecol. Res.* 5: 88–208.

Wang, T. 1993. *Synopsis on the classification of Mantodea from China.* Shanghai: Scientific and Technnical Literature Publishing House.

Wasserman, P. D. 1989. *Neural computing theory and practice.* New York: Van Nostrand Reinhold.

Webb, P. W. 1986. Locomotion and predator—prey relationships. In *Predator-prey relationships. Perspectives and approaches from the study of lower vertebrates*, ed. M. E. Feder and G. V. Lauder, 24–41. Chicago: University of Chicago Press.

Weiner, J. 1994. *The beak of the finch.* New York: Vintage Press.

Weis–Fogh, T. 1973. Quick estimates of flight fitness in hovering animals, including novel mechanisms of lift production. *J. Exp. Biol.* 59: 169–230.

Wenner, A. M. 1989. Concept–centered versus organism-centered biology. *Am. Zool.* 29: 1177–97.

Werner, E. E. 1994. Ontogenetic scaling of competitive relations: Size-dependent effects and responses in two anuran larvae. *Ecology* 75: 197–213.

Westwood, J. O. 1889 *Revisio Insectorum familiae Mantidarum.* London: Gurney and Jackson.

Wheeler, W. M. 1928. *Foibles of insects and men.* New York: Knopf.

White, T. C. R. 1993. *The inadequate environment.* Berlin: Springer-Verlag.

Whitham, T. G. 1977. Coevolution of foraging in Bombus–nectar dispensing Chilopsis: A last dreg theory. *Science* 197: 593–96.

Whitman, D. W. 1990. Grasshopper chemical communication. In *Biology of grasshoppers*, ed. R. F. Chapman and A. Joern, 357–91. New York: John Wiley.

Wickler, W. 1968. *Mimicry in plants and animals.* London: Weidenfeld & Nicholson.

Wickler, W. 1985. Stepfathers and their pseudo–parental investment. *Z. Tierpsychol.* 69: 72–78.

Wiener N. 1948. *Cybernetics.* Cambridge: MIT Press.

Wigglesworth, V. B. 1953. The origin of sensory neurons in an insect *Rhodnius prolixus* (Hemiptera). *Quart. J. Micros. Sci.* 94: 93–112.

Wilbur, H. M., and J. E. Fauth. 1990. Experimental aquatic food webs: Interactions between two predators and two prey. *Am. Nat.* 135: 176–204.

Wilkinson, R. S. 1984. An anonymous sixteenth century treatise on locusts. *Ent. Rec. J. Var.* 96: 34–35.

Williams, C. E. 1904. Notes on the life history of *Gongylus gongyloides*, a mantis of the tribe Empusides and a floral simulator. *Trans. Ent. Soc. London* 1904: 125–37.

Williams, G. C. 1966. *Adaptation and natural selection.* Princeton, N. J.: Princeton University Press.

Wilson, D. M. 1968. The nervous control of insect flight and related behaviour. In *Advances in insect physiology* 5, ed. J. W. L. Beament, J. E. Treherne, and V. B. Wigglesworth, 289–338. New York: Academic Press.

Wise, D. H. 1984. The role of competition in spider communities: Insights from field experiments with a model organism. In *Ecological communities: Conceptual issues and the evidence*, ed. D. L. Strong, D. Simberloff, L. G. Abele, and A. B. Thistle, 137–56. Princeton, N. J.: Princeton University Press.

Wise, D. H. 1993. *Spiders in ecological webs.* Cambridge: Cambridge University Press.

Wise, D. H., and J. D. Wagner. 1992. Exploitative competition for prey among young stages of the wolf spider *Schizocosa ocreata*. *Oecologia* 91: 7–13.

Wisser, A. 1987. Mechanism of wing rotation regulation in *Calliphora erythrocephala* (Insecta, Diptera). *Zoomorphology* 106: 261–68.

Wissinger, S. A. 1992. Niche overlap and the potential for competition and intraguild predation between size-structured populations. *Ecology* 73: 1431–44.

Wood, J. G. 1867. *Insect Life.* London: Bell and Daldy.

Wood, J. G. 1871. *The illustrated natural history.* London: George Routeledge & Sons.

Wood-Mason, J. 1898. On the presence of a stridulating apparatus in certain Mantidae. *Trans. Ent. Soc. London* 1898: 263–67.

Wootton, J. T. 1994a. Predicting direct and indirect effects: An integrated approach using experiments and path analysis. *Ecology* 75: 151–65.

Wootton, J. T. 1994b. The natural consequences of indirect effects in ecological communities. *Ann. Rev. Ecol. Syst.* 25: 443–66.

Wootton, R. J. 1979. Function, homology and terminology in insect wings. *Syst. Ent.* 4: 81–93.

Wootton, R. J. 1981. Support and deformability in insect wings. *J. Zool. London* 193: 447–68.

Wootton, R. J. 1990. The mechanical design of insect wings. *Sci. Am.* 26: 66–72.

Wootton, R. J. 1992. Functional morphology of insect wings. *Ann. Rev. Ent.* 37: 113–40.

Wordsworth, W. 1975. My heart leaps up when I behold (1807). In *The Norton anthology of poetry*, ed. A. W. Allison, 600–601. New York: W. W. Norton.

Yack, J. E., and B. I. Roots. 1992. The metathoracic wing–hinge chordotonal organ of an atympanate moth, *Actias luna* (Lepidoptera, Saturniidae): A light- and electron-microscopic study. *Cell Tissue Res.* 267: 455–71.

Yack, J. E., and J. H. Fullard. 1990. The mechanoreceptive origin of insect tympanal organs: A comparative study of similar nerves in tympanate and atympanate moths. *J. Comp. Neurol.* 300: 523–34.

Yager, D. D. 1989. A diversity of mantis ears: Evolutionary implications. In *Proceedings of the fifth international meeting of the Orthopterists' society*, ed. D. A. Nickle, 420. Madrid: Ministerio de Agricultura Pesca y Alimentacion.

Yager, D. D. 1990. Sexual dimorphism of auditory function and structure in praying mantises (Mantodea; Dictyoptera). *J. Zool. London* 221: 517–37.

Yager, D. D. 1992. Ontogeny and phylogeny of the cyclopean mantis ear. *J. Acoust. Soc. Am.* 92: 2421.

Yager, D. D. 1996a. Nymphal development of the auditory system in the praying mantis *Hierodula membranacea* Burmeister (Dictyoptera, Mantidae). *J. Comp. Neurol.* 364: 199–210.

Yager, D. D. 1996b. Serially homologous ears perform frequency range fractionation in the praying mantis, *Creobroter* (Mantodea, Hymenopodidae). *J. Comp. Physiol.* A: 178: 463–75.

Yager, D. D., and R. R. Hoy. 1986. The cyclopean ear: A new sense for the praying mantis. *Science.* 231: 727–29.

Yager, D. D., and R. R. Hoy. 1987. The midline metathoracic ear of the praying mantis, *Mantis religiosa*. *Cell Tissue Res.* 250: 531–41.

Yager, D. D., and R. R. Hoy. 1989. Audition in the praying mantis, *Mantis religiosa* L.: Identification of an interneuron meadiating ultrasonic hearing. *J. Comp. Physiol.* 165: 471–93.

Yager, D. D., and M. L. May. 1990. Ultrasound-triggered flight-gated evasive maneuvers in the praying mantis Parasphendale agrionina. II. Tethered flight. *J. Exp. Biol.* 152: 41–58.

Yager, D. D., M. L. May, and M. B. Fenton. 1990. Ultrasound-triggered, flight-gated evasive maneuvers in the praying mantis *Parasphendale agrionina*. I. Free flight. *J. Exp. Biol.* 152: 17–39.

Yager, D. D., and D. J. Scaffidi. 1993. Cockroach homolog of the mantis tympanal nerve. *Soc. Neurosci. Abstr.* 19: 340.

Yager, D. D., and J. D. Triblehorn. 1995. Compative neuroethology of ultra-high frequency hearing in praying mantises. In *Nervous systems and behavior. Proceedings fourth international congress of neuroethology*, ed. M. Burrows, T. Matheson, P. L. Newland, and H. Schuppe, 365. Stuttgart: Georg Thieme Verlag.

Yan, J., C. Xu, D. Yao, and Y. Li. 1981. Studies on the biometrics of six mantids from China. *Coll. Res. Publ. Chin. Acad. Forest.* 2: 67–74.

Young, D. 1991. *Nerve cells and animal behavior.* New York: Cambridge University Press.

Zabka, H., and G. Tembrock. 1986. Mimicry and crypsis: A behavioural approach to classification. *Behav. Processes* 13: 159–76.

Zack, S. 1978. Description of the behavior of praying mantis with particular reference to grooming. *Behav. Processes* 3: 97–105.

Zahavi, A. 1987. The theory of signal selection. In *Proceedings of the international symposium on biological evolution*, ed. V. P. Delfino, 305–25. Bari: Adriatica Edetricia.

Zahavi, A. 1991. On the definition of sexual selection, Fisher's model, and the evolution of waste and of signals in general. *Anim. Behav.* 42: 501–3.

Zänkert, A. 1939. Vergleichend-morphologische und physiologisch-funktionelle Untersuchungen an Augen beutefangender Insekten. *Sitzungsber. Ges. Naturforsch. Freunde Berlin* 1–3: 82–169.

Zarnack, W. 1972. Flugbiophysik der Wanderheuschrecke (*Locusta migratoria*). I. Die Bewegungen der Vorderflügel. *J. Comp. Physiol.* 78: 356–95.

Zarnack, W. 1983. Untersuchungen zum Flug von Wanderheuschrecken. Die Bewegungen räumlicher Lagebeziehungen sowie Formen und Profile von Vorder und Hinterflugeln. *Biona Rep.* 1: 79–102.

ADDITIONAL REFERENCES

Bragg, P. E. 1997. *An introduction to rearing praying mantids.* Derbyshire: Bragg.

Mukherjee, T. K., A. K. Hazra, and A. K. Ghosh. 1995. The mantid fauna of India (Insecta: Mantodea). *Oriental Insects* 29: 185–358.

Index

Page numbers in *italics* denote illustrations; illustrations in color gallery are listed as *fig.* followed by the figure number.

abdomen, 32–33, 37, 202–3, 206–7, 215, 218, 221, 254, 264, 272; development of, 265–69; dorsiflexion (flexion), 102, *103*, 186, 231; neuromuscular system, 268–69, *270, 271*, 273
abdominal bending, 82, 86
abdominal girth, 85
Acanthopinae, 39
Acanthops, 27, 30, 70, 71, 72, 74, 89, 286
Acanthops falcata, 70, 71, 72, 74, 89, 286, 297, fig. 5.1
acceptance angle, 116–18, 136–38
Acheta domestica, 273, 313
Achlaena, 21
Acontiothespis, 74, 79, 82, 83, 86, 88
Acontiothespis multicolor, 74, 82, 83, 86, 88
Acontista, 74
Acontistinae, 3, 27
Acontistini, 39
acoustic behavior 101–4, 109, 113. *See also* hearing
Acromantinae, 27, 36, 39, 74, 83
Acromantis insularis, 26
active escape, 184, 185, 290, 292
acute zone, 116–25, 137, 139
adaptive choice, 224
adaptive suicide, 49, 82, 84
Aelian, 5
aging, 255, 306–10
Aldrovandi, Ulisse, 6, 16
Ameles, 87, 315

Ameles desicolor, 87
Ameles heldreichi, 94, 105, 316
Ameles spallanzania, 186, *189, 191*
Amelinae, 25, 28, 36, 39, 40, 74, 313, 315
Amorphoscelidae, 29, 37, 38
Amorphoscelididae, 36, 38
Amorphoscelidinae, 36
Amorphoscelinae, 24, 25, 27, 28, 29, 36, 38
Amorphoscelis, 21, 281, 295
Amorphoscelis annulicornus, 26
Amorphoscelis annulipes, 23
Amorphoscelis elegans, 34
Amphecostephanus, 39
Anasigerpes, 33
Angela, 29, 33, 295, 297
Angela guianensis, 288, 293
Angelinae, 25, 27, 29, 36, 107
Antemna, 39
Antemninae, 40
antenna, 24, 25
antennal drumming: in food choice, 165; in sexual behavior, 230, 231
antlion larva, 61, 68
ant mimicry, 279, 289, 299
anuran amphibians, prey recognition by, 141
aposematism, 278–79, 289, 297, 298
apostatic selection, 281
approaching behavior, 146, 155–56
Apteromantis, 108
aptery, 107–8, 296
Archimantis, 125, 274

Archimantis latistyla, 125
artificial neural network, 170–72
Astape, 27, 34
Astape denticollis, 282, 293
Astollia chloris, 32
audiogram, 96, *97, 99, 106, 108, 110*
auditory bicyclops, 108–9, 111
auditory cyclops, 93–96, 97–99
auditory phylogeny, 110–11
auditory system, 93–113, 274; anatomy of, 93–96, *110*; —interneurons, 97, *98*, 100, 109, 111; —sensillae, 96; —tympanal nerve, 94, 96, 111–12; —tympanal organ, *94*, 96, 111–12; —tympanum, 94–96; development of, 99–100; evolution of, 109–13; —comparison with cockroaches, 111–13; —primitively earless, 105–6, *107*; —secondary loss, 106–8, 113; function of, 96–97, 100, 108–9, 113, 274; —directionality, 97–98, 102; —frequency range (*see* audiogram); —ultra–high frequencies, 103, 105; sensitivity of, 96, 99, 109 (*see also* audiogram); —correlation with wing length, 107–8, 113; serial homology of, 106, 109, *110*, 111–12; sexual dimorphism, 101, 107–8, 111, 113. *See also* hearing

back-and-forth rocking movements, 131–32, 295

background choice, 283, 284, 299
background matching, 281
bark mantids, 279, 282
Barrow, John, 7
Batesian mimicry, 277–81, 286, 289–90, 297
bats, 71, 101–4, 105, 277, 290
bees: bumblebees, 68; honey bees, 225
beetle, 68, 224; larva, 68
behavior, 37, 40; defensive (*see* defensive behavior); hierarchical organization, 224–50, *250*; interactions between, 238–45; satiated vs hungry mantids, 233, 240; sexual (*see* sexual behavior)
Belomantis, 30, 33
bestiary, 5
binocular: animals, 115, 121, 122, 125, 134–37; disparity, 122–24, 138, 207, 212, 213; interaction, 119, 122; mechanisms, 122–24, 136, 137, 139; neurons, 138; overlap, 136, 138, 139; vision, 114–40, 295
biological cycles, 37
Bistonbetularia, 282
bitrophic predators, 51, 57
Blattaria, 19
Blattarida, 19
Blattiformida, 19
Blattodea, 21, 37, 166
Blepharodinae, 36, 39
Blepharopsis mendica, 24, 183, *184*, 316
blood. *See* hemolymph
Bolivaria, 108
Book of the Dead, Egyptian, 3, 4
brachyptery, 30, 32, 107–8, 183, 288, 296
Brancsikia, 25, 26, 27, 32, 33, 39
breeding, 225, 253–54, 314–15
Brunneria, 25, 44, 318
Brunneria borealis, 44, 318
Bufo bufo, 141
Burmeister, H. C. C., 9

cages, 304, 312–13
Calamothespis, 21, 24, 27, 29
Caliridinae, 36, 40
Calospilota pulchra, 34
camouflage, 132, 183, 184, 281–86, 287
Campanotus acuapimensis, 289–90

cannibalism, 3, 6–16, 47, 49, 50, 51, 53, 319, 320, 63–64; in laboratory colonies, 313, 314, 315; prevention of, 101, 237; sexual, xi, xii 3, 7, 9–16, 49, 73–84, 86, 224
Cantipratanus, Thomas, 6
capture zones, 213–17. *See also* strike
Cardioptera, 29
Cardioptera brachyptera, 44
Cardioptera squalodon, 26
carrying capacity, 48, 58, 320; ontogenetically adjustable (OAK), 58
Catasigerpes, 70, 74, 83
Catasigerpes occidentalis, 70, 74, 83, 288, 293, fig. 13.5D
catching distance, 134
Catoxyopsis, 25
Caudatoscelis, 33
centipedes, 68
cerci, 19, 33, 37, 38, 87, 231
Chaeteessa, 27, 28, 31, 38, 106, 111
Chaeteessidae, 25, 27, 28, 33, 36, 37, 38, 105–6, 111
Chaeteessoidea, 38
Chalcoides aurata (coleoptera), 186
chantlitaxia, 278, 283
chemical defenses, 297
Chlidonoptera, 29, 33, 281, 287, 292
Chlidonoptera chopardi, fig. 13.10E
Chlidonoptera lestoni, 292, 297
Chloroharpax, 33
Chloroharpax modesta, 290
Choeradodidae, 39, 40
Choeradodinae, 36, 40
Choeradodis, 25, 101, 279, 286, 297
Choeradodis rhombicollis, 279, 286
Chopardiella, 29
Chopardiella latipennis, 30
Chorticocetes termifera, 273
chromosomes. *See* genotypes
Cilnia, 27
circadian rhythms, 225, 244; effects on behavior, 244, 247–49
classification. *See* taxonomy
claw-groove, 26, 27
Cliomantis, 27, 32
cockroach, 19, 111–13, 106, 166, 201, 213, 217, 221, 313
coloration, 33, 34

color change, 282–86, 287
color matching, 283–86, 287
color polymorphism, 255
communication, acoustic, 101, 103
community, 51, 54, 55, 59, 318
competition, 50–53
competitive exclusion, 52
compound eye, 24, 114–40
Compsothespinae, 27, 28, 29, 36, 40
Compsothespis, 21, 31
concealment: of head, 294–95; of legs, 295; of movement, 294; of wings, 295–97
convergence theory, 119, 120, 122
Coptopterix argentina, 296
Coptopteryx, 32, 70, 78
Coptopteryx gayii, 316
Coptopteryx viridis, 70, 78, 115, 197, 198
copulation, 84, 87–88, 314; during cannibalism, 49, 74–79, 82, 84; time of day, 70–71, 101; time of year, 70
correspondence problem, 123
countershading, 281
courtship, 93, 101, 224; possible displays by females, 81, 82; possible displays by males, 86
coxa, 26, 29, 36
Crematogaster, 289
Creobater urbana, 74
Creobroter, 23, 74, 102, *110*, 287, 313, 314, 315, 316
Creobroter gemmatus, 109, *110*
Creobroter pictipennis, 109
Creobroter urbana, 292
cricket, 101, 104, 125, 137, 192, 201, 205, 224, 313
crypsis, 183, 260, 278–86
Cursorida, 19
cuticle, 94; softening in histology, 324
Cuvier, Georges, 7, 8

Dactylopteryx, 28
damselfly larvae (*Pyrrhosoma nymphula*), 67
Danuria, 21, 26, 32, 39, 288–89, 295, 297
Danuria buchholzi, 288, 293
Danuria thunbergi, fig. 13.5E
DCMD. *See* descending contralater-

al movement detector (DCMD) system
death feigning, 293
Decembrus, Petrus Candidus, 5–6
defense of eggs, 293
defenses: primary, 277, 278; secondary, 277, 278
defensive behavior, 71, 101–4, 184, 225, 227, 235, 238, 252, 253, 258, 276–99; age-related changes, 227, 255, 257; age-related strategies, 272; morphological structures related to, 268, 274, 294–97; ontogeny of, 252–275, 294; plasticity of, 225, 238, 244, 265, 274, 298
defensive glands, 297
defensive mechanisms, 237, 258–65, 276–99
deflection marks, 290–91
deflection of attack, 290–91, 299
deimatic display (or reaction), 113, *110*, 185, 236, 262, 263, 287, 288, 291–93, 297, 299
demes, 50, 59
depth perception, 114–40
Deroplatyidae, 39
Deroplatyinae, 36, 39, 40
Deroplatys, 25, 39, 286, 313, 314, 316
Deroplatys desiccata, 22
Deroplatys truncata, 287, figs. 13.5A & 13.8F
descending contralateral movement detector (DCMD) system, 167–77
developmental time, 255, 305–6
diapause, 45, 316
dictuoptera, 19
dictyoptera, 19, 20, 21
Didymocorypha lanceolata, 24
diet, 313–14
disguise, 278
disparity, visual, 114, 122–24, 138
dispersal, 46–8, 50, 53, 59, 319
dispersion, 45
disruptive colors, 278, 281, 287, 292, 299
distance estimation, 114–40, 197, 207, 209
distance measurement, 114–40, 197, 200; absolute, 125, 203, 214, 215; relative, 125, 203, 214, 215
distance misestimation, 128, 129, 137
distribution, 45, 50, 320, 321
diversity, 105
drinking behavior, 166
Drosophila hydei sturdevant, 313
Drosophila melanogaster, 53, 186, 254, 313
Drosophila virilis, 303, 313
Dysaules himalayanus, 24
Dysaules longicollis, 24

ear. *See* auditory system
ecology, 43–60, 318–321
ecosystem, 51, 53, 54, 59
eggs. *See* ootheca
electrical brain stimulation, 224
electrophysiology, 96–110, 172–79, 210, 227
electroretinogram (ERG), 227, 245, 246, 247
embedding media, 324
Empusa, 28, 31, 132, 287, *296*, 316
Empusa fasciata, 105, *106*, 132
Empusa gongylodes, 295
Empusa spinosa, 24
Empusidae, 21, 25, 26, 27, 28, 29, 34, 36, 39, 40, 297
Empusinae, 27, 29, 34, 36, 39
Enicophlebia, 26
Enicophlebia hilara, 32
Enicophlebia pallida, 32
environment, 44, 311–13
Eomantis guttatipennis, 25
Epaphrodita, 29, 33
Epaphroditinae, 39
Ephestiasula, 74, 79, 81, 83, 86
Ephestiasula amoena, 74, 81, 83, 86
Episcopomantis, 24
Episcopomantis congica, 32
Epitenodera, 35
Eremiaphila, 28, 33, 105, 183, 281, 283, 290, 295
Eremiaphila brauerii, 105, *107*, 297, 312
Eremiaphila brunneri, 105, *107*, *184*, 312
Eremiaphila denticollis, 23
Eremiaphila numida, 30
Eremiaphila typhon, 30
Eremiaphilidae, 25, 27, 28, 29, 30, 32, 33, 36, 37, 38, 105–6, 111

Euantissa, 74, 79
Euantissa ornata, 74
Euchomenella heteroptera, 22
evolution, 43, 44, 51, 68, 109–13, 132, 139, 212, 213, 253, 294
evolution of prey recognition, 166–67
extinction, 46
eye, regions of, 115–17, 128, 136–38
eye grooming, 225, 227, 228, 231, 232, 233, 234, 241
eyespot, 184, 292

Fabré, J. H., 9, 10, 69, 224
facet diameter, 116, 118, 136
fecundity, 44, 61, 66, 82, 84–85
feeding behavior, 44, 49, 229
femoral brush, 26, 28, 232
femoral spines, 27, 200, 211
femur, 26, 29, 36, 197, 198, 200, 207, 209, 217
fertilization, 82, 84, 88
figure-ground discrimination, 130
fire melanism, 283, 299
fitness, 49, 61, 85, 320
fixatives, 323–24; for electron microscopy, 254, 323–24; for light microscopy, 323
flash coloration, 290
flesh flies (blow flies), 313
flight, 102, *103*, *104*, 132–34, 183–93, 290; kinematics, 189; speed, 102, 139, 185
floral simulators, 287
flower lures, 287
flower mantids, 279, 287–88, 292
fly detector model, 141
food, 3, 5–8, 11, 14, 16–18, 113–14, 225; choice, 313–14, 225; indirect herbivory, 49; limitation, 44, 45, 48–52, 320
food aversion. *See* taste aversion
foraging: behavior, 44, 48, 49, 51, 62; optimal, 44, 48, 49, 51; strategy, 44, 48, 49, 61–68
foreleg, 17, 21, 36, 102, 197. *See also* prothoracic legs
foreleg length (%FL), 197–200
forewing. *See* wing
fovea, 114, 116, 118, 161, 196, 227, 234; sexual dimorphism in, 235

Galapagos, 44
Galapagos solitaria, 44
Galepsus, 282, 295
Galepsus toganus, 283, 288, 293
gamete development, 70
generalist predator, 44, 51
genicular spine, 30, 37
genitalia, 33, 37, 40, 87
genotypes, 34, 37, 40
Geomantis, 108
Gesner, Konrad, 6
global warming, 46
Gongylus, 26, 27, 30, 101, 287
Gongylus gongylodes, 28, 29
Gonypetella, 24
grass mantids, 279, 295
growth efficiency, 44
growth rate, 67
Gryllus, 35
Gryllus campestris, 273
guild, 50, 51, 52. 53, 59
Gyromantis, 28

Haania, 26
Haaniinae, 27, 36
habitat, 53, 105
Hagenomyia micans, 68
Hagiotata hofmanni, 30
Haldwania liliputana, 105
Harpagomantis discolor, 292
hatching, 46, 51–53, 55, 303, 319
head, 21–25, 35; grooming, 48, 231, 238, 240, 243 (*see also* eye grooming); movements, 120, 129–32, 226. *See also* saccadic (head) movements
hearing: in courtship, 101; in nymphs, 99–100; in predator avoidance, 101–4. *See also* auditory system
Hemiempusa capensis, 289, 295, fig. 13.2D
hemolymph, 322–23
herbivore load, 57
Hestiasula, 26, 29, 74
Hestiasula brunneriana, 74, 109, 316
Hestiasula castetsi, 27
Heterochaeta 24, 39, 289, 295, 297, 313
Heterochaeta strachani, 316, figs. 13.5F & 13.10A
Heteronutarsus, 29, 30
Heterovates, 29

hierarchical organization of behavior, 224, 245, 250
Hierodula, 74, 80, 82, 83, 86, 87, 88, 136, 137, 314, 315, 322
Hierodula bipapilla, 316
Hierodula coarctata, 83, 86, 88
Hierodula crassa, 61, 143
Hierodula membranacea, 74, 80, 83, 86, 87, 88, 94, 99, *100*, *106*, 136, 137, 273, 316
Hierodula patellifera, 74, 83, 87, 88
Hierodula tenuidentata, 74, 82, 83, 86
Hierodulae, 199
hindwing. *See* wing
histology, 322–26; dehydration, 325; embedding, 324–25; fixation, 323–24; staining, 324–26
Holst, Erich von, 17
Homo sapiens, 307
Hoplocorypha, 21, 33, 295
Hoplocorypha nigerica, 296
houseflies, 61, 313
housing. *See* cages
Howard, L. O., 12, 69
Humbertiella, 74, 83, 86
Humbertiella ceylonica, 316
Humbertiella similis, 74, 83, 86
humidity, 304, 312
hunger hypothesis, 61
hunger level, 62, 225, 229, 231, 233; effects on behavioral hierarchies, 229–50; effects on predatory behavior, 61, 165
Hymenopodidae, 24, 26, 27, 29, 36, 39, 40, 74, 83, 111, 297
Hymenopodinae, 25, 34, 36, 39, 74, 83, 107, 111
Hymenopus, 107, 297
Hymenopus coronatus, 23, 29, 34, 281, 283, 287, 293, 313, 316
Hypsicorypha, 28

identifying mantids, 315–16
Idolium diabolicum. 287, 292, 299. *See also Idolomantis diabolica*
Idolomantis, 30, 40
Idolomantis diabolica, 25, 26, 28
Idolomorpha, 295
Idolomorpha lateralis, 289, *296*
Idolum, 297
indirect effects, community, 55, 56
individual selection, 69, 278
inhibition: feed-forward, 167–68;

lateral, 118, 167–68
interneurons, 97, *98;* visual, movement-sensitive, 167–77
interommatidial angle, 116–18, 119, 136–38
intraguild predation, 51, 53, 54
Iridopterygidae, 39
Iridopteryginae, 25, 27, 28, 34, 36, 40, 105
Iridopteryx, 31
Iris, 35, 71, 72, 73, 74, 78, 81, 82, 83, 85, 87, 88, 183, 189, 190, 192
Iris oratoria, 71, 72, 73, 74, 78, 81, 82, 83, 85, 87, 88, 183, *185*, 186, 189, *190*, *191*, 316, 318
Ischnomantis gigas, 21, 33
Isoptera, 19, 20, 21,

Juelez, 158
jumping range, 128, 129
jumps, 290
Junodia, 26

keystone effect, 57, 59
Kirby, William, 7

lacewing, 101, 104, 138
Latrodectus hasselti (redback spider), 84
leaf mantids, 279, 286–87
leaping, flight–assisted, 186, 189
learning, 134, 140, 165
legs, 264; mesothoracic (*see* mesothoracic legs); metathoracic (*see* metathoracic legs); prothoracic (*see* prothoracic legs); raptorial (*see* prothoracic legs)
Leptocola, 27, 295, 297
Leptocola phthisica, 289
Leptocola stanleyana, 25, 29
LGMD, 167–77
Lichtenstein, Anthony A. H., 17
life cycle, 44–48, 70
life history, 45–48, 304, 311; in laboratory colonies, 305–9, 311
life span, 305, 311
Ligaria, 28
light cycle, 312
lighting, 312
light sensitivity, 118
limiting resource,
Lindeman's ratio, 297
Linnaeus, 5, 35

Litaneutria, 74, 315
Litaneutria minor, 74, 105, *108,* 316
Liturgusa, 30, 282
Liturgusidae, 39, 40
Liturgusinae, 28, 33, 36, 74, 83, 105
locust, 125, 128, 129
Locusta migratoria, 191, 273
long-bodied grass mantids, 107, 286, 288–89
long-bodied stick mantids, 107, 286, 288–89
longevity, 64. *See also* life span
Lorenz, Konrad, 17
lunge, 134, 196, 198–200, 205, 207, 210, 217
lunge distance, 199, 201, 207, 220

Macracanthopus, 28
Macrodanuria elongata, 30, 105
Macromantis, 21, 26, 27, 28
Maculatoscelis, 34
Majunga, 31
mandibles, 25
Mantidae, 27, 29, 36, 39, 40, 74, 83, 111
Mantinae, 26, 27, 29, 33, 34, 36, 40, 74, 83
Mantis, 7, 27, 30, 35, 45, 48, 50, 52, 53, 55, 56, 58, 70, 71, 72, 73, 74, 78, 83, 84, 85, 86, 87, 88, *112,* 282, 290, 318–20, 322. *See also* Mantis religiosa
Mantispa styriaca, 137–39
Mantispidae, 21
Mantis religiosa, 32, 33, 34, 48, 50, 52, 53, 55, 56, 58, 70, 71, 72, 73, 74, 78, 83, 84, 85, 86, 87, 88, *95, 96, 97, 98,* 101, 102, *110,* 115, 118–20, 124, 129, 131–35, 137, 186, 189, *188, 192,* 273, 283, 286, 293, 312, 316, 318–20
Mantodea, 19, 27, 36, 38
Mantodidae, 37, 38
Mantoida, 31, 38, 106
Mantoida brunneriana, 23
Mantoida tenius, 21
Mantoidea, 38, 39
Mantoididae, 25, 27, 28, 29, 32, 36, 37, 38, 105–6, 111
marginal value theorem, 68

mate choice, 49, 69, 84–85, 88
mating, in laboratory, 80, 314
mating behavior, 12, 15, 69–89, 314. *See also* sexual behavior
mechanoreceptors, abdominal, 274
Mellierinae, 40
Mesopteryx robusta, 25
mesothoracic legs, 29–30, 37
mesothorax, 29–30, *107,* 108–9
Metallyticidae, 25, 27, 28, 30, 32, 34, 36, 37, 105–6, 111
Metallyticus, 31, 32, 37
Metallyticus semiaeneus, 290
Metallyticus splendens, 313
metathoracic legs, 29–30, 37
metathorax, 29–30, 93–110, *112*
Metilia, 27, 28
Metoxypilus, 26, 27
Metoxypilus spinosus, 25
milkweed bug, 151
mimicry, 132, 277–81
Miomantis, 27, 70, 71, 76, 83, 101, 102, 282, 313, 316
Miomantis aurea, 289, figs. 13.2F & 13.8E
Miomantis caffra, 76
Miomantis ehrenbergi, 105, 311
Miomantis natalica, 71, 105
Miomantis paykullii, 70, 76, 83, 103–4, 105–6, 283–85
misting, 314
Mittelstaedt, Horst, 17, 18, 130
Moffett, Thomas, 6
molting, 313
molting disaster, 315
monocular: animals, 115, 121, 122, 125, 134–37; mechanisms, 137
mortality, in laboratory colonies, 304–9, 315
motion parallax, 114, 115, 130, 134, 140
movement tricks, 128, 129
Müllerian mimicry, 277, 297
Musca, 313
Musca domestica, 61
museums: contact addresses, 317; with mantis collections, 105, 317
Myrmecomantis, 27
Myrmeleon bore, 68
Myrmeleon formicarius, 68

natural selection, 44, 53, 277
Negromantis modesta, 281
Nemotha metallica, 34
Neomantis australis, 286, fig. 13.10F
neuromuscular system, 255
neuronal circuits, development of, 273
niche, 43, 44, 46, 51, 52, 54, 57, 318
nocturnal activity, 71, 101, 104; attraction to lights, 101; flight, 101, 104, 183; pheromonal signaling, 101
nonpredatory ingestive behavior, 165
nutrition, effects on behavior, 246–47
nymph, 45–48, 52–56, 255–56, 287, 289, 311–15, 319, 320

object movements, 128–30
object-oriented, 125–28
ocellus, 24, 107, 227, 292; age-related changes, 255, 256, *257*
Odontomantis micans, 22
Oecyphylla longinoda, 289–90
Oligonychinae, 27, 28, 29, 36, 78, 83, 111, 313, 314, 315, 316
Oligonyx, 28, 78, 83, 86
Oligonyx insularis, 78, 83, 86
ommatidia, 116–19, 136–38, 161; corresponding, 119; optical characteristics, 116–18
ommatidial angle, 116–18, 136–38
ommatidial aperture, 116–18, 136–38
Omomantis, 28
Omomantis zebrata, 281
ontogeny, 51, 99–100, 253, 294
oocytes, 70
ootheca, 20, 21, 45, 48, 49, 55, 66, 293, 314–15, 318, 319
opimacus, 6
opisthognathous, 282, 286–89, 296
Opsomantis, 108
optokinetic (optomotoric) behavior, 235, 238, 239, 243
optokinetic head movements, 225, 234, 235, 236, 237
Orthodera, 31, 73, 78, 125
Orthodera ministralis, 23, 78, 125

Orthodera novaezealandiae, 73, 78
Orthoderella, 21, 295
Orthoderella ornata, 288, 295, 296
Orthoderidae, 39
Orthoderinae, 32, 36, 40, 78
orthognathous, 21, 296
Orthopterodida, 19
osmolarity, 322–323
oviposition, 45, 52, 314–15
ovipositor, 33, 87
Oxyelaea, 34
Oxyophthalma engaea, 24
Oxyothespinae, 30, 36
Oxyothespis, 24, 33
Oxyothespis longipennis, 289, fig. 13.10C
Oxypilinae, 28, 36, 39, 74, 111
Oxypilus, 34, 39, 74, 86, 281
Oxypilus hamatus, 74, 86, 296

Panorpa germanica (Mecoptera), 193
Panurgica, 286
Panurgica compressicollis, 286, 289, fig. 13.5B
Parablepharis, 39
Paramantis, 26, 35
Paramantis prasina, 292
Paramorphoscelis, 295
Paramorphoscelis gondokorensis, 288
Paraoxypilinae, 24, 26, 27, 28, 36, 38
Paraoxypilus armatus, 28
parasites, 59
parasitization, 293
Parasphendale, 102, 314, 316
Parasphendale agrionina, 103–4, 185, 189
Parastagmatoptera unipunctata, 143, 199
Paratenodera angustipennis, 305
Paratenodera sinensis, 199
Paratoxodera, 27, 29
Paratoxodera cornicollis, 26, 30
Parhierodula coarctata, 290. *See also Hierodula coarctata*
Paroxyophthalmus, 35
peering, 129–32, 140; amplitude, 129, 130; jump behavior, 125–30; movements, 129–32, 140; velocity, 129, 130
Periplaneta americana, 111–13, 313
Perlamantinae, 25, 27, 28, 32, 38
permits, 49

pest control, 59
phallomeres, 87
phasmids, 183
Phasmomantis sumichrasti, 316
phenology, 51–53
pheromone, 45, 49, 71–72, 101, 297
Philbrick, John and Helen, xi
Photininae, 27, 29, 30, 36, 40, 78, 111
photoreceptors, *116, 118*
Phthersigena, 28
Phyllocrania, 29, 33, 40, 286, 287, 297
Phyllocrania paradoxa, 25, 287, 293, 295, 297, fig. 13.2C
Phyllotheliinae, 36
Phyllovates, 30
Phyllovates cornutus, 281
phylogeny, auditory, 110–11
physiological saline, 322–23
Physiologus, 5
plant-part mimicry, 289
Pliny, 5
Plistospilota, 21, 96
Plistospilota guineensis, fig. 13.8B
Pnigomantis medioconstricta, 25
Poiret, 8, 12
Polymorphism: environmental, 282–86, 299; genetic, 282, 299
Polyspilota, 33, 73, 76, 125, 130, 132, 281, 282
Polyspilota aeruginosa, 33, 73, 76, 96, 293, 316, fig. 13.8A
Popa, 26, 28, 295
Popa spurca, fig. 13.5C
Popa undata, 288
population, 45–48, 50, 58, 59, 318, 319; density, 45–48, 71, 318, 319; regulation, 47, 58, 318
position-detecting neurons, 122
postembryonic development, 99–100, 134–37, 253, 305–6
praying posture, 294–95
predator, 101
predator-specific defenses, 253, 277, 294
predator evasion, auditory, 101–4
predator load, 57, 58, 59
predators, sit-and-wait, 43–60, 61–68
Presibylla elegans, 32
prey, 3, 5–8, 11, 14, 16–18, 313:

capture behavior, 194–23, 224, 226; sizes, 51–53, 61; types, 142–43
prey consumption, 61; effects on fecundity, 44–46, 48–49, 66
prey recognition, 141–79, 294–97; recognition algorithm, 141
primary defenses, 277, 278, 281–90
prognathous, *282, 286,* 296
Prohierodula, 34, 281, 282
Prohierodula coarctata. See Hierodula coarctata
Promiopteryx, 24
pronotum, 21, 25–26, 37
proprioceptive control, 130, 132, 140, 206–7, 211–13, 222
prothoracic legs, 26–29, 209, 210, 213, 264; in ontogeny of defensive behavior, 264
prothoracic tibial flexion reflex (PTFR), 210–12
prothorax, 35, 196, 198–200, 202–3, 206–7, 212–13, 215–16, 218, 222. *See also* pronotum
Pseudacanthops, 24, 26, 29
Pseudacanthops spinulosa, 25
Pseudaletia unipuncta (armyworm moth), 72
Pseudocreobotra, 34, 281, 287, 292, 297
Pseudocreobotra ocellata, 283, 289, figs. 13.8D & 13.10G
Pseudocreobotra wahlbergi, 109, 283, 295, 316
Pseudoharpax virescens, 281, 290, 296
Pseudomantis, 26
pseudopupil, 116, 120, 139
Pseudostagmatoptera, 31
punctuated equilibrium, 278
Pyrgomantis, 33, 282, 288, 295, 297
Pyrgomantis nasuta, fig. 13.2B
Pyrgomantis pallida, 283, 293

Rathet, Ilyse, 48
Rau, Phil and Nellie, 13, 14
Raubad, Etienne, 12
rearing, 303–4, 311–17
regenerated legs, 29
regulations, 316
reiterative learning hypothesis, 134

resolution, 116–18
respiratory rates, 68
retaliation, 293
retinal disparity, 122–24, 138
retinal flow field, 132, 134
retinal image, 128–30, 205, 213; motion, 128–30, 156, 207; speed, 129–30, 155
rhabdome diameter, *116*
Rhombodera, 40
Rhomboderella, 40
Riley, C., 12
Rilling, Susan, 17, 18
Rivetina, 33
Rivetina fasciata, 316
rocking movement, 131, 132, 295
Roeder, Ken, 12–18, 69, 224
Rosenhof, J. A. Rösel von, 6, 7, 8, 224

saccadic (head) movements, 120–22, 226, 234, 239–40. *See also* optokinetic head movements
sampling, 56, 319, 320
Schistocerca gregaria, 273
Schizocephala, 25, 34
Schizocephala bicornis, 21, 24
Schizocephalinae, 21, 25, 27, 28, 36, 105, 017
searching image, 279, 281, 292
secondary defenses, 277, 278, 290–93
SEG. *See* subesophageal ganglion
selection: green-beard, 278; individual, 69, 278; kin, 278; natural, 44, 53, 69, 200, 277; stabilizing, 277
senescence, 309–10
sensilla, 254, 268, 274
sensorimotor coordination, 224, 225, 273
sex ratio, 49, 50, 64, 81
sexual approaches and mounts, 72–73, 85–87
sexual behavior, 45, 49, 50, 101, 227, 229, 230, 231, 238, 314–15. *See also* mating behavior
sexual competition: between females, 45, 81; between males, 84, 88
sexual dimorphism, 21, 34, 35, 73; as a factor in rearing, 311, 313
sexual maturation, 70

short-bodied grass mantids, 286, 288–89
short-bodied stick mantids, 286, 288–89
Sibylla, 29, 78
Sibylla limbata, 287
Sibylla pretiosa, 78
Sibyllidae, 39
Sibyllinae, 21, 29, 36, 40, 73, 78
side-to-side movement, 124–32
signal selection, 277
simple eyes, 24. *See also* ocellus
Sinomantis, 286
Sinomantis denticulata, 283, 286
Solygia, 25, 26
sound production, 93, 101
spatial orientation, 114–40
spatial resolution, 116–18
spatial vision, 114–40
special resemblance, 278, 281, 286–89, 297, 299
Spence, William, 7
spermatheca, 88
spermatophore, 70, 87, 88, 232, 327
spermatozoa, 87, 88
sperm precedence, 84, 88
Sphodromantis, 24, 26, 70, 73, 76, 80, 82, 83, 86, 187, 282, 290, 313, 314, 315, 322. *See also Sphodromantis lineola*
Sphodromantis aurea, 316
Sphodromantis lineola, 22, 26, 70, 73, 76, 80, 82, 83, 86, 118–20, 196, 201, 202, 211, 215–21, 225, 227, 231, 272, 282–84, 289, 290, 292, 294, 316, fig. 13.10B; behavioral hierarchies in, 225, 227, 231; prey recognition by, 141–79; rearing of, 303–10
Sphodromantis viridis, 97, 253, 255, 272, 304, 305, 316; defensive behaviors, 253–75; developmental time, 255
spiders, 43, 49, 51, 53–54, 56, 67, 318, 319
stability, community, 59
Stagmatoptera, 40, 83, 86
Stagmatoptera biocellata, 83, 86, 115, 134, 136, 199, 272, 274, 291, 292
Stagmatoptera hyaloptera, 316
Stagmatopterinae, 40
Stagmomantis, 32, 45, 49, 69, 70,

71, 72, 73, 76, 80, 81, 83, 84, 87, 88, 314, 316, 318
Stagmomantis carolina, 13, 45, 69, 70, 71, 76, 81, 83, 87, *108*, 305, 318, 273
Stagmomantis limbata, 49, 70, 71, 72, 73, 76, 80, 81, 83, 84, 87, 88, 305, 318, figs. 5.2–5.5
startle display, 71, 185, 291–93
starvation, 45, 311
starvation tolerance, 68
Statilia apicalis, 281
stationary objects, 125–30
Stauromantis, 25
Stenophylla, 27, 28, 29
Stenophylla cornigera, 24
Stenophyllinae, 24, 40
Stenopyga, 29
Stenovates, 78, 79, 289
stereopsis, 114, 122–24, 139, 140
stereoscopic, 114, 122–24, 139, 140
stereoscopic area, 124
stick mantids, 279, 295
stigma, 31
stimuli triggering defensive behavior, 274
stimulus configuration, 254. *See also* visual stimulus
strike, 194–223; capture zones, 213–17; distance, 72, 134–36, 197, 200, 203, 207, 209, 214–15; elapsed time, 207; free-ranging mantids, 196, 200, 202–11; high, 202–9, 211; hit, 200, 201, 205, 207, 209, 213; horizontal plane, 213–22; immobilized mantids, 197–98, 211; low, 202–9, 211; midsagittal plane, 119, 120, 202–12; by movement-restricted mantids, 196, 197; orientation to prey, 202, 215; prey angle, 200, 203–4; proprioceptive cues, 206–7, 211–13, 222; pterothorax-centered space, 213, 222; range, 135, 136, 201, 203–5, 213–17; retinal image speed (of prey), 207; steering, 197, 200, 213–17; stop, 209, 210; tibial closure (*see also* prothoracic tibial flexion reflex), 207, 209, 210
styles, 33, 37
subesophageal ganglion, 11, 14, 15

subgenital plate, 33, 37
supernormal stimulus, 290
supra-anal plate, 32, 37
survival rates, 307
survivorship, 44, 47, 48, 53, 320

Talmud, 5
Tarachodes, 27, 71, 72, 78, 83, 282, 295
Tarachodes afzelii, 71, 72, 78, 83, 289, 293, 297, figs. 13.2A & G & 13.8C
Tarachodinae, 24, 27, 28, 30, 33, 36, 39, 40, 78, 83
Tarachodula pantherina, 289, 293
tarsus, 29, 20
taste aversion, 151, 160
Taumantis, 102, 314
Taumantis ehrmanni, 105–6, 316
taxonomy, 35–40
tegmina, 30, 37
temperature, 312
Tenodera, 35, 70, 71, 72, 73, 76, 78, 80, 81, 82, 83, 86, 87, 88, 281, 282, 315, 322. *See also Tenodera aridifolia sinensis*
Tenodera angustipennis, 48, 49, 51–53, 61–68, 71, 76, 83, 293, 318–20
Tenodera aridifolia sinensis, 44–53, 55, 56, 58, 64, 65, 70, 71, 72, 73, 76, 78, 80, 81, 82, 83, 86, 87, 88, 102, 115, 118, 126–28, 132, 135–38, 199–202, 209, 211, 225, 227, 228, 231, 255, 272, 274, 288, 293, 303, 305, 312, 313, 316, 318–20; behavioral hierarchies in, 224–50; prey recognition by, 144, 149
Tenodera australasiae, 117, 118, 121, 125
Tenodera superstitiosa, 70, 76, 83, 289, 293, fig. 13.10D
terrarium, 313. *See also* cages
tethered flight, 102, *103*, 192
thanatosis, 293
Theopompa, 282
Theopompella, 33, 282

Theopompella westwoodi, fig. 13.2E
Theopropus, 28, 34, 107
Theopropus elegans, 292, 293, 303, 316
Thespidae, 39, 40
Thespinae, 26, 28, 32, 34, 36, 111, 313
Thesprotia, 27, 28, 29
Thesprotia infumata, 34
Thesprotiella, 28
threat display, 184. *See also* deimatic display
tibia, 28, 30, 37
Tinbergen, Niko, 17
Tithrone, 39, 74, 83
Tithrone roseipennis, 34, 83
Topsell, E., 6
Toxodera, 24, 25, 27, 29
Toxoderidae, 39
Toxoderinae, 27, 29, 30, 34, 36, 39, 40
tracheal sac, auditory, 94, 96, 100, 112
Triaenocorypha, 28
triangulation, 119, 120, 134
trigger zones, 119
trochanter, 26, 29
trophic cascade, 53, 57

ultrasound, 93–105, 109, 111, 113

vagility, 50, 59
Vates, 40
Vates multilobata, 29
Vatidae, 39
Vatinae, 24, 26, 29, 30, 34, 39, 40, 73, 78, 83, 107
vision, 72, 73, 298; spectral sensitivity, 118; spectral sensitivity curves, *248*
visual: control, 128, 130, 132, 137, 138; deprivation, 114, 135–37; experience, 114, 137; field, 115, 117, 122, 134, 136–38, 202, 215–16, 222; fixation, 119, 234
visual stimulus: antiworm, 146–67; computer generated, 141–67; effects of: —brightness, 151; —contrast, 149, 160, 177; —direction, 161, 163–64; —geometry, 148, 254; —leading edge, 151; —location, 161, 173, 176; —size, 146–51, *175*, 254; —speed, 151–56, 174–75, 235–37, 243–44; eliciting head movements, 225: —light, 228–29; —mechanically driven, 141–60; —patterned, 225; —subthreshold components of, 158–59, 160–61; —super sign, 158; —worm, 146–67
visual tracking, 146

warning coloration, 289
water, visual recognition of, 151, 166
website, 317
Wheeler, William, 14
wing, 30–32, 99, 101–4, 132, 183–93, 258, 259; base, 185, 265, 266; claval furrow, 185, 187, 188, 191; clavus, 185; development, 258; downstroke, 186; elastic-inertial interactions, 190; flexion line, 185, 187, 190; folding, 32, 185, 186; forewing, 30, 31, 185, 186, *187*, 265, 266; hindwing, 31, 32, 186, *187*, 265, 266; jugal fold, 32, 185; kinematics, 189; loss of, 296; reduced, 296; upstroke, 190
winglessness. *See* aptery; brachyptery
wing vein, 30–32, 187
Woodson, Erika, 50

Yersiniops, 44, 315
Yersiniops solitarium, 107, *108*, 312
Yersiniops sophronicum, 44

Zimmerman, C. A., 9
Zoolea, 26

Library of Congress Cataloging-in-Publication Data
The praying mantids / edited by Frederick R. Prete . . . [et al.].
   p.   cm.
   Includes bibliographical references (p.).
   ISBN 0-8018-6174-8 (alk. paper)
   1. Mantidae. I. Prete, Frederick R., 1948–   II. Title.
QL508.M4P7   1999   CIP
595.7'27—dc21   99-044360